Understanding Communications Networks

– for Emerging Cybernetics Applications

Revised Edition

RIVER PUBLISHERS SERIES IN COMMUNICATIONS

The "River Publishers Series in Communications" is a series of comprehensive academic and professional books which focus on communication and network systems. Topics range from the theory and use of systems involving all terminals, computers, and information processors to wired and wireless networks and network layouts, protocols, architectures, and implementations. Also covered are developments stemming from new market demands in systems, products, and technologies such as personal communications services, multimedia systems, enterprise networks, and optical communications.

The series includes research monographs, edited volumes, handbooks and textbooks, providing professionals, researchers, educators, and advanced students in the field with an invaluable insight into the latest research and developments.

For a list of other books in this series, visit www.riverpublishers.com

Understanding Communications Networks
– for Emerging Cybernetics Applications
Revised Edition

Kaveh Pahlavan
Worcester Polytechnic Institute, USA

Published 2021 by River Publishers
River Publishers
Alsbjergvej 10, 9260 Gistrup, Denmark
www.riverpublishers.com

Distributed exclusively by Routledge
4 Park Square, Milton Park, Abingdon, Oxon OX14 4RN
605 Third Avenue, New York, NY 10158

First published in paperback 2024

Understanding Communications Networks – for Emerging Cybernetics Applications, Revised Edition / by Kaveh Pahlavan.

Routledge is an imprint of the Taylor & Francis Group, an informa business

Publisher's Note
The publisher has gone to great lengths to ensure the quality of this reprint but points out that some imperfections in the original copies may be apparent.

While every effort is made to provide dependable information, the publisher, authors, and editors cannot be held responsible for any errors or omissions.

ISBN: 978-87-7022-586-1 (hbk)
ISBN: 978-87-7004-316-8 (pbk)
ISBN: 978-1-003-33991-5 (ebk)

DOI: 10.1201/9781003339915

"to my life time partner, Farzaneh,

my children, Nima, Nasim, and Shek,

my grandchildren, Roya, and Navid, and

my academic son Prashant".

Contents

Instructor’s solution available on River Publishers’ website:
https://www.riverpublishers.com/book_details.php?book_id=919

Preface

Communications networks have emerged as a multidisciplinary diversified area of research over the past several decades. From traditional wired telephony to 6G and 7G cellular wireless networking industry and from wired local area networks and IEEE 802.3 Ethernet to wireless access to the Internet with IEEE 802.11 Wi-Fi, and IEEE 802.15 Bluetooth, ZigBee and ultra-wideband (UWB) technologies profoundly impacting our lifestyle. At the time of this writing, the wireless infrastructure of over a billion Wi-Fi access points and several hundred thousand cellular wireless base stations enable, several billions of smartphones, billions of desktops, and laptops as well as hundreds of billions of Internet of Things (IoT) devices to connect to the Internet from anywhere, at any time. The popularity of smartphones enabling the fusion of computers, networking, and navigation for location aware multimedia mobile networking and emergence of machine learning industry has fueled the emergence of the smart world affecting communications, trades, health, transportation, financing, education, and all aspects of our lives. In response to this growth, universities and other educational institutions must educate and prepare the task force in understanding these technologies to keep up with this amazing growth of the industry.

Communications networking is a multi-disciplinary field of study. To understand this industry and its technology, we need to learn several disciplines and develop an intuitive feeling of how these disciplines interact with one another. To achieve this goal, we describe important communications networking standards and applications, logically classify their underlying technologies, and give detailed examples of successful technologies with the fundamental science behind them. The selection of detailed technical material for teaching in such a large and multidisciplinary field is very challenging because the emphasis of the technology shifts in time and the skills of the instructors are limited to their research experiences in certain aspects of this gigantic industry. The author began his teaching of wireless networks by introducing the concept of wireless office information networks in the mid-1980s in a variety of conferences by comparing spread spectrum technology and optical communications as alternatives for implementation of physical (PHY) layer of these networks. Later, he integrated details of multipath indoor radio frequency (RF) propagation and the design of medium access control (MAC) for wireless local area networks (WLANs), commercially known as Wi-Fi, and wrote his first textbooks with Dr. Allan H. Levesque on the topic in 1995. This book was an expanded version of his research at the center for wireless information network studies (CWINS), Worcester Polytechnic Institute (WPI), MA, USA. The contents of courses he taught around this book were a description of measurement and modeling of multipath radio propagation in-depth and the design and performance evaluation of PHY and MAC alternatives for Wi-Fi and cellular networks. Then, he broadened the scope of his teaching to computer science departments and published his 2002 book with Prof. Prashant Krishnamurthy, University of Pittsburgh, PA, USA. In that book, more details of standards were included, and depths of analytical concepts in communications and signal processing were reduced to increase the circle of audience and include multi-disciplinary students. As a next step, he included wired networks into the teachings of fundamentals of communications networks in his 2009 book with Prof. Krishnamurthy. Then, he began an inclusion of new emerging wireless positioning systems, such as Wi-Fi and UWB positioning, which were the focus of his research in the 2000s, into his teachings and published the 2013 book with Prof. Krishnamurthy on wireless access and localization. At the time of this writing, both wireless networking and wireless positioning are mature technologies, and he teaches them in two separate courses: one on indoor geolocation science and technology based on his 2019 book and another on fundamentals of communications networks and applications, based on this book. These new books are focused on emerging cyberspace applications of communications networks in wireless positioning, gesture and motion detection, authentication and security, and intelligent spectrum management, which are his current areas of research and scholarship.

The emergence of wireless access and dominance of the Ethernet in wired local area networking (LAN) technologies shifted the innovations in networking toward the physical layer and characteristics of the medium. There is a need for a textbook that integrates all the aspects of current popular communications networks and applications together and places an emphasis on the details of physical layer aspects. In this book, we pay attention to the physical layer while we provide fundamentals of communications networking technologies and applications which are used in wired and wireless networks designed for local and wide area operations. The book provides a treatment of the wired IEEE 802.3 Ethernet and Internet, 2G–6G wireless networks, IEEE 802.11 Wi-Fi, and IEEE 802.15 Bluetooth, ZigBee, and UWB technologies as well as an overview of emerging cyberspace applications of communications networks in gesture-, motion-detection, and authentication and security. The novelty of the book is that it emphasizes on physical communications issues related to the formation and transmission of packets and characteristics of the medium for transmission in a variety of networks and cyberspace application development. The structure and sequence of material for this book was formed in lecture series by the author at the graduate and undergraduate school of the WPI, Worcester, MA, USA entitled "Introduction to WANs and LANs," "Introduction to Communications and Networks," and "Introduction to Wireless Networks." The author has also taught shorter versions of the course focused on the IEEE 802.3 Ethernet and IEEE 802.11 Wi-Fi at the University of Oulu, Finland, and the University of Science and Technology of Beijing, China, where he has held visiting international professorships.

The book is organized as follows. Chapter 1 provides an overview of communication networks dividing communications networking technologies into core and access networks. Chapter 2 is another overview chapter describing an operation of core networks. These core networks are public switched telephone networks (PSTN) evolved for connection-based telephone networks and the Internet evolved for packet switching. The emphasis of Chapter 2 is on the description of addressing, Quality of Service (QoS), wireless infrastructure, and interconnecting elements: bridges, switches, and routers. Chapters 3, 4, and 5 provided the technical background on characteristics of the medium, physical layer information transmission, and methods for MAC. In these chapters, we provide an analytical background needed to understand details of popular standards for communications networks. In the remainder of the book, we describe details of popular communication networking standards. Chapter 6 is on IEEE 802.3 standard for the Ethernet, the dominant standard for local networking. In this chapter, we present line-coding techniques that have enabled Ethernet to increase its data rate from legacy 10 Mbps to 100 Gbps. Chapter 7 is devoted to the IEEE 802.11 and the Wi-Fi technology explaining the importance of this technology in the emergence of the third industrial revolution and the "information age." In this chapter, we show how spread spectrum, orthogonal frequency division multiplexing (OFDM), multiple-input-multiple-output (MIMO) antenna systems, and mmWave technologies enabled Wi-Fi to increase its data rate from 2 Mbps to Gbps. Chapter 8 describes the IEEE 802.15 standard and its popular low energy technologies such as: Bluetooth, ZigBee, and UWB. In Chapter 9, we begin by the cellular telephone technologies evolving from 1G frequency division multiple access (FDMA), to 2G time division multiplex access (TDMA), 3G code division multiplexing (CDMA) for circuit switching connection to the PSTN. Then, we present 4G, 5G, and 6G technologies for packet-switched all Internet Protocol (all-IP) wireless metropolitan area networking. Chapter 10 is devoted to the deployment of wireless networks and technical issues associated with that such as cell planning and spatial throughput analysis. Chapter 11 provides an overview of emerging cyberspace applications benefitting from RF propagation cloud created by wireless devices. These applications are in wireless positioning and location intelligence, gesture and motion detection, and authentication and security.

The author believes that the most difficult part of the book for the students is Chapters 3–5 which provide a summary through the mathematical description of numerous technologies for analysis of the medium and design of effective PHY and MAC layers. Other parts of the book appear mathematically simpler but carry more details of how systems work. To make the difficult parts simpler for the students, an instructor can mix these topics as appropriate. The author believes that this is an effective approach for enabling the understanding of the fundamental concepts in communications networks to the students. Therefore, depending on the selection of the material, depth of the coverage, and background of the students, this book can be used for undergraduate or first-year graduate courses in Electrical and Computer Engineering, Computer Science, Information Science, Robotics, or Biomedical Engineering in one course or a sequence of two courses.

Materials presented in the book draw on previous writings of the authors with Dr. Allan H. Levesque and Prof. Prashant Krishnamurthy in 1995, 2002, 2005, 2009, and 2013. The earlier books were based on the pioneering research in Wi-Fi and Wi-Fi positioning at the CWINS. In the past few decades, communications networking research has grown to its maturity and this book is prepared to teach undergraduate and first-year graduate students, material needed to understand the evolution of standards and technologies in communications networks and the science and engineering behind them. The author has not directly referenced our referral to several resources on the Internet, notably Wikipedia. While there are people who question the accuracy of online resources, they have provided him with quick pointers to information, parameters, acronyms, and other useful references which helped him to build up a more comprehensive and up-to-date coverage of standards and technologies. He does acknowledge the benefits of these resources.

Since much of the material in this book builds from past research work of the students at the CWINS and the author's previous academic books, the author is pleased to acknowledge the students' and colleague's contributions to his growth in understanding of fundamentals of communications networks enabling him to prepare this book for teaching and training of the task force for the future of this industry. In particular, the author would like to thank Prof. Prashant Krishnamurthy of the University of Pittsburgh for being his co-author in the previous books in 2002, 2009, and 2013. The material presented in this book draws significantly from the material in those books. However, in multidisciplinary books, we need to revise the presented material frequently because of the direction of the industry and the importance of the material to teach shift rapidly in time. In this revision, prior family and work commitments precluded Prof. Krishnamurthy from participating in this book as a co-author. Without benefit from the author's past material published with him in the previous books, the author would not be able to complete the manuscript on time and with its current quality of presentation. The author also would like to express his appreciation to Dr. Allen Levesque, for his contributions in other books with the author in 1995 and 2005, which has indirectly impacted in the style of writing, formation of thoughts, and the methodology to present the material in this book. Most of the MATLAB codes presented in the book and the solution book for the problems at the end of the chapters are prepared by the author's teaching assistants, in particular, Zhuoran Su and Julang Ying, the author's current lead Ph.D. student. The author also owes special thanks to the Defense Advanced Research Projects Agency (DARPA), National Institute of Standards and Technology (NIST), National Science Foundation (NSF), Department of Defense (DoD), Department of Homeland Security (DHS), United Technology and Skyhook in the United States, as well as Finnish Founding Agency for Technology and Research (TEKES), Nokia, Electrobit, Sonera, and the University of Oulu in Finland, whose support of the CWINS research program at WPI enabled graduate students and the staff of CWINS to pursue continuing research in this important field. Writing of all of the new material in all of the author's books have flowed out of these sponsored research efforts.

The author has dedicated his income of the book to the educational fund of his grandchildren, Roya and Navid Kablan, for the love and energy they gave him to continue his research and scholarship in these late years of his career. The authors also would like to thank Dr. Rajeev Prasad and Junko Nakajima of River Publishers for their assistance and useful comments during various stages of the production of the book and for her help during the manuscript proofs.

List of Figures

List of Tables

List of Abbreviations

1-7G	first-seven-generation
ACF	autocorrelation function
ACK	acknowledgment
ADPCM	adaptive differential PCM
AES	advanced encryption standard
AI	artificial intelligence
AKL	Atwater Kent Laboratory
AMI	alternate mark inversion
AMPS	Advanced Mobile Phone Services
ANSI	American National Standards Institute
AOA	angle of arrival
AODV	ad-hoc on-demand distance vector
AoL	America online
AP	access point
ARIB	Association of Radio Industries and Businesses
ARP	address resolution protocol
ARPANET	advanced research projects agency network
ARQ	automatic repeat request
AS	autonomous system
ASK	amplitude-shift keying
ATIM	announcement traffic indication map
ATM	asynchronous transfer mode
AuC	authentication center
AWGN	additive white Gaussian noise
BAN	body-area network
BCH	Bose–Chaudhuri–Hocquenghem
BER	bit-error rate
BGP	border gateway protocol
BI	backoff interval
B-ISDN	broadband ISDN
BLE	Bluetooth low energy

BNC	Bayonet Neill–Concelman
BPDU	bridge protocol data unit
BPSK	binary phase shift keying
BRAN	broadband radio access network
BS	base station
BSA	basic service area
BSC	base station controller
BSS	basic service set
BSS	BS subsystem
BTS	base transceiver system
Cat-i	category-i
CBC	cipher-block-chaining
CCA	clear channel assessment
CCK	complementary code keying
CCMP	cipher-clock-chaining MAC protocol
CDDI	copper distributed data interface
CDM	code-division multiplexing
CDMA	code-division multiple access
CEPT	Committee of the European Post and Telecommunications
CFI	canonical format indicator
CFP	contention-free period
CH	corresponding host
CHAP	challenge handshake authentication protocol
C/I	carrier-to-interference ratio
CIR	channel impulse response
CM	connection management
COST	Co-operative for Scientific and Technical Research
CPS	cell-tower positioning systems
CRC	cyclic redundancy check
CSI	channel state information
CSL	coordinate sample listening
CSMA	carrier sense multiple access
CSMA/CA	CSMA with collision avoidance
CSMA/CD	CSMA with collision detection
CT	cordless telephone
CTS	clear-to-send
CW	contention window
CWINS	Center for Wireless Information Network Studies
DA	destination address

DARPA	Defense Advanced Research Projects Agency
DARPAnet	Defense Advanced Research Projects Agency Department Network
dB	decibel
dBm	decibel with respect to millimeter
DBPSK	differential BPSK
DCF	distributed coordination function
DCLA	DC level adjustment
DDS	digital data service
DECT	digital enhanced (formerly European) cordless telephony
DES	data encryption standard
DFE	decision feedback equalizer
DFIR	diffuse IR
DHCP	dynamic host configuration protocol
DHS	Department of Homeland Security
DIFS	DCF inter-frame space
DLL	data link layer
DM	medium rate data packet
DNS	domain name service
DOA	direction of arrival
DoD	Department of Defense
DQPSK	differential QPSK
DS	direct sequence
DS-UWB	direct sequence UWB
DSL	digital subscriber line
DSSS	direct sequence spread spectrum
DS-UWB	direct sequence UWB
DV	device
EAP	extensible authentication protocol
EDGE	Enhanced Data for Global Evolution
EGP	external gateway protocol
EIA/TIA	Electronic/Telecommunication Industry Association
EIR	electronic identity registration
EIRP	equivalent isotropic radio propagation
EO	end office
ERP	exterior router protocol
ESA	extended service area
ESME	emergency services message entity
ESN	emergency services network

ESP	encapsulated security payload
ESS	extended service set
ETSI	European Telecommunication Standards Institute
e-UTRAN	evolved UMTS terrestrial radio access network
EU	European Union
FA	foreign agent
FCC	Federal Communication Commission
FCS	frame correction sequence
FDD	frequency-division duplexing
FDDI	fiber distributed data interface
FDM	frequency-division multiplexing
FDMA	frequency-division multiple access
FEC	forward error correction
FEXT	far end cross talk
FFT	fast Fourier transform
FH	frequency hopping
FHSS	frequency-hopping spread spectrum
FM	frequency modulation
FMS	Fluhrer, Mantin, Shamir
FSK	frequency-shift keying
FTP	file transfer protocol
FTP	foiled twisted pair
GFSK	Gaussian frequency-shift keying
GGSN	gateway GPRS support node
GMSK	Gaussian minimum shift keying
GP	gap period
GPRS	general packet radio service
GPS	global positioning system
GSM	Global System for Mobile Communications
GTP	GPRS tunneling protocol
HA	home agent
HCI	human-computer interaction
HDLC	High-level data link control
HF	high frequency
HFC	hybrid fiber cable
HFSS	high frequency structure simulation
HILI	higher level interface
HIPERLAN	high-performance LAN
HLR	home location register

HPN	home phone networking
HSS	home subscriber server
HTTP	hypertext transfer protocol
IANA	Internet Assigned Numbers Authority
IAPP	inter-AP protocol
ICMP	Internet control message protocol
IEC	International Electrotechnical Commission
IETF	Internet Engineering Task Force
IFFT	Inverse FFT
IFS	interframe spacing
IGP	interior gateway protocol
IHL	Internet header length
IHT	internet header length
IMSI	international mobile subscriber identity
IMT-2000	International Mobile Telecommunications beyond the year 2000
IoT	Internet of Things
IP	Internet protocol
IPv4	IP version four
IPTV	Internet protocol television
IRIM	intelligent radio with interference management
IR	infrared
IrDA	IR Data Association
IRP	interior router protocol
ISDN	integrated service data network
ISI	intersymbol interference
IS-IS	intermediate system to intermediate system
ISM	industrial, scientific, and medical
ISO	International Standards Organization
ISP	Internet service provider
ITU	International Telecommunications Union
IV	initialization vector
LAN	local-area network
LANE	LAN emulation
LBT	listen-before-talk
LED	light-emitting diode
LF	low frequencies
LFSR	linear feedback shift register
Li-Fi	visible light communication

LLC	logical link control
LOS	line-of-sight
LS	least square
LSAP	logical service access point
LSTM-RNN	long short-term memory regressive neural network
LTE	long-term evolution
MAC	medium access control
MAN	metropolitan area network
MB-OFDM	multiband OFDM
M-BOK	multiple bi-orthogonal keying
MCM	multicarrier-modulation
ME	mobile equipment
MedRadio	Medical RadioCommunication Services
MH	mobile host
MIMO	multiple-input–multiple-output
MLME	MAC layer management entity
MM	mobility management
MME	mobility management entity
MMF	multimode fiber
mmWave	millimeter wave
MPDU	MAC protocol data unit
MPLS	multiprotocol label switching
MRC	maximal ratio combining
MS	mobile station
MSC	mobile switching center
MSK	minimum shift keying
MSS	mobile subscriber station
MTP	message transport protocol
MTU	Maximum transmission unit
NAP	network access provider
NAV	network allocation vector
NB	normal burst
NEXT	near end crosstalk
NIC	network interface card
NIST	National Institute of Standards and Technology
NGN	next generation networks
NMT	Nordic Mobile Telephone
NNI	network–network interface
NOMA	non-orthogonal multiple access

NRZ	non-return to zero
NSF	National Science Foundation
NSP	network service provider
NSS	network and switching subsystem
OAM	operation, administration, and maintenance
OC	optical carrier
OFDM	orthogonal frequency division multiplexing
OFDMA	orthogonal frequency division multiple access
OLOS	obstructed line-of-sight
OM	optical multimode
OS	optical single
OSI	Open Systems Interconnection
OSPF	open shortest path first
PAM	pulse amplitude modulation
PAN	personal-area network
PBX	private box exchange
PC	personal computer
PCF	point coordination function
PCM	pulse code modulation
PCS	personal communication service
PCF	point coordination function
PDF	probability density function
PDCP	packet data convergence protocol
PDN-GW	packet data network gateway
PG	processing gain
PHY	physical
PIFS	PCF inter frame space
PLCP	physical layer convergence protocol
PLME	PHY layer management entity
PLW	packet length width
PMD	physical medium dependent
PN	pseudo noise
POTS	plain old telephone service
POP	point of presence
PPM	pulse position modulation
PRB	physical resources per block
PSD	photo sensitive diode
PSF	packet-signaling field
PSTN	public switched telephone network

PTT	Post, Telegraph, and Telephone
QAM	quadrature amplitude modulation
QoS	quality of service
QPSK	quadrature phase-shift keying
RADIUS	remote authentication dial-in user service
RF	radio-frequency
RFCOMM	radio frequency communication
RFID	Radio frequency identification
RIP	routing information protocol
RIT	receiver initiated transmission
RJ	registered jack
RL	relay
RLC	radio link control
RNC	radio network controller
RRC	radio resource control
RRM	radio resource management
RSN	robust security network
RSS	received signal strength
RSSI	received signal strength indicator
RSVP	resource reservation protocol
RTLS	real-time localization system
RTP	real-time transport protocol
RTS	request-to-send
RZ	return to zero
SA	source address
SAP	service AP
SAW	surface acoustic wave
SCCP	signaling connection control part
SDH	synchronous digital hierarchy
SD	starting delimiter
SDMA	space division multiple access
SFD	start of the frame delimiter
SGSN	serving GPRS support node
S-GW	serving gateway
SHF	super-high frequency
SIFS	short IFS
SIM	subscriber identity module
SIMO	single-input–multiple-output
SIP	Session initiation protocol

SIR	signal-to-interference ratio
SISO	single input–single output
SME	station management entity
SMF	single-mode fiber
SMS	short messaging system
SNAP	sub-network access protocol
SNR	signal-to-noise ratio
SOHO	small office and home office
SONET	synchronous optical network
SS	spread spectrum
SS-7	Signaling System number 7
SSH	security shell
STA	spanning-tree algorithm
STC	space–time coding
STM	synchronous transport mode 1
STP	shielded twisted-pair
STS	synchronous transport signal 1
SYNC	synchronization
TB	tail bit
TCM	trellis-coded modulation
TCP	transmission control protocol
TDD	time-division duplexing
TDM	time-division multiplexing
TDMA	time-division multiple access
TIA	Telecommunications Industry Association
TIA/EIA	Telecommunication/Electronic Industry Association
TIM	traffic indication map
TKIP	temporal key integrity protocol
TMSI	temporary mobile subscriber identity
TOA	time of arrival
TP	twisted pair
TSN	transitional security network
TTL	Time to Live
UE	user equipment
UDP	user datagram protocol
UHF	ultra-high frequency
UMTS	Universal Mobile Telecommunications System
UNI	user-network interface
UNII	unlicensed national information infrastructure

U-TDOA	uplink-time difference of arrival
UTP	unshielded twisted-pair
UWB	ultra-wideband
VC	virtual connection
VHF	very-high frequency
VLAN	virtual LAN
VLF	very low frequency
VLR	visitor location registration
VoIP	voice-over-IP
WAN	wide area network
WARC	World Administrative Radio Conference
WBAN	wireless body area networking
WDM	wavelength-division multiplexing
WEP	wired equivalent privacy
Wi-SUN	wireless smart utility network
WLAN	wireless LAN
WPA	Wi-Fi protected access
WPAN	wireless personal area network
WPI	Worcester Polytechnic Institute
XML	extensible markup language

1

Overview of Communications Networks

1.1 Introduction

Technological innovations since the first industrial revolution in almost two centuries ago changed our labor-driven agricultural economy of a few millenniums to an industrial economy revolving around innovations and technology. Historians often divide the industrial revolution into eras to explain their impact on our lives. During the era of the first industrial revolution (1760–1840), we transferred from wood to charcoal as the source of energy, then an invention of the steam engine resulted in a revolution in transportation (railroads trains, steamboats, etc.), textile (fast clothing machines), and in metallurgy (hard iron for bridges and railroads). The academic engineering education in US began with an emergence of Civil Engineering curriculum, first at the West Point Military Academy (1802) and then at the Rensselaer Polytechnic Institute, Troy, New York, USA (1824).

During the second industrial revolution (1860–1940), we transferred from coal to petroleum for energy, and it followed by major technical innovation, combustion engine for cars, airplanes, a liquid rocket for access to space, steel for long bridges and tall buildings, **telegraph, and telephone for telecommunication, wireless telegraph for short messaging, and radio-television broadcasting,** electricity for light to work at dark, assembly lines for fast manufacturing, and new chemical engineering products such as petroleum, rubber, fertilizers, synthetic dye to enable numerous innovative applications touching all aspects of our lives. During the 1860s, after the conclusions of the civil war in the United States, engineering education had a major second jump with curriculum expansion from Civil Engineering into Mechanical Engineering, Electrical Engineering, and Chemical Engineering. A series of new academic institutions, Massachusetts Institute of Technology in Boston and Worcester Polytechnic Institute in Worcester Massachusetts, and Stevens Institute of Technology in Hoboken, New Jersey, implemented these new curriculums and began training a task force for the future of these industries. This tradition of engineering and science curriculum exploded in the next few decades resulting in numerous new colleges and universities in engineering and science across the United States, securing the foundation of a prosperous economy of this country in the 20th century and beyond.

The third industrial revolution (1990–2010) is transferring the energy from petroleum to renewable energy and has enabled the **information age** with an integration of computation and communications with instantaneous wireless access to these services, anytime and anywhere. The computer communications networking industry has enabled the Internet and data sciences for implementation of search engines such as Google or, Amazon, social networks such as Facebook and LinkedIn, and online shopping with eBay and many other services. The wireless access and localization technologies with Wi-Fi, cellular networks, and GPS, and wireless positioning combined with machine learning algorithms and cloud computing have enabled the Internet of Things (IoT) and the emergence of the Smart World (2010–present), which is referred to as the fourth industrial revolution. The fourth industrial revolution has already enabled intelligent transportation for Uber and DiDi, cyber–physical systems for robotic and smart heath, and many other applications. In the third and fourth industrial revolutions, engineering education is going through another "transformation" from the traditional specialty-focused curricula to "multi-disciplinary" curricula and "inter-disciplinary" research directed toward innovation and entrepreneurship, complemented with a project-based learning approach for delivery of engineering and science curriculum.

In this book, we intend to educate students interested in understanding of the communications networks and applications, the fundamental technologies that enable this industry to deploy its infrastructure, popular standard that makes communication devices communicate with one another, and the intelligent cyberspace applications that have emerged from radio frequency (RF) cloud of wireless communications devices. We begin by introducing the elements of communications network, followed by a historical perspective on the evolution of communications networks. Then, we introduce the role of government regulation and its impact on the evolution of communications networking standards protocols and enabling technologies. The last section of the chapter is devoted to a description of the structure of the book.

1.2 Elements of Communications Networks

Communications networks have evolved to interconnect telecommunication devices over a geographical area to share information generated by an application in the device. Figure 1.1 illustrates the abstraction of this general concept. The information source can be a human being creating an electronic signal on a telephone device that uses a telephone network to transfer that information to another geographical location. The information source could is a video stream from a video camera or sensor data to a robot that is sent through a networking interface card to a local area network (LAN) or the Internet to be delivered to another networking device in a geographically separated location. The sensor data, for example, could be

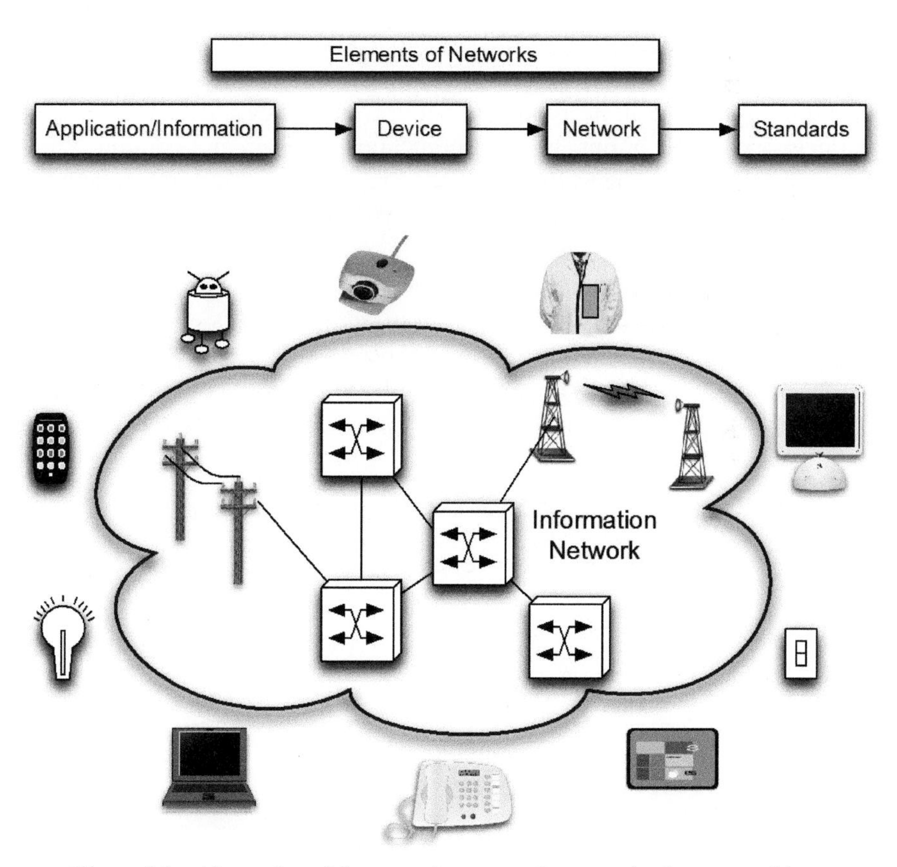

Figure 1.1 Abstraction of the general concept of communications networking.

used for remotely navigating the robot. The information could be a simple on–off signal generated by a light switch in one location to be transferred by a communication interface protocol to another location to turn a light bulb on. What is common among all of these examples is an *application* that needs the *transfer* of a certain amount of *information* from a location to another, a *network* that can carry that information, and an *interface device* that shapes the information to a format or protocol suitable for a particular networking technology.

Figure 1.2 shows a diagram of the elements affecting information networks and the relationships among them. Information generated by an application is delivered to a communication device that uses the network and delivers that information to another location. When the network includes multiple service providers, the interface between the device and the network should be *standardized* to allow communication among different network providers and various user devices. Standardization also allows multi-vendor operation so that different manufacturers can design different parts of the network. Applications, telecommunication devices, and communication networks evolve in time to support innovations that enable new applications. These are the new applications that fuel the economy and the progress in the quality of life over time. For example, the introduction of iPhones and iPads opened a new horizon for hundreds of thousands of new applications in the past few years. The evolution of these devices was enabled by the availability of reliable wireless mobile data cellular services as well as Wi-Fi and Bluetooth technologies defined by standards organizations for communications. These applications are changing how we work, eat,

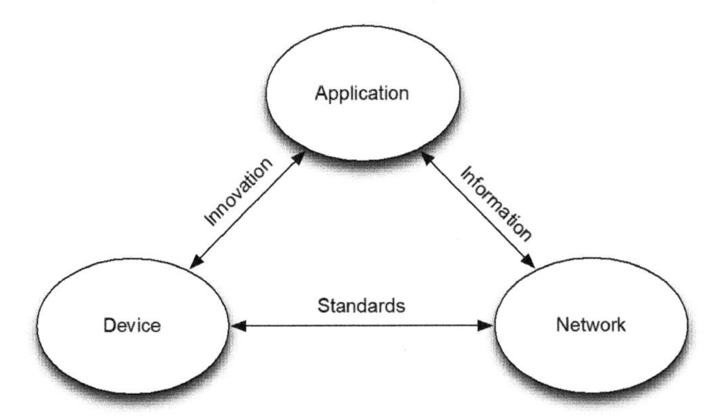

Figure 1.2 Elements of information networking.

and socialize; so, in fact, they are instrumental in the evolution of our humanity.

1.2.1 Classifications of Communications Networks

Communication networking is a huge worldwide industry involved in many technologies, standards, and applications. To study such a vast industry, we need to classify them, and these classifications can be done in a variety of groupings. Figure 1.3 provides an overview of the most popular networks and their applications.

The backbone of the communications networks is the core network that supports the connection-based circuit switched public switched telephone network (PSTN) and the connection-less packet switched Internet services for long-haul transmission. Then, we have access medium to connect to the core, private branch exchanges (PBX), Ethernet, Wi-Fi, cellular, and IoT. The privately owned, circuit switched PBX networks and Ethernet are the intermediate private networks in the corporate environment that connects to the PSTN and Internet, respectively. Ethernet is also used in homes to connect the Wi-Fi home router to the backbone fiber lines bringing these services to home. Wi-Fi is the wireless Ethernet to support portability in these environments. Corporate Wi-Fi consists of multiple access points with a deployment plan. Wi-Fi at home is often a multi-band router with several access points at different frequencies to optimize coverage and accommodate

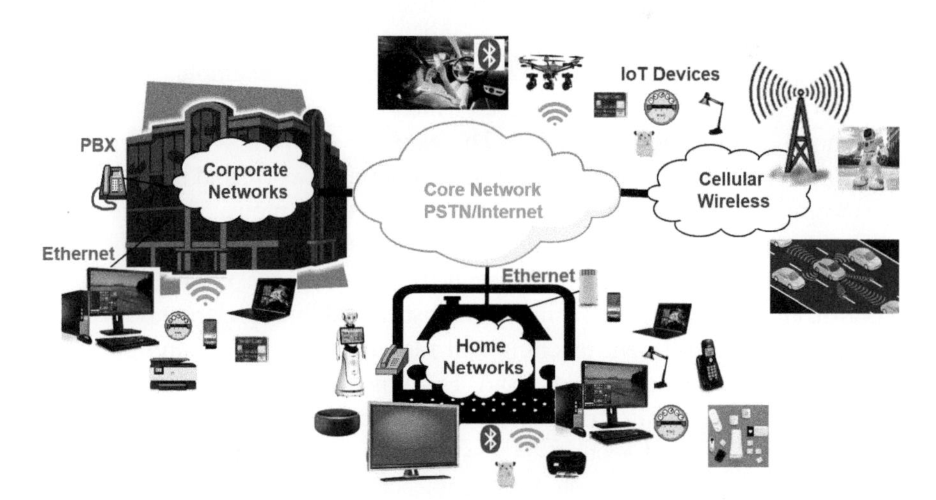

Figure 1.3 Overview of popular networks and applications.

a variety of devices with a huge difference in demand for the data rate. Cellular network base stations are deployed outdoors to cover metropolitan urban and suburban areas with mobility support to accommodate mobile services to the vehicles. IoT network is to connect the sensors in many different indoor and outdoor settings and they use Bluetooth, ZigBee, ultra-wideband (UWB), and other technologies to connect to the Internet, often through Wi-Fi in indoor residential and office setting and cellular wireless in wider outdoor areas. These classifications are instrumental in understanding communications networking technologies and applications for research and development as well as educational purposes, and we follow them in presenting communications networking technologies in this book. Different applications connect to both PSTN and the Internet and they have private local PBX networks for telephone services as well as local Ethernet networks for distributed computing and Wi-Fi for mobile computing. They also have remote access for virtual networking and the merger of different segments of the network to the main facilities. These subnetworks have different security issues because access to different computational facilities, information storage, and software services is restricted.

1.2.2 Examples of Corporate Networks

Corporate networks are perhaps the most complex of all local access networking technologies in terms of diversity in devices and technologies that are integrated in them. Figure 1.4 shows a simplified version of the corporate computational networking facilities at the Worcester Polytechnic Institute (WPI), Worcester, MA, USA, which we use as an example. This figure does not include the PBX network, which has a simple infrastructure for telephone networking with PSTN. The example network at WPI consists of the core WPI backbone network, the distribution network, and infrastructures for access to services from inside and outside the campus. The core of this infrastructure is the main router to connect to the outside world and the cloud, plus computer service facilities for security, wireless access management and control, and servers. These facilities are distributed through larger networking switches among different teaching and research laboratories and buildings, for example, Atwater, Fuller, and Higgins, hosting different academic departments and administrative offices. The corporate Wi-Fi infrastructure with a higher level of security is a part of this secure access infrastructure allowing all faculty, staff, and students access to their needed local and external computational and data storage facilities. In addition, we have a

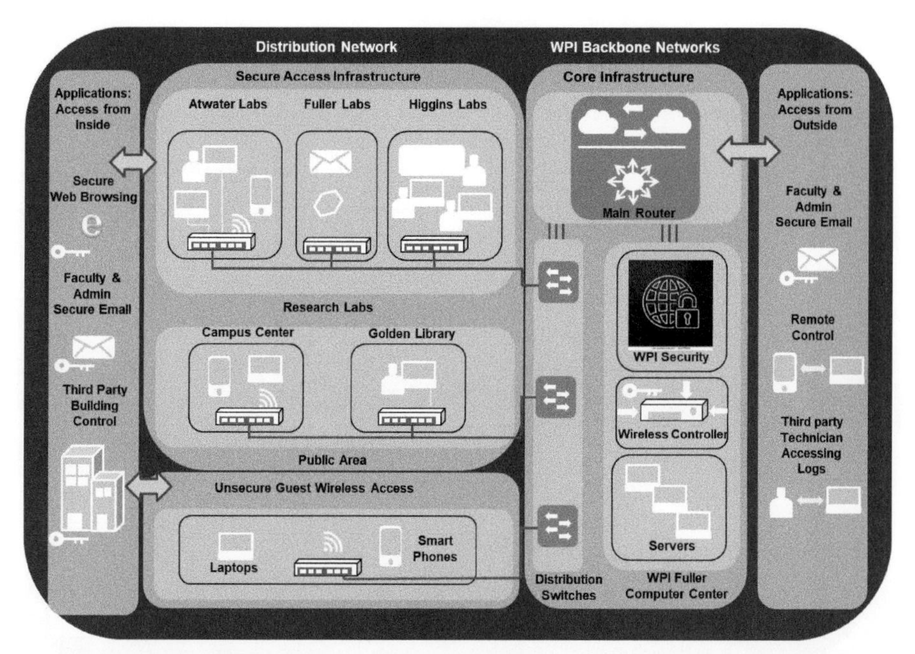

Figure 1.4 A simplified overview of the corporate computational networking facilities at the Worcester Polytechnic Institute, Worcester, MA, USA.

separate unsecured Wi-Fi sub-network for visitors' access as guests. The access to applications such as web browsing, email, and third-party building control and management services from inside the campus are all provided by top security. Access to these applications from outside the campus support secure email system for faculty and administration as well as remote control with mobile devices and third-party technician accessing log files for computer and communications software running inside the campus.

Figure 1.5 shows a typical floor in one of the buildings, the Atwater Kent Laboratory, with details of the networking devices in a research laboratory, a teaching laboratory, a lecture hall, an office, and a repair shop, and four Wi-Fi access points covering the floor. A fiber connection to the top distribution switch in the core infrastructure of Figure 1.4 brings a 10 Gbps Ethernet cable to the switch located on the second floor of this building, and all the wired equipment and computing devices on that floor in the second and third are connected to that switch. In addition, four multi-band Wi-Fi access points cover the floor for the faculty, staff, and students with a secure wireless connection and an unsecured connection to guests visiting the facility. All

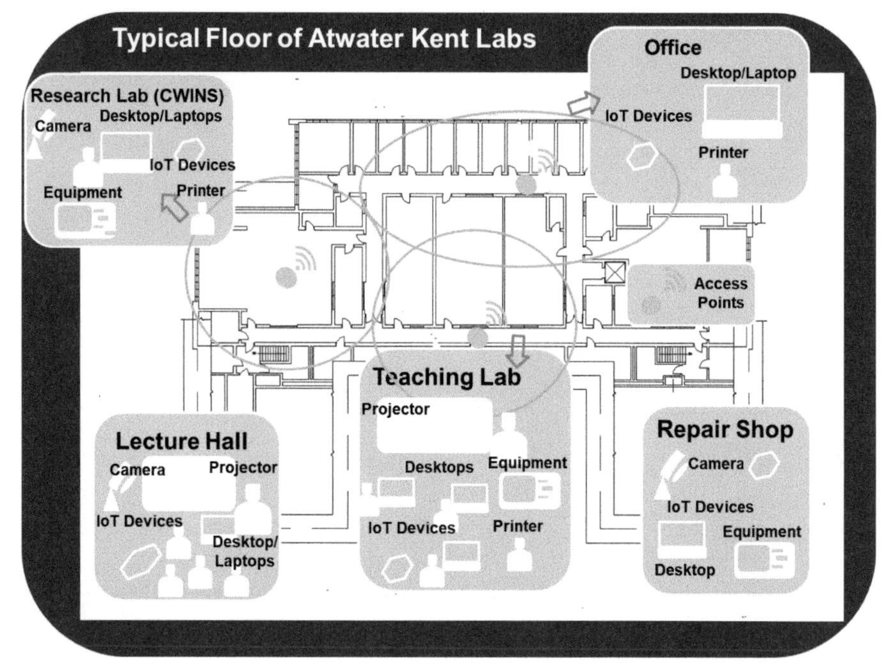

Figure 1.5 Layout of a typical floor at Atwater Kent Laboratory, WPI, and communication networking devices in that floor.

working equipment is equipped with IoT devices to control the lighting and the temperature. Other than that, each facility has desktops, printers, cameras projectors, and lab equipment that connect to Ethernet or Wi-Fi.

Figure 1.6 shows a typical networking research testbed in the CWINS laboratory on the third floor of the Atwater Kent Laboratory, WPI. A group of iBeacons, a group of cameras, several smartphones, and a group of laptops and desktops form a local network with a Wi-Fi access point controlled by a server for information exchange sessions. A group of Raspberry Pi small computers emulates malicious nodes to steal information or disrupt the operation of this testbed for emulation of a local IoT for body area networking (BAN) inside a hospital for patient monitoring. The actual network for the experience with details of the operation scenario is shown in Figure 1.6 [Yi18].

Figure 1.7 shows a typical hospital network focused on IoT devices and the BAN that connects the network to the cloud. The IoT sensors include body mounted health monitoring sensors, environment characteristics sensors, cameras, and smartphones for the patient and health care staff. The

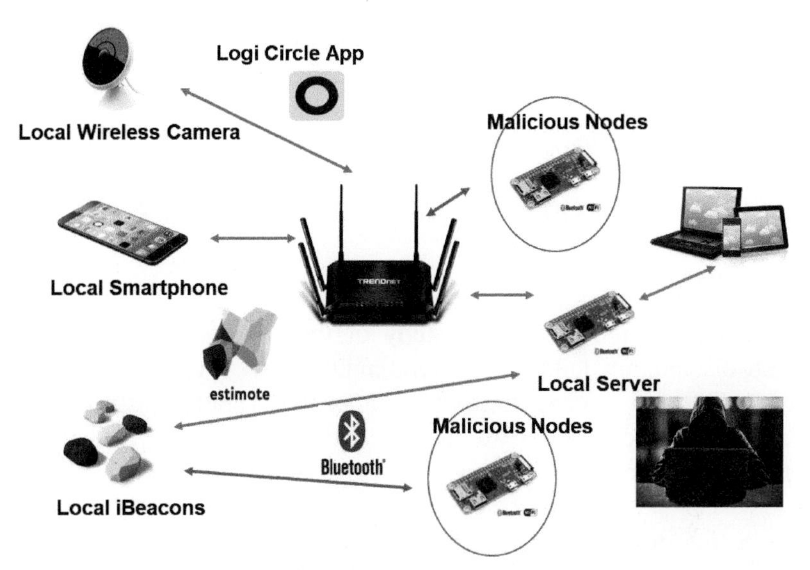

Figure 1.6 The laboratory testbed inside the CWINS research lab to emulate an environment for cyberspace security for a hospital room BAN [Yi18].

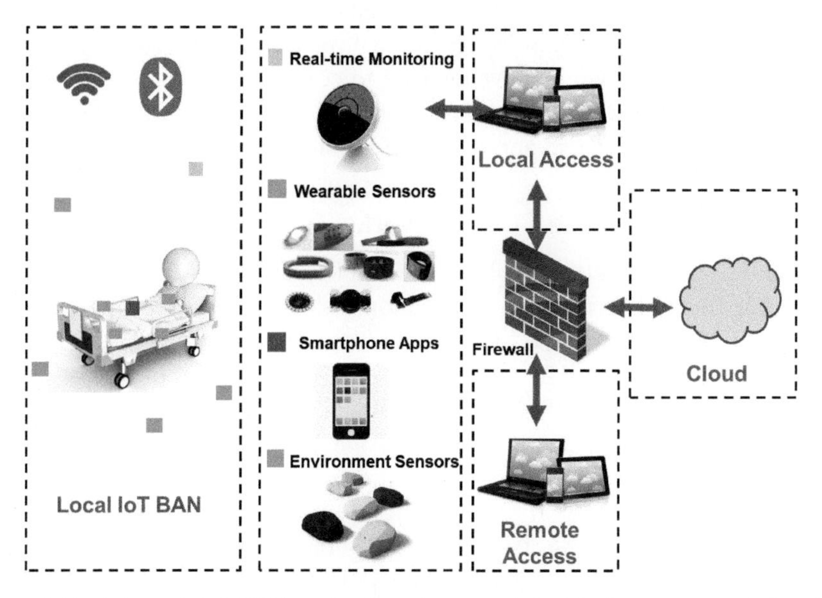

Figure 1.7 A typical IoT BAN applications in a hospital with devices connected to the network and the network layout to support secure operation of the network.

network supports secure local and remote access with different privileges set by the needs of the health care service and health care provider requirements. Since sensitive health related data is stored in the cloud, a firewall protects the security of operation.

1.3 Evolution of Communications Networks – A Historical Perspective

In this section, we provide a historical perspective for the evolution of communications networks. The study of the history of the evolution of a multi-disciplinary field such as communications networking enables the reader to develop a deeper understanding of the overall industry and role of technological innovations in the emergence of that industry. A holistic view of such a vast topic needs logical classifications, and as we explained in Section 1.2.1, we can classify today's communications networks into many different manners for different applications. For example, we can divide modern communications networks into a home, corporate, and cellular mobile networks connecting to the core PSTN and Internet (Figure 1.3). In this section, we divide them from a historical point of view into the core networks, indoor networks for local and personal area networking, and cellular wireless for comprehensive outdoor coverage in urban and suburban areas. Since communications networks have evolved to connect communication devices, this approach allows us to present a historical path for the evolution of different technologies and applications for communications networking.

1.3.1 Evolution of the Core Networks

Figure 1.8 illustrates the evolution of applications, devices, and networks to enable them. The first communication networking device that enabled a popular application was the Morse pad for the telegraph application that was invented in the 1830s. The telegraph was the very first application of short messaging systems (SMS). The communication network for the telegraph needed human operators familiar with the Morse code to route the text message along with long wire segments of the telegraph network from source to destination exchange offices. Therefore, we can consider the telegraph as the first packet switching data communication network using Morse pad as the communication device and operators as human routers benefitting from digital transmission over communication mediums. The Marconi's trial for

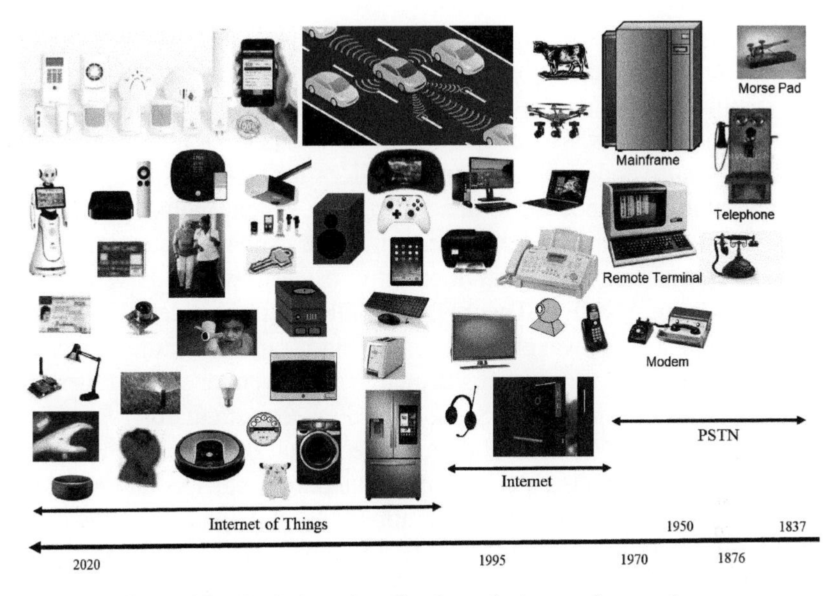

Figure 1.8 Evolution of applications, devices, and networks.

wireless telegraph over the Atlantic Ocean in the turn of the 20th century made telegraph also the first wireless communications network of the world.

The invention of the telephone in 1876 was the start of analog telephone networking and the so-called PSTN. The early-day networks used operators to manually switch the telephone connection among different available lines at the exchange offices to establish a connection line for the end users to begin their analog voice communication session. By the 1950s, the PSTN had more than ten million customers in the US and those interested in long-haul communication issues also needed other services to solve their problems. The end users were connected to the PSTN core with twisted-pair analog lines and the core long-haul transmission on the network was also analog. To provide flexibility and ease of maintenance and operation of the PSTN, the core network gradually changed to digital switches and digital wired lines connecting switches together. A hierarchy of digital lines called the T-carriers evolved in the US as the standard for trunks to connect switches of different sizes together, but the access remained analog. The analog user voice arriving at this core digital network was sampled at 8 Ksps with 8 bits per sample to form a 64 Kbps data stream for each direction of the telephone conversation. The PSTN with an analog telephone line and a core digital network exploded along with the world and through the 20th century.

With the popularity of optical communications in the 1980s and their ability to support much higher transmission rates, there was a need to extend the existing long-haul carrier hierarchies. At the same time, AT&T was broken up and there was a need to develop a standard for long-haul hierarchy so that the multivendor manufactured devices for different companies could cooperate with one another. In this setting, the synchronous optical network (SONET) standard hierarchy started in North America later in EU under the name synchronous digital hierarchy (SDH), which has only minor differences with SONET. The SONET/SDH hierarchy defines a common signaling standard for wavelength, timing, and frame structure for long-haul transmission in the core networks to enable multi-vendor and multi-providers operation, and it unified the US, EU, and Japanese standards for worldwide communications. SONET/SDH defines the format of the truck transmissions with data rates substantially higher than standard T-carriers and provides for better support of operation, administration, and maintenance of the network. SONET/SDH can be used for the encapsulation of earlier digital transmission standards.

Another advancement in the PSTN was the development of PBXs as privately owned local telephone networks for large offices. A PBX is a voice-oriented LAN owned by the end organization itself, rather than the telephone service provider. This small switch allows the telephone company to reduce the number of wires that are needed to connect all the lines in an office to the local office of the PSTN. This way, the service provider reduces the number of wires to be laid to a small area where large offices with many subscribers are located. The end user also pays less to the telephone company. The organization has thus an opportunity to enhance services to the end users connected to the PBX.

In the 1920s, Bell Laboratories conducted studies to use the PSTN facilities for data communications. In this experiment, the possibility of using analog telephone lines for transferring transoceanic telegrams was examined. Researchers involved in this project discovered several key issues that included the sampling theorem and the effects of phase distortion on digital communications. However, these discoveries did not affect applications until after World War II when Bell Laboratories developed voice-band modems for communication among air force computers in air bases that were geographically separated by large distances [Pah88]. These computer communication modems soon found their way into commercial airlines and banking industries resulting in the associated private long-haul data networks. These pioneering computer communications networks consisted of a central computer and a bank of modems operating over 4-wire commercial grade leased telephone lines to connect several terminals to the computer. In the late 1960s, the highest data rate for commercial modems was 4800 bps. By the early 1970s, with the invention of quadrature amplitude modulation (QAM), the data rate of 4-wire voice-band modems reached 9600 bps. In the early 1980s, trellis coded modulation (TCM) was invented that increased data rates to 19.2 kbps and beyond.

In parallel with the commercial 4-wire modems used in early long-haul computer networks, 2-wire modems emerged for distance connection of computer terminals. The 2-wire modems operated over standard 2-wire analog telephone lines and they were equipped with dialing procedures to initiate a call and establish a data communication session on a dial-up analog line. These modems started at data rates of 300 bps. By the early 1970s, they reached 1200 bps, and by the mid-1980s, they were running at 9600 bps. These 2-wire voice-band modems would allow users in the home and office having access to a regular telephone to develop a data link connection with a distant modem also having access to the PSTN. Voice-band modems using 2-wire telephone connections soon found a large market in the residential and small office remote computer access (telnet) and the technology soon spread to several popular applications such as operating a facsimile machine or credit card verification device. With the popularity of Internet access, a new round of gold rush for higher speed modems began, which resulted in 33.6 kbps full-duplex modems in 1995 and 56 kbps asymmetric modems by 1998. The 56 kbps modems use dialing procedures and operate within the 4 kHz voice-band, but they directly connect to the core pulse code modulation (PCM) digital network of the PSTN that is like digital subscriber lines (DSLs). DSLs use the frequency band between 2.4 kHz and 1.1 MHz to support data rates up to 10 Mbps over two wired telephone lines.

More recently, cellular telephone services evolved. To connect a cellular telephone to the PSTN, the cellular operators developed their own infrastructure to support mobility. This infrastructure was connected to the PSTN to allow mobile to fixed telephone conversations. Addition of new services to the PSTN demanded increases in the intelligence of the core network to support these services. As this intelligence advanced, the telephone service provider added value features such as voice mail, auto-dialing through network operators, call forwarding, and caller identification to the basic analog telephone service traditionally supported.

Data networks that evolved around voice-band modems connected a variety of applications in a semi-private manner. The core of the network was still the PSTN, but the application was for specific corporate use and was not offered privately to individual users. These networks were private data networks designed for specific applications and they did not have standard transport protocols to allow them to interconnect with one another. Another irony of this operation was that the digital data was first converted to analog to be transmitted over the analog access to the PSTN at arrival to the core. It was again converted to digital 64 Kbps PCM format for transmission over a core, and at the other end, it was turned back analog to get to the end user to be reconverted to the original digital stream. To avoid this situation, starting in the mid-1970s, telephone companies started to introduce digital data services (DDS) that provided 56 Kbps digital service directly delivered to the end user. The DDS services later emerged as integrated service digital network (ISDN) standard for services providing two 64 Kbps voice channel and a 16 Kbps data channel to individual users. Asynchronous Transfer Mode (ATM) core protocol was used in SONET/SDH backbone of the PSTN for long-haul transportation of the ISDN data. Penetration rates of ISDN services were not as expected, but the second generation Pan European digital cellular widely adopted a wireless version of the ISDN. These technologies are largely superseded and integrated into IP technology for long-haul transmission.

In the early 1970s, the rapid increase in the number of terminals in offices and manufacturing floors was the force behind the emergence of LANs. LANs provided high-speed connections (greater than 1 Mbps) among terminals facilitating sharing printers or mainframe computers from different locations. LANs were providing a local medium specifically designed for data communications that were completely independent of the PSTN. By the mid-1980s, Ethernet emerged among several successful LAN protocol alternatives and LANs were installed in most large offices and manufacturing floors connecting their computing devices. Another important and innovative event in the 1970s was the implementation of advanced research projects agency network (ARPANET), the first packet switched data network connecting 50 cities in the USA. This experimental network used routers rather than the PSTN switches to interconnect data terminals. The routers were originally connected via 56 Kbps digital leased lines from the telephone company and the ARPANET interconnected several universities and government computers around a large geographical area. This network was the first packet switched network supporting end-to-end digital services. This basic network was later upgraded to higher speed lines and numerous additional networks. To facilitate a uniform communication protocol to interconnect these separated networks, the transmission control protocol/Internet protocol

(TCP/IP) evolved which allowed LANs as well as several other public data networks to interconnect with one another and form the Internet. In the mid-1990s, with the introduction of popular applications such as telnet, ftp, e-mail, and web browsing, the Internet industry was created. Soon, the Internet penetrated the home market, and the number of Internet users became comparable with that of the PSTN creating another economic power, namely computer communications applications, that competed with the traditional PSTN. The IP-based Internet provides a cheaper solution than circuit switched operations and it could better accommodate the integration of an exponentially growing number of data applications (middle part of Figure 1.8).

At the turn of the 21st century, we witnessed an explosion of the wireless networking industry. The Wi-Fi infrastructure began to cover indoor areas for tetherless access to the Internet and cellular wireless networks in outdoors, first provided access to the PSTN and then to the Internet and they nurtured the emergence of smartphones close to the end of the first decade of this century. Smartphones were the first telephones equipped with cellular chipsets as well as Wi-Fi and Bluetooth chipsets. Now Internet could connect to ever-growing devices benefiting from wireless technologies using these three technologies and through the smartphone, and what we call IoT emerged (left side of Figure 1.8). The IoT is enabled by wireless connection to the Internet, mostly through intervention from smartphones and sometimes directly. The backbone wireless networks to enable IoT are Wi-Fi for indoor areas and cellular wireless for outdoors with Bluetooth and ZigBee as intermediary networks to establish the connections. We divide these technologies into "local and personal area networks" and "wireless cellular networks" to present the story of their evolution in the next two sub-sections.

1.3.2 Evolution of Local and Personal Area Networks

The LAN industry emerged during the 1970s to enable sharing of expensive resources like printers and to manage the wiring problem caused by the increasing number of terminals in offices. These applications for local indoor networking was for distance coverage inside the buildings and they could support data rates well beyond that of the digital service over long-haul PSTN. By the early 1980s, three standards were developed: Ethernet, supported by Digital Equipment Corporation, Xerox, and Intel, Token Ring, presented by IBM, the giant of the computer industry at that time, and Token

Ring, promoted by Hewlett Packard, which was focused on manufacturing market. The data rates of all these LAN alternatives were 10 Mbps. Ethernet evolved as the most successful and finally became the trend for the local wired computer networking.

The need for higher data rates in LANs emerged from two directions: (1) there was a need to interconnect LANs that are located in different buildings of campus to share high-speed servers and (2) computer terminals became faster and capable of running high-speed multi-media applications. To address these needs, several standards for higher data rate LANs emerged in the market. The first fast LAN operating at 100 Mbps was the Fiber Distributed Data Interface (FDDI) that emerged in the mid-1980s as a backbone medium for interconnecting other LANs. In the mid-1990s, 100 Mbps fast Ethernet emerged which was based on the success of the FDDI. In the late 1990s, the Gigabit Ethernet was introduced followed by 10 and 100 Gbps Ethernets and beyond. At the time of this writing, Ethernet is penetrating into the core network and SONET/SDH long-haul transmission. These advances in accommodation of higher data rates and longer coverages materialized based on the design of more efficient transmission over a variety of wired media and the adoption of high-speed transmission techniques using complex coding techniques.

During the 1990s, the Wi-Fi technology emerged as wireless Ethernet and nurtured mobile computing to connect laptops, in homes and offices, to the Internet. In the 2000s, wireless personal area networking (WPAN) emerged and introduced Bluetooth followed by ZigBee technologies for wireless sensor networks that could virtually connect the Internet to everything to create the IoT (left side of Figure 1.8).

The Wi-Fi as a wireless LAN (WLAN) was focused on increasing the data rate in an area of up to 100 m range and Bluetooth and ZigBee on low energy for *ad hoc* and sensor networking. The Wi-Fi technology began with data rates of 1–2 Mbps and evolved to Gbps. Bluetooth and ZigBee originally began aiming at long battery life from few days to go to a few months and years. The Wi-Fi technology was discovering higher data rates using wideband and UWB technologies and WPAN technologies were focused on transmission technologies with lower energy consumption to stretch the life of the batteries. These technologies became the core of modulation and coding techniques forming the wireless industry. The popularity of Wi-Fi in home, small offices, and corporate areas created a huge infrastructure of over a billion access points around the world. The RF signal radiated from these access points by their beacons carries the location of the area of the access point and their electronic addresses. Since access points are installed in fixed locations, if we know the location of an access point, we can locate a device that can read the beacon of the access point. This simple and neat idea created the Wi-Fi positioning industry in the first decade of the 21st century. More recently, we use the signal from Bluetooth of smartphones to find the social distance. These are location-related cyberspace applications taking advantage of the RF cloud radiating around the devices transmitting beacons such as Wi-Fi access points or Bluetooth low energy (BLE). The RF cloud then entered the arena of other popular cyberspace application in gesture, activity, motion detection, and authentication and security during the 2010s. It is expected that these applications revolutionize the human–computer interfacing in the 2020s [Pah19].

1.3.3 Evolution of Wireless Cellular Networks

The cellular networks began with the analog cellular telephone to provide wireless mobile access to the PSTN telephone services. The technology for the analog cellular first generation (1G) systems was developed at the AT&T Bell Laboratories in the early 1970s. However, the first deployment of these 1G systems took place in the NORDIC countries under the name Nordic Mobile Phone (NMT) about a year earlier than the deployment of the advanced mobile phone services (AMPS) in the US. Since the US was a large country, frequency administration and other regulations in the process were slower; so it took a long time for the deployment. The digital cellular networks in NORDIC countries started with the formation of the Groupe Special Mobile or GSM standardization group, later renamed as Global Service Mobile. The GSM standards group was originally formed to address international roaming, a serious problem for cellular operation in the European Union (EU) countries. The standardization group shortly decided to go for a new digital time division multiple access (TDMA) technology because it could allow integration of other services to expand the horizon of wireless applications [Hau94]. In the US, however, the reason for migration to digital cellular was that the capacity of the analog systems in major metropolitan areas such as New York City and Los Angles had reached its peak value and there was a need for increasing the capacity in the existing allocated bands. Although, NORDIC countries, led by Finland, maintained the highest rate of cellular penetration in the early days of this industry, the US was, by far, the largest market. Meanwhile, with the huge success of digital cellular telephony, all manufacturers worldwide started working on the 3G cellular IMT-2000 wireless networks. The 1G cellular offered analog voice and the 2G cellular provided circuit switch voice and data like ISDN. The GSM also introduced SMS, the first packet switched cellular network overlaid on the control and signaling channel. The connection-based networks could be considered as two networks, one for signaling and control to establish and maintain the call using short bursts of data and one for transferring information after the call is established. The SMS in the original GSM sent the SMS over the signaling and control channel, not the user traffic channel. The 3G IMT-2000 aimed at 2 Mbps data link for packet switching on top of the telephone service. This jump in the demand for high data rate packet switching was motivated by the popularity of the Internet for home applications in the mid-1990s and its explosive exponential growth ever since its inception. While 3G was focused on the quality of the delivered circuit switched voice application as the mainstream of revenue, it did a very drastic shift in accommodation of packet switching. Not only the 3G aimed at supporting 2 Mbps packet data services, but it also defined a core IP-based infrastructure enabling mobile devices wireless access to both PSTN and the Internet. This core is referred to as the 3G partnership project (3GPP) and was adopted by all major service providers and has become the core for the following generations of cellular networks. After the 3G WCDMA, cellular networks became like Wi-Fi adopting OFDM and MIMO transmission techniques for optimization of packet switching and the Internet access. This movement first began with worldwide interoperability for microwave access (WiMAX) that could be thought of as an extension to the WLAN technologies for outdoor antenna deployment, followed by long-term evolution (LTE) as the fourth generation (4G) cellular wireless data network.

The LTE provided an enhancement more suited to fragmented frequency bands used by cellular networks compared to WiMAX. LTE-advanced was the next standard emerging in this area aiming at maximum data rates on the order of a gigabit per second. The basic physical layer of these technologies is like the physical layer of Wi-Fi that better suits emerging data applications. The implementation of the medium access control is adjusted to the specifics of cellular networks that are designed for comprehensive wide area coverage, a fragmented bandwidth allocation, and support of a large number of users from one antenna location. The transmission novelty of 5G cellular was in mmWave communications with UWB wireless data communication using pulse transmission, adoption of smaller cells, and massive MIMO antenna arrays enabling multiple streaming as well as capabilities for accommodating

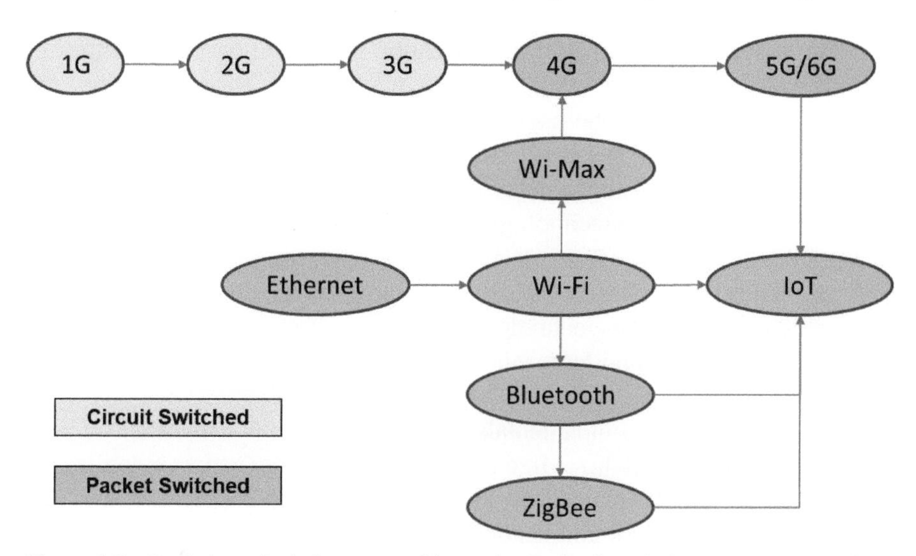

Figure 1.9 Evolution of wireless networking technologies for wireless access to the PSTN and the Internet.

IoT devices. This trend of reduction in cell size, expansion of antenna arrays to support narrower beams to increase the number of streams, and resorting to higher frequencies in Thera Hertz (THz) continues to the evolving 6G and 7G technologies. Another trend for these technologies is the intelligence spectrum management for efficient use of spectrum to enable intelligent radio with interference management.

Figure 1.9 provides a summary of the evolution of wireless technologies, the cellular telephone networks, Wi-Fi, Bluetooth, and ZigBee for wireless access to the PSTN, the Internet, and the IoT. The upper parts of the figure illustrate the evolution of standards for wireless access to the PSTN core and the lower part is related to wireless access to the Internet core. The 6G is evolving around intelligent spectrum sharing and the application of machine learning algorithms to spectrum management and regulations.

1.4 Standards, Regulations, and Technologies

The core information networks consist of communication links and interconnecting elements connecting users worldwide. This infrastructure is owned by numerous national and international organizations, and they need standards to define the interfaces between the elements of the network. These standards define the format of packets for different available mediums for communications to enable multi-vendor operation, to manufacture the interconnecting elements, and to design the medium to interconnect. We also need standards for the users to access the core network with wires or wireless technologies. The increasing number of fixed, portable, and mobile wireless applications on different communication devices demand a variety of standard wireless access technologies operating on different frequency bands. These frequency bands are regulated by government agencies and innovation in using these bands has an outstanding impact on the evolution of the wireless communications technologies as well as navigation, environmental, surveillance, and transportation industries, which are sharing the frequency bands with the wireless communications networks.

To understand the fundamentals of communications networks, we need to have an idea about how frequency administration and standardization organizations operate, as well as an understanding of the technologies that are recommended by communications standards.

1.4.1 Intelligent Radio with Spectrum Management

Frequency bands are regulated by national agencies such as the Federal Communication Commission (FCC) in the US. Figure 1.10 is a reprint of the U.S. frequency allocations for different radio spectrums covering all applications. This regulation covers a frequency range of 3 kHz to 300 GHz shown in logarithmic scale with six segments. The frequency range 3–30 KHz is called very low frequency (VLF) and it covers wavelengths of 100–10 Km,[1] and for that reason, it is also called myriametric (10Km) wave spectrum. This band is used for a few radio navigation and military applications. The 30–300 kHz frequency band, called low frequencies (LF), has wavelengths of 10 m to 1 Km, and for that reason, it is also called kilometer wave spectrum. Its most popular usc is in aircraft navigation. The 300 KHz to 3 MHz is the medium frequency band, hosting AM radio broadcasting. The 3–30 MHz band is the high frequency (HF) band, hosting amateur radio and citizen bands. The 30–300 MHz band is very-high frequency (VHF), with popular FM radio TV broadcast applications as well as air traffic control. The ultra-high frequency (UHF) band of 300 MHz to 3 GHz with a wavelength of 1 m to 1 cm is the most popular in wireless cellular, Wi-Fi, Bluetooth, and ZigBee technologies as well as GPS. The wireless

[1]The frequency, f, and the wavelength λ are related by $\lambda = c/f$, where $c = 10^8 \mathrm{m/s}$ is the speed of light is free space.

Figure 1.10 Reprint of the US radio spectrum allocations chart, FCC, Department of Commerce.[2]

industry is now discovering 3–30 GHz super-high frequency (SHF), 30–300 GHz mmWave bands, as well as 0.3–3 Terahertz (THz) for the future of wireless communications. These bands are in the middle of microwave and infrared lights and are popular in astronomy, outdoor applications, and in X-ray imaging in indoor areas. After these frequencies, we enter visible light and infrared optical communication, which have been subject to a variety of innovative applications for wireless communications. Infrared was one of the options of the legacy Wi-Fi and visible light communication (Li-Fi) has been popular for WPAN applications.

Technologies that are discussed in this book include PSTN and the Internet core networks, and wired LANs, which do not get affected by FCC decisions directly. The wireless access to the core with outdoor metropolitan area cellular wireless or the wireless indoor local and personal area networks are highly affected by FCC regulations. The wireless indoor local and personal area networks technologies emerged after the release of an instrument, scientific, and medical (ISM) unlicensed bands in 1985, and they could not otherwise. Licensed bands are like a privately owned backyard. The

[2]https://www.ntia.doc.gov/files/ntia/publications/2003-allochrt.pdf

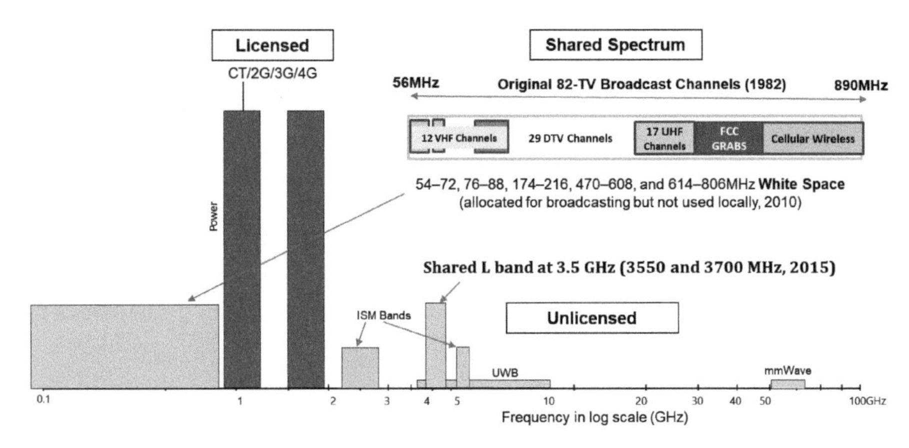

Figure 1.11 Samples of licensed and unlicensed bands spectrums in the US.

owner of the band needs to invest a substantial amount of money and effort to obtain permission for using that band in a certain geographical area. These bands usually allow higher transmission power, but they are more restricted in the size of the bandwidth. Unlicensed bands are like public gardens; users of these bands have access to a wider bandwidth but with restrictions on the transmission power.

Unlicensed bands were the first innovation in intelligent spectrum management allowing different technologies to use the same band without any coordination. In 2010, FCC introduced the white space bands, which were unused TV station bands (56–890 MHz) in coverage wholes of TV station coverages. To avoid interference, TV stations are operated in geographically separate areas. Further, because of population density shifts and shifts in habitats of residence in different areas, there are geographical areas in which we have channels that are not utilized. These unused spectra between TV stations in different areas are called white spaces and it provides a valuable opportunity for wireless mobile applications for capacity expansion and with innovative new technologies. In 2010, the FCC adopted final rules to allow unlicensed radio transmitters to operate in the white spaces for spectrum sharing.

The shared spectrum bands indeed overlay an unlicensed operation in a licensed band giving a priority to the existing licensed operation. In 2015, FCC allowed secondary use of 3.5 GHz band (between 3550 and 3700 MHz), for its current primary use, offshore radar operations by the US Navy, and cellular vendors rushed to develop equipment that benefits from this opportunity to increase the capacity of wireless cellular networks. Providers and other organizations will be granted access using a three-tier priority allocation structure: (1) incumbent users such as the US Navy, (2) LTE providers and other organizations that pay license fees for the right to share, and (3) general users in an unlicensed format of operation.

Figure 1.11 illustrates several examples of licensed, and shared bands in the US that are used for different generations of cellular wireless networks as well as several unlicensed bands for wireless local and personal area networks. These are licensed frequency bands around 1 and 2 GHz as well as unlicensed ISM bands at 2.4 and 5.2 GHz, UWB at 3.6–104 GHz, mmWave bands at 57–64 GHz, white space sharing with TV bands, and L bands at 3.5 GHz shared with three levels of priorities. Other bands of interest for cellular network providers are (1452–1492 MHz), which is assigned for 5G in some European countries, Iridium Phones bands (1616–1626.5 MHz), and Inmarsat bands (1525 and 1646.5 MHz). GPS operates at 1176.45, 1227.60, 1381.05, and 1575.42 MHz, and there are concerns for interference to that important technology from wireless communications. This concern by GPS community affected FCC decision-making on unlicensed UWB templates and it has been debated for cellular radio operation in neighboring bands.

These examples provide an insight on complexity, bandwidth, frequency of operation, and rules for the operation of the bands. At the time of this writing, the cellular wireless networks mainly rely on expensive licensed bands for outdoor applications. This industry is discovering intelligent spectrum management in existing bands and operation at THz, where design faces challenges but bandwidth are abundant. Because of the lack of accurate models for RF cloud interference in the smart world with trillions of IoT devices, the existing bands are utilized inefficiently and the research community is discovering intelligent interference management science and technology for the design of next generation intelligent radio with interference management (IRIM).

1.4.2 Standards Organization for Communications Networks

Standards define interface specifications between elements of a network infrastructure allowing a global multi-vendor operation, which facilitates the growth of the industry. Figure 1.12 provides an overview of the standardization process in communications networking. The standardization process starts in a special interest group of standards developing body such as the Institute of Electrical and Electronics Engineers (IEEE 802.11)

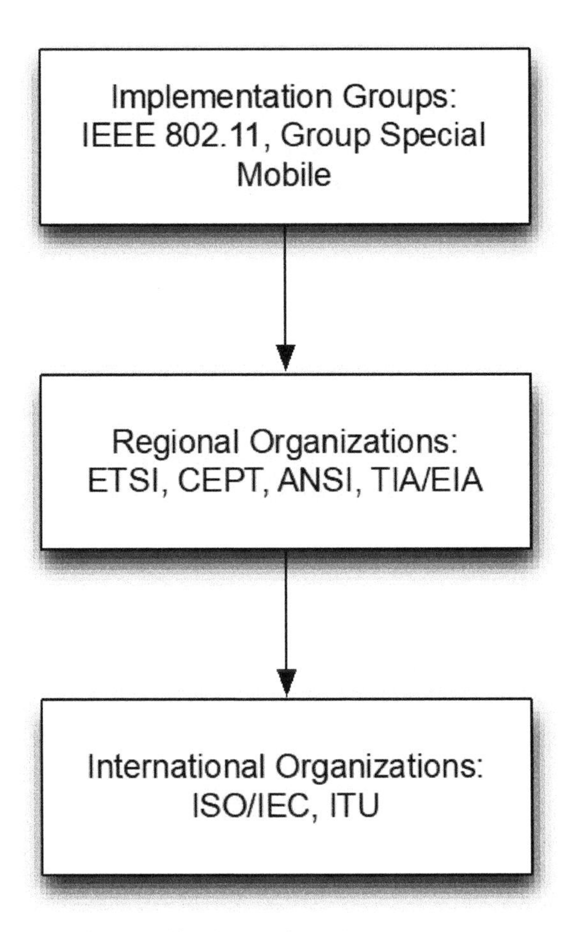

Figure 1.12 Standard development process.

or Global System for Mobile (GSM) communications, which defines the technical details of networking technology as a standard for operation. The defined standard for implementation of the desired network is then moved for approval by a regional organization such as the European Telecommunication Standards Institute (ETSI) or the American National Standards Institute (ANSI). The regional recommendation is finally submitted to world level organizations, such as the International Telecommunications Union (ITU), International Standards Organization (ISO), or International Electrotechnical Commission (IEC), for final approval as an international standard. There are several standards organizations involved in information networking. Table 1.1

Table 1.1 Summary of important standard organizations and government agencies for information networking.

FCC (Federal Communication Commission): The frequency administration authority in the US.

IEEE (Institute of Electrical and Electronics Engineers): Publishes 802 series standards that turned to popular Wi-Fi, Bluetooth, ZigBee, and UWB technologies.

GSM (Global System for Mobile): Special group defined 2G TDMA standard sponsored ETSI.

ATM (Asynchronous Transfer Mode) Forum: An industrial group worked on a standard for ATM networks for ANSI and ITU.

IETF (Internet Engineering Task Force): Publishes Internet standards that include TCP/IP and SNMP.

EIA/TIA (Electronic/Telecommunication Industry Association): US national standard for North American defined specification for TP, cable, fiber, and American cellular wireless.

ANSI (American National Standards Institute): A US national organization that has accepted 802 series and ATM and forwarded them ISO.

ETSI (European Telecommunication Standards Institute): Published GSM, HIPERLAN, and UMTS.

CEPT (Committee of the European Post and Telecommunication): Standardization body of the European Posts Telegraph and Telephone (PTT) ministries. Co-published GSM with ETSI.

IEC (International Electrotechnical Commission): Publishes jointly with ISO.

ISO (International Standards Organization): Ultimate international authority for approval of standards.

ITU (International Telecommunication Union, formerly CCITT): International advisory committee under United Nations. The telecommunication sector formalizes SONET/SDH for long-haul transmission and local, wired, and wireless access to core networks. The radio sector works on international frequency administration.

provides a summary of the important standards playing major roles in shaping the information networking industry, which is discussed in this book.

The most important standard developing organizations for technologies described in this book are the IEEE 802-series standards for personal, local, and metropolitan area networking. The IEEE is the largest engineering organization in the world, which publishes numerous technical journals and magazines and organizes numerous conferences worldwide. The IEEE 802 community is involved in defining standard specifications for information networks. The number 802 was simply the next free number IEEE could assign to a committee at the inception of the group in February 1980, although "80-2" is sometimes associated with the date of the first meeting. Regardless of the ambiguity of the name, the IEEE 802 community has played a major role in the evolution of wireless information networks by introducing IEEE 802.11, commercially known as Wi-Fi, and IEEE 802.15, which introduced Bluetooth, ZigBee, and UWB technologies, which are discussed in detail in this book.

Another important standard developing organization is the Internet Engineering Task Force (IETF) which was established in January 1986 to develop and promote Internet standard protocols around the TCP/IP suite for a variety of popular applications. In the 1990s, the ATM Forum was an important standard developing group trying to develop standards for connection-based fixed packet length communications for the integration of all services. This philosophy was in contrast with Internet/Ethernet networking using connection-less communications with variable and long length packets and it lost its momentum.

The Telecommunication/Electronic Industry Association (TIA/EIA) is a US national standards body defining a variety of wire specifications used in LANs, metropolitan area networks (MAN)s, and WANs. The TIA/EIA is a trade association in the US representing several hundred telecommunications companies. The TIA/EIA has cooperated with the IEEE 802 community to define the media for most of the wired LANs used in fast and gigabit Ethernet. TIA/EIA also defines cellular telephone standards such as Interim Standards (e.g., IS-95) or cdmaOne second generation (2G) cellular networks and the IS-2000 or CDMA-2000 third-generation (3G) cellular telephone networks. ETSI and the Committee of the European Post and Telecommunications (CEPT) were the European standardization bodies publishing wireless cellular networking standards, for different generations of wireless cellular networks.

The most important international standards organizations are ITU, the ISO, and the IEC, and are all based in Geneva, Switzerland. Established in 1865, ITU is an international advisory committee under United Nations and its main charter includes telecommunication standardization and allocation of the radio spectrum. The Telecommunication Sector, ITU-T, has published, for instance, the Integrated Service Data Network (ISDN) and wide area ATM standards, as well as different generations of wireless networks. The World Administrative Radio Conference (WARC) was a technical conference of the ITU where delegates from member nations of the ITU met to revise or amend the entire international Radio Regulations pertaining to all telecommunication services throughout the world. ISO and IEC are composed of the national standards bodies, one per member economy. These

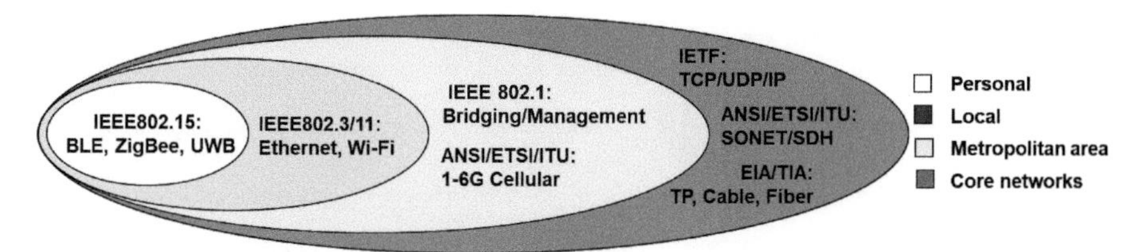

Figure 1.13 Overview of standards organization technologies and their area of coverage.

two standards often work with one another as the ultimate world standard organization. Established in 1947, ISO nurtures worldwide proprietary industrial and commercial standards that often become law, either through treaties or national standards. The ISO seven-layer model for computer networking was one of the prominent examples of ISO standards. The IEC started in 1906 and it is a non-governmental international standard organization for "electrotechnology" which includes a vast number of standards from power generation, transmission, and distribution to home appliances and office equipment, to telecommunication standards. The IEC publishes standards with the IEEE and develops standards jointly with the ISO as well as the ITU. Figure 1.13 shows a summary of standards organizations and technologies that we present in this book and categorizes them into their intended areas of coverage.

1.4.3 Communications Networking Technologies

Networks connect applications at end users to one another, communication technologies carry the information produced by the application over a guided or wireless medium (Figure 1.14). Standards organizations recommend the architecture of the network, characteristics of communication medium, and specification of communications links. The architecture of the network defines the elements of the network and their functionalities as they relate to other elements and the overall networking objectives. Characteristics of the guided medium define expected attenuation of the communication line for a given distance and expected radiation from the medium to interfere with the environment. Characteristics of wireless medium define models for the propagation of the signal for deployment and interference analysis. Specification of communication link defines protocols to communicate over the medium. Vendors design communication network equipment and manufacture the guiding mediums according to the recommended standards

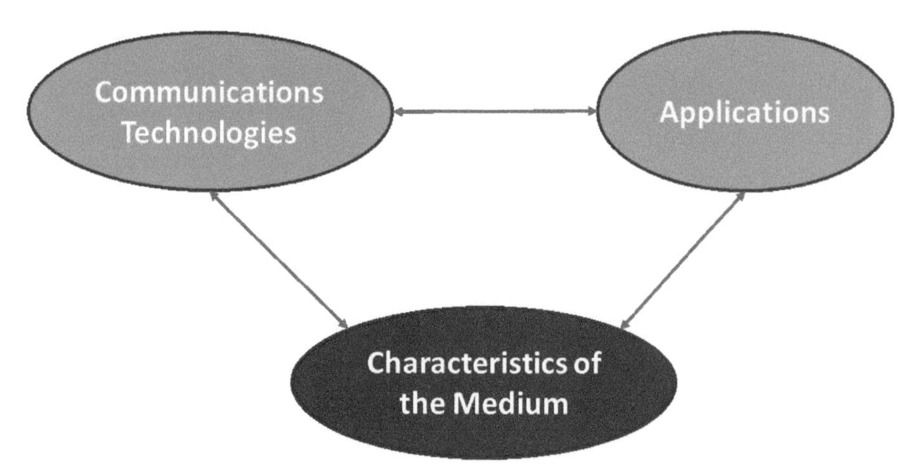

Figure 1.14 Applications, communications technologies, and characteristics of the medium.

specification. Service providers use the equipment and guiding media to deploy their networks to support services to communication devices.

Definition of communication protocol involves considerable details, and it is often defined in multiple layers, each layer receiving service from the layer below and providing service to the layer above. For example in Figure 1.15a, a layer that defines how to control error-free packet flows, data link layer, provide a service for the layer above it that integrate packets from different sources, network layer, and it provides a service to the layer below it, the physical layer, which sends the data bit-by-bit over the medium. In the 1970s, when PSTN was the dominant core network and the Internet was experimentally emerging to unify the definition of layered protocols for communications, OSI defined the open systems interconnection (OSI) a reference model for layer communications (Figure 1.15(a)). In 1975 CCIT (now ITU-T) defined the Signaling System Number 7 (SS-7) protocol stack for control and management of the worldwide PSTN core following OSI reference model. The OSI is a conceptual reference model attempting to relate characteristics of communication functions of the telecommunication PSTN and Ethernet/Internet enabling computer communications.

OSI reference model provides a broad view of the network operation protocols. However, the actual protocol stack defined by standards organizations follows their own technical needs. Figure 1.15(a) shows the OSI seven layers as a reference as it is compared with the Internet engineering task force (IEFT) protocol layers (Figure 1.15(b)) for a wide area datagram communications within the Internet core network and the IEEE

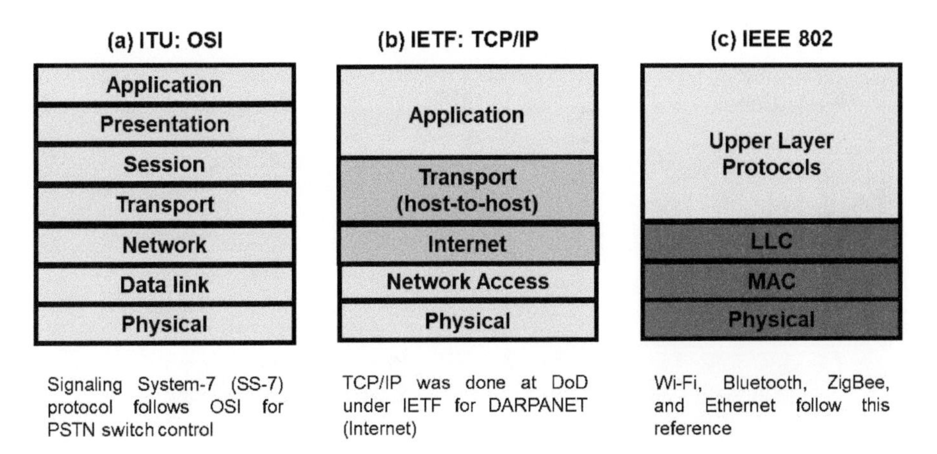

Figure 1.15 Three popular reference models for protocol stack. (a) OSI. (b) TCP/IP. (c) IEEE 802.

802 community reference model (Figure 1.15(c)) for local area networking. Regardless of how we define the layers by name, in practice, layers are a bunch of headers or trailers added to the application data. Using an example simplifies the explanation of the purpose of layers and how they get embedded in communication protocols.

Figure 1.16 shows an example of packet headers added to a data packet using the 802.11 framing and TCP/IP networking protocol stacks. The user data is encapsulated in a TCP/IP packets with two headers. The maximum TCP segment size (MSS) for communications is 65495 bytes, if the user application packet is longer, the TCP protocol needs to manage the fragmentation and reassembly at the end. In addition, TCP manages multi-port operations over a single end-to-end IP connection. The IP header carries the transmitter and receiver addresses along the segments of a communication network to deliver the packet to the intended destination. The maximum transmission unit (MTU) is the MSS plus overhead for TCP/IP headers. The packet is then delivered to LLC layer, which adds sub-network access protocol (SNAP) heather as well as the LLC header. SNAP identifies the hardware port connected to the network and LLC ensures sanity and flow of the transmitted packet over the PHY. The PHY header has a physical convergence layer (PLCP) header mapping the MAC packet data unit (MPDU) into the specific PHY format to make it ready for packet transmission over the PHY. When the MTU is larger than the MPDU, either LLC or MAC should support service for another

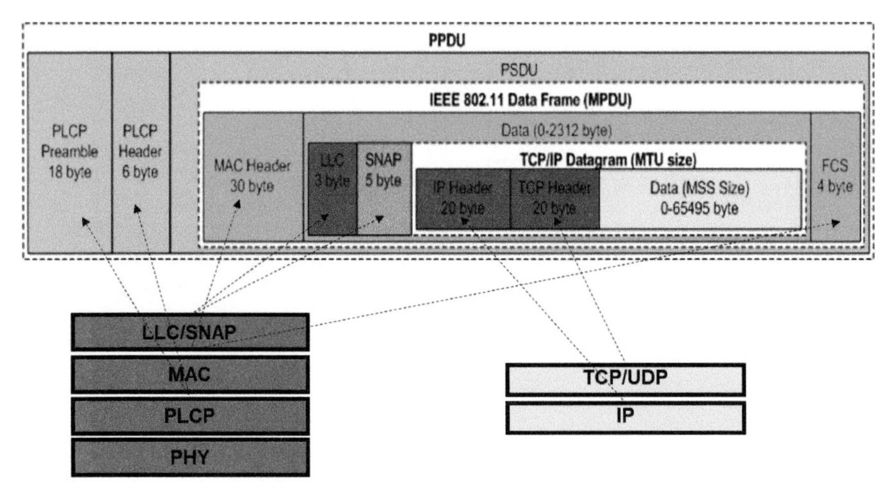

Figure 1.16 Example of Internet packet using IEFT TCP/IP protocol and an IEEE 802.11 protocol stack to connect to the Internet.

fragmentation and reassembly. These layer functioning headers are overhead for transmission of the application data and they are different for different protocols. The services provided with these layers enable a packet from multiple software applications to reach multiple hardware ports of the end users through different segments of the communication network and maintain the communication session.

When a segment of the TCP packet arrives at LLC/MAC, the MAC segment is formed and sent through an access network to the core network to be delivered to the device for reassembly and reconstruction of the flow. This transaction involves several segments of the local access network as well as the core network using different mediums and protocols for communications. The interconnecting elements of the network connecting different segments of the network modify the packet format to the recommended protocol and send that to the next connecting element. The core datagram with source and destination address is preserved in all these transmissions in packet-switched networked. In circuit switched transmission, a virtual path is established from the source to the destination, and data is transferred over that path during the communication session.

1.5 Structure of the Book

Communications networking is a very complex multi-disciplinary systems engineering discipline. To describe these networks, we need to divide the

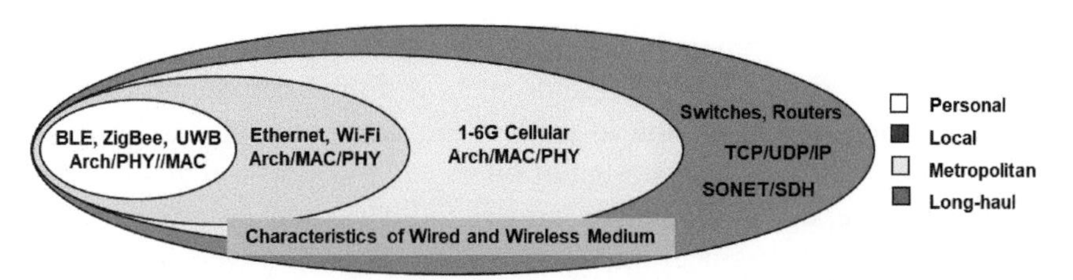

Figure 1.17 Overview of technologies for understanding communications networks in different areas.

details into several categories to create a logical organization for presentation of important material. Figure 1.17 provides an overview of technologies for understanding communications networks in different areas. Personal area networks are designed for communication networking among devices collocated in a room (<10 m). BLE, ZigBee, and UWB technologies have evolved for these areas. A student interested in learning details of these technologies need to learn the details of architecture (packet format and topology), MAC protocol defining how to share the medium, the PHY a protocol describing how the bits are transmitted over the medium, and characteristics of wireless multipath propagation inside a room to learn how to deploy these wireless networks and how to analyze their interference with other networks. Ethernet and Wi-Fi are designed for covering inside buildings (<100 m); to learn these technologies, we need to find out their architecture, MAC, and PHY, as well as characteristics of the wire for Ethernet and RF multipath propagation inside a building for Wi-Fi. For understanding the cellular networks, covering metropolitan areas, we need the same details for technologies as well as RF multipath propagation in urban and sub-urban areas. All these technologies are designed to access the core network for world wide web applications to support IoT access to sensors, telephone services to smartphones, and high-speed data connections to personal and mobile computing platforms. The core networks, Internet and PSTN, supporting circuit switched and packet switched services consists of bridges, switches, and routers, as interconnecting elements, and transmission lines to connect them for long-haul transmission. To understand the core networks, we need to learn (1) differences among bridges, switches, and routers, (2) protocols for long-haul packet transmission, such as SONET/SDH for PSTN and TCP, user datagram protocol (UDP), and IP for the Internet, and (3) characteristics of medium that connects them (copper, fiber, wireless).

To make the book suitable for teaching in an engineering or science curriculum, we need to present useful quantitative and analytical examples that are relevant for comparative evaluation of these systems. In textbooks and associated courses in specific fields such as digital communications or signal processing, it is common to present the details of the derivation of techniques for transmission or design of filters and transforms that are useful for processing the signal. Multi-disciplinary fields such as communications networking, robotics engineering, or bioengineering, however, *use* results of several other disciplines and merge them for the creation of a new field. The analytical examples used for students in such fields are more diversified and they must be selected carefully to avoid excessive complexity and yet remain useful and non-trivial to carry educational value. Therefore, clear organization of the presented material and depth of discussion on the variety of issues is needed because the multi-disciplinary nature of the material plays a significant role in the properness of the book.

We have organized this book as follows. Chapter 1 provides an overview of communication networks dividing communications networking technologies into core and access networks. Chapter 2 is another overview chapter describing an operation of core networks. These core networks are PSTN evolved for connection-based telephone services and the Internet evolved for packet switching services. The emphasis is on the description of interconnecting elements: bridges, switches, and routers. Chapters 3, 4, and 5 provided the technical background on characteristics of the medium, physical layer information transmission, and methods for medium access control. In these chapters, we provide the analytical background needed to understand details of popular standards for communications networks. In the remainder of the book, we describe details of popular communication networking standards. Chapter 6 is on IEEE 802.3 standard, commercially known as the Ethernet, the dominant standard for local networking, which is gradually penetrating the wide-area core networks. In this chapter, we present line-coding techniques that have enabled Ethernet to increase its data rate from legacy 10 Mbps to 100 Gbps, and we present the MAC layer details and a method to analyze its performance. Chapter 7 is devoted to the IEEE 802.11, commercially known as Wi-Fi, by explaining the importance of this technology in the emergence of the third industrial revolution and the "information age." In this chapter, we show how spread spectrum, orthogonal frequency division multiplexing (OFDM), multiple-input-multiple-output (MIMO) antenna systems and mmWave technologies enabled Wi-Fi to increase its data rate from 2 Mbps to Gbps. Chapter 8 describes the IEEE 802.15 standard and its popular low energy technologies such as: Bluetooth, ZigBee, and UWB. In Chapter 9, we begin by the cellular telephone technologies evolving from 1G frequency division multiple access (FDMA), to 2G TDMA, 3G code division multiple access (CDMA) focused on circuit switched services for mobile telephone and connection to the PSTN. Then, we present 4G, 5G, and 6G technologies for packet switched all-IP wireless metropolitan area networking. Chapter 10 is devoted to the deployment of wireless networks and technical issues associated with that, such as cell planning and spatial throughput analysis. Chapter 11 provides an overview of emerging cyberspace applications benefitting from RF propagation cloud created by wireless devices. These applications are in wireless positioning and location intelligence, gesture and motion detection, and authentication and security.

Assignments

Questions

1. Explain the role of communications networks in the emergence of information age.
2. How do the applications, devices, and radio propagations relate to one another?
3. How is a wireless network different from a wired network? Explain at least five differences.
4. What is the difference between a licensed and an unlicensed band? Give one example of a wireless technology standard that operates in licensed and one that operates in unlicensed bands.
5. What is a white space spectrum and how is it related to the TV spectrum?
6. What is the concept of the shared spectrum at 3.5 GHz and how does it differ from unlicensed bands?
7. Explain why standardization is important for the wireless wide, local, and personal area networks.
8. What are the main differences among the characteristics of technologies needed for wireless access to the PSTN and the Internet?
9. How does a standard evolve for wireless networks and what type of organizations are involved in the process?
10. What are the differences between connection-based and connection-less services? Give an example application for each of the two services.
11. What is WPAN? What is the difference between WPANs and WLANs? Name the two major standardization organizations writing the draft standards for these networks.

12. What was the first transmission technique used for telecommunication? Was it voice or data? Analog or digital?
13. What is IoT and how does it relate to core and access wireless networks?
14. What is the purpose of SONET/SDH standard?
15. What are the differences between IEEE 802.15 and IEEE 802.11 standards in terms of applications that they target, the data rate and battery life of the devices, and the targeted coverage area for their operation?
16. What are the differences between Wi-Fi and 4G/5G data services in terms of regulation for frequency of operation, data rate, coverage, and cost charging mechanism?

Project 1

Search chapter 1 and the Internet (IEEE Explore, Wikipedia, Google Scholar, ACM Digital Library) to identify one area of research and one area in business development which you think are the most important for the future of the communications networking industry. Give your reasoning why you think the area is important and cites at least one paper or a website to support your statement.

2

Overview of Core Networks

2.1 Introduction

Core networks are the backbone infrastructure of communication networks with universal coverage. The importance of core communication networks is in their ability to enable the information to be distributed worldwide with the speed of light. As we explained in Chapter 1, the first core network was developed for the telegraph digital messaging application to exchange short messages (like SMS) worldwide. However, the PSTN network, which later evolved for analog telephone application, became dominant and assimilated the telegraph messaging transmission in its infrastructure. The reason for the growth of PSTN was the ability of the telephone network to penetrate homes, as compared with the telegraph, which was an office-based networking infrastructure. Since the number of homes is orders of magnitude larger than a number of offices, PSTN industry grew into a much larger industry and it became more cost efficient to send the telegraph over the PSTN. This way, the PSTN became the dominant core network of the 20th century until the emergence of the Internet in the mid-1990s.

In the mid-20th century right after the second World War, the computer communication industry emerged for military, banking, airline reservations, and other applications also using PSTN infrastructure, originally designed for analog telephone application. Computers were using voiceband data communication modem technology as an interface between the computer data and the existing PSTN core network [Pah88]. PSTN is designed to enable communication between analog telephone devices worldwide. Access to the network was through a 4 kHz analog line and it remained that way up to the recent time when in some countries fiber lines began to replace them. However, the core of the PSTN, which was originally analog too, turned to digital in the mid-20th century by replacing analog lines in the core with digital lines carrying telephone signals in digital form with 64 Kbps pulse code modulation (PCM) encoding. To use this infrastructure for computer communications, voiceband modems were converting the digital computer data into analog voice-like signals to access the PSTN network using another digital coding technique with a variety of data rates. The analog modem signal with embedded digital data was treated like a telephone analog signal sampled at 8 Ksps with 8-bits per sample to form the 64 Kbps digital stream of the user in the core digital network of the PSTN. At the receiver end, the 64 bps data were converted to 4 kHz analog and sent to the user. The user then converted the analog signal with embedded digital data with the modem to recover the originally transmitted computer data. The highest data rate of the modems was 19.2 Kbps, but the network was carrying that with 64 Kbps! To explain this strange situation, one must understand that the major cost of wired networking is laying down the wires. When they are established, it is very difficult to change because it needs international coordination on standards and physical change of a huge network worldwide. The very expensive network infrastructures for PSTN evolved for analog telephone application, then the core evolved to digital because it was cost-effective and involved in only a few companies, but the access remained analog because it was not cost-effective and involved billions of individual connections to homes. Each core network carries numerous individual users while home access belongs to an individual user. In summary, the dominant core network in the second half of the 20th century was the PSTN with a 64 Kbps digital voice channel per user in the core and an analog access channel with 4 kHz band per user. Digital data communication over this channel was through a modem interface converting the digital data into a voice-like analog signal to access the core network.

The popularity of the Internet and its penetration into the home market occurred at the emergence of the 21st century, where the Internet penetrated homes. The internet also provides a worldwide market and generates incomes exceeding that of PSTN with a less cost per user. Internet was an evolution in computer communication originally building on voiceband modems. The industrial voiceband modems for computer communications were 4-wire modems, while in home, we had 2-wire connections to the core PSTN. Therefore, the early Internet service providers (ISPs) such as America online (AoL) were using 2-wire voiceband modems with a maximum data rate of 9.6 Kbps. Unlike voice that always has the same bandwidth per user, the data applications demand ever-increasing data rates. As a result, a few new communication technologies emerged to support wideband signaling over voiceband telephone channels. The leaders in these technologies were digital subscriber line (DSL) for voiceband modems and cable modems benefitting from cable-TV wirings. These technologies increased the data rate to over 1 Mbps. Later, fiber lines reached homes to support Gbps connections. The core networks had already moved toward optical lines to mesh the backbone. The access technology evolved to the dominance of wireless access using Wi-Fi and cellular networks, which support portability and mobility as well. However, core networks remained the same as before with more demand on cost reduction and lower delay in delivery of packets. The analog circuit switched voice communications are now obsolete and all the networks at the core and access use packet data communications. In this chapter, we provide an overview of the technologies and standards for core networks. In Chapter 3, we discuss characteristics of the wired and wireless medium, before we introduce the details of wired and wireless access to the core network, which is far more complicated than the core network itself.

2.2 Element of Core Networking

Core networks are mesh networks with point-to-point connections among multiple interconnecting devices. The main asset of these networks is the offices for the deployment of the interconnecting elements of the mesh networks and underground and overground wiring facilities connecting interconnection devices of the mesh network together. Laying down wires under or over the ground is expensive and needs to be regulated by the governments because it involves digging lands, which are property of the nations. Traditionally, in the US and some other countries, these wirings are installed, owned, and maintained by private companies such as AT&T or GTE (now Verizon). In some other countries, government agencies, such as ministries of Post, Telegraph, and Telephone (PTT), own and administer this infrastructure.

The main technologies involved in the design of core networks are the design and manufacturing of interconnecting devices and the design and manufacturing of interconnecting lines. To support multi-vendor operation standardization committees, define the interface protocols for communication between the elements. Manufacturers of interconnecting devices and lines follow these protocols to facilitate plug and play operation with multi-vendor designs. The main two core networks, PSTN and the Internet follow different packet forwarding methods demanding different technologies for interconnecting devices. Transmission over the lines follows the same standard protocols. We first introduce the packet forwarding issues and then give an overview of standards for transmission over the lines.

2.2.1 Principle of Packet Forwarding

Figure 2.1 shows the basic principles of packet transmission over a network. Packets from terminal A are destined to terminal B through a set of interconnecting elements to network the transmission media. Each interconnecting network has a few options to forward the packets raising two fundamental questions for the design of the interconnecting elements.

1. Who should decide the appropriate route for the packet, the terminal, or the interconnecting elements?
2. Should all packets from one user go through the same route or each may take a different route?

In bridges, switches, and routers used for traditional and popular local and wide area networks such as Ethernet, PSTN, and Internet, decision making for packet forwarding is transparent to the terminal. These techniques are generally divided into a datagram and virtual circuit switching techniques. Figure 2.2 illustrates the basic concept of the datagram and virtual circuit packet switching. In datagram transmission used for connectionless transmissions in Ethernet bridges and Internet routers, each packet finds its own route and they are reassembled in the last stage of transmission in the proper order. In virtual circuit switching used in connection-based networks such as asynchronous transmission mode (ATM), the network first defines a route for the packet delivery during the connection phase and then all packets follow the same route while the connection is alive.

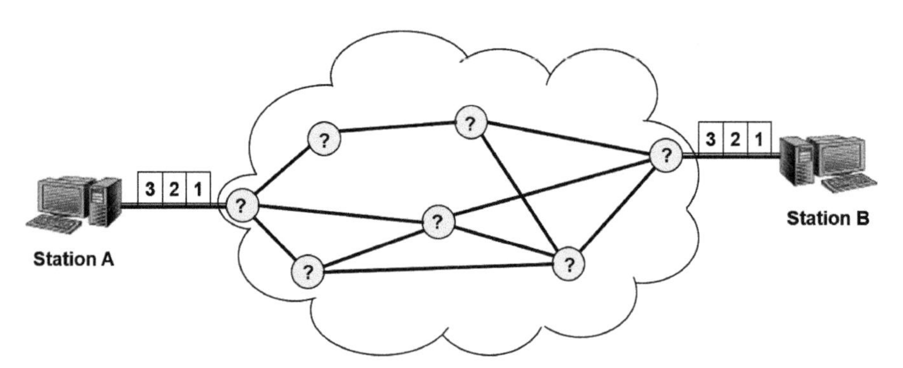

Figure 2.1 Packet forwarding problem.

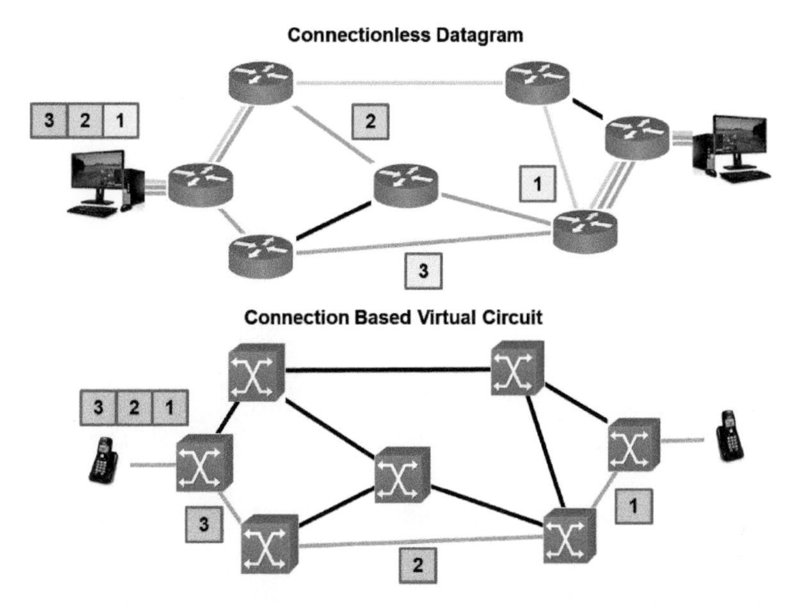

Figure 2.2 Packet switching approaches. (a) Connection-less datagram. (b) Connection-based virtual-circuit.

In the *ad hoc* and sensor networks, the terminal is involved in the routing of the packet. The traditional interconnecting elements of the PSTN were switches because it is a connection-based switching network. Later, these switches turn into virtual switches implemented on ATM that is a core protocol used in the synchronous optical network (SONET)/synchronous digital hierarchy (SDH) long-haul backbone of the PSTN. The ATM technology has largely been superseded in favor of next-generation all-IP networks by encapsulating these packets into IP packets for transmission over the Internet. The interconnecting elements on the Internet are routers. At the time of this writing, switches and router are the basic elements for interconnecting the public core network and the privately owned enterprise corporate networks. In corporate-networks, switches are used to connect computers to and from the network and routers are used to connect multiple networks. For long-haul transmission in core networks, routers establish software controlled connections packet by packet, while switches transfer the data on a pre-assigned hard wired path. Switches establish faster connections and are more expensive; as a result, they are used only when the link is sensitive to delay, for example, for video streaming among Hollywood studios.

Table 2.1 Connection-based virtual circuit switching versus connectionless datagram (adopted from [Tan03]).

Issue	Datagram networking	Circuit-switched networking
Circuit setup	Not needed	Required
addressing	Each packet contains full source and destination addresses	Each packet contains a short VC address
State information	Routers do not hold state information about connection	Each VC requires routing table for all connections
Routing method	Each packet is routed independently	Route is pre-selected during call setup
Effects of failure	None, except for packet lost during the crash	All VCs that pass through the switch terminate
User QoS assurance	Challenging	Easy
Network congestion	Probable	Avoided by call blockage

Table 2.1 adapted from [Tan03] summarizes the traditional comparison between packet-switched datagram subnet and virtual-circuit-switched subnets. Circuit-switched PSTN services require call establishment to establish a communication session and PSTN provides that service; in a datagram, we can establish the connection through transmission control protocol (TCP) transport, but the network is not involved in establishing and tearing off a session. Datagrams information packets carry the full source and destination addresses all the way throughout segments of the network. Circuit-switched network carries a short address of the virtual connection (VC) between the two ends of each transmission link. The interconnecting elements for VC must maintain a table for routing each incoming VC to an outgoing VC. The interconnecting elements for the datagram, the routers, should read the sources and destination addresses of a packet from an incoming port and determine the most suited output port. In interconnecting switches for VC, a table is established during the call setup. In circuit-switched network, if a switch is dead, the table disappears, and the connection is lost. If a router is dead, the packet is delivered through another route, making Internet a fault-tolerant core. In a circuit-switched networks, we can allocate resources at call establishment and maintain that throughout the session to provide a certain level of quality of service (QoS). Maintaining QoS on the Internet is more challenging. The core network supporting datagram faces challenges in congestion control, while in VC, we may block the incoming traffic to maintain traffic flow for the rest.

PSTN uses virtual circuit switching elements for packet switching and Internet includes hubs, LAN bridges, LAN switches, and routers. We discuss evolution of PSTN switching technologies in Section 2.3 and routers, LAN switches, and bridges in Sections 2.4 and 2.5. Here, we continue our discussion by providing an overview of transmission standards for lines connecting interconnecting elements.

2.2.2 Transmission Standards for Core Networks

For long-haul transmission in core networks, multiplexing is used where multiple data streams are combined into one signal to share the expensive long-haul transmission resources. From a different perspective, multiplexing divides the physical capacity of the transmission medium into several logical channels, each carrying a data stream. The two most basic forms of multiplexing over point-to-point connections are time-division multiplexing (TDM) and frequency-division multiplexing (FDM). In optical communications, FDM is referred to as wavelength-division multiplexing (WDM). Multiple streams of digital data can also use code division multiplexing (CDM) which has not been commercially successful over wired networks. Variable bit rate digital bit streams may be transferred efficiently over a fixed bandwidth channel like TDM utilizing statistical multiplexing techniques such as ATM. If multiplexing is used for channel access, it is referred to as medium access control (MAC) in which case TDM becomes time-division multiple access (TDMA), FDM becomes FDMA, CDM becomes CDMA, and statistical multiplexing into something like carrier sense multiple access (CSMA). Multiplexing techniques are simple, and they are part of the physical layer of the network. We discuss them in this section. MAC protocols are more complex, and we describe them in Chapter 6.

In core networks, the end user's telephone line carrying customer data typically ends at the remote concentrator boxes distributed in the streets where these lines are multiplexed and carried to the central switching office on significantly fewer numbers of wires and for much longer distances than maximum reach of the customer's end line. Fiber multiplexing lines, such as ATM protocols, were commonly used as the backbone of the network which connects home wired phone lines with the rest of the core and carries Internet data provided by other lines. As a result, the core network has been the driver for the development and standardization of most multiplexing standards for long-haul transmission. In the early 1960s, the first multiplexing system used for PSTN telephony was FDM. Figure 2.3 illustrates this early multiplexing system. This system multiplexed 12 subscribers, each with a 4-kHz bandwidth signal in frequencies between 60 and 108 kHz. Figure 2.3

time
60KHz
108KHz
4KHz
amplitude
f0 f1 f2
f10 f11
frequency

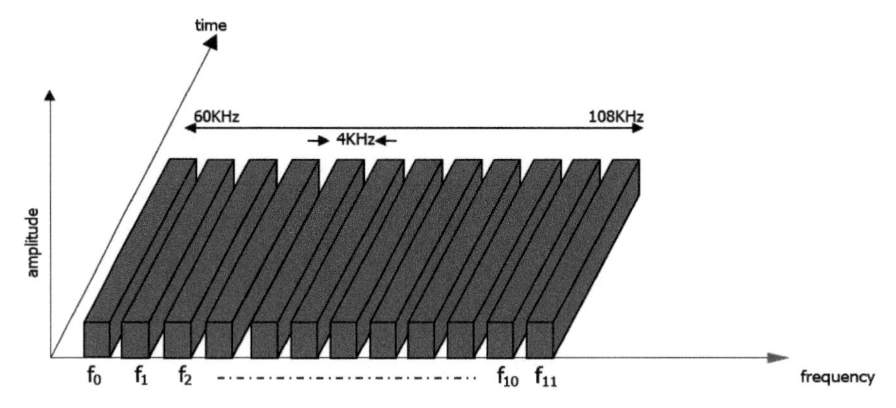

Figure 2.3 FDM used in telephone networks multiplexing 12 subscribers each with 4-KHz bandwidth in frequencies between 60 and 108 KHz.

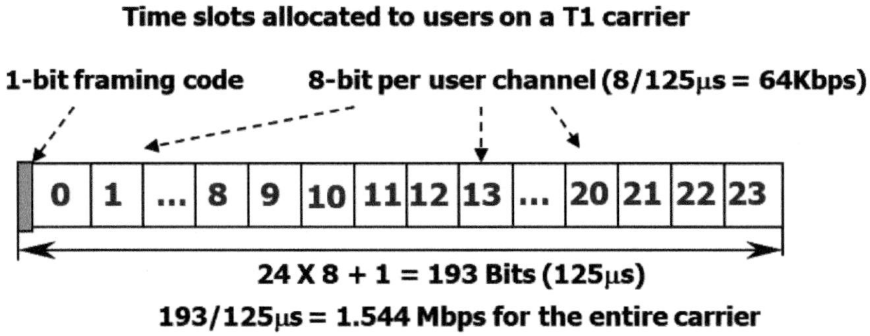

Time slots allocated to users on a T1 carrier
1-bit framing code
8-bit per user channel (8/125μs = 64Kbps)
0
1
...
8
9
10
11
12
13
...
20
21
22
23
24 X 8 + 1 = 193 Bits (125μs)
193/125μs = 1.544 Mbps for the entire carrier

Figure 2.4 T1 carrier for multiplexing 24 digitized voice signals each carrying 64 Kbps for a total 1.544 Mbps.

shows the original bandwidths for each subscriber, the bandwidths raised in frequency, and the multiplexed channel. The first TDM system was developed for PSTN telephony applications as well. This system carried 24 PCM encoded digitized voice calls, every 64 Kbps, over 4-wire copper lines at a rate of 1.544 Mbps. Figure 2.4 shows the 193-bit frame used in a T1-carrier – there are 8 bits per channel for $24 \times 8 = 192$ bits and 1 bit known as the frame bit. Each 8 bit consists of 7 bits representing a sample of the voice and 1 bit for control signaling. Each frame is transmitted in 125 ms to support a bit rate of 8(bits)/125(ms) = 64 Kbps per channel and an effective data rate of 7(bits)/125(ms) = 56 Kbps. The remaining 8 Kbps is used to carry the signaling information to set up telephone calls. A T1 carrier has a higher layer hierarchy to support higher data rates. Figure 2.5 shows the T-carrier hierarchy – every four T1 carrier results in a T2 carrier with the rate of 6.312 Mbps, each seven T2 carrier forms a T3 carrier which carries 44.736 Mbps, and six T3 carriers form a T4 carrier at 274.176 Mbps. Although the Japanese followed the American carrier system, later, Europeans developed their own TDM hierarchy for telephony which combines 30 channels rather than 24 channels to form the basic stream. The next three layers of hierarchy in the European system each is four times larger than the previous level.

With the popularity of optical communications in the 1980s and their ability to support much higher transmission rates, there was a need to extend the existing TDM hierarchies. At the same time, AT&T was broken up and there was a need to develop a standard TDM hierarchy so that the multi-vendor manufactured devices for different companies could cooperate with

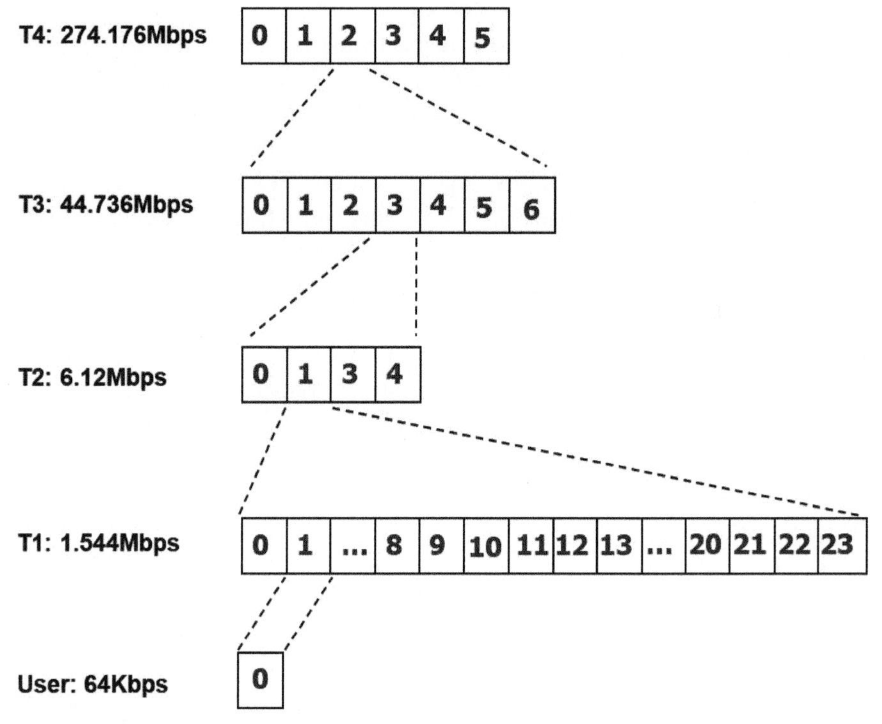

Figure 2.5 The legacy T-carrier hierarchy for long-haul TDM transmission of PSTN voice streams in US and Japan.

one another. In this setting, the SONET standard hierarchy started in North America.

Later, the International Telecommunication Union (ITU) became involved and the name SDH was selected as the name of the standard (which has only minor differences with SONET). The SONET/SDH hierarchy defines a common signaling standard for (wavelength, timing, frame structure, etc.) for multi-vendor operation, unify the US, the EU and Japanese standards, defines data rates higher than T-carriers, and provides for support of the operation, administration, and maintenance (OAM) of the network. SONET/SDH can be used for encapsulation of earlier digital transmission standards and they can be directly used to support ATM or SONET/SDH packet mode of operation providing for generic and all-purpose Ethernet and IP transport containers for multimedia information.

In a manner like the T1 carrier, the overhead, and the payload of the SONET streams are interleaved and all frames take 125 ms. Here, the basic frame, called synchronous transport signal one (STS-1)[1] or optical carrier one (OC-1), is 810 octets in size and the frame is transmitted with 3 octets of delimiter indicating the start of the frame followed by nine 87 octets of payload each having 3 octets of overhead, followed by 84 octets of user data. If we align all 87 rows, we have a block of 9×87 octets of payload in which the first three overhead columns appear as a contiguous block, as does the remaining 84 columns of user data. With this format, if we extract one octet from the bitstream every 125 ms duration of a frame, this gives a data rate of 8 (bits)/per (125 ms) = 64 Kbps representing a telephone user. This is very useful for the fast switching of low speed streams embedded in extremely high data streams without any need to understand or decode the entire frame. The basic frame in SDH is called synchronous transport mode one (STM-1) which has similar properties as OC-1/STS-1 with minor differences. These simple formats allow the design of relatively simple devices to take SDH data for a specific user and insert that in a SONET stream. This simple structure allows the implementation of faster switches operating at higher data rates. Again, like the T1 hierarchy, SONET and SDH have their own TDM hierarchy which is shown in Table 2.2. SONET defines four layers, photonic layer, section layer, line layer, and path layer. The photonic layer uses simple line coding for physical layer transmission that we will explain in Section 4.2.1. This layer also specifies the optical fiber type for the communication channel. The other three layers correspond to the data link

[1]This is the name for electrical rather than optical signal.

Table 2.2 SONET/SDH hierarchy of data rates and frame format.

SONET optical	SONET electrical	SDH level electrical	Payload bandwidth (Kbps)	Line rate (Kbps)
OC-1	STS-1	STM-0	48,960	51,840
OC-3	STS-3	STM-1	150,336	155,520
OC-12	STS-12	STM-4	601,344	622,080
OC-24	STS-24	STM-8	1,202,688	1,244,160
OC-48	STS-48	STM-16	2,405,376	2,488,320
OC-96	STS-96	STM-32	4,810,752	4,976,640
OC-192	STS-192	STM-64	9,621,504	9,953,280
OC-768	STS-768	STM-256	38,486,016	39,813,120
OC-1536	STS-1536	STM-512	76,972,032	79,626,120
OC-3072	STS-3072	STM-1024	153,944,064	159,252,240

layer of OSI model correction of transmission error, protocol to establish and terminate the link, and to control the flow of data.

With the popularity of Ethernet local area networking standard on the Internet, the Ethernet frame format which was originally designed for 10 Mbps local data traffic began defining higher data rates up to hundreds of Gbps inviting application of this frame as an alternative to SONET/SDH frames for long-haul networks. We will discuss this movement of Ethernet from the local area to long-haul wide area networking in Chapter 8, when we provide the details of the Ethernet standards and technologies. At the time of this writing, typical world manufacturers of long-haul interconnecting elements, switches, and routers, such as Cisco, Ericsson, or Siemens, have multiple ports to accommodate SONET/SDH, ATM, and Ethernet connections for long-haul transmission and for corporate networking. Popularity of transmission protocols began with an application to PSTN during the second half of the 20th century and then they got carried into the Internet in the 21st century.

2.2.3 Quality of Service (QoS)

As a packet is delivered to the network until it arrives at the destination depending on the type of network, several impairments occur which affect the quality of the received information. The quality of the received information is usually measured by several characteristics. The throughput can be sustained between two endpoints that reflected by a variety of measures such as maximal data transfer rate or bandwidth, average data rate, peak data rate, or minimum data. The delay time of the packet to transport from the source to the destination caused by transmission facilities and

Table 2.3 Quality requirements for popular applications.

Application	Bandwidth	Delay	Jitter	Reliability
Email	Low	Low	Low	High
File transfer	Medium	Low	Low	High
Web browsing	Medium	Medium	Low	High
Audio streaming	Medium	Low	Medium	Low
Video streaming	High	Low	High	Low
Telephony	Low	High	High	Low
Video conferencing	High	High	High	Low

interconnecting elements. Variation in end-to-end transit delay which is referred to as delay jitter. The reliability of the network in delivering the packets or packet loss which is the ratio of the number of undelivered packets to the total number of sent packets. Different applications have different sensitivities to these performance measures. Table 2.3 shows the relation between these performance measures and several popular applications. The access methods and interconnecting elements of different networks have evolved around specific applications, which was the main reason to generate income in an industry. Interpretation of QoS and mechanisms to control and maintain a certain level of QoS in connection-based PSTN and the Internet are different.

QoS in Connection-Based PSTN:

In the connection-based circuit-switched telecommunication industry, QoS was sometimes defined as the ability of a network to have some level of assurance that its traffic and service requirements can be satisfied. To support QoS, all layers and every network element should cooperate to guarantee several features, such as time, to provide service, voice quality, echo level, and connection loss rate. A subset of telephony QoS is the grade of service requirements related to the blockage and outage probability. The term QoS is sometimes used as a quality measure rather than referring to the ability of the network to reserve resources. The QoS is not the same as the ability of the network to support a high bit rate, low error rate, and low latency. It is the ability to support a certain level of assurance for traffic requirements. To define the level of QoS, the traditional wired telephone and video industry used subjective measures based on the user perceived performance and the mean opinion scores which reflect the cumulative effects of all system imperfections affecting the service. This approach uses collective human opinion in the assessment process to determine objective measures such as delay, throughput, and error rate for designing the network. As an example, the wired telephone network industry required measures such as 1% packet loss and 150 ms end-to-end delay for telephone services to satisfy the mean opinion of scores. The switches and transmission facilities in the PSTN were designed to support this level of QoS for wired telephony application using 64 Kbps PCM encoding techniques. When it came to cellular phone applications, the limitation of bandwidth forced the industry to resort to different encoding techniques with data rates around 10 Kbps. Lower data rate and fading characteristics of the radio channel reduced the expected QoS from the wireline quality traditionally used in PSTN over the past century. This experience showed that religious adherence to QoS specifications can be compromised. To integrate data, application into connection-based ATM technology evolved, which had an elaborate framework to plug in QoS mechanisms. Later, using ATM for local networks to connect devices became restricted to carrying live professional uncompressed video and audio in environments demanding low latency and very high QoS such as the professional media production industry. However, ATM is still used in connecting some core network elements worldwide. In the turn of the 21st century, the popular commercial applications and the main bulk of the local and core network connections tend to delve further into connectionless Ethernet/Internet environment for most applications.

QoS in Connectionless Internet:

In connectionless packet-switched networks for computer networking, QoS refers to resource reservation control mechanisms rather than the achieved service quality. This is due to the fact that the packet-switched networks were originally designed for connectionless-packet-based data applications such as file transfer, telnet, or email for which user's satisfaction is much less sensitive to a continual QoS needed for two-way telephone conversations. Data applications recover the packet loss at higher layers, and in most legacy applications, they are not that sensitive to the delay. As a result, the legacy Ethernet was designed without any provisioning for QoS. The Ethernet original MAC packets do not have any field for priority assignment and the MAC mechanism is contention-based having no control over the delay. Some discontinued legacy LAN technologies, such as fiber distributed data interface (FDDI) or Token Ring, had priority bits in their MAC packets. These bits were designed to differentiate users, for example, to give higher priority to the network manager, but not to ensure the certain level of quality to the users and they never became popular in practice. Indeed, early bridges connecting different LAN protocols were eliminating priority assignments when they were connected to long-haul networks because that priority could not be carried to the core Internet. When the Internet was first deployed, it was not designed to support QoS guarantees due to the limitation in router computing power and it operated at "best effort." Although the IP addressing header has allocated four "Type of Service" bits in each message, but, like priority bits in some of the legacy LANs, these bits were also ignored in practice. As real-time streaming multimedia applications such as voice over IP (VOIP) and IP-TV complemented traditional Ethernet/Internet, applications type of service bits in IP and other available resources in LANs were exploited for traffic provisioning. As we explained earlier, the need for QoS in Ethernet/Internet networks became more important when streaming became popular in resource limited cellular networks. In summary, in packet-switched networks, QoS is the ability to provide different priority to different applications, users, or data flows, or to guarantee a certain level of performance to a data flow, and although there are mechanisms to implement priorities between packets, there is no mechanism to ensure certain quality.

As we mentioned earlier, in the packet-switched environment of the Ethernet/Internet, the QoS can be provided by over-provisioning a network so that interior links are considerably faster than access links, e.g., VoIP over Ethernet or Wi-Fi. With the emergence of broadband services, this relatively simple approach is practical for many applications such as audio and video streaming for which high jitters can be compensated with large buffers or VoIP under low load. Under high traffic load conditions, e.g., cellular networks, the delivered QoS for applications such as VoIP degrades significantly and the use of a QoS mechanism in the network would allow significantly higher user satisfaction and more balanced traffic among the subscribers. Application of QoS mechanisms becomes more important as we encounter lower bandwidth wireless applications such as satellite IP [Kot04]. To support QoS in packet-switched networks in the early days, the integrated service "IntServ" philosophy of reserving network resources was used. In this model, applications used the resource reservation protocol (RSVP) to request and reserve resources through routers in the network. Using these methods, core routers for a major service provider needed to accept, maintain, and tear down numerous reservations (and maintain this information for many flows) which did not scale well with the rapid growth of the Internet and the notion of designing core routers at the highest possible packet switching rates. As a result, a second approach called differentiated service "DiffServ" emerged as an alternative. In the DiffServ approach, packets are marked according to the type of service they need. Routers supporting DiffServ use these marks to form multiple queues for different priorities.

In the Internet part of the WAN-type traffic bits and the LAN, priority bits are used to mark the packets. Router manufacturers provide different capabilities for configuring this behavior to include the number of queues supported, the relative priorities of queues, and bandwidth reserved for each queue. Packets carrying VoIP, for example, are assigned to the highest priority queues; control packets will receive a default portion of the bandwidth, and the leftover bandwidth is allocated to best effort traffic.

2.3 User Addressing in Core Networks

Modern networks have evolved around connection-based voice-oriented PSTN and the data-oriented connectionless Ethernet/Internet networks. Addressing scheme for these two networks is quite different, and PSTN is a circuit-switched network and Ethernet/Internet is a packet-switched network. In circuit-switched environments, network assigns a virtual path for transmission during the call establishment and all the traffic packets are transmitted through that path. Therefore, address identifies a destination that is used during call establishment. In packet-switched network, we have MAC addresses for LANs and local networking and IP addresses for WAN operation. Wireless access to these networks adds a new dimension of complexity to this environment because the connection to the network is mobile and it connects to the network from different access points or base stations (BSs). Another dimension of complexity is the handling of multicast and broadcast. Internet has embedded multicast and broadcast features which are desirable for several modern applications. To extend these features of Internet to PSTN, we need to find new approaches for scaling these features. In the rest of this section, we provide an overview of the issues involved in addressing with more details of how these issues are handled in a modern network.

The core of the Internet is interconnected through Routers which are less expensive and more flexible than PSTN switches. The addressing and transportation of messages on the Internet is through IP addressing and TCP/user datagram protocol (UDP), which are completely different from Integrated Services Digital Networks (ISDN) addressing. The cost advantages and flexibility of the Internet to integrate multimedia traffic increased the size of the Internet industry, and, today, it has integrated the core traffic of the PSTN in itself, Internet, the same way that the PSTN core network integrated the telegraph core networking in the early 20th century. What differentiates the PSTN from the Internet for the end user is the access technologies and the addressing methods. The majority of the PSTN telephone connection-based traffic arrives through a wireless connection using cellular mobile telephone technology and the telephone calls use 10-digit ISDN addressing. The user application of the data traffic for the Internet is identified by MAC address of the hardware and the TCP/IP connection address.

2.3.1 Addressing End Users in PSTN

In the late 20th century, the interconnecting element in PSTN was digital switches capable of establishing a connection-based line between two devices worldwide. The addressing in PSTN is the ISDN addressing assigning different fields to countries, cities, and neighborhoods. The legacy connection-based circuit-switched network is the PSTN which was basically used for the voice application, and later in ISDN networks, it provides an integrated voice and data service. ISDN services never gained the popularity that they expected; however, digital cellular networks used a modified version of this technology for the implementation of 2G TDMA cellular networks with an integrated service. The ISDN addressing is the core for this type of communication. Virtual packet switching networks such as ATM networks also have their own addressing schemes for handling the packets in the core. In the rest of this subsection, we provide a fundamental overview of the addressing techniques used in these networks.

In connection-based networks, a terminal needs to forward its message to the destination terminal and, for that reason, it passes the destination address to the first node in the network so that the network can find the best route and establish the connection shown in Figure 2.2(b). This process involves five steps:

1. **Setup time** in which the terminal indicates to the network that it needs to establish a connection with a given address.
2. **Call processing** in which the route is determined and the connection to the destination terminal is established.
3. **Alerting** in which the network informs the calling party that the destination terminal is ringing.
4. **Connecting** in which network sent back a message to the calling party indicating how the intended destination has answered the call.

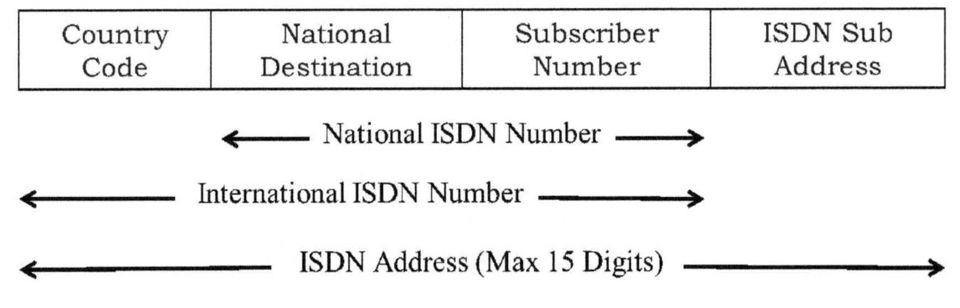

Figure 2.6 General format of an ISDN address.

5. **Release** in which either the source or the destination indicates that the call is to be terminated.

Digital cellular networks, from 2G GSM onwards, use connection-based networking and ISDN addressing. Figure 2.6 shows the general format of the ISDN addresses. The international ISDN number carries a country code assigned by ITU, telecom authorities in different countries distribute the number groups to cities and services with a national destination, and service providers assign the subscriber number to the end users. The international ISDN number has a maximum length of 15 digits. An ISDN address consists of an ISDN number plus a subaddress with a maximum length of four digits. The subaddress is transparent to the public network and can be used for the implementation of additional private features such as an extension number. Figure 2.7 shows the typical hierarchy of telephone switches consisting of regional, sectional, primary, toll, and local offices. The ISDN number 358-294-48-5463 is for an office with PBX number 358-294-48-5000 for which 358 specifies Finland regional center, 294 specifies the section switch in Oulu Finland, and 48-5436 is a subscriber number in an office connected to the PBX 5000 in Oulu region. Other POTS and cordless telephones in the area have their own direct lines.

Mobile Addressing in PSTN:

As shown in Figure 2.7, in PSTN infrastructure, one can track an ISDN address geographically to the home or office location and the network uses the number to route the call using the interconnecting elements of the network. When the telephone terminal user picks the phone and dials the destination number, the end office (EO), which is the last switch in the network establishes the telephone connection to the network and the toll center establishes a route for the traffic to the exchange office at the

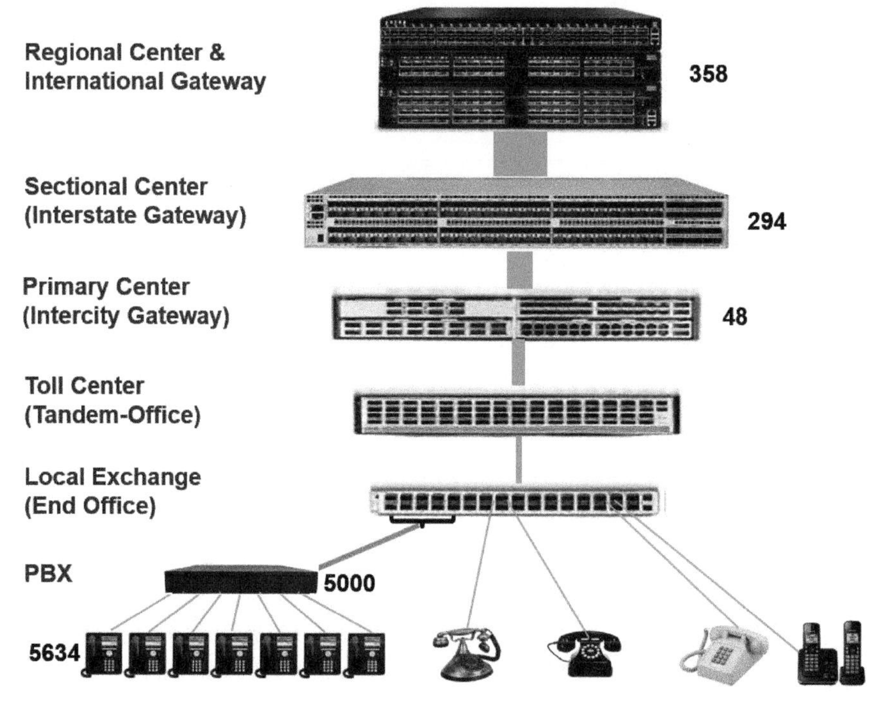

Figure 2.7 Example for PSTN switching hierarchy and how ISDN number addresses a telephone.

destination with SS7 protocol and toll office at the other end establishes a connection to the destination device through the associated EO. Then traffic is transferred through the physical or virtual dedicated traffic channel. The fixed service user profile identifying the location of the service holder and terms of the contract is stored in a location registration database at the network for service and billing purposes.

When cellular wireless networks were introduced, the fixed connection to the PSTN changed to a mobile connection that can connect to the network from any location and the details of the service contract for billing information changed to a different format. To implement the mobile telephone service standardization organization published a reference model with elements enabling mobile connection to the PSTN. Figure 2.8 shows the fundamental blocks of these reference models. We have BSs to establish the connection to the end user; this is an EO capable of handling mobile rather than fixed operation. A base station controller (BSC) handles resource

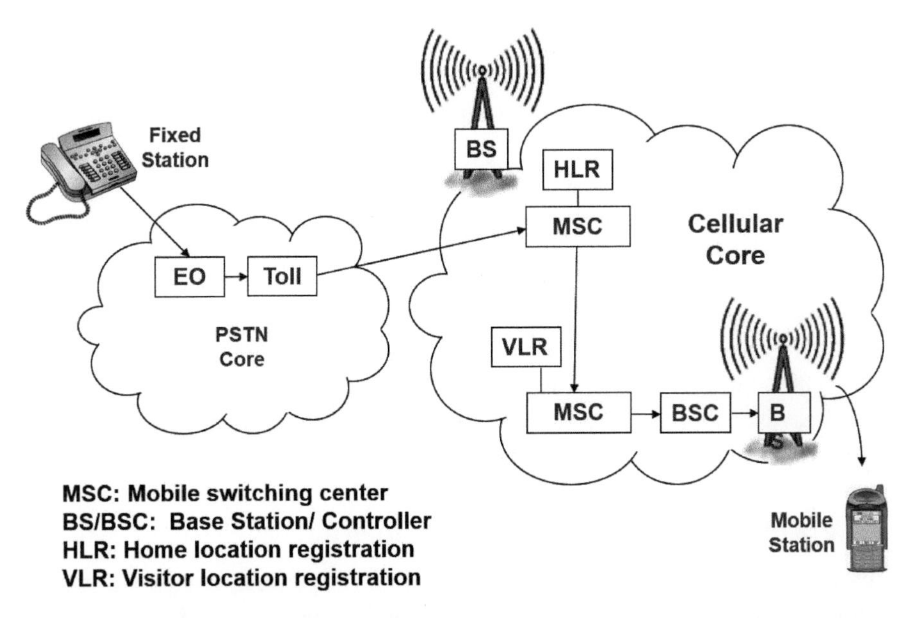

Figure 2.8 Core cellular telephone infrastructure enabling mobile ISDN addressing over PSTN and an example of fixed to mobile call establishment.

allocation by distributing a limited number of wireless resources among mobile stations (MSs) and managing hand-offs when a user changes its BS that connects to the network. The mobile switching center (MSC), like Toll Center in PSTN, manages the call and billing information plus identification of authentication of subscriber and the user equipment. The MSC has two databases, home location registration (HLR) and visitor location registration (VLR). When a service provider establishes a contract with a customer, it sells a number that is kept in the HLR for contractual monitoring and services. An MS is registered in an HLR database by its permanent address. When the mobile moves from the home to connect as a visitor to another part of the cellular core, it receives another address for the visitor database. The HLR and VLR databases communicate with one another to allow call forwarding to the new location. In cellular phones, the actual address is not broadcasted on the air, and each time that a user connects to the network, it receives a temporary address for communication through the air. This is to protect the user from fraudulent connections. Therefore, mobile services have a virtual permanent number, like actual fixed service numbers, but they have several other temporary addresses for communication in the core as well as for the wireless communications. Figure 2.8 shows more details of a fixed station that initiated a call to an MS. The fixed station connects to the core PSTN through EO and the toll center attempts to establish a connection to home MSC. The home MSC redirects the call to visiting MSC to establish a wireless link to the MS through the BSC and the BS. When mobile initiates a call, the visiting MSC manages the call directly. In addition to ISDN address of a cellphone, the cellphone hardware has an electronic identity registration (EIR) number and the subscriber identity module (SIM) card inserted in the cellphone has a four-digit security code. EIR helps to identify stolen cellphones and SIM card key allows authentication of ownership of the ISDN number to avoid fraud. The SIM card security key is also used in ciphering the wireless traffic to avoid eavesdropping. These features were first introduced in 2G GSM digital cellular to counter the large analog cellular fraud industry.

2.3.2 Addressing End Users on the Internet

Internet is the public network connecting numerous smaller computer networks to create a platform for worldwide packet switching using the Internet protocol (IP). The Internet allows the implementation of applications running over millions of smaller networks owned by the government, private sectors, academic institutions, and industrial organizations. The Internet service is provided by ISPs which connect to network access points (NAPs) and from there to the worldwide Internet. Devices connecting to the Internet are computing devices running numerical applications with a variety of communication interfaces. The unique soft address of the device on the Internet is identified by the IP address, and the hardware address of communication hardware interfaces is identified by the MAC address. Each IP address can transport multiple applications with different TCP or UDP addresses and each MAC address may connect with multiple hardware section through logical link control (LLC) addresses. With these embedded address diversities for communication over the Internet, a computing device can handle complex operations with multiple software applications and different hardware interfaces.

IP Addressing for the Internet:

The IP or layer 3 or Internet software address is an address identifier for an electronic device allowing it to network with other devices through the Internet. All Internet network devices such as routers, switches, computers, printers, IP telephones, and IP-TVs have their own address that is unique

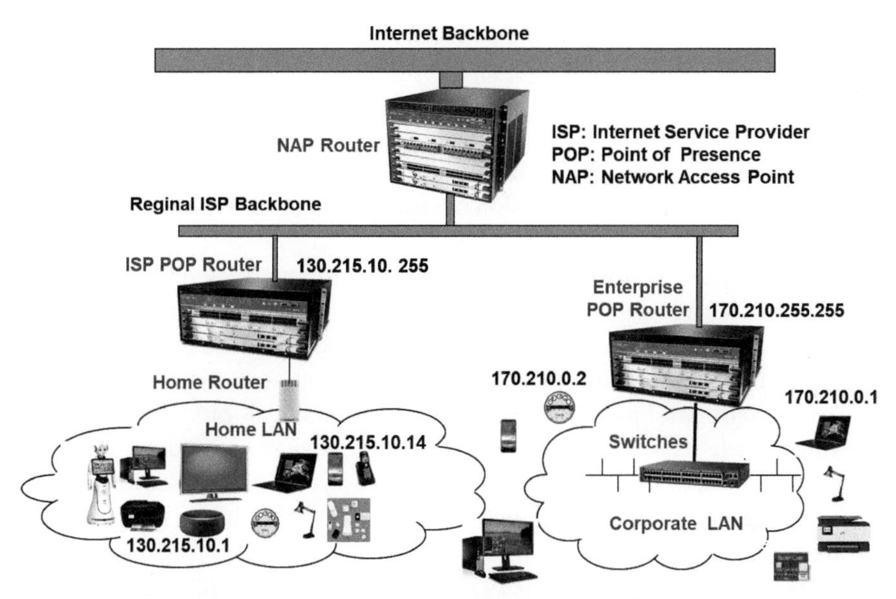

Figure 2.9 Example for Internet routers hierarchy and how IP addresses point to a device in a LAN using Ethernet or Wi-Fi.

within the scope of the specific network. IP addresses are created and distributed by the Internet Assigned Numbers Authority (IANA) which allocates super-blocks to different regions to be divided into smaller blocks for different ISPs and enterprises, as shown in Figure 2.9. The small blocks of IP addresses are then assigned dynamically by the network administrator of each organization to the individual terminals. As a result, parts of the IP address is used to locate an IP device, and from that device, one can find the target device and interact with it. Therefore, like ISDN addresses, IP addresses have a hierarchical geographically selective structure which helps routing the message toward the destination address through several independent interconnecting element, but unlike ISDN used in fixed or mobile telephone, they do not identify a fixed connection to a location or a specific mobile terminal. The popular IP version 4 uses 32-bit addresses which are usually represented as four decimal number equivalents of the four blocks each with the length of eight digits. As compared with the MAC addresses, the IP addresses provide geographical selectivity while the geographical distribution of MAC addresses are random and IP address is a software address that is changed while MAC address is burned into a network adapter card (NIC) hardware.

As an example, the IP address 130.215.10.14 in Figure 2.9 represents 10000010, 11010111, 00001010, and 00001110. A 32-bit address results in close to 4 billion (232 = 4,294,967,295) different possibilities. Considering that at the time of this writing, we have approximately 6 billion cell phones and more than a billion fixed Internet connections, this number is small, and we need a longer address. IP version 6 with 128-bit addresses was introduced to solve this problem.

IP addresses have two parts which identify the network and the node in the network. There are five coded classes of addresses and a subnet mask to differentiate the network address from the node address. Class of the address and the subnet mask determine which part belongs to the network address and which part belongs to the node address. The first 4 bits can be used to determine each class. Figure 2.10 shows the five classes of address for IP v4 and different fields for each class. Using Figure 2.10, one can determine the class of the address and from which network and node address for classes A, B, and C.

In the IP address 130.215.10.14 (Figure 2.10), the first three digits "130" indicate that this is a class B address, and, by default, the first two sets 130.215 represent the network address and the second set 10.14 the address of the device. In other words, the address for the main router in the network or the network address is 130.215.0.0. If we set the node address to all "1"

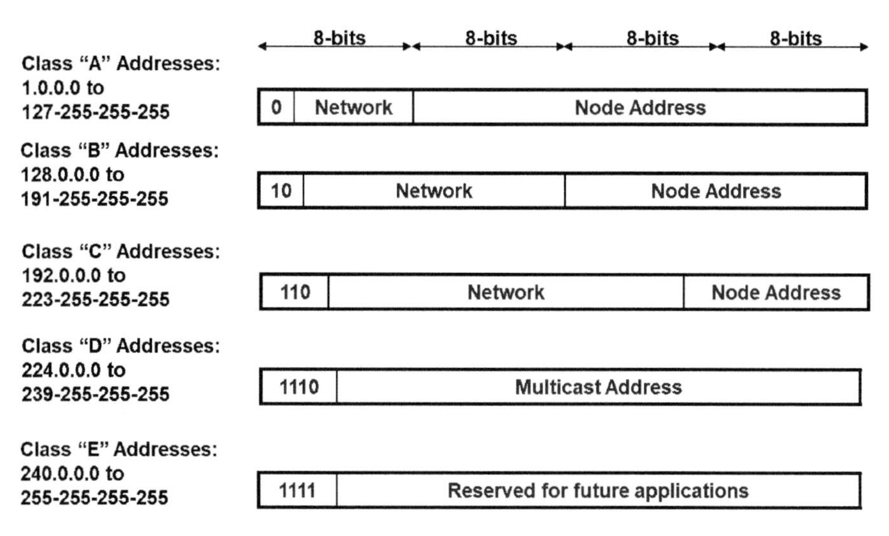

Figure 2.10 Overall structure of an IP address and different classes.

which means 130.215.255.255, it signifies a broadcast for that network which delivers the message to all nodes of the network. Addresses beginning with 127 (**01111111**) are reserved for loopback and internal testing on a local machine. For example, if you ping **127.0.0.1**, it always works because it points at yourself.

IP Multicast provides for one to many messaging, but it takes place in an IP infrastructure. This approach utilizes the network infrastructure efficiently by allowing the source to send a packet only once while it is delivered to many destinations. The nodes in the network manage the packet switching so that it arrives in multiple destinations without any prior knowledge of the specific destinations. IP Multicast is implemented based on the IP Multicast group address, a multicast distribution tree which is created by the destination terminals. The IP Multicast group address is used by sources and the destination terminals in different ways. Source terminal uses the group address as the IP destination address. The destination terminal uses the group address to inform the network nodes that it is interested in accepting the broadcast. IP multicast is ideal for applications such as distance learning where several widely distributed users demand a wideband service such as the class video. Implementation of IP multicast requires much more complex operation at the interconnecting element and its complexity increases as the problem scale to a huge number of users.

Mobile IP Addressing:

Consider one of the Ethernet LANs in Figure 2.9; each node is identified by a MAC address in the LAN and with an IP address in the network. IP addresses are not permanent, but MAC addresses are permanent. How can we keep track of the mapping between the two addresses? The simplest approach coming to the mind is that we keep a table in the Routers which maps the MACs to IP addresses, but updating such a table is difficult because when we need the mapping, we are not certain that the table is updated. The more common approach is using the so-called address resolution protocol (ARP) which broadcasts a question in the LAN asking who owns the IP of the arriving packet. The terminal that has the address releases the MAC address to the network and the packet gets delivered to the correct device. The mobile IP standard defined by Internet Engineering Task Force (IETF) benefits from ARP protocol to allow a device to maintain a unique IP address while connecting to the Internet through different subnetworks.

Unlike mobile addressing for cellular networks, mobile IP is not designed only for wireless connection to the network. Mobile IP allows wired or

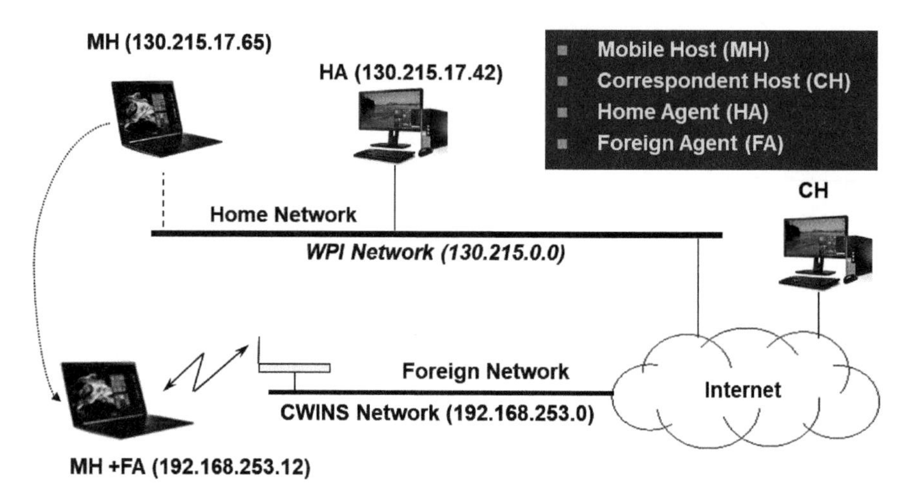

Figure 2.11 Basic concept of mobile IP operation showing network entities involved.

wireless connection to the network from any place in the network. Figure 2.11 shows the basic element for the implementation of the mobile IP protocol. We have a home network in which a mobile host (MH) is registered, and its IP address **130.215.17.65** is assigned; there is a visiting or foreign network to which the MH desires to connect, and a corresponding host (CH) which desires to send a message to the MH with the address taken from home network not knowing that the MH is currently connected to a foreign network with address **192.168.253.12**. To enable the network to provide this type of mobility to the host terminal, the IETF recommends the addition of two agents in the home and the foreign networks. The MH has two addresses, a **permanent address 130.215.17.65** for the home network and **care of address 192.168.253.12** associated with the visiting foreign network. The home agent (HA) stores information about MH and its permanent address in the home network. The foreign agent (FA) stores information about MH visiting its network and advertise care-of address (these databases have similar functionalities as HLR and VLR in cellular networks, Figure 2.8).

Figure 2.12 shows how a terminal sends a packet to a wireless MH terminal registered at the home network but visiting the foreign network. The CH sends its packet for the MH using the home address. This packet is intercepted by the HA which uses ARP protocol to find if MH is still connected to the home network; if the MH MAC address is not connected, the HA uses the IP table to tunnel the packets to the MH's care-of address.

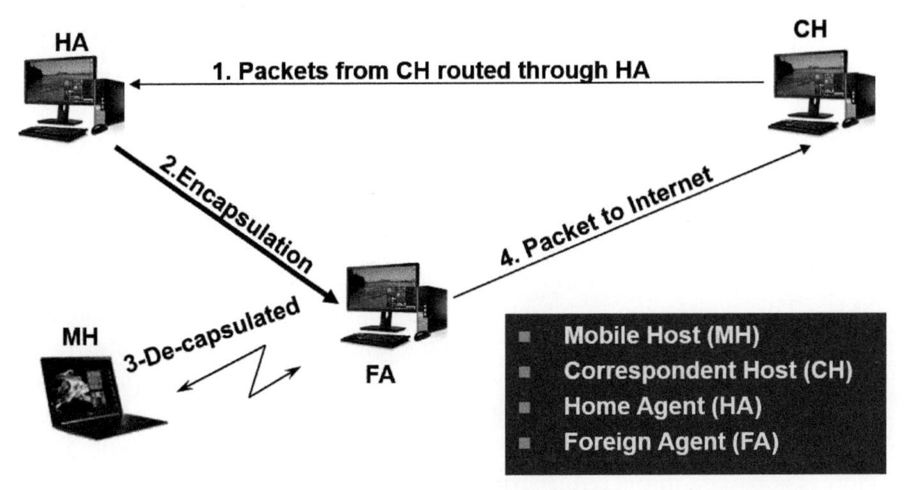

Figure 2.12 Basic principle of mobile IP protocol for a wireless laptop.

Tunneling operation adds a new IP header with the care-of address while keeping the original IP header in the packet. Upon the arrival of the packet at FA the packet is decapsulated at the end of the tunnel to remove the added IP header and delivered to the MH. The mobile IP protocol benefits from ARP protocol to identify the device for the delivery. When foreign MH wants to send a packet, it simply sends the packet directly to the CH through the FA. Mobile-IP protocol is most suitable for hybrid handoff for wireless data services among wireless network infrastructures, such as hybrid handoff between wireless cellular networks and Wi-Fi [Pah00]. For VoIP handoff between cell towers, mobile IP is not fast enough to accommodate transitions and, for that reason, it is not adopted by 3GPP to handle those types of delay-sensitive conversational streaming applications for 3G cellular and beyond. Note that the FA database can reside in MH (Figure 2.11), but to demonstrate the decapsulation process more clearly, in Figure 2.12, we have shown them as separate devices.

MAC Addressing in LANs:

Connectionless networks carry the information using datagram packets. In datagram, we expect each packet to carry source and destination address so that an interconnecting element such as any bridge or a router can forward the packet toward the destination. Two addresses have evolved for the datagram, the MAC address introduced for LANs and the IP address

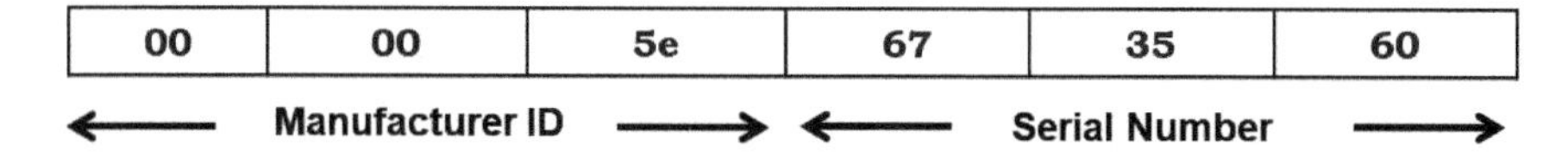

Figure 2.13 An example 48-bit IEEE MAC address represented numbers hexadecimal identifying manufacturer and the serial number of an NIC.

introduced by the Internet community. The MAC or layer 2 or Ethernet hardware address is a unique address identifier for NIC to connect to the LANs. This addressing technique was originally introduced in the Ethernet and it uses 48 bits[2] assigned to manufacturers of the NICs by the IEEE organization. As compared with the IP addresses, the IP addresses provide the geographical selectivity while geographical distribution of MAC addresses is random and IP address is a software address that is changed while MAC address is burned into a NIC hardware.[3]

As an example, the MAC address 00:00:5e:67:35:60 represented in 12 hex letters each carrying 4 bits. The first three pairs of octet numbers represent the manufacturer and the next three the serial number (Figure 2.13). This 48-bit address space contains potentially 2^{48} or 281,474,976,710,656 possible MAC addresses. A NIC is uniquely identified by an MAC during the manufacturing process and products of a manufacturer are randomly distributed all over the world. As a result, the MAC address uniquely identifies the hardware wherever it is, but it does not carry information on the geographical location of the terminal.

As compared with ISDN addresses, the maximum 15-digit ISDN number is more than three times the potentially 2^{48} possible MAC addresses. ISDN assigned by the manufacturer carries geographical information, but MAC addresses do not. MAC addresses are enabled for multicast and broadcast in addition to traditional ***unicast*** addressing. In ***multicast***, a single packet transmitted from a source can address several destinations in the network, and in ***broadcast*** mode, that packet arrives in all destinations in the network. Broadcast and multicast features utilize network infrastructure efficiently by requiring the source to send a packet only once when it is delivered to many destinations. The nodes in the network take care of replicating the packet

---

[2] An unpopular 16 bits option for MAC address also exists and it is sometimes used in sensor networks.

[3] Using MAC spoofing technique, it is possible to change the MAC address on most of the hardware designed recently. In this approach, a locally administered address is assigned to a device by a network administrator to override the burned MAC address.

to reach multiple destinations only when it is necessary. A MAC multicast or broadcast address is used by the source and the destination terminal to exchange information. These addresses are useful for the implementation of user groups in a LAN.

The MAC address for the unicast operation has a zero as the least significant bit of the most significant byte of the address. If the least significant bit of the most significant byte is set to 1, it is a multicast address that reaches several destinations enabled for multicasting with that number. Packets sent to a multicast address are received by all stations on a LAN that have been configured to receive packets sent to that address. Packets sent to the broadcast address carry an all one address. All terminals in a local area network receive these packets. In hexadecimal, the broadcast address would be "FF:FF:FF:FF:FF:FF." Ethernet, Wi-Fi, Bluetooth, Zigbee, and several other networks use MAC addresses to identify a NIC. A multifunction device such as a smartphone has two MACs for Wi-Fi and Bluetooth but one IP and one ISDN address for connection to the Internet and PSTN.

Four Addresses in Ethernet/Internet Packets:

Figure 2.14 shows the overall frame format for the user MAC packet data unit (MPDU) in the Ethernet/Internet datagram packets before delivery to the PHY for digital transmission. By adding TCP/IP headers (each 20 bytes) to the user application data with a variable length (0–65,495 bytes), a TCP/IP packet is formed to be delivered to the LLC/MAC to form an IEEE 802 packet for transmission over a variety of PHY. The TCP packet header contains the source and destination addresses of TCP port (16 bits each), which allows 2^{16} = 56,536 ports to connect to an IP connection. Each port can carry a different application packet. The TCP protocol is an acknowledge-based protocol to ensure the streaming of the application data, which is suitable for file transfer types of applications demanding error free transport. An alternative to TCP is the UDP that delivers the same services without acknowledgment for faster processing of a sequence of application data for live streaming applications such as VoIP or Internet protocol television (IPTV). The IP header carries the packet through the Internet with another set of source and destination addresses (32-bits v4 and 64-bits v6), identifying the end user devices.

To connect the end devices to the Internet, a local network forms an IEEE recommended MPDU for transmission of alternative PHY options (e.g., Ethernet or Wi-Fi). Before adding the MAC, header, and trailer, we have the LLC header (8 bytes), which also carries source and destination addresses each 8-bit long. The addresses identify the so-called logical

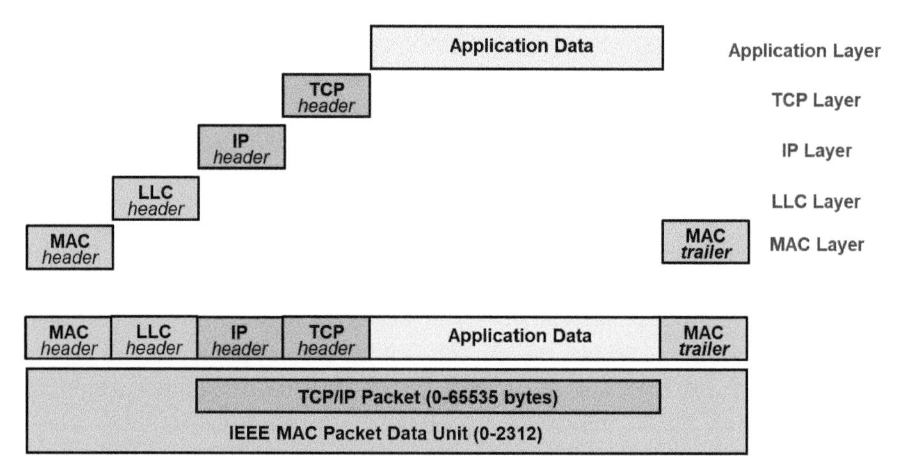

Figure 2.14 General format of an MAC packet data unit in Ethernet/Internet.

service access point (LSAP) allowing alternative services for logical end-to-end transmission links over the MAC of an IEEE LAN. LLC defined by the IEEE 802.2 allows connection-based and connectionless services with and without acknowledgment. The connectionless services allow another layer of multicast broadcast to the terminal. In practice, all products use unacknowledged connectionless services. The two layers of connectionless LLC and MAC addressing each with three options for unicast, multicast, and broadcast allow nine different addressing techniques in a LAN.

In the wireless local network when we need to address a WLAN, we need two MAC addresses – one to identify the access point and the other to identify the wireless terminal. As we will see in Chapter 7 on 802.11 WLAN, the MAC frame of this standard has four address fields which are used to identify the access point and the mobile terminal. When a mobile node with WLAN connection moves from coverage of one access point to another, it keeps its own MAC address, but the MAC address of the access point will change. This mobile MAC addressing mechanism allows implementation of local roaming among different points of connections for a terminal using a WLAN.

2.4 Evolution of Circuit Switches for PSTN

The most popular LAN interconnecting element, discussed in Section 2.5, is also referred to as switches. These network interconnecting elements evolved for datagram packet switching in LAN environment for today's Ethernet/Internet datagram environment. Another class of switches had already evolved for circuit-switched connection-based PSTN core, which is fundamentally different from the packet-switched connectionless in the LANs (see Table 2.1). In this section, we discuss the evolution of these wide area circuit switched for PSTN. The PSTN is also referred to as a circuit-switched network; so we may refer to these switches as circuit switching elements. The hierarchy of circuit switches in the PSTN was introduced in Figure 2.7; these are local, toll, primary, sectional, and regional switches to establish connection among peer switches as well as lower and upper layer switches with lower and higher transmission capability for shorter and longer distance coverage.

Originally PSTN was designed using analog voice connections through manual switchboards in the late 19th century. The manual switchboards were subsequently replaced by automated telephone exchanges in the early 20th century and later digital switches took over core connection for the PSTN in the mid-20th century. At the time of this writing, analog two-wire circuits are still used to connect to most wired telephones worldwide, but intermediate PSTN switches are digital physical or virtual circuits. The basic digital circuit in the traditional core PSTN has a data rate of 64 Kbps carrying a typical phone call from a source telephone terminal to a destination terminal. The analog audio sound is digitized at an 8 KSps using 8-bit PCM resulting in a 64 Kbps data stream. The digitized voice is then transmitted from one end of the PSTN network to another through a hierarchy of digital circuit switches. The call is switched using a signaling protocol known as signaling system number 7 (SS7) between the telephone exchanges under an overall routing strategy. With the popularity of wireless digital cellular networks and VoIP services, smartphone and VoIP phones are eliminating the analog telephone connections in the home and office networks and circuit switching turns to virtual circuit switching; however, the hierarchal transmission infrastructure remains the same.

SS7 is a set of signaling protocols to set up and tear down telephone calls. The SS7 signaling protocol is also used for number translation, prepaid billing mechanisms, short message services, and a variety of other services such as caller ID, call waiting, and voice messages, except the two-way voice traffic for conversation. The IETF has also defined level 2, 3, and 4 protocols that are compatible with SS7 but use an IP transport mechanism enabling VoIP telephone services. In principle, PSTN can be thought of as an overlay of a circuit-switched network used for traffic and a packet data network for signaling and control. After call establishment, each circuit cannot be used by other callers until the circuit is released and a new connection is set up. As a result, during the connection period, the QoS is maintained, but even when there is no traffic, the channel remains unavailable for other users. Technically speaking, these switches are designed for a constant bit delay during the connection, whereas interconnecting element in packet switching uses queues which may cause varying delays for arriving packets.

At the time of this writing, the PSTN and circuit switching in most part is integrated with the Internet in an all-IP core networks carrying all types of traffic, including circuit-switched telephone. The wired telephone connections at homes, which were once the source of prosperity of the telecommunication industry as perhaps the largest technical industry of the world, are diminishing and replaced by cellular wireless telephone services and VoIP. Starting from 2G digital cellular telephones, the end-to-end digital services evolved from ISDN.

2.4.1 ISDN for Integrated Circuit-Switched Data

In the 1970s, the telecommunications industry began to pay attention to digital services and the concept of ISDN emerged in the industry. In an ISDN network, voice and data are both treated as digital circuit-switched services. In an ISDN network, the access of the POTS, which was a two-wire analog line like the original access in the early days of the telephone industry, changes to digital access. In the POTS, the analog signal arriving from the customer would be digitized at the network, while in an ISDN network, the analog voice is digitized at the terminal to be integrated with the data. Circuit-switched data in the POTS for applications such as fax machines or telnet was modulated by voiceband modems to an analog signal and then it was transmitted over the PSTN with data rates of up to 9600 bps that was much less than the 64 Kbps dedicated to each line in the core. With the growth of the importance of data services, ISDN emerged to solve this problem. ISDN can deliver at minimum two simultaneous connections, in any combination of data, voice, video, and fax, over a single line. Multiple devices can be attached to the line and used as needed. At a top level, ISDN is a set of protocols for establishing and breaking circuit-switched connections with advanced call features for the user, such as caller ID, call forwarding, three-way conferencing, etc. The basic ISDN provides two traffic channels at 64 Kbps and one control channel at 16 Kbps, all using digital transmission. As compared with POTS for delay sensitive applications, ISDN provides a better quality voice with two simultaneous channels when both ends are ISDN. As a result, it found some niche applications such as the transmission of radio signals in a radio station, video conferencing, and many PBX systems to connect the end office of the PSTN through a T1 carrier. For call setup, control, and other administrative purposes, ISDN services use a separate dedicated signaling channel from the end node to the network while POTS did not. As compared with the voiceband modems used for data application, ISDN avoids analog transmission and multiple digital-analog-digital-analog conversions by providing a complete end-to-end solution with defined switching infrastructure. As compared with DSL modems, ISDN cards provide lower data rates at considerably lower complexity and a direct connection to the end office in the PSTN. The next step in ISDN standard was broadband ISDN (B-ISDN) [Min89] to achieve data rates on the order of 2 Mbps.

Although wired ISDN and B-ISDN never met the expected market and never achieved a large penetration, ISDN was the backbone of protocols for the 2G GSM digital cellular networks and 3G IMT-2000 cellular wireless networks successfully implemented the B-ISDN goal of 2 Mbps for the data rate IMT 2000. In fact, 2G digital cellular networks were the first popular application of an integrated voice and data service over a circuit-switched network in which digitization of the voice is taken place at the terminal and not at the network and the IMT 2000 adopted 3GPP all-IP backbone [Cha99].

2.4.2 Packet Switching Over Virtual Circuits

The early packet-switched computer networks were designed to connect remote data terminals and computers together. These networks were using 4-wire voiceband modems over dedicated analog leased telephone lines connected to the core digital PSTN. A Connection was dedicated and did not need dialing to establish and terminate the call, and rather than 2-wired, it was 4 wires allowing the modem to operate in two directions simultaneously. These networks emerged after the Second World War, first, for military applications to connect computers and terminals in remote airbases and shortly spread in airline reservation, banking, and many other commercial applications. The main technical challenge in that industry was the design of higher speed modems to save on the expensive cost of the leased lines. These networks were using high-level data link control (HDLC) or similar protocols for data link layer protocols with extensive coding to support reliable data transmission over an unreliable telephone connection. HDLC protocol provides connectionless and connection-based services with multipoint capabilities, but it was commonly used for point-to-point connections. The first real packet-switched network, X.25, emerged in that environment as the first international packet switching network in 1978. It had large global coverage during the 1980s.

X.25 protocol was a CCITT (now ITU-T) international standard for packet switching over virtual circuits provided by PSTN core and provided reliable data transmission with a data rate of 2.4 Kbps up to 64 kbps. These virtual circuits carry variable-length packets with up to 128-byte data over the core PSTN worldwide. Today, the dominant Ethernet/Internet packet switching technology is connectionless and it does not use a dedicated physical or virtual circuit where X.25 packet switching operates based on connection-oriented virtual circuits. In X.25, a connection is established, the data packets are transferred, and then the connection is terminated. This protocol allows virtual switching for multiple streams over a single line. The HDLC is adopted by ITU as the link control layer protocol for X.25 to ensure reliable end-to-end data flow and error control. This protocol, used commonly on legacy computer networks, provides connectionless and connection-based services with multipoint capabilities. Today, this protocol is used for point-to-point Internet connections such as transaction processing for credit card authorization.

With the widespread digital and optical links in the PSTN, transmission became very reliable and almost error-free which did not need the large overhead of X.25 and it was replaced by Frame Relay. X.25 checks point-to-point connections with error correction overall intermediate switches which caused large delay and reduced the throughput. Frame relay did not have error and flow control on the point-to-point links, and it provided an end-to-end connection with much higher throughput which is suitable for LAN connections. X.25 specified processing at layers 1, 2, and 3, while frame relay operated at layers 1 and 2 only, which has significantly less processing at each node. In addition, X.25 had a fixed bandwidth which wastes the throughput on low load, while frame relay dynamically allocates bandwidth during call setup negotiation to avoid throughput waste.

2.4.3 ATM for Virtual-Circuit Packet Switching

Figure 2.15 illustrates the overview of the evolution of switching techniques in PSTN to support circuit switch and packet-switched voice and data services. The legacy circuit-switched networks supporting POTS using

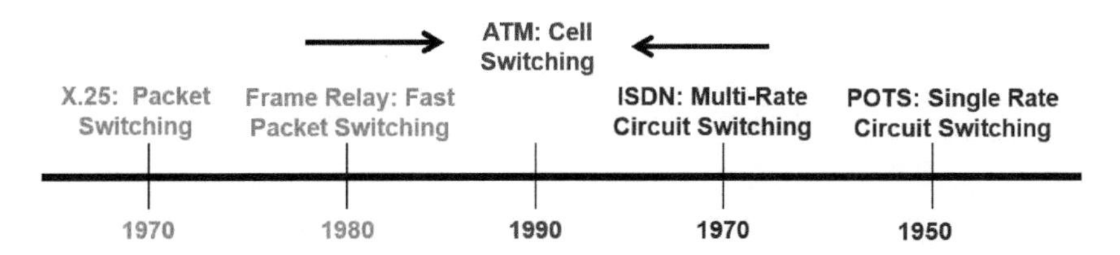

Figure 2.15 Evolution of ATM switches to integrate circuit-switched and packet-switched traffic.

analog connection to end user terminal evolved into ISDN using multirate circuit switching techniques using a connection-based virtual circuit. The packet-switched networks using datagram over virtual circuit-switched networks started with the lower speed X.25 and emerged into much faster frame relay. In the late 1980s, the next step in the evolution of this industry emerged as the asynchronous transfer mode (ATM) or cell switching technology. The objective of this technology was to design a single networking strategy that could support both delay sensitive applications such as real-time video and audio, usually transported over circuit-switched media, as well as data applications such as transport of image files, text, and email, usually carried over packet-switched networks. The standard was designed by a special group called ATM Forum and became an ITU-T standard. ATM was a virtual circuit technology, which uses fixed-length cell relay connection-oriented packet switching. ATM was the technology designated for implementation of the B-ISD services to carry both synchronous voice and asynchronous data services.

Figure 2.16 shows the original concept of the ATM networks. The encoded information traffic is broken into small 53 bytes packets with 48 bytes of information and 5 bytes of header (Figure 2.16(a)) to be transferred over the network (Figure 2.16(b)) with two basic interfaces, user to network interface (UNI), and network to network interface (NNI). This approach is different from other packet-switched networks such as Frame Relay, Internet, and Ethernet, in which variable sized *packets* or frames are used. Like Frame Relay and X.25, ATM is a connection-oriented technology, in which a logical connection is established between the two endpoints before the actual data exchange begins. ATM's ability to carry multiple logical circuits on a single physical or virtual medium was useful. Figure 2.16(b) provides an overview of the integration of PSTN and Internet applications using ATM networks. ATM envisioned a utopic end-to-end network to integrate everything into

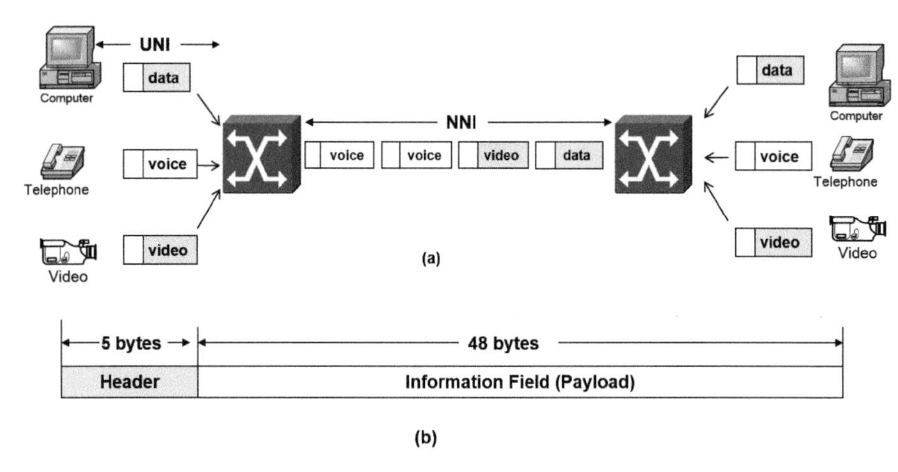

Figure 2.16 Principle of operation of an ATM network. (a) Integration of media into fixed packet length. (b) The format of the packet.

the PSTN and had attempted to define details of how wired and wireless connections can reach the end users. An overview of some of these details is available in [Pah09] for readers interested in history. ATM was designed for efficient integration of voice streams, which was the main income of the service providers at the time, and the emerging data applications which were expected to form the future. But this entire idea was before the commercial success of the Internet and its associated industry, which generated incomes exceeding that of voice telephony. At that time, the AT&T alone commanded a budget equal to the fifth economy of the world and that income was dominated by voice telephony application. Therefore, the design of the system pays more attention to the needs of the telecom industry and its popular telephony application. As a result, motivation for the use of short packets or *cells* was the control of the delay jitter in the integration of voice and data applications in a high speed link. At the time ATM was designed, 155 Mbps SONET/SDH with a payload of 135 Mbps was considered a fast optical core network link. At this rate, a typical full-length 1500 byte (12,000-bit) data packet from a LAN would take 77.4 μs to transmit. In a lower-speed link T1 carrier link at the rate of 1.544 Mbps that packet would take up to 7.8 ms. A queuing delay caused by a few of these randomly generated data burst packets between very short and regulated streaming speech packets causes excessive jitter bringing QoS of the voice below the accepted level by service providers. Designers of the ATM standard were convinced that to be able to provide short queuing delays for voice and carry large datagrams, they had to have fixed short packets or cells. As a result, ATM breaks both long data packets and voice streams into 48-byte pieces and adds a 5-byte switching header to each piece to form a cell. The header is used for transport over the switching fabric and reassembly at the destination. The choice of 48 bytes was a compromise between American 64-byte and European 32-byte proposals. 5-byte headers provided around 10% of overhead for switching information. By multiplexing these 53-byte cells instead of long data packets, ATM technology reduced the worst-case queuing jitter for voice applications to more than an order of magnitude resulting in a better QoS for the voice users.

The challenge for the ATM technology earlier was the unexpended exponential of the LAN industry and the emergence of wireless LAN. In the core of modern optical networks, a 1500-byte Ethernet packet takes 1.2 μs to transmit on a 10 Gbps link allowing reasonable chances to control the jitter for voice streaming packets and the consequent need for small cells. In addition, the cost of segmentation and reassembly hardware at those high speeds made ATM was less competitive for growing IP traffic. ATM forum responded by two unsuccessful attempts, LAN Emulation (LANE) to replace LANs [Mcd98], and wireless-ATM [Pah97, Liu98].

Today, ATM technology is only used in some of the backbones of high-quality high-speed applications such as video transfer in low latency environments. The advantage of virtual circuits is to support delay sensitive real-time applications. Datagram packet switching over connectionless networks in the original Ethernet/Internet networks was best-effort networks lacking such mechanisms. Technologies such as multi-protocol label switching (MPLS) and the RSVP created virtual circuits on top of datagram networks. MPLS provides services like ATM for variable length packets, and it is sometimes referred to as ATM without cells. Modern supper high-speed routers for long-haul and international connections with terabit per second (Tbits/s) speeds do not require these technologies to be able to forward variable-length packets because the transmission time of the packets are so short that their associated queuing delay causing jitters do not harm the QoS of time sensitive streaming applications, such as VoIP or IPTV, significantly.

2.5 Evolution of Internet Interconnecting Element

Figure 2.17 provided a simplified description of the Internet infrastructure and its relation to interconnecting elements and protocols. One can view the entire

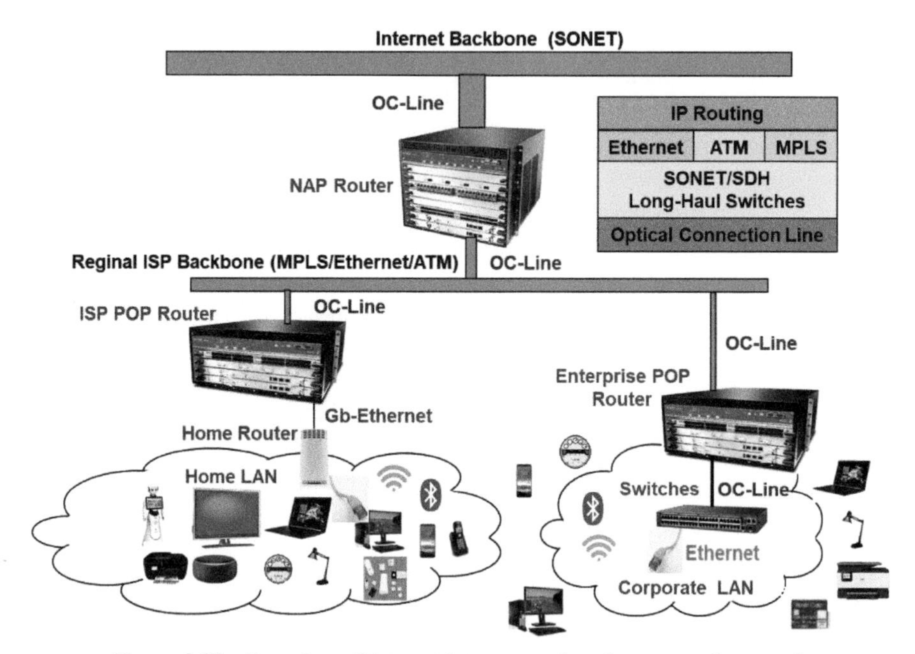

Figure 2.17 Overview of Internet interconnecting elements and protocols.

network as many homes and corporate or enterprise LANs connecting in local areas LAN switches and connecting to wide areas through routers. The core of the LAN connections follows Ethernet protocol to form packets with IP headers to communication with other LANs through the Internet. Wi-Fi traffic is dominant for local distribution, but it is bridged to Ethernet protocol at arrival to the core of the wired LAN to connect to the LAN switches or router to the Internet. The connectionless computer networking datagrams are labeled by TCP header and the connection-oriented data such as VoIP or IPTV is labeled with UDP heather (Figure 2.14). The physical infrastructure of this Ethernet/Internet network uses a variety of links over the optical connections (OC) over fiber lines following SONET/SDH with Ethernet, MPLS, and even ATM protocol to support the diversified applications with different Q0S requirements.

Considering Figure 2.17, the first natural question which comes to mind is "why so many options to interconnect networks?" The simple answer to this question is that "no one was able to solve all issues" or may be "networking industry evolved that way". Long-haul telecommunication TDM switches for voice traffic in PSTN came first. Then ISDN and ATM switches came to integrate the data with the voice. These switches are fast hardware equipment designed for connection-based services with the support of QoS, but service per user is expensive. ATM was designed to answer our question, but it failed to integrate everything and that was the last attempt of the telecommunication industry for a utopic solution. In the computer communication industry, bridges, which are also called LAN switches, interconnect LANs using hardware MAC addresses which do not scale well, but are fast and inexpensive. Routers scale well using IP software addresses and can handle data traffic at a reasonable price for long-haul communications, but still they are more expensive than LAN bridges and they cannot guarantee the QoS.

Internet was originally designed to connect privately owned corporate LANs, connecting employee computers, together. The interconnecting elements in LANs are hubs, bridges, or LAN switches and the LAN itself connects to the Internet via a router. In this section, we provide an overview of these interconnecting elements of the Internet. We begin with LAN interconnects and then we discuss routers connecting the LANs to form the Internet.

2.5.1 Bridges and LAN Switches

In the evolution of the IEEE 802 standards for LANs, we see repeaters, hubs, bridges, and switches defined as interconnecting devices. Repeaters were defined for legacy Ethernet operating with a bus topology over a single thick cable, enabling all computers to connect to the cable with a vampire connector. The maximum length of the legacy Ethernet cable was 500 m and four repeaters could extend the coverage to 2.5 Km. Later, Ethernet migrated from the bus topology to centralized star topology with a hub managing connection among computer terminals up to 100 m away. The centralized topology was better suited for hierarchical expansion of the network coverage. A hub is a network LAN hardware device for connecting multiple LAN devices together to create a segment of the local network. Typically, a hub has multiple ports and the signal introduced at the input of any port appears at the output of every port except the original incoming port. In a sense, it is a repeater of the arriving packet in all ports. The Ethernet hubs for the early versions of this standard were also engaged in carrier sensing and sending a jamming signal at the time of collision. Today, hubs are mostly replaced by bridges or LAN switches (see Chapter 7 on the Ethernet).

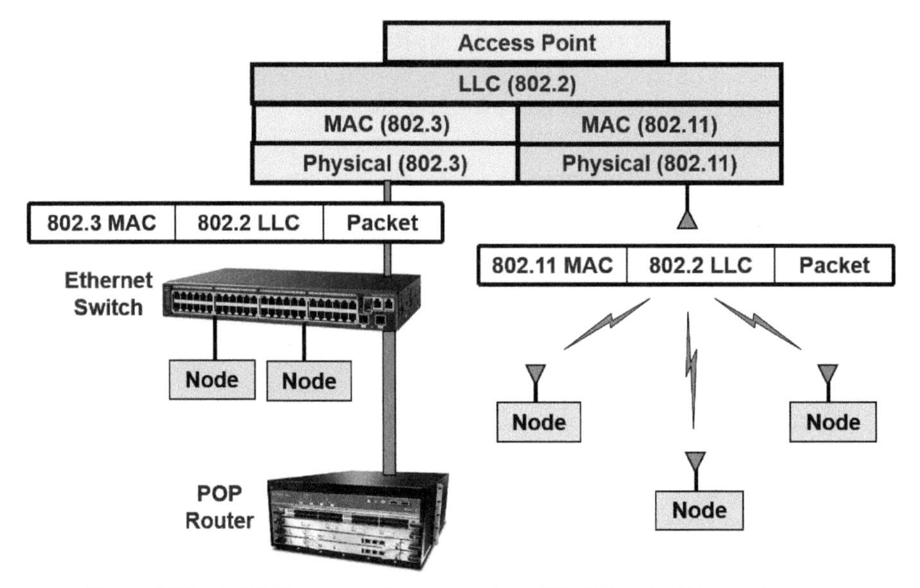

Figure 2.18 A Wi-Fi access point connecting a WLAN to the Ethernet/Internet.

A bridge is a device that connects two networks using the same or different data link protocols. Since in OSI model, the data link layer is the second layer of the protocol stack, they are sometimes referred to as layer 2 devices. One of the early applications of the bridges was to connect two separate segments of a network, such as two LANs in two different buildings of a corporate, with a point-to-point link. Another traditional application of bridges was to connect two different LANs with different MAC and PHY protocols, such as IEEE 802.3 Ethernet and IEEE 80.5 Token Ring or the IEEE 802.11 Wi-Fi. As shown in Figure 2.18, a single 802.11 network interface card in the AP is connected to the backbone Ethernet/Internet network. The AP enables each Wi-Fi device in the subnetwork access to the Ethernet/Internet as well as to other devices in the subnetwork. The AP changes IEEE 802.11 protocol stack to the IEEE802.3 protocol stack and vice versa. From this standpoint, it can be referred to as a bridge between two different LAN protocols. The AP here can also function as a router in assigning IP address or implementing different IP protocol services and consequently referred to as a router. Wireless APs sometimes connect two clusters of users, for example, located in two different buildings, to one another. In such applications, WLAN technology is used to bridge two LANs, and it is sometimes referred to as a wireless bridge.

The traditional bridges were hardware devices operating at LAN speed using the MAC address (rather than IP address) to direct the packet to the appropriate segment of the network. Although the function of a bridge was to connect separate LANs, together, it was also used to connect two LANs using the same protocol. Bridges use the MAC addresses of the nodes residing on each LAN segment of the network and allow only the necessary traffic to pass through them. A bridge can also filter out certain traffic and prevent it from passing through. When a packet is received by the bridge, the bridge determines the destination and source segments. If the segments are the same, the packet is filtered (dropped); if the segments are different, a packet is forwarded to the appropriate segment. In addition, bridges prevent unnecessary and corrupted packets from spreading throughout the network by dropping them. For example, a Token Ring LAN packet carries priority and has token packets for the operation of the LAN which are not used in the Ethernet. A bridge connecting a Token Ring to an Ethernet LAN eliminates the priority bits and prevents token packets to enter the Ethernet segment of the network. Bridges improve the overall throughput by distributing the shared traffic into different shared domains, increase security and resistance to failure by partitioning the traffic, and increase overall throughput and geographical coverage by using several LANs. A LAN switch is a bridge connecting more than two LAN segments together preventing unnecessary traffic to enter each segment. The word LAN bridge and LAN switches are sometimes used interchangeably to address the same entity.

A bridge or a LAN switch receives packets on one port and retransmits them on another port and it does not start retransmission until it receives a complete packet. As a result, stations on either side of a bridge can transmit packets simultaneously without causing collisions. A fixed bridge directs every packet toward all ports. A learning bridge examines the destination address of every packet to determine which port the packet should be directed to using a table that has been built by analyzing previous arriving packets in different ports. This approach to bridging increases efficiency because the bridge avoids unnecessary transmission of the packets. Learning bridges are very common in the industry.

Bridges are devices which are connecting different LANs together. Since IEEE 802 community has developed all popular LAN standards, popular bridge standards are also developed by IEEE 802 community. Figure 2.19 shows an overall perspective of the organization of the IEEE working groups of the IEEE 802 project. The IEEE 802.1 is a general group working on issues common to all other 802 LAN standards, and it is concerned with several

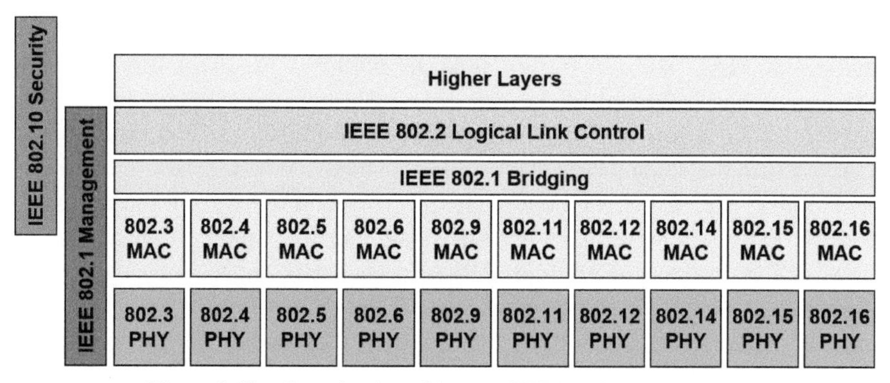

Figure 2.19 Organization of legacy IEEE 802 Standard Series.

issues related to all LANs including internetworking among 802 LANs, metropolitan area networks (MAN)s, and other wide area networks. Other concerns of the IEEE 802.1 include defining the architecture, link security, network management, and higher interface to higher layer protocols and layer 2. IEEE 802.2 defines the LLC protocol which is also used for all other standards. Starting from IEEE 802.3 Ethernet, traditionally, each number was associated with a specific MAC technique with several physical layer options. For example, as described in Chapter 6, the IEEE 802.3 standard defines a single CSMA/CD MAC and several physical layers operating at different data rates over a variety of media. The most important IEEE recommendations for bridges are defined under 802.1 group. The most popular IEEE 802.1 recommendation related to bridges is the IEEE 802.1D which defines the operation of the transparent bridges and spanning tree algorithm for interconnecting bridges. Another interesting 802.1 recommendation is IEEE 802.1Q defining the operational environment for the virtual LAN (VLAN).

IEEE 802.1D Transparent Bridges and STA:

The word "transparent" is used for the IEEE 802.1D bridges because bridging does not involve any transaction with the stations connected to the bridge; so they are transparent to the user terminals. These bridges have an automated address finding mechanism that builds up a table based on the address of the arriving packets to each port. The IEEE 802.1D transparent bridges have packets exchanged within themselves to establish the routing protocol known as the "spanning tree" algorithm (STA). The traditional IEEE 802.1D bridges were designed to support protocol conversion between legacy LAN technologies such as Ethernet, Token Ring, and FDDI. These bridges were

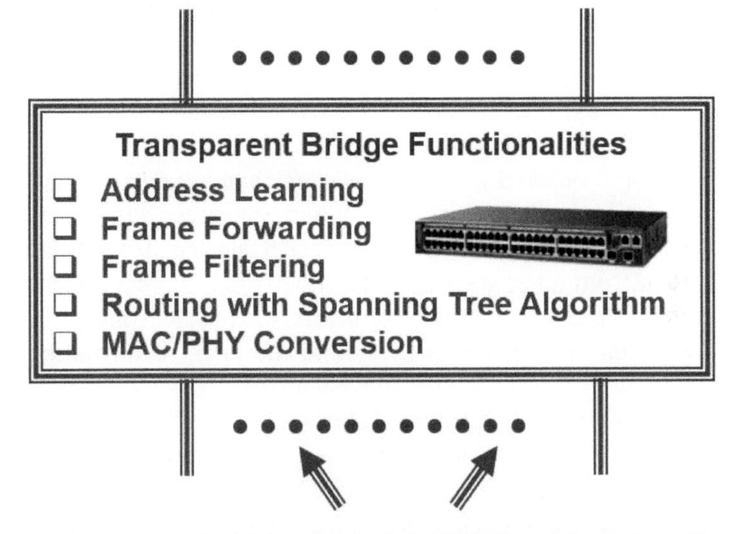

Figure 2.20 Overall structure of the IEEE 802.1D Transparent bridges.

filtering control packets associated with different LAN technologies and adjusted for differences in MAC priorities.

Figure 2.20 shows the general structure of the traditional IEEE 802.1D transparent bridges. A bridge receives and sends three different types of packets in each port, user terminal's data frames from the connected LANs, control packets such as tokens used for the operation of certain LAN technologies such as Token Ring, and bridge protocol data unit (BPDU) which is the protocol for communication among different bridges. Each port of the bridge has a separate MAC address to connect to different LANs. The functionality of the bridge is to *learn* about the terminals connected to each port, *forward* frames to appropriate ports, *filter* unwanted frames, execute *routing* algorithm, and *convert* MAC and PHY layer protocols of the packets to fit the associated LAN technology connected to each port.

During the *learning* process, a transparent bridge reads the source address of the arriving packets and creates a table to associate them with the port number they are arriving from. The arriving information is time stamped so that it can get refreshed (e.g., around every 300 s) to manage for changes in network nodes. During the *forwarding* process, the bridge looks at the destination address of the arriving packet and if it matches the addresses associated with an existing port, the bridge forwards the packet to that port. If the destination address is not known, the bridge floods all ports except the source port and blocked ports. If the source and the destination address are the same, the packet is removed. The filtering process eliminates control packets such as token or ring maintenance packets used in Token Ring LANs to prevent them from other segments of the network. If ports are connected to LANs using different technologies or media, the physical and MAC layers are *converted* at the bridge. The IEEE 802.11D uses STA for routing which prevents the circulation of packets inside a network.

Local networks support broadcast and multicast addressing; therefore, bridges have the capability of flooding all their ports for broadcast or multicast messages. When a bridge flood a packet over all ports to many networked LANs connected to that bridge there are chances that other bridges redirect the packet to the original bridge and create a loop circulating the packet again and again. We use an example to further explain this concept. Figure 2.21(a) shows the topology of a network with seven LANS, A, B, C, D, E, F, and G and five bridges, 1, 2, 3, 4, and 5 connecting them for different purposes; each bridge is shown as a node and each LAN as a connection between two nodes. Each bridge has at least two ports to connect a pair of LANs and can have more to connect three or more LANs. In Figure 2.21, bridges 1, 2, 3, and 4 each have 3 ports and bridge 5 has 2 ports. The topology of this subnetwork can create a few loops to circulate packets among bridges. This situation is caused because we have some redundant paths to connect at least one link for all LANs to connect to all others. Figure 2.21(b) offers a solution to this problem by keeping four paths, shown by solid lines, and three redundant paths, identified by dashed lines. If the redundant dashed connections are blocked, there is no loop in the network, while all LANs are connected to each other.

The IEEE 802.1D recommends an STA which automatically breaks loops in the network and maintains access to all LAN segments. This algorithm

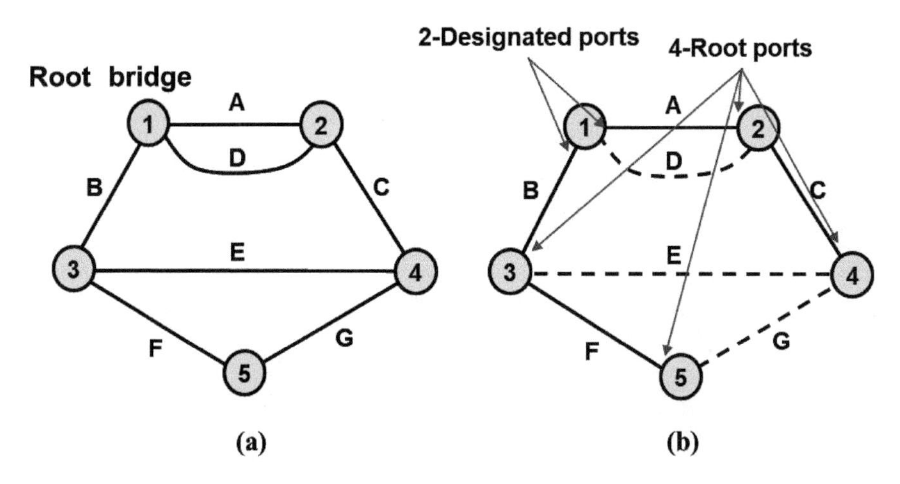

Figure 2.21 An example subnetwork topology with five LANs and seven bridges connecting them. (a) Network topology. (b) Blocked paths selected by STA.

is implemented in several stages using a set of messages exchanged among the bridges using the BPDU protocol. Each BPDU is issued from a bridge regularly (e.g., every 2 s) and it is broadcasted to all other bridges. A BPDU carries the ID of the original bridge and its port as well as its root path cost through different LANs. The algorithm operates in a few steps to form the tree that connects all LANs and blocks ports connected to redundant paths to prevent the loops. To start the algorithm, the STA defines a *route bridge* and assigns *root ports* and *route path cost* for all other bridges to connect to the root bridge. The network administrator can assign a specific bridge as the route bridge or there are procedures in the standard to find it automatically. In the next step of implementation, the STA decides on the lowest cost path through LANs to connect to the source bridge to any other bridge port. The bridge port connecting bridge ports to the root bridge at the lowest cost is named as *root port* for that bridge and the port that connects the source bridge to that bridge is called a *designated port* of the source bridge. As the last step, the source bridges block ports that are not a *root ports* and other bridges block all ports that are not a *designated port* to eliminate loops. This blockage is for the flow of data from the terminals, but it does not block the packets exchanged among the bridges using BPDU; therefore, communication among bridges is alive, allowing them to adjust to changes in the network topology in time. Figure 2.21(b) shows the complete picture for all the root ports, designated ports, and the blocked ports and the resulting implementation of the STA for the network shown in Figure 2.21(a). To select a root port when we have two candidates with the same cost, one of them is selected randomly. If two bridges have the same cost to connect to the LAN to the root port, the bridge with a lower identification number is selected. A more detailed description of STA is available in [Pah09].

The spanning tree protocol was replaced by the rapid STA introduced by IEEE 802.1w and adopted by IEEE 802.1D in 2004. The STA was a passive algorithm waiting for time to pass for information collection. Rapid STA actively sends inquiry packets seeking information from neighboring switches which results in faster convergence. The algorithm also has a faster root detection procedure, digs for backup ports in case forwarding ports failed, and uses other practical tricks to speed up the process. Other modifications to STA have been made to make it applicable to VAN environments.

IEEE 802.1Q Virtual LAN:

A VLAN allows a group of terminals to form a subnet regardless of their physical location. Figure 2.22(a) illustrates an example of the formation of a VLAN among a terminal connected to a LAN and a LAN located in another place in the network. Implementation of a VLAN is based on the attachment of a tag to the MAC address which carries a unique address to identify a VLAN group.

The format of the tag is defined by the IEEE 802.1Q and bridges complying with this standard have means to read the tag and identify the VLAN address. The VLAN tag is then used by the bridge to direct the packet to specific segments of the network which are a part of the VLAN. Figure 2.22(b) shows the general format of a MAC packet and the insertion of a VLAN tag that carries the VLAN address. A VLAN operates in the same way as a physical LAN to group end stations together even when they are not located on the same LAN segment. Using the VLAN feature, a network manager can configure the network through software control rather than physical relocation of terminals. A VLAN allows the creation of a broadcast domain using hardware MAC addressing mechanisms at the so-called layer 2. In a larger network, switch ports can be assigned to a VLAN allowing packets to be forwarded and flooded only to stations in the same VLAN. VLANs allow the creation of segmentation services traditionally provided by routers in LAN configurations. A VLAN forms a logical network, and packets destined for stations that do not belong to the same VLAN must be forwarded through a routing device using an IP address. The STA can be implemented to a VLAN. More details of the implementation of VLAN on

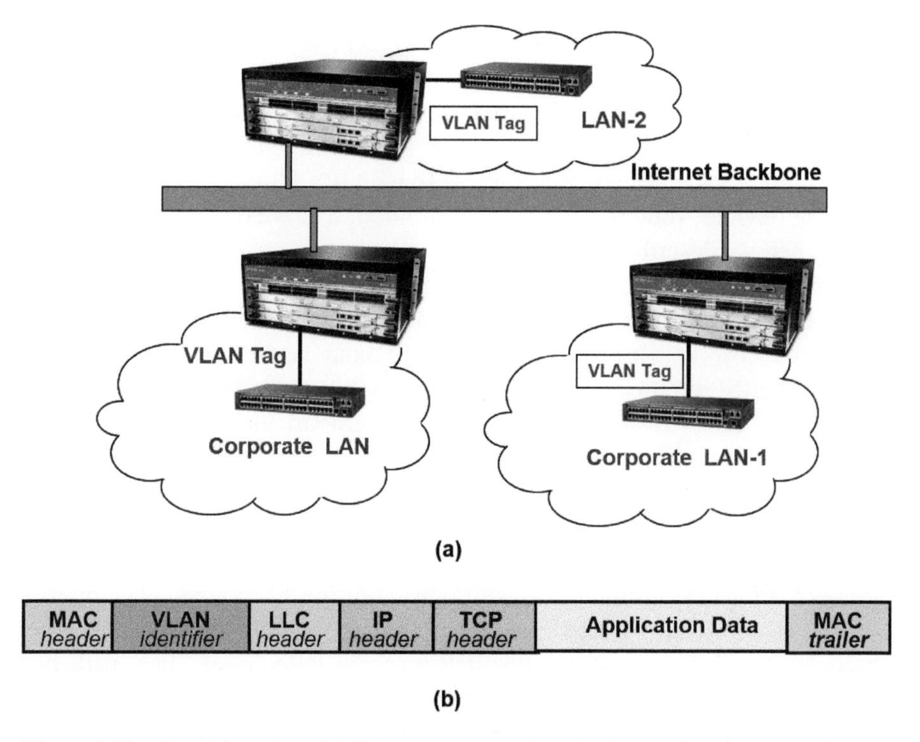

Figure 2.22 Formation of a VLAN between geographically dispersed networks. (a) Overall architecture of VLAN inside the Internet. (b) Packet format for implementation.

the Ethernet packets and how the address is managed are described in Section 6.4.1, where we discuss how Ethernet packets are modified to accommodate additional services such as VLAN.

2.5.2 Routers Protocols

An IP router partitions a network into subnets and directs the arriving traffic destined for IP addresses toward a segment of the network using a routing algorithm that optimizes the cost. The cost for intelligent forwarding and filtering is usually calculated using the speed of the network. This protocol filtering usually takes more time than packet filtering used in LAN bridges. Routers examine data as it arrives, determine the destination address for the data, and use routing algorithms to determine the best way for the data to continue its journey. Unlike bridges and switches, which use the hardware to determine the destination of the data, routers use the software network address to make decisions. This approach makes routers more functional than bridges or switches, but it also makes them more complex and expensive. IP addresses are soft addresses, and the address of a device may change in time; when a packet arrives in a LAN, it needs to know the MAC addresses which uniquely identifies the hardware. To map IP address to MAC address of different devices, routers use the ARP protocol that we discussed in Section 2.3.2 when we described Mobile IP.

Figure 2.23 is another view of the same concept of Figure 2.17 to provide a simplified model for the operation of the Internet in connecting home and corporate networks through ISP and network service provider (NSP). The ISP provides wide-area connectivity among different enterprises and home networks and NSPs connect the ISP to the rest of the world. Each ISP organization manages a group of connected routers exchanging information using a common protocol. This group of routers is referred to as an autonomous system (AS). The routers inside an AS domain use interior gateway protocol (IGP) to communicate with each other and exterior gateway protocol (EGP) to communicate with outside of the AS domain. The interior router protocol (IRP) adopts routing protocols to carry the packet toward the destination. The most popular IRP protocol is the open short path first (OSPF) which uses Dijkstra's shortest path algorithm. The most popular ethernet ring protocol (ERP) is the border gateway protocol (BGP). BGP is the core ERP routing protocol of the Internet operating on TCP/IP protocol. It allows negotiation between bordering routers to learn if another AS is willing or able to carry traffic of an AS. BGP is fundamentally a distance vector protocol in which distances are defined based on political, security, or economic considerations. These policies are manually configured in the BGP routers and they are not part of the protocol. ISPs must use BGP to establish routing between one another.

Different sizes of routers have emerged for a variety of applications. Smaller routers are often found in small homes and small offices, while the large routers are found in ISP core networks, academic and research network facilities, and networks for large businesses. One way to classify different routers, shown in Figure 2.23, is to divide them according to their application into *access*, *distribution*, and *core routers*. *Access routers* are small routers used by small subnetworks at customer sites such as a branch office of an enterprise, a small office, or a home office to connect to ISP services such as IP over cable, DSL, or fiber optic links. These small subnetworks do not need hierarchical routing of their own and are designed for minimal cost. These routers or gateways commonly use network address translation

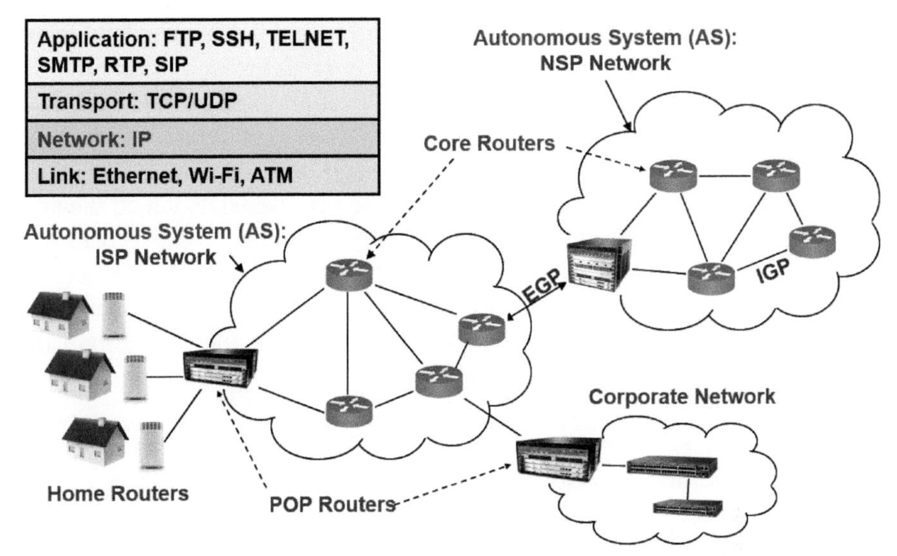

Figure 2.23 Internet protocol stack and TCP/IP. (a) Packet headers. (b) Operation for terminals and routers.

instead of routing algorithms to support the connection. Therefore, instead of connecting local computers to the remote network directly, these routers make all local computers appear to be a single computer connecting to the ISP. Today, most of these routers integrate Wi-Fi access for the devices and carry several other ports to connect to fixed desktops in the proximity of the router.

Distribution or point of presence (POP) routers connect access routers to the rest of the Internet. These routers collect and police multiple streams of data from different access routers to assure the QoS for the user and to enforce that across the wide-area networks. In corporate-networks POP routers connect multiple buildings of a campus or distributed buildings of large enterprise locations. As a result, distribution POP routers have large memories, complex algorithms, and processing intelligence, and multiple connections to the wide-area networks to support diversified resources for transmission. *Core routers* provide a backbone to interconnect the distribution POP routers from multiple buildings of a campus or large enterprise locations. These routers process a huge amount of information and they tend to optimize for high bandwidth.

TCP and UDP Transport on the Internet:

As shown in Figure 2.23, the IP routing protocol is based on a four-layer reference model by the IETF consisting of an application layer, transport layer (TCP/UDP), the network layer, and link layer (IP). Several protocols at a higher level use different options for lower layer protocol to accomplish their objectives. For example, an application layer protocol such as FTP, SSH, TELNET, or SMTP for implementation of file transfer, remote connection, or email have transport option protocols such as TCP and RTP/SIP for implementation of VoIP or video conferencing can use UDP, which then can be sent over IP-v4 or IP-v6 network protocol using a link layer protocol such as Ethernet, Wi-Fi, or even ATM. The TCP/UDP protocol provides an end-to-end connection between the two applications and IP protocol is used for routing the protocol. The IP protocol is so important for the Internet that sometimes the entire suite is referred to as the TCP/IP protocol suite or an all-IP network.

IP protocol is designed for datagram delivery over the Internet using packets of information also referred to as datagram. Information packets are short sequences of binary data consisting of a header and a body. The body contains the application data, and the header describes the destination and some additional bits to be used by the routers on the Internet to pass the packet along a sequence of routers toward the destination. Since, in a datagram, different packets may take different paths, the packets may arrive out of sequence. In case there was congestion in a specific router, the IP may discard a packet, causing a packet loss in a stream of packets. Since packets sometimes go through LAN bridges or other interconnecting devices which sometimes flood the ports, we may have packet duplications at the destination. Therefore, a stream of packets generated by an application may arrive at the destination in the wrong order and with missing or duplicate packets. The TCP protocol is an end-to-end connection-based protocol operating on the arriving packets to provide a reliable delivery service that guarantees to deliver a stream of packets sent from a host to a destination without packet duplication or packet loss. TCP also allows multiplexing several applications over an IP stream. In a sense, the functionality of the TCP is like data link layers because they both are designed to provide reliable communications. Data link layer protocols are working over a specific physical link while TCP operates over a randomly varying sequence of links with the possibility of delivering packets in duplication or out of sequence. Therefore, like popular data link layer, protocols such as connection-based FTP, security shell (SSH), or SNMP, the TCP protocol is a connection-based protocol using congestion or flow control as well as error control mechanisms.

Since TCP protocol is optimized for accurate delivery rather and is not focused on timely delivery, sometimes, it results in relatively long delays in the orders of seconds. This delay is caused while the TCP protocol waits for out-of-order messages or retransmissions of lost messages. As a result, TCP protocol is not particularly suitable for real-time applications such as VoIP in which delays more than a hundred milliseconds cause undesirable disturbances on the QoS while losing up to 1% of the packets does not have any significant impact on the QoS for the user. For such applications, the connectionless UDP is usually replacing the TCP protocol. Application layer protocols such as the real-time transport protocol (RTP) can run over the UDP transport layer. Session initiation protocol (SIP) can establish the signaling protocol for call establishment. RTP and SIP are protocols enabling the implementation of VoIP and Internet telephony for voice and video calls that are developed on UDP transport. TCP protocol provides reliable in-order delivery of a stream of bytes, making it suitable for traditional Internet applications such as file transfer, email, or web access. TCP is the transport protocol that manages the individual conversations between web servers and web clients. In these applications, TCP divides the long messages into smaller segments, and it is responsible for controlling the size and rate at which messages are exchanged between the server and the client.

TCP has been optimized for wired networks in which lost packets are caused by congestion mostly in the core network. In wireless links, packets are also lost because of multipath and shadow fading as well as handoff in the wireless access portion of the network. The congestion control mechanisms used to remedy the packet loss in long-haul networks with relatively long window sizes are not suitable for the packet losses in short distances due to wireless channel impairments. Modern wireless networks implementing VoIP use different approaches to address this problem by

Ipv4 and Ipv6 for Routing the Datagram Over the Internet:
The IP protocol is a data-oriented network layer protocol used for communicating datagrams or packets across a heterogeneous packet-switched network. The packets with IP protocol headers were originally designed to be encapsulated in data link protocols such as Ethernet. The service of an IP header to a lower layer protocol with its own MAC layer address is to provide for a communicable unique global addressing amongst computers. The current and most popular version of the IP protocol is IP version four (IPv4) originally defined in 1981. IPv6 is the proposed successor to IPv4 whose most prominent change is the modification of 32-bit addresses (approximately 4 billion) to 128-bit addresses (approximately 3.4×10^{38}).

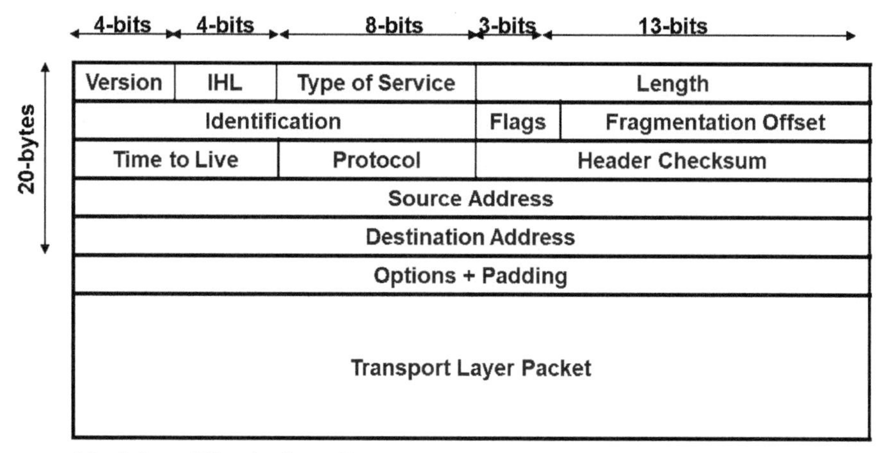

Figure 2.24 Frame format for the IP-v4 packets with the details of the header.

Figure 2.24 shows the general format of an IP-v4 packet and the details of its header. The length of the header is 20 bytes in which 4 bytes are used to identify the address. As compared with the IEEE 802 MAC addresses that were 26-byte long with 6-byte addresses. The main difference between the two addresses, however, comes from the fact that MAC addresses are randomly distributed while the IP addresses carry certain order and distribution technique that allows reasonable tracing of the address into geographical locations. In the header, the first four take advantage of the acknowledgment packets at the MAC layer of Wi-Fi or other wireless technologies. Bits associate with the version field that identifies the version of the IP protocol. The next four bits are used for internet header length (IHT) that specifies the length of the header in 4-byte (32 bits) segment lengths. Most of the time, there is no option or pad leaving the value of IHL at 5 for a header length of $5 \times 4 = 20$ bytes. The type of service field is 8 bits, and it is reserved for the implementation of priority for a given packet. This type of priority is like the priority of the LANs which at most can serve if the interconnecting bridge or router has multiple queues with different priorities. The next 16-bit identifies the length of the packet including the header in bytes. Therefore, the maximum length of an IP packet is $2^{16} - 1 = 65,535$ bytes. The link layer for carrying IP packets such as Ethernet often has a lower limit for packet length which calls for fragmentation and reassembly mechanisms.

The 16-bit identification field is used to identify the length of the fragments of a long packet and it is the same for all fragmented packets. The 3-bit flags field indicates if the packet is fragmented so that, at the destination, the packet can be reassembled. The 13-bit fragment offset shows the sequence number for a fragmented packet in 8-byte (64-bit) units. This way, the fragmented packets can be reassembled in the destination terminal and routers do not get involved in the reassembling process. The 8-bit Time to Live (TTL) field is used in practice to count the number of hubs to discard packets that are trapped in routing loops. The value can be set to any value up to the limit of 255 and each router reduces counts of this number one time down until this field represents a "0" for which the router drops the packet. The default value of the time to leave field commonly used in practice is 64. The 8-bit protocol field identifies the transport layer protocol that the packet should be delivered to. For example, Protocol field 6 identifies TCP and Protocol field of 17 signifies UDP transport.[4] The 16-bit Checksum field is calculated by dividing the entire header into 16-bit blocks and adding them in a simple binary form to generate a checksum. Since the generation of checksum includes the destination address, if the check sum does not match, the packet can be discarded, leaving it to TCP layer to recover that information. As compared with cyclic redundancy codes (CRC)s used for IEEE 802 MAC packets, the coding technique used for IP headers is simpler but less powerful. The source and destination addresses follow the IP addressing format. The options field is used for security, source routing, record routing, time stamping, and other features. The padding field is designed to make sure that the length of the packet is a multiple of 32 bits.

Comparing the IP-v4 header with the header of the Ethernet LANs, the IP header addresses provide for destination traceability, means for fragmentation and reassembly of the packets, a timer for discarding wandering packets, and a few bits for priority. These features are exploited by routing algorithms operating in the routers to direct a packet toward its destination. Using an IP header data from an upper layer protocol is encapsulated inside one or more datagrams without any circuit setup and it is delivered to its destination host with which it may not have previously communicated, and for that reason, we call it a connectionless protocol. This approach is quite different from PSTN in which we establish a connection before a phone call goes through and we refer to it as a connection-based protocol. The TCP protocol also provides an

[4] Details for other protocols is available at www.iana.org

end-to-end connection, but that connection is for reliable packet transmission and does not specify a physical or VC on the network elements as PSTN does.

The IP protocol works based on encapsulation of lower layer packets, which enables networking over heterogeneous data link layer environments such as Ethernet, Wi-Fi, or even ATM. Using IP protocol to connect two computers, makes no difference to the upper layer protocols in which the data link layer network is connected to the individual computers. Each of these data link layer networks can have its own method of addressing with a corresponding need to resolve IP addresses to data-link addresses. This address resolution is handled by the ARP. IP provides an unreliable best-effort delivery service which may result in corrupted, lost, or duplicated packets; the only reliability assurance of the IP protocol is that the header is error-free with a very high probability using a checksum. The primary reason for the lack of reliability is to reduce the complexity of the routers and allow them to discard packets when necessary. Perhaps, the most complex aspects of IP are IP address assignments and routing.

Figure 2.25 shows the format of the IP-v6 packets. The overall header is 40-byte compared with 20-byte IP-v4 header. The address is 128 bits

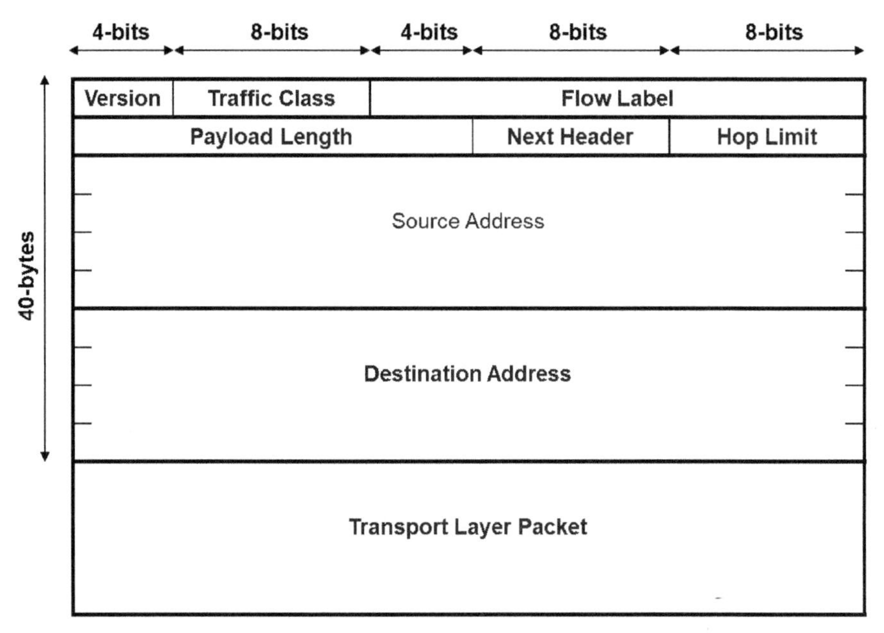

Figure 2.25 Frame format for the IP-v6 packets with the details of the header.

rather than 32 bits, allowing a practically unlimited number of addresses to accommodate the envisioned future networks in which every light bulb can have an IP address. Another feature of the IP-v6 is that the header compression option which can reduce the overhead. The 4-bit version field is the same as IP-v4. The 8-bit Traffic Class is an extended version of the Type of Service field in IP-v4. The 20-bit Flow Label is reserved for flow control and together with the Traffic Class field is used to implement QoS over the Internet. The 16-bit Payload Length field is the same as the Length field in IP-v4 with the difference that the length in byte in IP-v6 excludes the length of the header. The 8-bit Next Header field replaces the IP-v4's Options and Protocol fields. Normally, this field shows the transport layer code, but if there is a need for an option, that is also indicated in this field. In case there is an option, that field follows the header. The 8-bit Hop Limit is the same as the TTL field of IP-v4 with a more appropriate name used in practice.

The main change brought by IPv6 is a much larger address space that allows greater flexibility in assigning addresses. A large number of addresses allows a hierarchical allocation of addresses that may make routing and renumbering simpler. Since using IP-v6 has an impact on the design of the routers and end stations, the transition from IP-v4 to IP-v6 was much less successful than expected. Figure 2.26 provides an overview of the different functionalities of the network layer used within a router. As we discussed earlier in this section, IP is a routable protocol used for adjusting the packet lengths and support structured addressing schemes for routing among different subnetworks. The IP protocol basically supports host-to-host datagram services in a system of interconnected networks. Routers also need to communicate between themselves for control purposes and with the host terminals, for example, to report an error in datagram processing. The Internet Control Message Protocol (ICMP) messages are used to implement these functions. The objective of ICMP control messages is to provide feedback about problems in the communication environment, but it does not fix the problem and IP packets will remain unreliable. When a reliable transmission is needed, the higher level is responsible to provide reliable communication procedures and they can use ICMP messages. The IP address is used by routing algorithms, such as RIP, OSPF, and intermediate to intermediate (IS-IS), to prepare a routing table inside each router. The routing information protocol (RIP) was one of the most used IRP allowing routers to dynamically adapt to changes of network connections using a distance-vector algorithm. Although RIP is still actively used, it is generally considered to have been

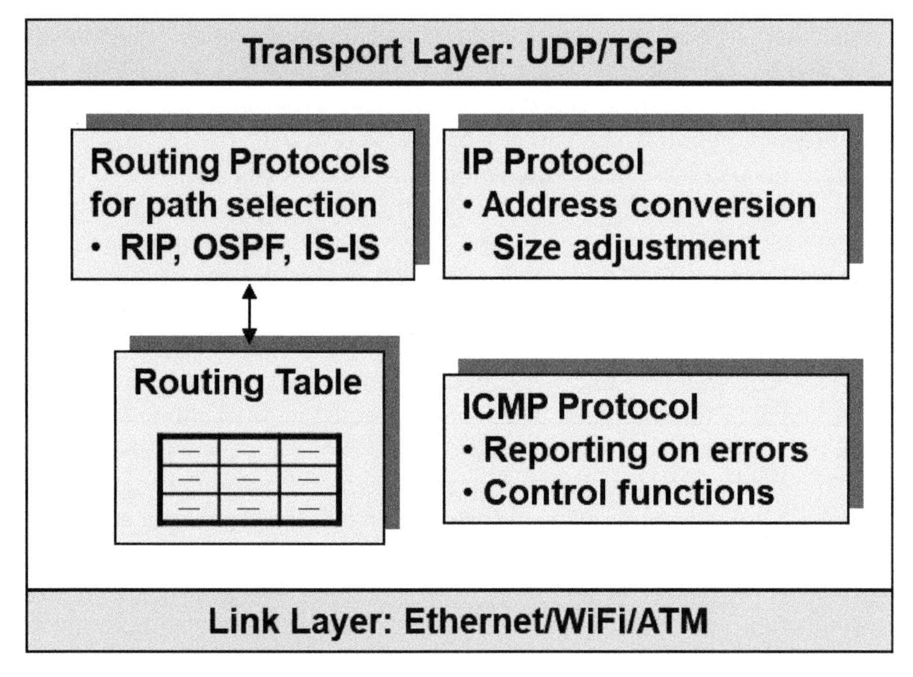

Figure 2.26 Overall view of the functionalities provided at the network layer.

made obsolete by routing protocols such as OSPF and IS-IS using the link-state algorithm. In the next section, we discuss algorithms used for routing datagram on the Internet.

2.5.3 Routing Algorithms

A routing protocol is a protocol that specifies how routers communicate with each other to exchange information that allows a router to find out the next router, they have to deliver an arriving packet so that the packet is directed toward its destination address. This communication is necessary so that routers can learn the network topology and its changes throughout the time. Since each router can only communicate directly with its immediate neighbors, a routing protocol provides a means for sharing this information with distant routers so that ultimately each router has knowledge of the entire network topology. Therefore, the routing protocol can be defined as protocols operating at the network layer to disseminate topology information among routers. The routing protocols differ from one another by the way they determine preferred routes from a sequence of hop costs and other preference factors as well as the method they use to handle routing loops. Routing protocols rely on the addressing of the routable protocols such as IP to implement their algorithms. There are two major types of routing algorithms: distance-vector routing algorithms and link-state routing algorithms.

In link-state algorithms also known as short path first algorithms, each router floods portions of the routing table that describes the states of its own links to all routers in the entire network allowing each router to build a picture of the entire network in its routing tables. In distance-vector algorithms, also known as Bellman–Ford algorithms, each router sends all or some portion of its routing table only to its neighboring routers. In other words, link-state algorithms send small updates everywhere, while distance-vector algorithms send larger updates only to neighboring routers. Routers using link-state algorithms need larger memories and longer computational time than routers using distance-vector algorithms increasing the implementation costs of the router. However, link-state algorithms converge faster, scale better, and create a smaller number of routing loops than distance-vector algorithms.

Routing algorithms use a metric to describe the "cost" of a certain route and based on the cost, they decide how to direct a packet toward a certain node. The metric can be a combination of the number of hubs to destination, the rate of the links which govern the delay for transmission of the packet, or a value that is assigned by an administrator to discourage the use of a certain route. Path length, reliability of the path, delay, bandwidth, communication cost, and the load of the link has been used as metrics in different routers. The alternative to using routing protocols is that the network administrator manually enters the route information. Manually entering routes is time-consuming, subject to human error, and demands manual reconfiguration when the network topology changes. This approach is referred to as static routing and it is sometimes used in very small networks.

Distance-Vector Routing Algorithm:

With the distance-vector algorithm, each router calculates the distance and direction of all routers in the network by exchanging information only with the neighboring routers. The distance or cost of reaching a destination is determined by using mathematical calculations comparing the costs of directing packets toward different directions or vectors. A different implementation of the algorithm may use simple hop count or

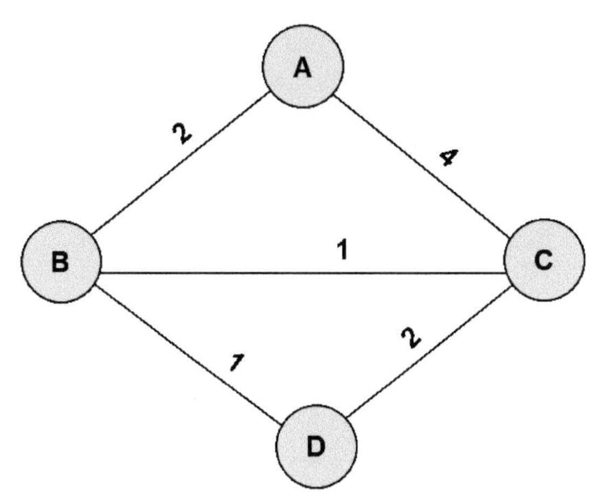

Figure 2.27 A simple network with four routers.

other information such as data rate, traffic density, and economic or political preference factors to calculate the costs. Neighboring routers using the same distance-vector protocol updates their tables periodically to adjust to changes in the network. The frequency with which routers send route updates depends on the routing protocol, and it is usually between 10 and 60 s. At each update, the entire routing table of a router is sent to all its neighboring routers. Other routers check the received information against the existing routing table to make appropriate changes if necessary.

Figure 2.27 shows a simple network example with four routers and five links interconnecting them for a better understanding of the algorithm. At the start of the operation, each router knows only about the immediate neighbors and the cost of connection with them, and based on that, each router forms its own table. Figure 2.3 shows how the table in each of the routers is formed and then updated iteratively with incoming information from reported tables from neighboring nodes. The four tables on the left side of Figure 2.28 shows the table associated with the start of the operation. In the first iteration of the algorithm, all routers broadcast these tables to all of their immediate neighbors: A to B and C; B to A, C, and D; C to A, B, and D; and D to B and C. In the second iteration, each router uses the tables arrived from the neighbors to update its own table and find the shortest path to get to all other routers.

The second column of the tables in Figure 2.28 shows the updated tables after the completion of the second iteration. To go over the details, consider

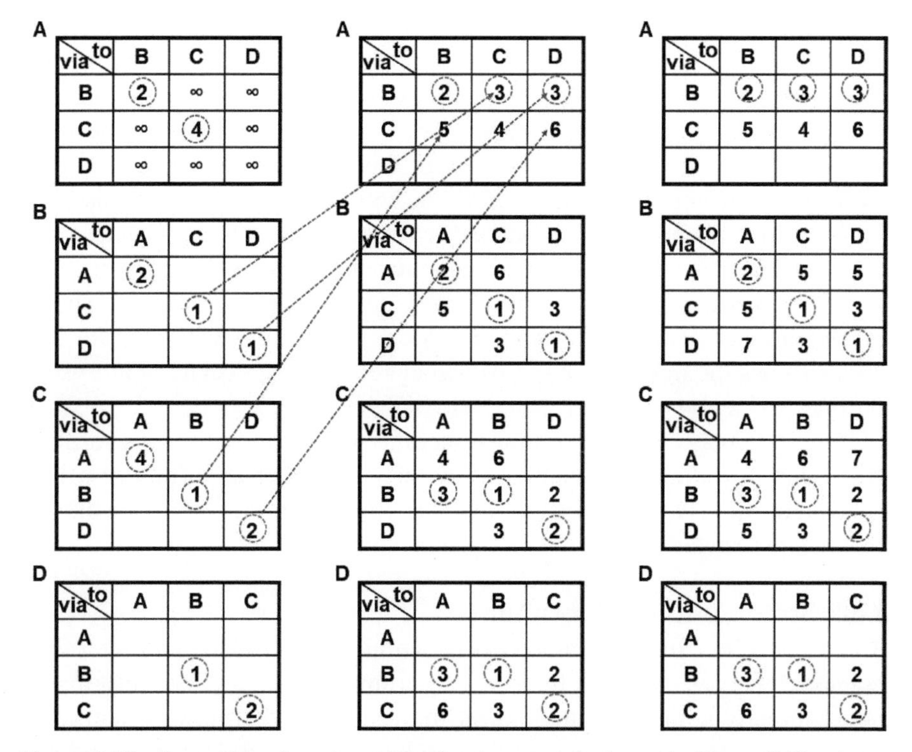

Figure 2.28 Formation of routing tables for the network shown in Figure 2.29 using a distance-vector protocol.

updating process for a table of the router A, a neighbor of the B and C routers, during the second iteration. Using the information arriving in the table from router B, router A discovers that (1) there is another router, D, which was unknown to A previously and (2) router B is connected to C as well. Now A can complete all elements of the row of distances via B. If A wants to approach C via router B, the distance is the sum of the distance between A and B and distance between B and C or 2 + 1 = 3. Similarly, the distance between A and D is 3 if the next router is B. Then router A uses information from router C to complete the second row of its own table associated with distances between A and other routers when the connection is made through router C. Since the distance between A and C was 4, the route going from A to C via B is 4 + 1 = 5 and the distance between A and D when connection is through C would be 4 + 2 = 6. These details are also shown in Figure 2.28. Similarly, packets from A, C, and D are used to update the table for router

B and the information from A, B, and D to update the table for router C and the information from routers B and C to update router D. In each of the four tables in the middle column of the tables in Figure 2.28, the circled number under each column of a table shows the minimum distance to connect to a specific router. The third column of four tables shows the results obtained from the third iteration of the algorithm with the same approach used for previous updates. In this iteration, tables associated with B and C make some changes and tables for A and D remain unchanged. After the third iteration, all tables stay the same and converge to their final value before any change is made in the network. This way, the information in the final table for each router shows the next router and its associated shortest distance for a packet destined to any other router.

The distance-vector algorithm sometimes causes routing loops resulting in a count-to-infinity problem. An alternative routing algorithm is a link-state algorithm. We next describe link-state protocols which provide another alternative to avoid distance-vector protocol problems.

Link-State Routing Algorithm:
The distance-vector algorithm sends the entire routing table of a router to all connected neighbors. In the link-state algorithm, the router sends information about directly connected links to all the routers in the network and every router builds the map of the entire network topology with link costs and uses that map to determine the shortest paths for routing packets. Each router in a network using link-state protocol broadcasts link-state advertisements informing all other routers about what networks it is connected to. This way, all routers share the same information database which shows all connections for all routers.

Using the same database and the same algorithm, all routers can build the topology of the network for themselves. When building the network topology in each router is complete, like distance-vector protocol, routers update each other periodically. However, a link-state protocol needs much less frequent updates than link-state protocols. Updates are also sent when a change in the network topology is detected. This feature combined with the fact that routers hold maps of the entire network results in a much faster convergence for the link-state protocol as it is compared with the distance-vector protocol. As compared with a computational complexity of the distance-vector protocol, routers using link-state protocol require higher computational speed to calculate the tables and much larger memory to store the entire table. A router using distance-vector protocols only maintain a small database of the neighboring routers and form the table with much simpler calculations.

In the link-state protocol, the only information passed between the routers is the information used to construct the connectivity maps. This information is processed by every router in the network to construct the map of the connectivity of the network in the form of a graph showing all connections and costs among the nodes in the network. Then, each node uses the graph to independently calculate the best next hop for every possible destination in the network. The collection of the best next hops forms the routing table for a router. In practice, a link-state protocol is implemented in a hierarchical structure dividing network into smaller areas so that each router does not need to store the map of the entire network. When the table of costs is completed, the topology of the network is settled and each router uses Dijkstra's algorithm to process the tables and find the shortest paths. We demonstrate this process with an example.

Figure 2.29(a) shows a network topology and Figure 2.29(b) shows the formation of the iterative algorithm to find the shortest paths for router A in the graph to all other routers. The algorithm starts with forming the first row of the table for which the node set contains only the source router, {A}. The following columns of the first row show the path to the destination and the distance from the source router. Node D has no direct connection to the set; so the distance is shown by infinity. The closest node to the set is node B with distance 2. In the next iteration, the node-set includes the shortest distance node and expands to {A, B}. The rest of the columns of row two are then filled by the closest distances to this node set and their associated paths. Nodes C and D are the neighbors of the {A, B} node set; the minimum distance to C using A-B-C path is 4 and the minimum distance to D using A-B-D is 3. Therefore, we add the closer neighbor D to our set to form the new node-set {A, B, D} to be used in the next iteration for calculation of the third row of the table. The two nodes outside this set are C and E with minimum distances of 3 and 6, respectively. Therefore, the next row is formed for node-set {A, B, C, D} for which the distance to node E will change from A-B-D-E with distance 5 to A-B-C-E with distance 4. In the next iteration, E is added to the set making it {A, B, C, D, E} which includes all the nodes. The last row of the table in Figure 2.29(b) shows all the paths and minimum distances from A to all other nodes.

This procedure creates a tree containing all the nodes in the network with the source node representing the *root* of the tree. The shortest path from that node to any other node and the sequence of routers to get to that node are

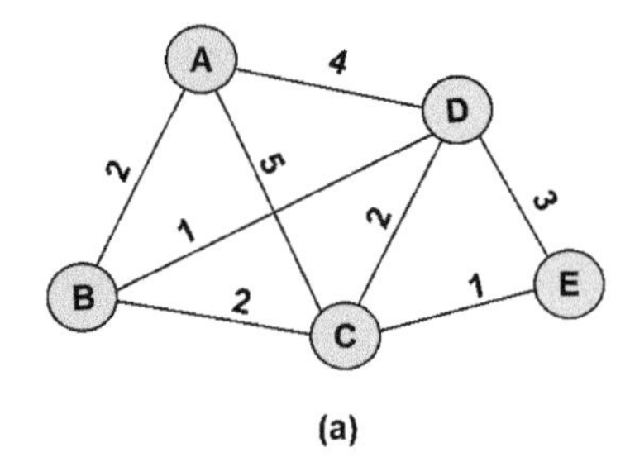

Destination / Nodes Set	B Path	B Distance	C Path	C Distance	D Path	D Distance	E Path	E Distance
{A}	A-B	2	A-C	5	A-D	4	-	∞
{A,B}	A-B	2	A-B-C	4	A-B-D	3	-	∞
{A,B,D}	A-B	2	A-B-C	4	A-B-D	3	A-B-D-E	6
{A,B,C,D}	A-B	2	A-B-C	4	A-B-D	3	A-B-C-E	5
{A,B,C,D,E}	A-B	2	A-B-C	4	A-B-D	3	A-B-C-E	5

(b)

Figure 2.29 Example of link state routing algorithm. (a) Topology of the network. (b) Table construction to find the shortest path for router A.

identified by the algorithm. These paths identify the nodes one traverses to get from the root of the tree to any desired node in the tree. In link-state protocol, each router determines its own routing table and shortest-path tree independent from other routers. If for some hardware or software reason, two routers start with different maps, it is possible to have scenarios in which routing loops may occur, but this event is much more unlikely than distance-vector protocol. Examples of link-state routing protocols using Dijkstra's algorithm are Open Shortest Path First (OSPF) and IS-IS protocols. OSPF is an IGP; it runs directly over IP, and it has its own reliable transmission mechanism. IS-IS is also an IGP protocol, but it is neutral regarding the type of network addresses and runs over the data link layer.

2.6 All-IP and Next Generation Core Networks

In the early 1990s, the traditional cores for networking were the PSTN and the Internet. The PSTN had evolved to support circuit-switched services for telephone application and Internet supported datagram packet switching services for computer data applications: web browsing, file transfer, and email. We began this chapter in Section 2.1.1 and compared these services in Table 2.1. At that time, cellular service providers began deploying their own infrastructure to support cellular wireless telephone services for connection to the PSTN and Wi-Fi emerged as a technology for wireless local access to the Internet. After the success of the Internet to penetrate the home market in the mid-1990s, computer data applications began to grow exponentially and the demand for expansion of the Internet core increased. At that time, one could differentiate core PSTN and Internet with their interconnecting elements, the telephone switches, and packet data routers. Switches carry the traffic over pre-assigned virtual routes with fixed short packets with short address overhead. The more expensive hardware implementation of switches made them faster and more in control of delay for enforcing a QoS. The less expensive software implemented routers were slower because paths were not pre-assigned, and addresses were longer, and processing the packet and finding the best next step caused delays.

In 1997, the IETF formed the MPLS group to define a circuit-switched like IP routing protocol, which soon became a very popular technology for core connecting elements of the independent ISPs. Around the same time, in 1998, the 3GPP partnership was formed and it began working on an all-IP core for cellular wireless infrastructure to support up to 2 Mbps packet-switched service on top of the circuit-switched telephone for the ITU International Mobile Telephone for the year 2000 (IMT-2000) for what we call the 3G cellular networks. The 4G and 5G cellular networks also evolved over the 3GPP all-IP core that supports packet-switched networks focused on increasing the data rates and circuited switched services demanding certain QoS. In 2005, ITU began thinking of next generation networks (NGNs) to integrate MPLS and 3GPP networks in a unified core. The NGN is currently evolving around cloud and edge computing.

2.6.1 The Multi-Protocol Label Switching

The MPLS transfers IP packets over a virtual circuit technology which has been designed to work with routers. Traditionally, a part of the long-haul IP traffic, sensitive to QoS, was carried through the virtual circuit-switched ATM, which has better handling of the delay. Figure 2.30(a) shows more details of this transmission approach. A variable-length IP packet is broken into short, fixed length ATM cells at the egress to the long-haul network. The short ATM packet, each carrying a label addressing the next switch, S_i, on the path, are carried through the network with a controlled delay because the ATM switches operate at layer 2 or link layer and they switch the path by reading the label address, S_i, and directing that according to a preexisting table to the next destination, S_{i+1}. This needs a connection establishment process to identify the path and a segmentation and reassembly of the packet at two ends and that makes it good for longer communication sections such as telephony and video conferencing with VoIP or and IPTV, but it is not the best solution for datagram services. In IP packet forwarding using layer 3 routers, shown in Figure 2.30(b), the IP address of the packet remains unchanged throughout the transmission. In this approach, the router processes IP address of each packet to find its next routing address using an IP address table lookup and a routing algorithm. This process takes more computational time at the router than label switching table lookup in ATM switches, resulting in longer delays which are not desirable for delay sensitive real-time applications such as VoIP or IP-TV. In MPLS, shown in Figure 2.30(c), each packet carries a virtual circuit an identifier, called *label*, enabling virtual circuit switching over the IP router network. This process eliminates the computational time for fragmentation and reassembly at two ends as well as delays associated with long address reading and execution of routing algorithms at the router. In MPLS shown in Figure 2.30(c), each packet carries a virtual circuit an identifier called *label*, like short ATM addresses, enabling virtual circuit switching over the IP routers network for variable length long IP packets. This process eliminates the long address search in the packet as well as execution of a routing algorithm for each packet because the route path has been pre-assigned, and the label address is short.

Since it is a combination of layer 3 datagram routing and layer 2 virtual circuit switching, it is also referred to as layer 2.5 switching. MPLS handles labels like ATM virtual circuit identifiers. An IP packet arrives at an MPLS-enabled network at location A. The packet is forwarded through an MPLS network or *domain* between A and B. As the IP packet arrives at the first router of the MPLS-enabled network, the egress router uses the source and destination addresses of the packet to establish a virtual circuit for transmission between locations A and B. The address of the next router in the virtual circuit is then added as a part of the MPLS header to the packet. Upon the arrival of the packet in the next router, the label is updated by swapping the address of the next router. The last router in the MPLS domain removes the MPLS header and delivers the IP packet to location B.

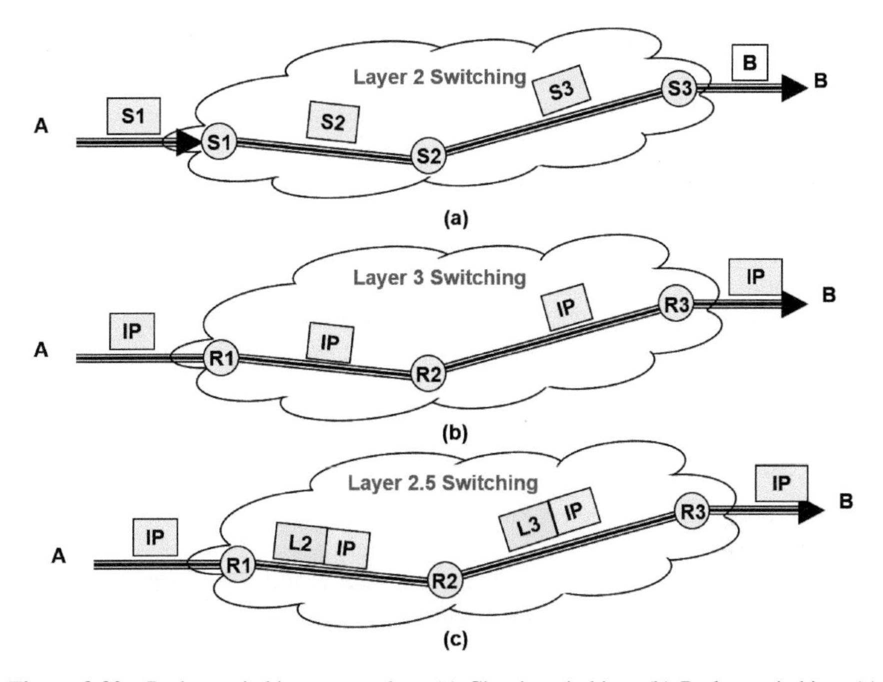

Figure 2.30 Packet switching approaches. (a) Circuit switching. (b) Packet switching. (c) Label switching.

Figure 2.31 shows the implementation details of the MPLS header. The header is 32 bits long and it is inserted between the network layer and link layer headers carrying layer 3 (IP) and layer 2 (MAC) routing addresses. The label address takes 20 bits of the 32-bit header. The first 3 bits after the label provides a QoS to handle eight different QoS classes. The next bit "S" indicates that MPLS has a stacked hierarchy or not. This allows

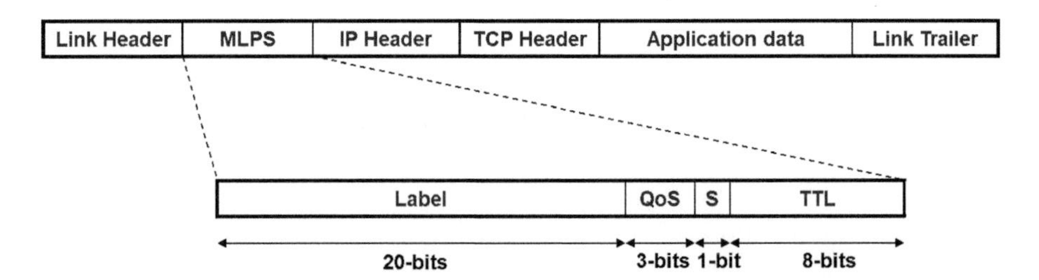

Figure 2.31 Details of multi-protocol label switching (MPLS) header.

implementation of multiple headers for multi-MPLS networks stacking on top of one another. In other words, using this bit, a packet within an MPLS domain can enter another MPLS domain and carry more than one MPLS label. The "S" or bottom of the stack bit indicates whether a label is the last in the stack or not. The last 8-bit field in the MPLS header is the TTL field. Each MPLS router decrements this field and discards packets when the value of this field, initially set by the ingress router, reaches zero. The purpose of this field is to provide a means to avoid packets from indefinite looping in case a circular virtual circuit was created by mistake. Just like any other virtual circuit-switched network such as ATM, handling multicast in MPLS is a challenging task.

2.6.2 3GPP and Next Generation Core Networks

Cellular networks evolved for wireless connection to the PSTN that supported QoS in the core network for the telephone application. We can consider 1G analog FDMA cellular, 2G TDMA digital cellular, and the 3G CDMA cellular networks as voice-oriented telephone networks. The analog 1G was strict circuit-switched extension of PSTN with the analog access like the plain old telephone services (POTS). The 2G TDMA success, the GSM standard, provided circuit-switched data services for the first practical ISDN services over the traffic channel as well as the first packet-switched SMS services over the control plain using SS7 [Hau94]. The access to the GSM was a wireless ISDN type service; while wired ISDN services had difficulties to penetrate end user market, GSM was the first end-to-end popular voice and data digital service over the circuit-switched core of the PSTN penetrating a successful worldwide market. QoS for the GSM core was embedded into the PSTN design with ATM long-haul transport with ISDN protocol. To maintain QoS in the wireless access, a high-quality speech coding technique at 13 Kbps was adopted and the effects of handoff in call drops were studied carefully so that the end-to-end user experience become compatible with the wired telephone services.

The 3G CDMA access to PSTN emerged to increase the circuit-switched the capacity of cellular networks in handling more telephone calls, but it was also the first to support packet-switched services of up to 2 Mbps for mobile applications enabling Internet access at a decent rate for many popular computer applications such as email, file transfer, web browsing, and video streaming. Another important development emerging at 3G cellular was the formation of the 3G Partnership Project (3GPP) consortium of several standards organizations to create an all-IP infrastructure for the cellular networks. Shortly after the success of MPLS and 3GPP in the late 1900s, the idea of integration of the two technologies in the unified core network, the next generation networks (NGN) was introduced by ITU and it appeared in the literature [Kni05]. The general idea is to design a unified core network that transports data, voice, and video by encapsulating these into IP packets and, for that reason, it is also referred to as all-IP network. The NGN supports fixed and mobile services end-to-end. With the growth of cloud computing, the idea of mobile edge computing emerged which shifted the cloud close to the edge to reduce the delay in access for the mobile user services (a good survey of this technology is available at [Abb18]).

In wired access to the Internet, by simple allocation of additional bandwidth to the user one can assure the QoS. This is easily manageable for Ethernet connections with an extremely higher data rate than 64 Kbps. For wireless connections with Wi-Fi, where usually a few user devices share an access point, this is still manageable because Wi-Fi operates in unlicensed bands which are plentiful in bandwidth. Therefore, for wired and wireless local area networks, implementation of VoIP or IPTV was not a hurdle. However, the cellular wireless data access from 4G and beyond needed to pay more attention to QoS, and it became an important factor in the design of these networks. Regulating QoS enables wireless cellular networks to accommodate a large number of applications with diversified wireless communications needs for integration of the Internet of Things (IoT) traffic with the traditional streaming and datagram applications traffic. In 5G and beyond, the industry began to classify a large number of applications for mobile devices into different categories like Table 2.4, to accommodate a wider set of applications and diversified needs data rate, latency, mobility, density, and power [Zha19].

The standards organization involved in the design of next generation or the next step in the evolution of wireless mobile cellular networks define an expected performance for these networks based on the foreseeing wireless applications to connect to the Internet. The fundamental measures for the performance of a wireless access to the Internet is the user experienced data rate and latency and that is enabled by the abilities of the network to support these services in mobile platforms, in home, and the office in densely populated areas, as well as to the growing population of "Things" sensing the environment to enable the smart world. An example of this process is reflected

Table 2.4 Design goals for QoS in wireless mobile cellular.

Data rate	**Latency**	**Mobility**	**Density**	**Power**
Tactile/Haptic Internet	Industrial Internet	Travel Internet	Internet of Everything	Internet of bio-nano things
Ultra-high-definition video for holography and virtual reality inside vehicle	Fully automated driving with car traffic coordination capability	High-speed trains, planes, satellites, submarines, boats/ships	Smart home–office networking to accommodate densely populated urban canyons	BAN for boy mounted sensors and sensors inside the body
Beyond Tbps	**10–100 ms latency**	**1000 Km/h**	**10^7 devices/Km^2**	**100 times less**

in [Zha19], which defines Internet applications into tactile/haptic, industrial, travel, everything, bio-nano things. Then, by defining typical applications, associate a non-existing technical feature in cellular networks to enable that application. Table 2.3 provides a summary for emerging applications of wireless mobile cellular networks by classifying them into different logical categories each with a distinct technical requirement assuming a user experience data rate of 1 Gbps. With these features, we need up to 5–10 times increase in bandwidth efficiency and a 100 times smaller power efficiency at the PHY technology. Based on the overall picture of technical demands from applications, as shown in Figure 2.32, one can sketch the capabilities of 5G and justify the need for 6G cellular wireless mobile networks to support these emerging applications. This trend creates an interactive dialog between the standardization organization, service providers, and the vendors manufacturing the equipment and communication devices [Kni05]. The general idea is to design a unified core network that transports data, voice, and video by encapsulating these into IP packets and, for that reason, it is also referred to as all-IP network.

Assignments

1. What are the differences among ISDN, MAC, and IP addressing techniques?
2. What are broadcast and multicast addressing and how do they get implemented on the Internet?
3. What is ARP and how does it work?

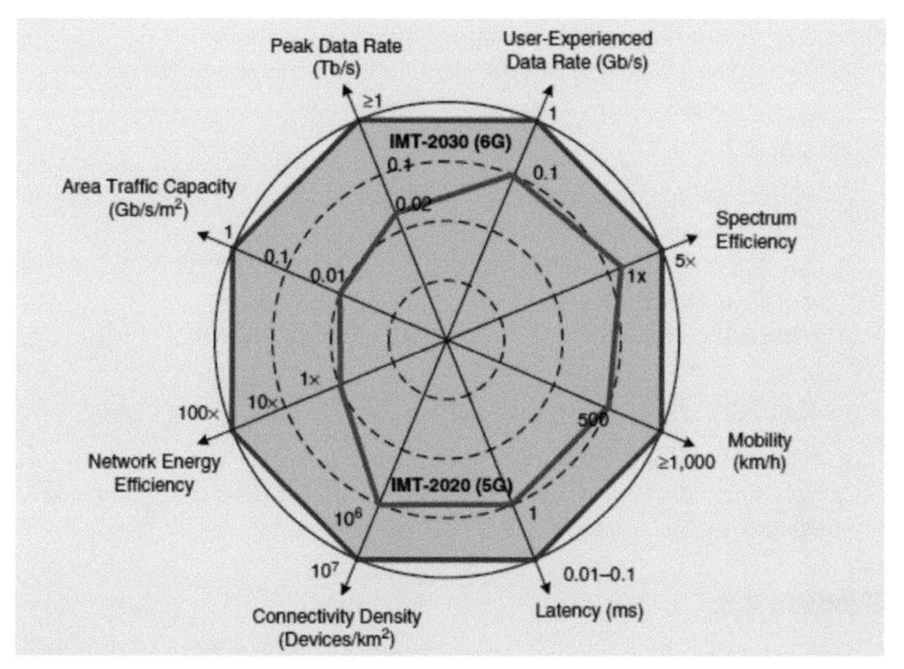

Figure 2.32 A comparison among technical features of the 5G and 6G cellular networks [Zha19].

4. In Figure 2.12, how does HA intercept the packets sent for the MH when it is away from the home network?
5. Why is PSTN better suited than the Internet for the implementation of QoS?
6. What is the role of IEEE 802.1Q in the implementation of VLAN tags?
7. What are the similarities and differences between IEEE 802.1D transparent bridges and a dumb bridge only changing protocol to connect two LANs with different protocol?
8. What is the meaning of filtering in 802.1D bridges?
9. How can one assign a cost to a connection between two bridges or two routers?
10. What was the purpose of IP-v6 and how does it differ from IP-v4?
11. How does mobile-IP work?
12. Draw the details of the MPLS tag and explain what its significance is.

Instructor's solution available on River Publishers' website:
https://www.riverpublishers.com/book_details.php?book_id=919

Problem 2.1

1. What is the maximum number of addresses using IEEE 802 format for MAC addresses?
2. How many different serial numbers can be assigned for the MAC address by a specific manufacturer?
3. What is the broadcast MAC address in the IEEE format?

Problem 2.2

1. What is the maximum number of IP-v4 addresses?
2. How many different Class A, B, and C IP addresses can be produced?
3. What is the maximum number of IP-v6 addresses?

Problem 2.3

1. A network on the Internet has a subnet mask of 255.255.240.0. What is the maximum number of hosts it can handle?
2. Give the broadcast IP address for this network.
3. Give the loopback address for this network.

Problem 2.4

We want to assign IP addresses at 198.16.0.0 to three organizations requesting 5000, 3000, and 6000 IP addresses.

1. Give the first and last IP assigned to each of the three organizations.
2. What arc the masks used by each of these organizations to filter their traffic?

Problem 2.5

1. In Figure 2.21, assume the root bridge is LAN number "5" and identify destination ports and root ports for the graph.
2. Reconstruct the topology by showing the blocked links with dashed lines (like Figure 2.21(b)).

3. Compare your new graph with that of Figure 2.21(b) and comments are similarities, differences, in appearance as a graph and functionality as a part of a communications network.

Problem 2.6

For the direct graph shown in and link costs are shown in Figure P2.1, construct the least-cost routing algorithm table using OSPF algorithm when B is the source node.

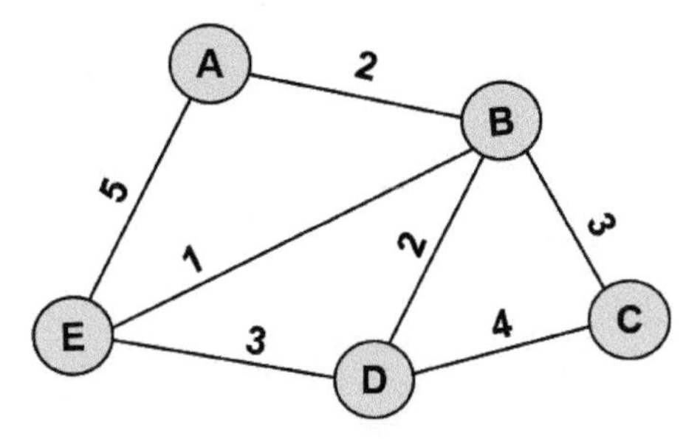

Figure P2.1 A local network with seven LANs and five bridges.

3

Characteristics of the Medium

3.1 Introduction

Communications networking technologies have evolved around signaling a set of waveforms or digital communication symbols with fixed time duration on the wired and wireless mediums. These symbols are alphabets of communication over the medium. With a sequence of these alphabets, we create sentences and the language for communications. The medium has a restriction on the duration of the symbols and, consequently, the rate of transmission of the symbols that are referred to as the bandwidth of the medium. Users always demand on higher data rates, and that motivates researchers to find digital communications and signal processing methods to increase the alphabet for digital communication or increase in bandwidth of the medium by going from twisted-pair telephone line to cables and fiber for wired communications or shifting to higher frequencies in wireless communications, where wider bandwidths are available. Except for the data rate, another important factor for communications networks is the coverage, the length of the cable for wired communication, and the area of coverage for a wireless antenna. Communication devices have a practical limitations on power transmission, and that restricts the coverage of the technology. Any medium for communications has a certain path loss in distance. For a fixed transmitted power after certain distances, the received signal strength becomes close to the background noise and the receiver cannot differentiate symbols from each other, causing erroneously received data unacceptable to the user. Research is needed to find ways for the medium to decrease the relation between path loss and distance to increase the coverage of a digital communication link. Communication networking science and technology is focused on discovering the relationship among data rate, bandwidth, characteristics of the medium, and analysis of the noise at the receiving devices.

In this chapter, we begin with a simple review of the fundamentals of communications to relate the medium bandwidth to the achievable data rate and power requirement (Section 3.2). Then we present characteristics of guided media, twisted pair (TP), coaxial cables, and fiber optics in Section 3.3, followed by characteristics of the wireless medium in Sections 3.4 and 3.5. Since propagation of signals on the wireless medium is complicated and more detailed, we have devoted two sections to explain it. The reader is referred to [Tho06] for more details on properties of copper and fiber media and [Pah05] for a rigorous treatment of wireless channels.

3.2 Fundamentals of Digital Communications

In Section 1.3, we explained a historical perspective for the evolution of the PSTN as the dominant core of the communications networks in the 20th century and the fact that as of this time of writing, wired connections to the PSTN carries the obsolete analog plain old telephone service (POTS). However, Bell Laboratories researchers fed from the income of the POTS services laid the foundation of data communication networking that is the subject of this book. Foundation of data communication science and technology is in the digital transmission of the symbols over the wired and wireless medium. We begin this part with a historical perspective of the evolution of data communication over the telephone network and the emergence of information theory followed by an overview of the wired and wireless medium before we discuss them in detail in the remainder of this chapter.

3.2.1 Digital Communications – A Historical Perspective

Before the 1990s, most data communication was carried out by voice-band modems over analog telephone networks, and the modem business had revenues on the order of a billion dollars [Pah88], which was negligible if it was compared with that of the POTS. However, for several decades, voice-band data communication was an active area for both scientific investigation and commercial product development. The data modem field had been the scene not only of theoretical advances but also of early commercial application of many signal processing algorithms such as automatic equalizations of the medium behavior, high-speed transmission, and coding techniques, and various synchronization methods. These advances found the field of digital communications and enabled Ethernet, cable modems, DSL modems, and Wi-Fi as well as cellular wireless communications to achieve high-speed digital communication over different mediums. The history of digital communications methods falls quite naturally into a few chronological eras [Pah88].

From roughly 1919 to the mid-1950s, work grew out of a need for transmission of low-speed telegraphic information over the PSTN. As early as 1919, the transmission of teletype and telegraph data had been attempted over both long-distance landlines and transoceanic cables. During these experiments, it was recognized that data rates would be severely limited by signal distortion arising from the nonlinear phase characteristics of the telephone line. This effect, although present in analog voice communication, had not been noticed previously due to the insensitivity of the human ear to phase distortion. From these studies came techniques, presented around 1930, for quantitatively measuring phase distortion and for equalizing lines with such distortion. This work apparently resolved the existing problems of that time, and little or no additional work appears to have been done until the early 1950s. In this era, research efforts were focused primarily on basic characterization of the lines and the basic theories of data communications. The Nyquist studies on limits for a sampling rate of analog signals for digital transmission were one of the most important scientific results of that era. The maximum data transmission rate over voiceband telephone lines in this era was approximately 100 bits/s.

The advent of the digital computer around the 1950s and the resulting military and commercial interest in large-scale data processing systems led to a new interest in using telephone lines for transmitting digital information. The government need for transmitting aircraft radar data to central locations resulted in the development of Bell System data communication products for voiceband networks, and these products soon found their way into commercial applications. The highlight of Bell Laboratories scientific contributions in this era was the discovery of Shannon–Hartley bound for digital communications and the emergence of information theory. In the early 1960s, the commercial importance of voice-band modems was being established. Magazines such as Fortune, Business Week, and US News and World Report reflected this growing commercial interest, and numerous vendors responded with products. With the FCC's Carterfone Decision in 1968, which opened the switched network to the interconnection of non-Bell modems, the growth of data communication products increased even more rapidly. As a result, the demand for higher-speed data communications increased, and it became desirable to attempt a more efficient utilization of the telephone channel.

Starting in about 1954, researchers at both the MIT Lincoln Laboratory and Bell Telephone Laboratories began to investigate more efficient utilization of the telephone network. These and other studies were continued by a moderate but ever-increasing number of people during the late 1950s. By 1960, numerous modulation techniques for obtaining higher rates (over 500 bits/s) had been proposed, built, and tested. However, these systems were still quite poor relative to what many people felt to be possible. Because of this, and due to a growing interest in the application of coding techniques to this problem, work continued at a rapidly increasing pace. In this period, carrier and timing recovery techniques were quite primitive and included the use of pilot tones or a significant carrier component. In addition, since the transmitted data was not scrambled, automatic gain control circuits were often unreliable because of possible extended idle periods between bursts of data. For these reasons, noncoherent modulation techniques, which did not need phase synchronization between a transmitter and a receiver, such as frequency-shift keying (FSK), were the central focus for most of the modems at that time. In the early 1960s, the states of the art in phase lock loops enabled transmission of electrical symbol with both phase and amplitude and in the late 1960s, the quadrature amplitude modulation (QAM) was invented and implemented for 9600 bps modems for voiceband data communications at Codex Inc., Mansfield, MA, USA. During the 1970s, the maximum practical speed remained at 9600 bits/s, but extensive research efforts by those working in this area improved methods of implementing modems and affected investigations in other areas, such as adaptive filtering, digital signal processing, and design of integrated circuits. This era led to significant reductions in size and power requirements for modems and laid the groundwork for the next generation of voiceband modems.

The 1980s saw the exploitation of improvements in telephone lines, introduction of the Trellis Code Modulation (TCM) that doubled the rate of QAM digital communications over the telephone channels to achieve 19.2 Kbps, and introduction of orthogonal frequency division multiplexing (OFDM) as an alternative to TCM for 19.2 QAM. Today, OFDM/QAM is the core of wireless communications for Wi-Fi and cellular mobile wireless networks and TCM and adaptive filters are the foundation of the Gbps Ethernet over cooper. The next innovation in transmission techniques from digital communications and information theory communities were the emergence of space-time-coding (STC), and multiple input multiple output antenna (MIMO) systems in the early 2000s, which enabled multiple streaming in wireless networks.

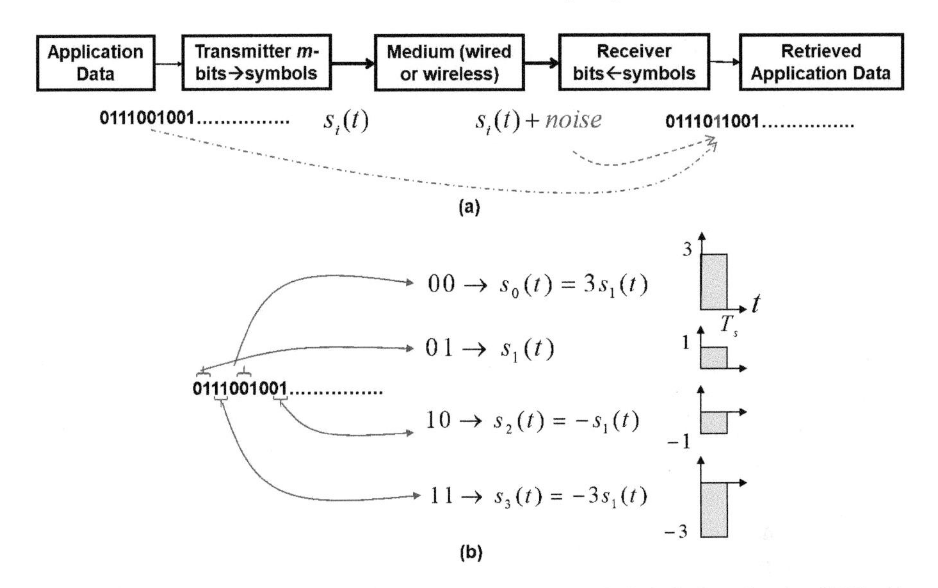

Figure 3.1 Overview of transmission and reception of digital data in the PHY. (a) Relation between data, communication symbols, noise, and error. (b) Example of a digital communications system.

Chapter 4 is dedicated to the details of these digital communications technologies. This chapter is devoted to the transmission medium, but before we discuss the detail of the transmission medium, we give a brief overview of the noise sources in digital communications, followed by an introduction to the Shannon–Hartley equation to relate the medium bandwidth to data rate, transmitted power, and background communication noise. We think with that background the reader will appreciate the details of characteristics of the medium in a more practical context.

3.2.2 Data Rate, Medium Bandwidth, and RSSI

Figure 3.1(a) shows a general block diagram for different components involved in a communications link for digital information transmission. The application data encoded into a bitstream with arriving bit ***data rate*** of $\boldsymbol{R_b}$ bits per second (bps) and each m-bits of the stream is encoded into one of the symbols in a set of analog waveforms $\{s_i(t);\ i = 1, 2, ..., M\}$ before transmission over the medium. Symbols are transmitted every T_s seconds referred to as the symbol duration or symbol interval. The symbol transmission rate is $R_s = 1/T_s$ symbols per second (sps). The maxim symbol transmission rate over a medium is the ***medium bandwidth***, $\boldsymbol{W = R_s = 1/T_s}$. The waveform representing the symbols passes through the channel which may make changes to the shape of the signal and disturb it with additive noise at the receiver. The received symbol is then processed at the receiver which extracts the best processed estimate of the transmitted bits associated with the received symbol to reconstruct the transmitted data stream. During the detection process, because of the noise, we may make errors that are not desirable for the users. Scientists and engineers design symbols for transmission and model characteristics of the medium to achieve the highest data rate with minimum errors. Figure 3.1(b) shows an example of a digital communication system with four square symbols with amplitudes of $\{-3, -1, 1, 3\}$ and duration of T_s. In this figure, $m = 2$, and $M = 2^m = 4$. Stream of data is divided into 2-bit blocks and each 2-bit is mapped to one of the four symbols or waveforms for transmission over the medium. The bandwidth of the medium is $\boldsymbol{W = R_s = 1/T_s}$ and the user data rate of the stream is $R_b = 2R_s = 2W$.

The received power of a symbol is the average of the energy of the symbol

$$E_{s_i} = \int_{-\infty}^{\infty} |s_i(t)|^2 \, dt.$$

For example, the received powers from the first and last symbols in Figure 3.1(b) are $9T_s$ and T_s, respectively. The average received energy per symbol, $\bar{E}_s$, is the average of the energy from all symbols. For example, in Figure 3.1(b), if all symbols are transmitted with equal probability, it is $\bar{E}_s = 5T_s$. The average received power, or the ***received signal strength indicator (RSSI)***, is the average energy over transmission time

$$\text{RSSI} : P_r(\text{Watts}) = \frac{\bar{E}_s(\text{Joules})}{T_s(\text{seconds})}. \tag{3.1a}$$

If the amplitudes in Figure 3.1 were in Volts, the energy would be in Joules and power in Watts. As an example, the average power of the symbols in Figure 3.1(b), if they are sent with equal probability, is 5W.

We often describe power in logarithmic form with respect to a reference. The most popular description in logarithmic form in communications networks is the dBm that is 10-time rations of the power measured in mW:

$$P_r(\text{dBm}) = 10 \log \left[P_r(\text{mW}) \right]. \tag{3.1b}$$

As an example, the maximum transmitted power allowed by the Wi-Fi is 100 mW (20 dBm) or maximum power for Bluetooth is 1 mW (0 dBm). The minimum RSSI for Wi-Fi could be -90 dBm and that for Bluetooth could be -75 dBm. The difference between the transmitted power and the minimum RSSI is the maximum path loss or link budget of the communication medium:

$$\text{Maximum Path} - \text{Loss} : L_{p-\max}(\text{dB}) = P_t(\text{dBm}) - \text{RSSI}_{\min}(\text{dBm}). \tag{3.1c}$$

3.2.3 Thermal Noise and Intersymbol Interference

As we have shown in Figure 3.1(a), the transmitted symbols arrive at the receiver with some noise that can cause unprecedented errors in the received data stream. In that regard, the noise can be the thermal noise produced by the receiver electronic circuitry or the intersymbol interference (ISI), caused by transmission medium characteristics. In multi-channel operation, we also have cross-channel and co-channel interferences, and we address them in the deployment of cellular wireless networks. There are other noise sources like atmospheric noise that are beyond the scope of applications discussed in this book. Figure 3.2 introduces the thermal noise and ISI that are important in digital communications because they can cause errors in detecting the correct transmitted symbols. Figure 3.2(a) shows a rectangular transmitted waveform as a typical alphabet for digital communications and the effects of additive thermal noise on the received symbol. It preserves the duration and shape of the transmitted symbol, but it adds noise to the amplitude of the signal. The thermal or Johnson–Nyquist noise is the electronic noise generated in the receiver electronics as a background for any filtering. The background thermal noise is a Gaussian distributed random variable with a variance (power) of

$$P_N = {\sigma_N}^2 = 4K_B T_K W = N_0 W, \tag{3.2a}$$

where T_K is the temperature in Kelvin (290 Kelvin degrees in normal room) and $K_B = 1.381 \times 10^{-23}\,\text{W/Hz/Kelvin degrees}$ is the Boltzmann's constant. Calculated in the dBm, we can derive the following simple equation for the power of the background noise:

$$P_N(\text{dBm}) = 10 \log N_0 W = 30 + 10 \log(4K_B T_K W) = -168 + 10 \log(\text{W}). \tag{3.2b}$$

The above equation for the calculation of the thermal noise in dBm is very useful for practical application to calculate the background noise level of different communication devices. For example, the bandwidth of

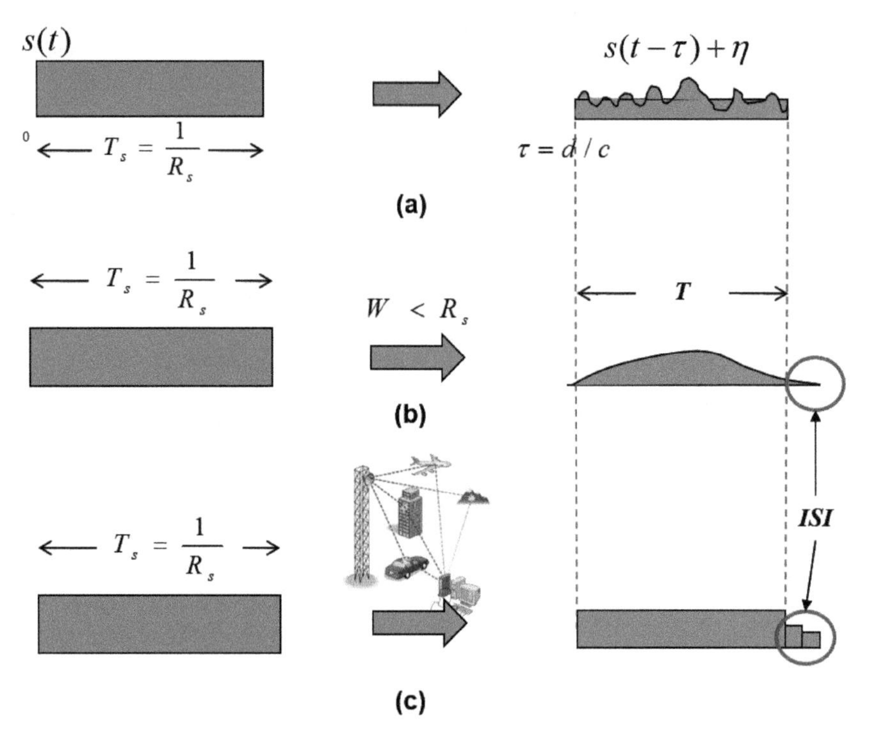

Figure 3.2 Different sources of noise for communications. (a) Additive thermal noise. (b) ISI caused by bandwidth limitation of the wired medium. (c) ISI caused by multipath.

traditional Wi-Fi devices is 20 MHz, and the typical background noise for these devices is

$$P_N(\text{dBm}) = -168 + 10\log(W) = -168 + 10\log(20 \times 10^6) = -95\,\text{dBm}\,.$$

For a ZigBee device with a bandwidth of 5 MHz, the background noise for this system is −98 dBm, for a Bluetooth with a bandwidth of 1 MHz, it is −108 dBm, and for a UWB system with a bandwidth of 2 GHz, we have −75 dBm background noise. Obviously, background noise reduces with the bandwidth.

Figure 3.2(b) illustrates the effects of medium bandwidth on the transmitted power and how the ISI is formed when we increase the symbol transmission rate beyond the bandwidth of the medium. The wired medium acts as a frequency filter; it transmits a waveform that has a wider symbol transmission rate, R_s ,than the bandwidth of the medium, *W*, and the shape of the transmitted waveform used as a symbol for communications will change. The rectangular pulse transmitted in Figure 3.2(b) will arrive as a pulse with a modified shape and a duration that is more than the original symbol duration, T_s. In communication applications, we transmit and receive symbols every T_s and when the received symbol stretches beyond its assigned time slot, it interferes with the neighboring symbol causing ISI. This noise, like background noise, can cause errors in the correct detection of the transmitted symbol. In additive noise communications, if we are not happy with the error rate at the receiver, we can increase the transmitted power to remedy the situation. In an ISI noise environment, we cannot remedy the situation with an increase in the power level because when we increase the power, we increase the ISI as well. Equalization of the channel effects with an adaptive filter at the receiver that inverses the filtering effects of the medium is a very effective method to control the ISI.

A wired medium is a guided communication link where the transmitted signal is guided to the receiver through one wired path. Wireless medium propagates the signal in many directions and the received signal arrives from multiple paths with different transmission attenuations and delays because the delay of arrival, τ, for a path is related to the length of the path, d, with $\tau = d/c$, where c is the speed of electromagnetic wave propagation that is approximately the same as the speed of light. Figure 3.2(c) shows how multipath arrivals with different amplitudes and delays of arrival cause ISI. Effects of ISI from multipath, like ISI for the wired channel, cannot be fixed by increasing the power and need equalizers at the receiver.

The signal-to-noise ratio (SNR) is a very important parameter in the design of alternatives for digital communications because different communications technologies operate at different received SNR. The SNR is basically defining the ratio of the receive signal power over the variance of the noise

$$\mathrm{SNR} = \frac{P_r}{P_N} = \frac{P_r}{{\sigma_N}^2} = \frac{P_r}{N_0 W}. \tag{3.3a}$$

The RSSI in dBm is what communication devices use for measurement of the received power $\mathrm{RSSI} = P_r(\mathrm{dBm}) = 10\log\left[P_r(mW)\right]$. SNR in dB is

$$\begin{aligned} \mathrm{SNR(dB)} &= P_r(\mathrm{dBm}) - P_N(\mathrm{dBm}) \\ &= \mathrm{RSSI} + 168 - 10\log(W) \end{aligned}. \tag{3.3b}$$

For example, if we read an RSSI of -80 dBm from a Wi-Fi device with 20 MHz band, the received SNR is

$$\mathrm{SNR(dB)} = -80 + 168 - 10\log(20 \times 10^6) = -80 + 94 = 14\,\mathrm{dB}.$$

If the source of noise is ISI, co-channel interference, or the adjacent channel interference, sometimes we call the SNR the signal-to-interference ratio (SIR), but the fundamental relations remain the same. The SNR relates the power to noise and the medium bandwidth in the case of additive noise channels.

In many of the applications, we have thermal noise and ISI noise and co-channel and cross-channel interferences. The SNR is then the ratio of the signal power to all interfering noises, P_I:

$$\mathrm{SNR} = \frac{P_r}{P_N + \sum P_I} = \frac{P_r}{N_0 W + \sum P_I}.$$

If additive noise is dominant and interference is negligible, we call the communication channel an additive noise communication channel. In the design of modems, we often assume additive noise channel to compare the performance of alternative solutions. If the interference dominates, we can refer to the system as an interference limited channel. Wireless networks are designed and deployed based on ISI, co-channel, and cross-channel interferences.

3.2.4 Shannon–Hartley Bounds for Information Theory

In our discussions thus far, we have introduced the data rate, R_b, medium bandwidth, W, and the SNR (representing the combined effects of the signal power and noise), as the fundamental elements of digital communications. The Shannon–Hartley equation provides a theoretical foundation to relate bounds on the achievable data rate for a given medium bandwidth and SNR (in linear form) by

$$R_b \leq W \log_2 (1 + \mathrm{SNR}) \quad \text{bits/s}. \tag{3.4}$$

This simple but powerful equation is the theoretical foundation for understanding the complex digital communications systems with simple explanations. The equation basically says you can achieve any data rate that you desire if you can have unlimited bandwidth or unlimited power. However, in practice, we have a limitation on the bandwidth of the medium as well as the received signal strength.

For example, in PSTN, the telephone wiring for analog telephone has the bandwidth of 4 KHz, and for a received SNR of 20 dB (100 times), the maximum achievable data rate is

$$R_b = W \log_2(1 + \mathrm{SNR}) = 4 \log_2(1 + 100) = 24\,\mathrm{Kbps}.$$

Or, in a reverse situation, if we want to design a 56 Kbps digital data service over the analog telephone line, we need SNR of

$$\mathrm{SNR} = 2^{R_b/W} - 1 = 2^{56/4} - 1 = 16,283(\sim 42\,\mathrm{dB}).$$

Digital subscriber line (DSL) achieves data rates of up to 10–30 Kbps over TP wirelines using 1.25 MHz wide band above the telephone band of the TP telephone lines; with a 30 dB (1000 times) SNR, they can achieve

$$R_b = W \log_2(1 + \mathrm{SNR}) = 1.25 \log_2(1 + 1000) = 12\mathrm{Mbps}.$$

DSL modems benefit from both bandwidth expansion and increase in SNR to achieve data rates above 10 Mbps.

3.2.5 Medium Characteristics: Wired vs. Wireless

The fundamentals of digital communications and Shannon–Hartley bounds provide an overview of theoretical foundations to relate data rate, medium bandwidth, signal power, and noise. The remainder of this chapter is devoted to the practical technical description of characteristics of the wired and wireless medium. A wired medium provides a reliable guided link that conducts an electric signal associated with the transmission of information from one fixed terminal to another. There are several alternatives for wired connections that include TP used in telephone wiring and high-speed local area networking, coaxial cables used for television distribution, and fiber optics used in the backbone of long haul networks. The guided media act as filters that limit the maximum transmitted data rate of the channel because of band-limiting frequency response characteristics. The signal passing through a guided media may also radiate outside of the wire to some extent that can cause interference to close by radio or other wired transmissions. These characteristics differ from one wired medium to another. Laying additional cables in general can duplicate the wired medium, and, thereby, we can increase the bandwidth.

Compared to wired media, the wireless medium is unreliable, has low bandwidth, and is of a broadcast nature. It, however, supports mobility due to its tetherless nature and it is free of wiring installation costs. Different signals through wired media are physically conducted through different wires, but all wireless transmissions share the same medium – air. Thus, it is the frequency of operation and the legality of access to the band that differentiates a variety of alternatives for wireless networking. These bands are licensed, unlicensed, or shared spectrum and are used for communications and radars for commercial and military applications. As the frequency of operation and data rates increase, the implementation cost (in hardware) increases, and the ability of a radio signal to penetrate walls decreases. The electronic cost difference has become insignificant with time, but in-building penetration and licensed versus unlicensed and shared spectrum frequency bands have become important differentiations. For frequencies up to a few GHz, the signal penetrates through the walls allowing indoor applications with minimal wireless infrastructure inside a building. At higher frequencies, a signal that is generated outdoor does not penetrate buildings and the signal generated indoor stays confined to a room. This phenomenon imposes restrictions on the selection of a suitable band for a wireless application.

For example, if one intends to bring wireless Internet service to the rooftop of a residence and distribute that service inside the house using other alternatives such as existing cable or TP wiring, he or she may select a metropolitan area wireless network equipment operating in licensed bands at several tens of GHz. If the intention is to penetrate the signal into the building for a direct wireless connection to a computer terminal, he or she may prefer equipment operating in the unlicensed ISM bands at 900 MHz or 2.4 GHz. The first approach is more expensive because it operates at licensed higher frequencies where implementation and electronics are more expensive, and the service provider has paid to obtain the frequency bands. The second solution does not have any interference control mechanism because it operates in unlicensed bands.

3.3 Characteristics of the Guided Medium

Wired media provide us an easy means to increase capacity – we can lay more wires where required if it is affordable. With the wireless medium, we are restricted to a limited available band for operation and we cannot obtain new bands or easily duplicate the medium to accommodate more users. As a result, researchers have developed several techniques to increase the capacity of wireless networks, to support more users with a fixed bandwidth. The simplest method, comparable to laying new wires in wired networks, is to use a cellular architecture that reuses the frequency of operation when two cells are adequately far from one another. Then, to further increase the capacity of the cellular network, one may reduce the size of the cells. In a wired network, doubling the number of wired connections will allow twice the number of users at the expense of twice the number of wired connections to the terminals. In a wireless network, reducing the size of the cells to half will allow twice as many users in one cell. Reduction of the size of the cell increases the cost and complexity of the infrastructure that interconnects the cells. Recently, by exploiting the rich multipath environment with MIMO techniques, expanding the capacity of wireless links has become possible.

The most common guided media for communications are TP wires, coaxial cables, optical fibers, and power line wires. TP wires, in shielded (shielded twisted pair (STP)) and unshielded (unshielded twisted pair (UTP)) forms are available in a variety of categories. They are commonly used for local voice and data communications in homes and offices. The telephone companies use category 3 UTP to bring POTS to customer premises and distribute the service on the premises. This wiring is also used for voice-band modem data communications, integrated service digital network (ISDN), digital subscriber loops (DSL), and home phone networking (HPN). The wired media in local area networks (LANs) are dominated by a variety of UTP and STP wires to support a range of local data services from several Mbps up to over a Gbps within 100 m of distance.

Coaxial cable provides a wider band useful for multi-channel FDM operation, lower radiation, and longer coverage than TP, but it is less flexible in indoor areas. The early LANs were operating on the so-called thick cable to cover up to 500 m per segment. To reduce the cost of wiring, the thin cable, sometimes referred to as Cheaper-LAN, which covers up to 200 m per segment, replaced thick cables. Cabled LANs are not popular anymore and TP wiring is taking over the LAN market. Another important application for cable is cable television that has a huge network to connect homes to the cable TV network. This network is also used for broadband access to the Internet. The cable-LANs were using baseband technology with data rates of 10 Mbps, while broadband cable modems provide comparable data rates over each of around a hundred cable TV channels. Today, in addition to traditional cable TV, cable modems are becoming a popular access method to the Internet and some cable service providers are offering voice services over the cable connections.

Fiber optic lines provide extremely wide bandwidth, smaller size, lighter weight, less interference, and very long coverage (low attenuation). However, fiber lines are less flexible and more expensive to install, and implementation costs of wave division multiplexing (WDM) and time division multiplexing (TDM) on fiber are more expensive than frequency division multiplexing (FDM) and TDM on wires or cables. Because of the wide bandwidth and low attenuation, optical fiber lines are becoming the dominant wired medium to interconnect switches in all long-haul networks. In LAN applications, fiber lines mostly serve as the backbone to interconnect servers and other high-speed elements of the local networks. Optical fibers have not found a considerable market for distributing any service to the home or office desktop.

Power line wiring has also attracted some attention for low-speed and high-speed home networks. The bandwidth of power lines is more restricted, the interference caused by appliances is significant, and the radiated interference in the frequency of operation of AM radios is high enough that frequency assignment agencies do not allow power line data networking at these frequencies. However, power lines have good distribution in existing homes because power plugs are available in all rooms. In addition, almost all appliances are connected to the power line, and with only one connection, a terminal can connect to the network, and the power supply. Existing power line networks either operate at low data rates at low frequencies, below the frequency of operation of AM radios, to interconnect evolving smart appliances or they operate at high data rates of up to 10 Mbps in frequencies above AM radio in support of home computing.

Wireless channels considered in this book are all for relatively short connections. For long haul transportations, these networks are complemented by wired connections. As we have discussed in this chapter, the average path loss in wireless channels is exponentially related to the distance and we often describe the path loss in terms of a fixed dB value per decade of distance (see Appendix 3.A for a definition of dB). The path loss in wired environments is linearly related to the distance. Therefore, wired connections are more power-efficient for shorter distances and wireless for longer distances.

The category 3 UTP wires used for Ethernet lose around 13 dB per 100 m of distance [Sta00]. Therefore, the path loss for 1 km is 130 dB. In free space, the path loss of a 1 GHz radio in the first meter is around 30 dB (with an omni-directional dipole antenna) and after that, the path loss is 20 dB per decade of distance. Therefore, the path loss at 100 m is 70 dB, and at 1 km, it is 90 dB. Indeed, for 130 dB path loss, the 1 GHz radio can cover 1000 km. Therefore, for short distances, the path loss with the wired transmission is smaller, but for long distances, path loss with radio systems is smaller. This is the reason why in long haul wired transmissions, repeaters are commonly used. Also, for the same reason, radio signals can provide very long distance satellite communications.

3.3.1 Twisted Pair

TP wiring is the most popular wiring used in distribution and access networks for end users. At home, TPs are used for access and distribution of telephone services and Internet access using DSLs. In offices, the original TPs were used for telephone distribution and a separate better graded TP is used for LAN connections.

In TP wiring, as the name indicates, two insulated copper conductors are twisted together for the purposes of canceling out electromagnetic interference from external sources and cross-talk from neighboring wires. The greater the number of twists, the greater the attenuation of cross-talk. TP wires are normally bundled in a sheath as a group of pairs differentiated by a standard color code. Original TPs first widely used by the telephone companies across the world did not have any shields between the pairs or around the bundle, and, today, they are referred to as UTP. Figure 3.3 shows a general picture showing details of TP wiring. Figure 3.3(a) shows how wires are twisted together, Figure 3.3(b) shows all possible details inside the sheath jacket of the wire, Figure 3.3(c) shows a picture of a piece of UTP wiring used for Ethernet wirings, and Figure 3.3(d) shows a registered jack (RJ-45) connector usually used with TP cables. If individual pairs in a bundle are shielded with metallic screens, the bundle is referred to as STP. If the bundle has a metallic shield, it is referred to as foiled twisted pair (FTP). The most

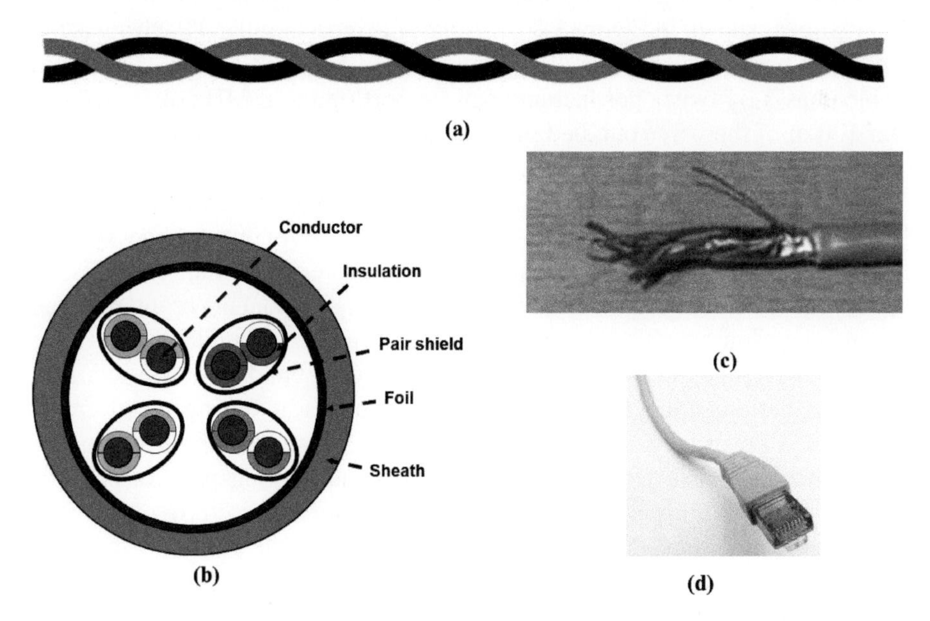

Figure 3.3 Twisted pair wirings. (a) Basic twisting concept. (b) Details inside a bundle of four pairs. (c) Photo of a sample twisted pair bundle. (d) RJ-45 twisted pair connector.

complicated TP wiring has both foil and shield and is referred to as S/FTP. Therefore, the characteristics of the TP wires depend on the number of twists, shielding complexity, and the length of the cable. For local distribution inside a home or office, the length of the TPs are less than 100m, and for local access to the network, it may go up to a couple of kilometers.

In the industry, different grades of TP wiring are referred to in terms of categories. The original TP cables first used in the American telephone network around the 1900s are now referred to as Category-1 (Cat-1) wiring. This wiring was previously used in POTS, ISDN, and for the doorbells. The Cat-1 TP, however, is not recognized by a TIA/EIA standard. TPs dominated the 20th-century wiring of the telephone and computer communications networks and several TP wiring options became standards for wiring for indoor and outdoor applications. The most popular TP wirings are Cat-3 and Cat-5. Billions of meters of these TP wirings around the world are installed mostly by telephone companies and are commonly used in telephone and Ethernet networks, respectively. Recently, TP wiring has been complemented by cable and fiber media which are described later in this chapter.

Telephone wiring had 25 color codes and the bundles of the telephone wiring can be as large as 25 pairs, but, today, the most popular bundles carry four pairs as shown in Figure 3.3(b). Today, the most popular TP wiring is the so-called voice grade Cat-3 UTP wiring defined by TIA/EIA 568-B standard which has 3–4 twists per feet and can support up to 16 MHz of bandwidth for 100 m of the wires bundled in four pairs. The main question we arrive at this point is how do we define the bandwidth?

As we show in Chapter 3, increasing the symbol transmission rate or the "bandwidth" of the system will increase the error rate of the system. This error rate is a function of the SNR of the received signal. The noise is either caused by the thermal noise of the receiver components or interference from other sources. The dominant source of noise in bundled TP wiring is the interference between the wires in the same bundle. In the industry, two terminologies are used for these interferences – the near end cross-talk (NEXT) and far-end cross-talk (FEXT) which are referring to the electromagnetic interference caused by cross-talk among parallel pairs of wire in a bundle. Figure 3.4 illustrates the nature of these cross-talks in a four-pair bundle of TP wires. The dominant cross-talk occurs at the ends of the cable where all protection on the wire is removed to connect the wire to the connector jack. Standardization organizations such as TIA/EIA specify the maximum allowed NEXT and FEXT for the 100 m TPs for

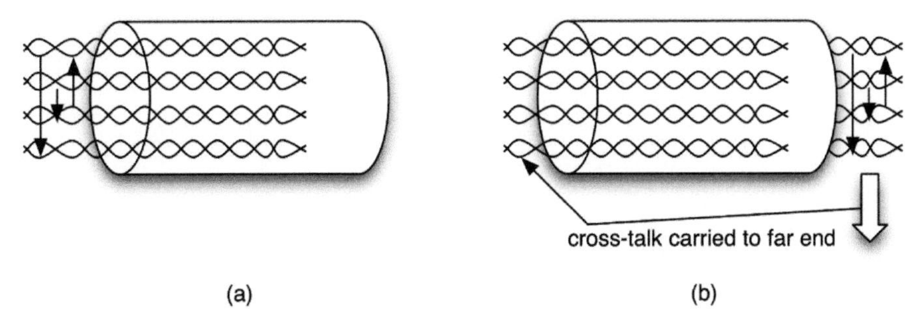

Figure 3.4 Cross-pair interference in twisted pair bundles. (a) Near-end cross-talk (NEXT). (b) Far-end cross-talk (FEXT).

different categories and manufacturers design their wires to comply with these specifications.

The path gain function, $G_P(f, l)$ of a Cat-3 TP as a function of the frequency, *f*, in MHz and the length of the cable, *l*, in *meters* can be approximated as

$$G_p(f, l) = -0.0235(\sqrt{f} + 0.1f)l\,. \tag{3.5a}$$

The relation between the power or path loss of a guided medium and the distance is linear, while the power loss in a conductor is related to the square root of the frequency. Deviation from ideal square root relation to frequency by adding a linear component is caused by dielectric isolation loss and other practical issues [Che97]. The solid line in Figure 3.5 shows this relation, Equation (3.5a), for frequencies up to 40 MHz for a Cat-3 cable with a length of 100 m used as the limit of the length for TP LAN options in the IEEE 802 standardization committees. As the frequency increases, the signals becomes weaker; at 16 MHz, the signal strength is approximately 13.2 dB weaker.

On the other hand, the NEXT interference for 100 meter Cat-3 can be approximated by:

$$NEXT(f) = -23 + 15 \log\left(\frac{f}{16}\right). \tag{3.5b}$$

The dashed line in Figure 3.5 shows the NEXT interference, Equation (3.5b), for a Cat-3 twisted wiring as a function of frequency. As the frequency increases, the interference is increased. At 16 MHz, the NEXT interference is approximately -23 dB. Therefore, the SNR which is the difference between the signal and interference levels at 16 MHz is 9.8 dB. As we show in

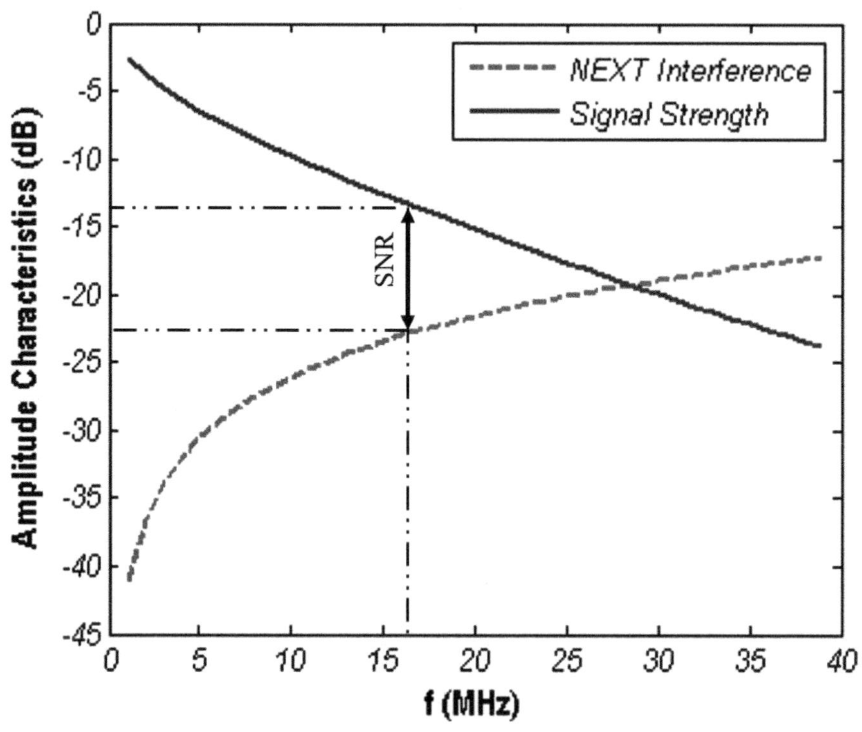

Figure 3.5 Amplitude characteristics of the Cat-3 twisted pair wires: the solid line is the 100 m cable attenuation, and the dashed line shows the interference.

Chapter 4, each transmission technique requires a different level of SNR to support a given data rate. Therefore, the cable path gain characteristics versus interference level govern the supported data rate of a transmission technique.

To improve the performance of a TP to support higher data rates, Cat-5 wiring was specified by the TIA/EIA standard 568-A. To implement this TP, we need 3–4 twists every inch rather than every foot in Cat-3. This arrangement reduces the power versus frequency and the amount of NEXT interference. The approximated equation for path gain can now be given by

$$G_p(f, l) = -0.02(\sqrt{f} + 0.01f)l \tag{3.5c}$$

and the NEXT interference by

$$\mathrm{NEXT}(f) = -32 + 15 \log\left(\frac{f}{100}\right) \tag{3.5d}$$

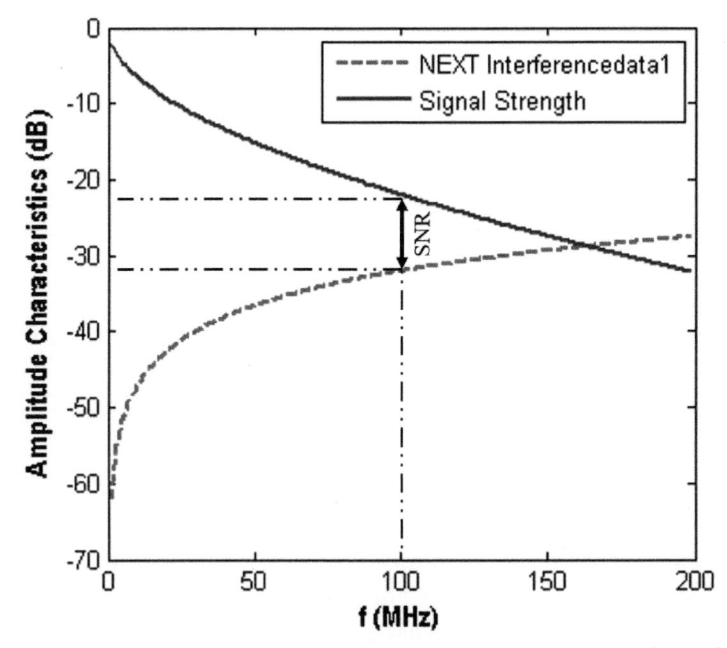

Figure 3.6 Amplitude characteristics of the Cat-5 twisted pair wires: the solid line is the 100 m cable attenuation, and the dashed line shows the interference.

Figure 3.6 shows the plots of the signal strength, Equation (3.5c), versus NEXT interference, Equation (3.5d). At 100 MHz, the signal strength is attenuated by 22 dB and the NEXT interference level is −33 dB which results in the SNR = 10 dB and performance at 100 MHz which is comparable with that of CAT-3 at 16 MHz. As we describe in Chapter 6, this characteristic of Cat-5 has been exploited by the IEEE 802.3 standardization committee to defined Ethernet options operating at 100 Mbps and above.

Based on the above discussions, in the computer communications literature, it is customary to state that the bandwidth of the Cat-3 UTP is W = 16 MHz and the bandwidth of Cat-5 UTP is W = 100 MHz. As the need for higher data rates at Gbps raised the bar, TIA/EIA standardization activities have defined higher quality TP options by specifying more details on FEXT as well as NEXT to guide manufacturers to design these cables. Today, the most popular of these TP options are enhanced Cat-5 (Cat-5e) with the same bandwidth as Cat-5 but more specified FEXT behavior, Cat-6 with W = 250 MHz bandwidth, and Cat-7 with up to W = 600 MHz of bandwidth. According to Shannon–Hartley bound for Cat-3 with 16 MHz bandwidth at SNR = 10 dB, we can achieve $R_b = 16\log_2(1+10) = 35\,\text{Mbps}$, while legacy Ethernet was 10 Mbps. Therefore, there should be better coding techniques to utilize Cat-3 TP and achieve higher data rates. Also, with 600 MHz bandwidth of Cat-7, we can achieve up to $R_b = 600\log_2(1+10) = 2.1\,\text{Gbps}$. If we use four pairs of a TP bundle instead of one pair and find coding methods that reduce the SNR requirements, we can achieve 100 Mbps with Cat-3 and 10 Gbps with Cat-7. As we show in Chapter 6, the IEEE 802.3 standards, established in 1980, with 10 Mbps recommendation over TP has gradually specified physical layer medium and transmission techniques to achieve data rates of up to 10 Gbps.

3.3.2 Coaxial Cables

Coaxial cables have been the most popular wiring used for home access for broadband multimedia services. Cable TV installations that started in the late 1960s were carrying multiple analog TV channels over the coaxial cables. More recently, the availability of such broadband access to home motivated the invention of broadband modems for Internet access, and, today, cable TV providers are adding telephone services to their cable connections to the home. In the office environment, the so-called "thick cable" was used first in the 1970s to implement legacy LANs such as Ethernet, Token Ring, and Token Bus that became recommendations by the IEEE 802 standardization community. The second-generation wired LAN industry resorted to the more flexible lower bandwidth "thin cables" in the mid-1980s to improve the wiring difficulties facing the "thick cables" before they finally resort to TP wirings in the early 1990s. Since the resistance in metallic conductors is inversely proportional to the diameter of the conductor, the thinner cables have more resistance causing a higher attenuation rate. The IEEE 802.3 recommendation for the maximum length of the less flexible thick cable for Ethernet was 500 m, while the more flexible thin cables were recommended to be used for distances up to 200 m. The recommended length of the most flexible TP wiring recommended by this standard is 100 m. In other words, the early day guided LAN networking community determined a practical compromise between a suitable length of coverage and a comfortable rigidity of the cable for the ease of wiring. The cost of installation and relocation of wiring for the LANs has been an important factor behind the growth of guided media

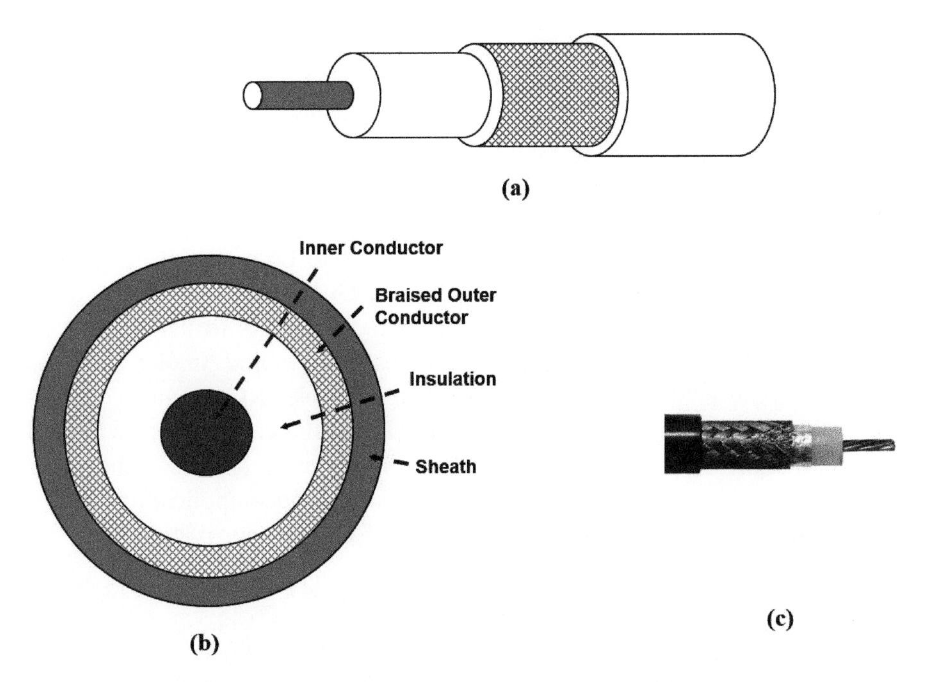

Figure 3.7 Coaxial cable wiring. (a) Basic coaxial concept. (b) Details inside a coaxial cable. (c) Photo of a sample coaxial cable.

LAN and the ultimate emergence of the wireless local area network (WLAN) industry.

Figure 3.7 provides the details of coaxial cable construction. A coaxial cable consists of a round conducting wire, surrounded by insulation, embraced by a cylindrical braided conductor, and covered by a final plastic sheath as a jacket. In ideal conditions, this arrangement keeps the electromagnetic field carrying the signal only in the space between the inner and outer conductors isolating the transmission of the signal from the outside electromagnetic signals in the air from different sources. Therefore, an ideal cable does not interfere with or suffer interference from external electromagnetic fields. The general design concept of the coaxial cable which controls the interference provides for a medium for high-frequency and longer-distance transmissions to distribute broadband signals such as multi-channel TV stations in neighborhoods.

As we explained in the previous section with TP wiring operation, the main source of transmission impairment is the interference caused by other transmission lines in the near and far end connectors. The connectors used in cables follow the same inner and outer connections, which control the harmful effects of radiations around the connectors. Therefore, the general propagation characteristics of the coaxial cables follow the same pattern as that of the path gain of the TP wirings while the NEXT and FEXT interference is almost eliminated.

The most recognized cable in the industry is the so-called RG-6 which is used for the distribution of cable TV signals to homes and many other applications in commerce. RG means Radio Guide and it was originally a military specification for cables carrying radio signals to the antenna or other parts of a radio system and it is not used anymore. Among a variety of RG coaxial cables, we use RG-6 as an example. RG-6 coaxial cables typically have a copper-coated steel conductor as the center metal wire and a combination of aluminum foil and aluminum braid shield as the outer conductor. Better grade cables used in professional video applications have a denser copper braid. The path gain of the RG-6 can be approximated by

$$G_p(f,l) = -0.0067 \times \sqrt{f} \times l \tag{3.6}$$

Figure 3.8 compares the amplitude characteristics of the RG-6 coaxial cable, in Equation (3.6), with the Cat-5 TP in Equation (3.5c), for a cable length of 100 m. For the same 22 dB loss observed at 100 MHz for the TP, we can have a coaxial cable with an order of magnitude wider bandwidth of more than 1 GHz to support up to $R_b = 1(\text{GHz}) \times \log_2(1 + 10) = 3.5\,\text{Gbps}$. If we keep the bandwidth at 100 MHz, the coaxial cable loss is 6.7 dB as compared with the TP loss of 22 dB. If we increase the length of the cable three times, we have $3 \times 6.7 = 20.1\ (\text{dB})$ which still provides better performance than Cat-5 TP.

Figure 3.9 compares the performance of the RG-6 cables with the Cat-3 voice grade TP wires. For the 100 m of distance used in IEEE 802.3 LANs and 10 MHz of bandwidth, Cat-3 has attenuation around 10dB. For the same attenuation, an RG-6 coaxial cable can run for 500m. This fact governs the assignment of 500 m for the original Ethernet using "thick cables" and the selection of 100 m as the length of the first TP star Ethernet. Coaxial cables have different levels of thickness (between 1 and 2.5 cm) and flexibility. The more flexible cables have a braided sheath and are usually used with thin copper wires. The thicker cables are less flexible but have thicker inner copper. Since the resistance of the conductors is inversely proportional to the thickness of the inner wire, the more flexible thin cables have shorter coverage. For example, the "thick cable" used in the original legacy Ethernet can cover up to 500 m, while the "thin cable" adopted later for the second

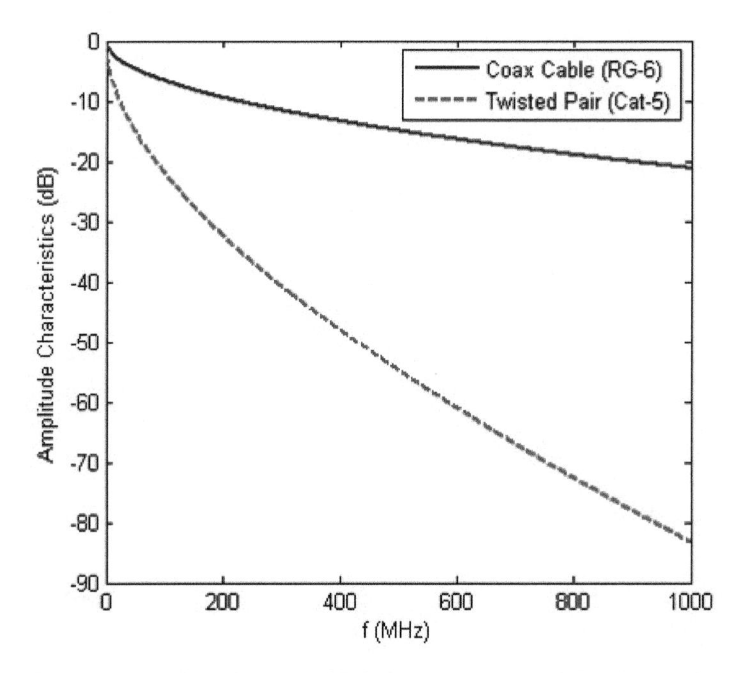

Figure 3.8 Amplitude characteristics of RG-6 coaxial cable and Cat-5 twisted pair wiring for a cable with length of 100 m.

generation of cabled Ethernet could cover up to 200 m. Connections to the ends of coaxial cables are usually made with radio frequency (RF) connectors. The most popular RF connectors used with cables are the Bayonet Neill–Concelman (BNC) connectors and F-connectors which are shown in Figure 3.10.

Therefore, for the same data rate, coaxial cable can cover a longer distance, between the towns, under the ocean, and for LANs, they can support much higher data rates. The first successful use of coaxial cables in communications networks was in long-haul telephone lines, then in the distribution of analog cable TV, followed by the use in the legacy Ethernet. In a long time, the local networking applications of coaxial cable became replaced by TP wiring because they were already available in residential and office environments, and the fact that TP is more flexible to snake wires in the building's walls and the emergence of wider bandwidth TP media. In long-haul communications networking, coaxial cables became replaced with Fiber which had wider bandwidth, smaller power loss per unit of distance, and lighter weight.

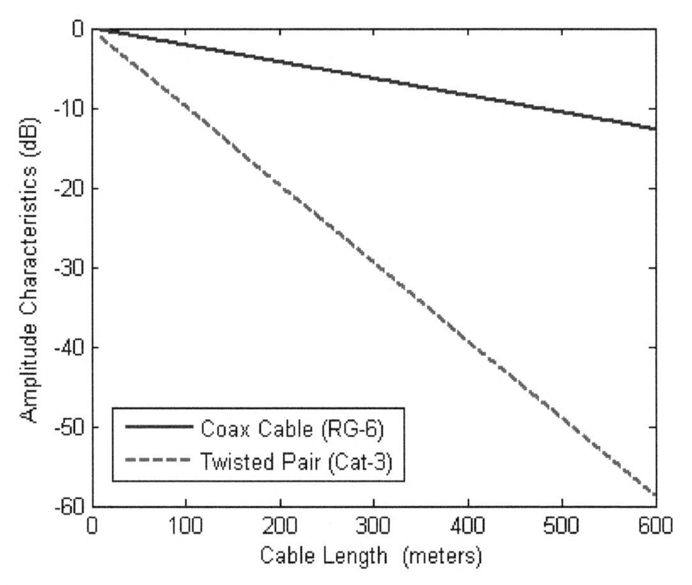

Figure 3.9 Amplitude characteristics of RG-6 coaxial cable and Cat-3 twisted pair wiring at the frequency of 10 MHz.

3.3.3 Optical Fiber

Optical fiber is the medium of choice for the backbones of modern networks, and it is rapidly penetrating the multimedia home access market worldwide, replacing the access by cable provided by the cable TV industry. In the office environment, optical fiber lines are connecting the backbone of LANs and many modern buildings lay optical fiber lines for fiber connections to the

Figure 3.10 Cable connectors. (a) BNC connector. (b) F-connector sample fiber bundle.

desktop computers as an alternative to current Cat-3 and Cat-5 wirings. Fiber lines provide for extremely higher transmission rates, longer coverage, lighter weight, smaller size, and they are free from RF interference.

The optical fiber cable uses glass or a plastic or a combination of the two fibers to confine and guide light along its length. The glass provides for lower optical attenuations, and it is used for longer-distance telecommunication applications. Figure 3.11 shows the general picture of optical fiber cables. The core of the fiber is used for guiding the light, the cladding is used to confine the light inside the fiber core, and a plastic jacket protects the cable. If the diameter of the core fiber is large, the confinement is based on total internal reflection. In such a case, the fiber is referred to as multimode fiber, and it is used for shorter distance communications of up to a couple of hundreds of meters. The single area of contact is larger, and the light is not directed. Inexpensive light-emitting diodes (LEDs) are used with multimode fibers to carry the signal over multiple paths. Single-mode optical fibers using thinner cores and sharper laser diodes are used for longer distances. Figure 3.12 shows the basic concept

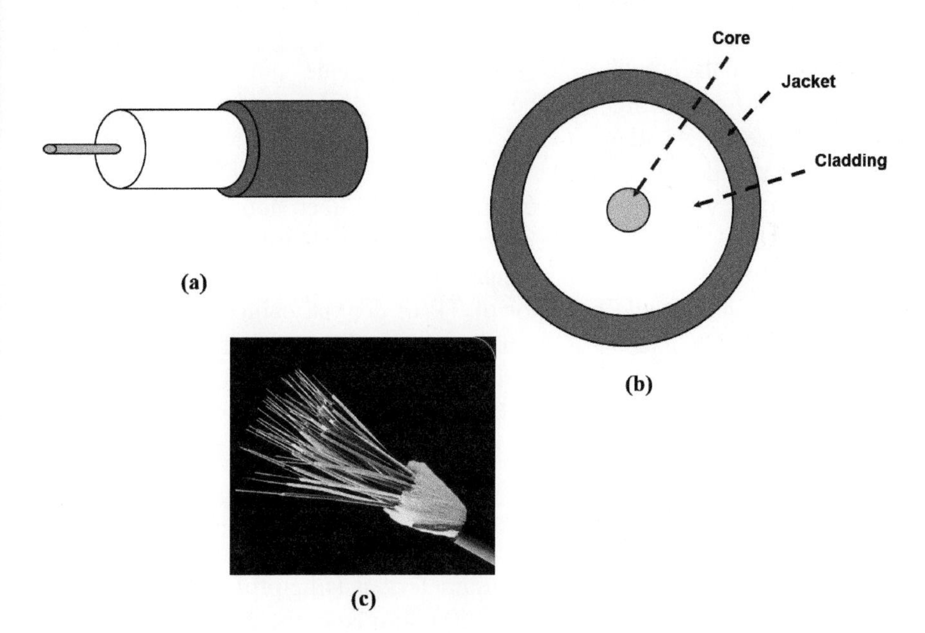

Figure 3.11 Optical fiber cable. (a) Basic concept. (b) Details inside a fiber line. (c) Photo of a sample fiber bundle.

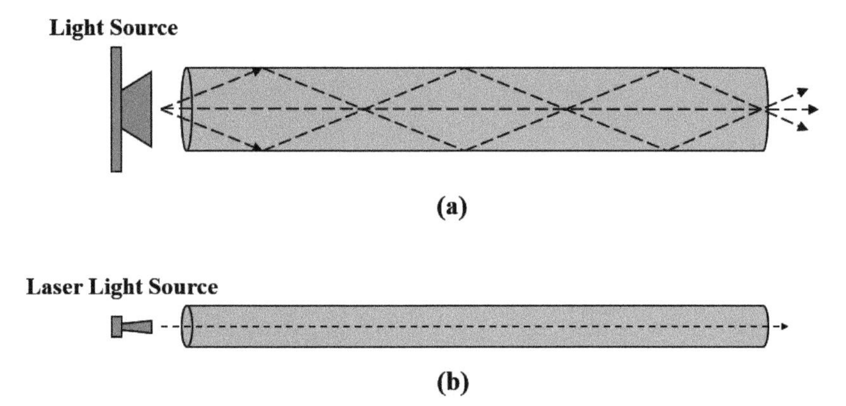

Figure 3.12 Light source and optical fiber transmission modes. (a) Multimode with multiple reflecting lights. (b) Single-mode with narrow fiber and laser light.

behind single-mode and multimode optical fiber communications. To design multimode and single-mode optical fibers operating at different wavelengths, several different designing techniques such as graded-index or step-index are used.

Optical signals are like electromagnetic signals, but the frequency of operation is extremely high (on the order of 10^{14} Hz). It is customary in the industry to use the wavelength $\lambda = c/f$ where f is the frequency of operation and c is the speed of light to identify a specific band. Figure 3.13 illustrates the gain versus wavelength of an ensemble of fiber optics cable material types. The bumps in the performance are caused by chemical reactions during the manufacturing process and cannot be avoided. This relationship does not follow the monotonic behavior of TP or coaxial cables and shows that certain wavelengths cause less attenuation providing a better opportunity for information transmission. The four wavelengths of 850, 1300, 1310, and 1550 nm is the most popular in industrial applications. The lower wavelengths are used for multimode fiber and the upper wavelengths for single-mode fibers.

The core of the fiber optic cables and their cladding has extremely small diameter sizes which are often expressed in microns or micro-meter. To have an intuition about the size of these diameters, it is helpful to point out that human hair has a diameter of around 100 microns. In the industry, fiber optic cable sizes are usually expressed by first giving the core size followed by

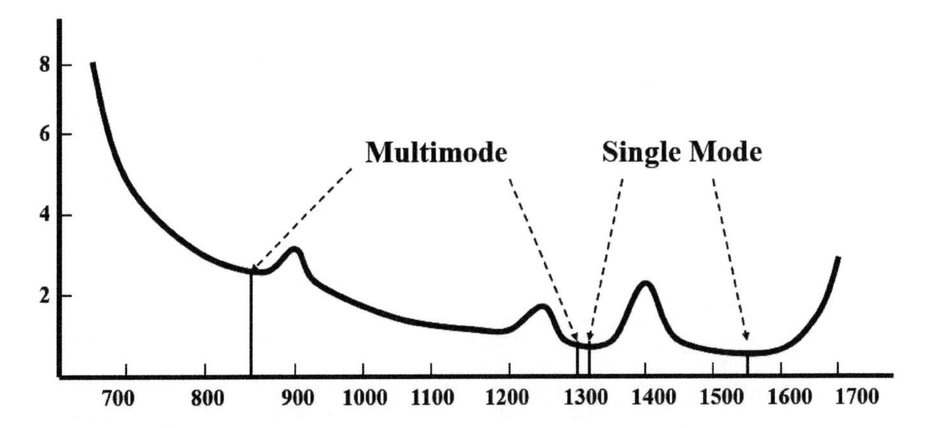

Figure 3.13 Attenuation versus wavelength for an ensemble of optical fiber lines.

the cladding size, for example, 50/125μm indicates a core diameter of 50 microns and a cladding diameter of 125 microns.

The wire specification standardization organizations identify the mode, diameters, wavelength, and attenuation as a guideline to the manufacturers. Table 3.1 provides *TIA/EIA* recommended path-loss and physical specifications for different types of fiber. The EIA/TIA specifies 3.5 and 1.5 dB attenuation per kilometer for 850 and 1300 nm multimode fibers, respectively. This specification is defined for both 50/125 and 62.5/125 μm core/cladding multimode optical cables. The same standardization committee specifies single-mode fibers with 9 μm core diameters at the wavelengths 1310 and 1550 nm to have an attenuation of 0.4 and 0.3 dB per kilometer. The TIA/EIA also specifies the connector loss and splice loss of the fiber optic cables to be 0.75 and 0.1 dB, respectively. These numbers can be used to specify the length of the fiber optic cable for a specific application. The main physical difference between the single and multimode, as shown in Table 3.1, is the thickness of the core.

Example 3.1: Optical Budget with an Optical Fiber in Gigabit Ethernet:

The IEEE 802.3 standardization committee for Gigabit Ethernet recommends a maximum fiber optic cable length of 700 km for the single-mode 9/125 μm diameters at 1550 nm. Using 0.3 dB/km loss recommended by the EIA/TIA, we have a cable loss of $700 \times 0.3 = 21$ dB for the cable. Adding connector

Table 3.1 Path loss for different types of fiber recommended by TIA/EIA.

Fiber type	Wavelength (nm)	Fiber attenuation (dB)	Connector loss (dB)	Splice loss (dB)
Multimode 50/125mm	850 13002	3.5 1.5	0.75	0.1
Multimode 62.5/125 mm	850 13002	3.5 1.5	0.75	0.1
Single MODE 9 mm	13102	0.4	0.75	0.1
Single mode 9 mm	15502	0.3	0.75	0.1

losses of $0.75 \times 2 = 1.5\,\text{dB}$ for the two connectors at the end of the cables results in a minimum power loss of 22.5 dB. A typical laser transmitter has -8 dBm transmitter power and its associated receiver has a minimum received signal power of -34 dBm for proper operation allowing a maximum power loss of $-8 - (-34) = 26\,\text{dB}$ for the medium. This arrangement recommended by the IEEE standard allows $26 - 22.5 = 3.5\,\text{dB}$ as a safety margin. Usually, there is a loss of 0.1 dB per splice. If we have five splices in this long cable, we will have a 3 dB safety margin. The difference between the transmitter power and the receiver sensitivity of 26 dB is referred to as the optical budget of the transmission system.

Example 3.2: Optical Fiber in 10 Mbps Ethernet:

The IEEE 802.3 committee for legacy Ethernet operating at 10 Mbps recommends using a 50/125 μm multimode fiber optic line at 850 nm. The path loss per kilometer for this fiber is 3.5 dB resulting in a total of $3.5 \times 2 + 0.75 \times 2 = 8.5\,\text{dB}$ for the cable and connectors.

Considering a 3.5 dB safety margin, a less expensive lower grade LED and photosensitive device with an optical budget of 12 dB can be used for the implementation of the system instead of the higher quality more expensive laser beams used in the previous example with an optical budget of 26 dB.

At the time of this writing, optical multimode (OM) fiber is widely deployed in short-range in-building communications. The OM1 legacy fibers at 62.5/125 dimensions could carry up to W = 500 MHz per km and latest OM5 wideband multimode can carry up to W = 3.5 GHz per km with multi-channel wave-division multiplexing [source: Wikipedia]. The more expensive optical single (OS) mode fiber has a theoretical infinite bandwidth, where the data rate limitation is imposed by the switching ability of the devices, and it is most popular in long-haul communications links for the core. Local access to the core is through wireless communications links that we discuss its fundamentals next. Quantum communication is an emerging field for communications for secure key wireless communications [PIR17].

3.4 Characteristics of Path Loss for Wireless Media

In the previous section, we analyzed the behavior of the guided media by relating the path loss to distance and frequency of operation. The basic relation between path loss, distance, and frequency of operation in the wireless medium and free space are also relatively simple. However, wireless networks such as cellular telephones or WLANs operate in urban and indoor areas where the signal is subject to multipath which causes fading in the received signal that complicates the channel behavior. Indeed, this behavior is so complicated and environment-dependent that we need to resort to statistical modeling. As a result, while we were describing the behavior of a guided medium with a simple path-loss equation that was deterministic and environment-independent, here, we need several complex statistical models to represent the behavior in different application environments. Since these models are mostly developed by telecommunication engineers with an electrical engineering backgrounds, a major question raised by a computer networking professional would be "why should we study radio propagation?"

An understanding of radio propagation is essential for coming up with appropriate design, deployment, and management strategies for any wireless network. In effect, it is the nature of the *radio channel* that makes wireless networks are far more complicated than their wired counterparts. Radio propagation is heavily site-specific and can vary significantly depending on the terrain, frequency of operation, velocity of the mobile terminal, interference sources, and other dynamic factors. Accurate characterization of the radio channel through key parameters and a mathematical model is important for predicting signal coverage, achievable data rates, specific performance attributes of alternative signaling and reception schemes, analysis of interference from different systems, and for determining the optimum locations for installing base station antennas.

The three most important radio propagation characteristics used in the design, analysis, and installation of wireless information networks are the achievable signal coverage, the maximum data rate that can be supported by the channel, and the rate of fluctuations in the channel [Pah05]. The achievable signal coverage for a given transmission power determines the size of a cell in a cellular topology and the range of operation of a base station transmitter. This is usually obtained via empirical *path*-loss models obtained by measuring the received signal strength as a function of distance. Most of the path-loss models are characterized by a distance-power or path-loss *gradient* and a random component that characterizes the fluctuations around the average path loss due to shadow fading and other reasons. For efficient data communications, the maximum data rate that can be supported over a channel becomes an important parameter. Data rate limitations are influenced by the multipath structure of the channel and the fading characteristics of the multipath components. This will also influence the signaling scheme and receiver design. Another factor that is intimately related to the design of the adaptive parts of the receiver such as timing and carrier synchronization, phase recovery, and so on is the rate of fluctuations in the channel, usually caused by movement of the transmitter, receiver, or objects in between. This is characterized by the *Doppler spread* of the channel. We consider path-loss models in detail in this section and provide a summary of the effects of multipath and Doppler spread in subsequent sub-sections of this chapter.

Depending on the data rates that need to be supported by an application and the nature of the environment, certain characteristics are much more important than others. We begin with free-space radio propagation in wide open areas for long-distance applications such as communications with satellites or airplanes as well as very short distance areas in the proximity of an RF device.

3.4.1 Friis Equation for Free Space Propagation

In previous sections, we discussed the relation between the guided media length, l, and the loss of the cable and provided equations to relate this relation to the frequency. In radio propagation literature, this is referred to as path-loss modeling. In path-loss modeling for the guided media, a signal is guided through the medium in the direction of the guide. In radio propagation, the signal propagates in the medium according to radio propagation characteristics of the antenna. The ideal antenna is an isotropic antenna radiating the signal in all directions with the same intensity. Most models for path-loss calculation for wireless medium develop their models for isotropic antennas and add an antenna gain factor relating the propagation to the antenna's radiation pattern. We start our path-loss modeling for wireless networks with path loss in the free space for an isotropic antenna. Here, rather than the length of the cable, we relate the loss to the distance, d, between the transmitter and the receiver. All our models for the guided media were

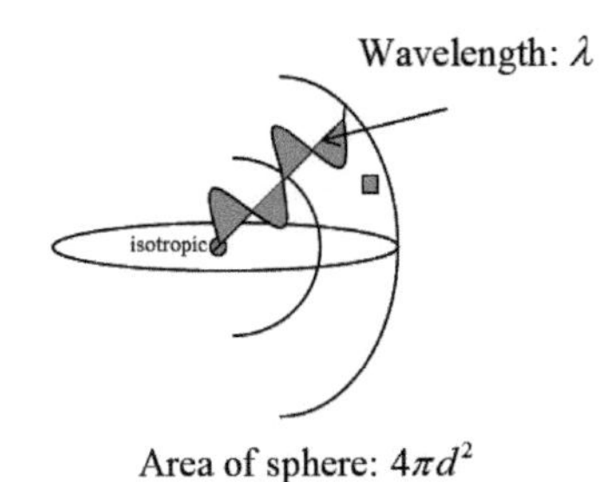

General antenna gain: $G = A \times \frac{4\pi}{\lambda^2}$

Antenna area for normalize received power: $A = \frac{\lambda^2}{4\pi}$

Normalizer received power on the sophere: $P_r = P_t \times \left(\frac{\lambda}{4\pi d}\right)^2$

Figure 3.14 Free space propagation and Friis equation.

relating the power loss in dB to the length of the cable, l, with a linear function. In the case of radio propagation, the relation between the path loss in dB and the distance is not linear anymore and it follows a logarithmic function.

Figure 3.14 illustrates the general concept of free space propagation and the Friis equation derived for that environment [Frii46]. Friis equation is well known in the antenna propagation literature for an ideal isotropic antenna with no heat loss, which radiates signal strength in all directions at the same rate the radiation intensity is given by$\lambda^2/4\pi$. At a sphere of radius d, the total radiated signal strength is divided by the area of the sphere, $4\pi d^2$, resulting in a path gain of

$$G_p(\lambda, d) = \frac{P_r}{P_t} = \left(\frac{\lambda}{4\pi d}\right)^2 \tag{3.7}$$

where $\lambda = c/f$ is the wavelength of the carrier, c is the speed of light in a vacuum (3×10^8 m/s), and f is the frequency of the radio carrier. Therefore, in the free space, Equation (3.7), signal strength falls with the square of the distance, and the power path loss after a distance d in meters is proportional to d^2 . The transmission delay as a function of distance is given by $\tau = d/c = 3d$ ns or 3 ns per meter of distance.

In the radio propagation literature, it is customary to express the channel attenuation behavior in terms of path loss which is the inverse of the path gain. Taking 10 times the logarithm of quantities in Equation (3.7), the path loss in free space, expressed in dB form, is

$$L_p = L_0 + 20\log_{10}(d)\,, \tag{3.8a}$$

where L_0 is the path loss in the first meter of distance, and it is given by

$$L_0 = 20\log_{10}\left(\frac{4\pi}{\lambda}\right). \tag{3.8b}$$

Equation (3.8a) shows that the path loss in the free space has a fixed component L_0 which increases as the frequency increases (wavelength decreases) and a second component which causes attenuation of 20 dB per decade (or 6 dB per octave) of the distance.

Example 3.3: Path Loss in Wireless LANs:

The transmitted power of the WLANs operating at 2.4 GHz is 100 mW (20 dBm). The wavelength of these devices is

$$\lambda = \frac{c}{f} = \frac{3 \times 10^8}{2.4 \times 10^9} = 0.125\,\mathrm{m} = 12.5\,\mathrm{cm}.$$

Therefore, the path loss in the first meter is

$$L_0 = 20\log_{10}\left(\frac{4\pi}{\lambda}\right) = 20\log_{10}\left(\frac{4\pi}{0.125}\right) = 40.05\,\mathrm{dB}.$$

The overall equation for the path-loss calculation of these devices in free space is then given by

$$L_p = L_0 + 10\alpha\log_{10}(d) = 40.05 + 20\log_{10}(d)\,.$$

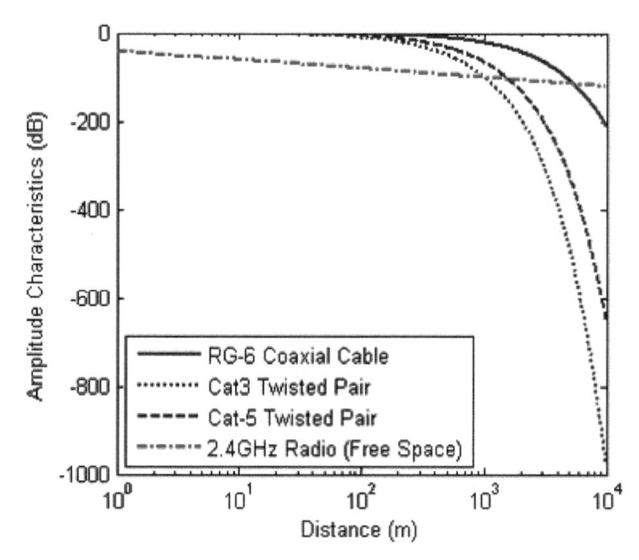

Figure 3.15 Attenuation versus distance for different guided media versus 2.4 GHz radio.

At 10 m, we have 40.05 + 20 = 60.05 dB loss and at one 100-m 80.05 dB path loss. Figure 3.15 shows the attenuation versus distance for RG-6 coaxial cable and Cat-3 and Cat-5 TP wires versus 2.4 GHz radio propagation in free space. At shorter distances, radio transmissions lose significant power, but over a longer distance, since the path loss in dB is exponentially related to distance, radio transmissions will perform better. Since the loss of the power in a wireless medium is logarithmically related to the distance when we show the distance on a logarithmic basis this relation is linear, this trend of presentation is common in the radio path-loss modeling literature. If we increase the distance exponentially, the loss in any guided media will appear as a water-fall curve that ultimately crosses the loss of the radio at a certain distance. At around 1 km, the path loss with a radio transmission becomes smaller than the loss over a Cat-3 wire, at 1.6 km smaller than the loss over a Cat-5 wire, and at 5 km better than the loss with an RG-6 coaxial cable. Therefore, wireless media have more power loss than the wired media at shorter distances from the transmitter but form a more power-efficient solution over longer distances. But over longer distances, we need to compensate for the heavy power loss either by a high power transmission or by designing more sensitive receivers. The difference between the transmitted power and the sensitivity of the receiver, which is sometimes referred to as the link budget for the cable-based modems are on the order of 20–30 dB, while in wireless media, it can easily go over 100 dB. For example, IEEE 802.11 devices transmit at 20 dBm (100 mW) and the receiver sensitivity is around

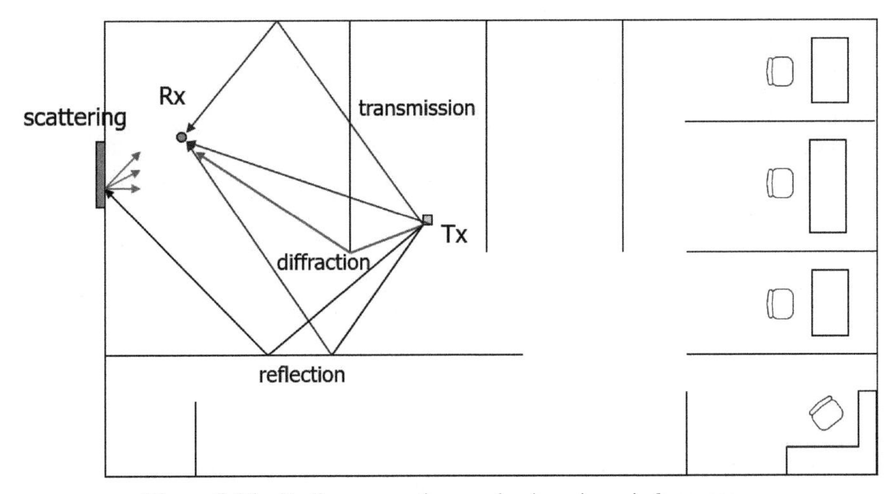

Figure 3.16 Radio propagation mechanisms in an indoor area.

−80 to −95 dBm providing for a link budget on the order of 100−115 dB. As shown in Figure 3.16, this link budget can support operations up to several kilometers in free space. However, most of us have experienced that if we are around 100 m from an Wi-Fi access point, we do not have much of a chance to save our connections. This is because of the extra attenuation caused by multipath arrivals from reflection off or loss through walls and objects in and around the transmitter and the receiver, which ultimately change the distance power gradient.

3.4.2 Path-Loss Modeling in Multipath

Radio signals with frequencies used in wireless networks have small wavelengths compared to the dimensions of building features so that electromagnetic waves can be treated simply as rays. This means that ray-optical methods can be used to describe the propagation within and even outside buildings by treating electromagnetic waves as traveling along localized ray paths. The fields associated with the ray paths change sequentially based upon the features of the medium that the ray encounters. To describe radio propagation with ray optics, three basic mechanisms are generally considered while ignoring other complex mechanisms. These mechanisms are illustrated in Figures 3.16 and 3.17 for indoor and outdoor applications, respectively.

Reflection and transmission: Specular reflections and transmissions occur when electromagnetic waves impinge on obstructions larger than the wavelength. Usually, rays incident upon the ground, walls of buildings, the ceiling, and the floor undergo specular reflection and transmission with the amplitude coefficients usually determined by plane wave analysis. Upon reflection or transmission, a ray attenuates by factors that depend on the frequency, the angle of incidence, and nature of the medium (its material properties, thickness, homogeneity, etc.). These mechanisms often dominate radio propagation in indoor applications. In outdoor urban area applications, the transmission mechanism often loses its importance because it involves multiple-wall transmissions that reduce the strength of the signal to negligible values.

Diffraction: Rays that are incident upon the edges of buildings, walls, and other large objects can be viewed as exciting the edges to act as a secondary line source. Diffracted fields are generated by this secondary wave source and propagate away from the diffracting edge as cylindrical waves. In effect, this results in propagation into *shadowed* regions since the diffracted field can reach a receiver, which is not in the line-of-sight (LOS) of the transmitter. Since a secondary source is created, it suffers a loss much greater than that experienced via reflection or transmission. Consequently, diffraction is an important phenomenon outdoors (especially in micro-cellular areas) where signals transmission through buildings is virtually impossible. It is less consequential indoors where a diffracted signal is extremely weak compared to a reflected signal or a signal that is transmitted through a relatively thin wall.

***Scattering*:** Irregular objects such as wall roughness and furniture (indoors) and vehicles, foliage, and the like (outdoors) scatter rays in all directions in the form of spherical waves. This occurs especially when objects are of dimensions that are on the order of a wavelength or less of the electromagnetic wave. Propagation in many directions results in reduced power levels especially far from the scatterer. As a result, this phenomenon is not that significant unless the receiver or transmitter is in a highly cluttered environment. This mechanism dominates diffused IR propagations when the wavelength of the signal is so high that the roughness of the wall results in extensive scattering. In satellite and mobile radio applications, tree leaves and foliage often cause scattering.

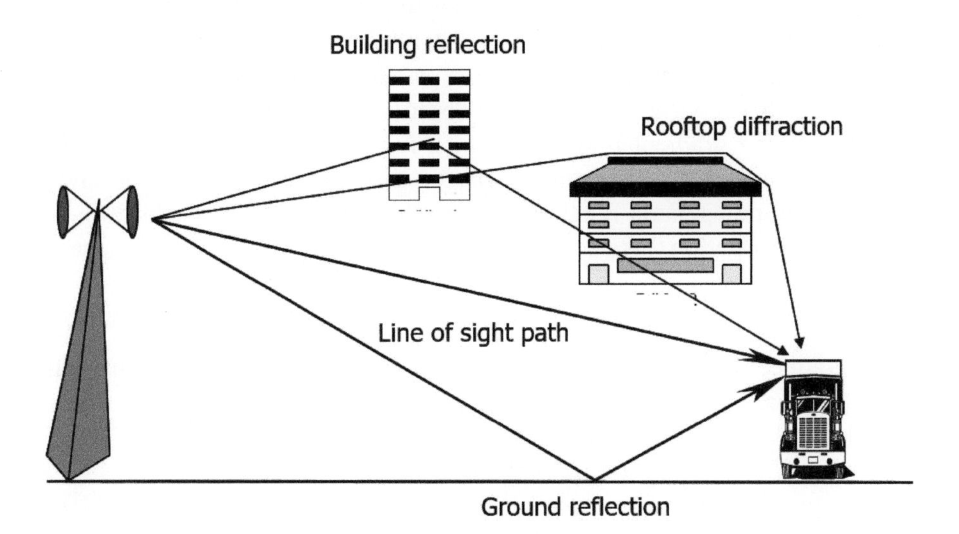

Figure 3.17 Radio propagation mechanisms in outdoor areas.

Calculation of signal coverage is essential for the design and deployment of wireless networks. In wired transmissions, we needed models for loss of the signal strength in the media to find the maximum length of the cable for effective operation. In wireless networks, signal coverage is influenced by a variety of factors, prominently, the radio frequency of operation and the terrain. Often, the region where a wireless network is providing service spans a variety of terrain. An operation scenario is defined by a set of operations for which a variety of distances and environments exist between a transmitter and the receiver. As a result, a unique channel model cannot describe radio propagation between the transmitter and the receiver, and we need several models for a variety of environments to enable system design. The core of the signal coverage calculations for any environment is a path-loss model that relates the loss of signal strength to the distance between two terminals at a given frequency. Using the path-loss models, radio engineers calculate the coverage area of wireless base stations and access points as well as the maximum distance between two terminals in an *ad-hoc* network. In the following, we consider path-loss models developed for several such environments that span different cell sizes and the terrain in the cellular hierarchy used for the deployment of wireless networks.

Path-Loss Model and Coverage:

In most environments, it is observed that the average received radio signal strength falls as some power of the distance α called the power-distance gradient or path-loss gradient. That is, the path loss after a distance *d* in meters is d^{α} . For the free space propagation described in the previous section, the distance power gradient is $\alpha = 2$ which indicates that the signal strength falls as the square of the distance between the transmitter and the receiver. Considering this general relation between the path loss and the distance, in the radio propagation literature, the path loss in dB is expressed by the general relation

$$L_p = L_0 + 10 \times \alpha \log_{10} d \tag{3.9a}$$

which reduces to Equation (3.8) for free space when we let $\alpha = 2$. In urban areas, typically $\alpha = 4$ is used for the calculation of the coverage of the cellular networks. In indoor areas, αcan take values from less than 2 up to 6. Equation (3.9) again presents the total path loss as the path loss in the first meter, L_0, plus the power loss relative to the power received at 1 m, $10 \times \alpha \log_{10}(d)$. For a one-decade increase in distance, the power loss is 10α dB, and for a one-octave increase in distance, it is 3α dB. For a free-space path, the power loss is 20 dB per decade or 6 dB per octave of distance as already discussed. In urban areas with $\alpha = 4$, the attenuation is 40 dB per decade or 12 dB per octave. The received power in dB is the transmitted power plus transmitter antenna gain in dB minus the total path loss L_p plus receiver antenna gain.

The RSSI, P_r, in a wireless device is calculated by subtracting the path loss, L_p, in dB from the transmitted power in dBm, P_t, plus the transmitter and receiver antenna gains, G_A, in dB:

$$P_r = P_t + G_A - L_p = P_0 - 10\alpha \log d. \tag{3.9b}$$

In this equation, $P_0 = P_t + G_A - L_0$ is the RSSI in 1 m distance from the transmitting device.

The path-loss models of this form are extensively used for the deployment of wireless networks. The coverage area of a radio transmitter depends on the power of the transmitted signal and the path loss. Each radio receiver has a power sensitivity, P_s, and it can only detect and decode signals with a strength larger than this sensitivity. Since the signal strength falls with distance, using the transmitter power, the path-loss model, and the sensitivity of the receiver, one can calculate the coverage that is the maximum distance for receiving a signal strength higher than the power sensitivity. We can calculate the maximum path loss that is acceptable for a device by finding the difference in transmitted power and the receiver sensitivity $L_p = P_t + G_A - P_s$. Substituting this maximum allowable path loss in Equation (3.9a), we have

$$L_p = P_t + G_A - P_s = L_0 + 10 \times \alpha \log_{10} d.$$

Solving this equation for *d,* we can find the coverage of a device from

$$d = 10^{\frac{P_t + G_A - P_s - L_0}{10 \times \alpha}}. \tag{3.9c}$$

Example 3.4: Coverage of a WLAN in Free Space and Urban Areas:

The transmitted power and the receiver sensitivity of a WLAN operating at 2.4 GHz are 20 dBm and -80 dBm, respectively. If the transmitter and the receiver antenna gains are 3 dB and 1 dB, the maximum acceptable path-loss or link budget is

$$L_p = P_t + G_A - P_s = 20\,\text{dBm} + 3\,\text{dB} + 1\,\text{dB} - (-80\,\text{dBm}) = 104\,\text{dB}.$$

The path loss at 1 m for 2.4 GHz calculated in Example 3.3 was $L_0 = 40.05\,\text{dB}$ using Equation (3.9c); the coverage of this system for free space,

$\alpha = 2$, is

$$d = 10^{\frac{104-40.05}{20}} = 1576\,\text{m}.$$

The coverage of the same system in urban areas with $\alpha = 4$ is

$$d = 10^{\frac{104-40.05}{40}} = 40\,\text{m}.$$

The substantial difference in the coverage in different areas is caused by the change of distance power gradient from 2 in free space to 4 in urban areas. Therefore, the distance-power gradient plays an important role in the calculations of the coverage of wireless networks. The value of this gradient changes drastically in different environments and is often calculated empirically.

Shadow Fading:

Depending on the environment and the surroundings, and the location of objects, the received signal strength *for the same distance* from the transmitter will be different. In effect, Equation (3.9) provides the mean or median value of the signal strength that can be expected if the distance between the transmitter and receiver is d. The actually received signal strength will vary around this mean value. This variation of the signal strength due to location is often referred to as *shadow fading* or *slow fading*. The reason for calling this shadow fading is that, very often, the fluctuations around the mean value are caused due to the signal being blocked from the receiver by buildings (in outdoor areas), walls (inside buildings), and other objects in the environment. It is called slow fading because the variations are much slower with distance than another fading phenomenon caused due to multipath that we will discuss later. It is also found that shadow fading has less dependence on the frequency of operation than multipath fading or fast fading as will be discussed later. The path loss Equation (3.9a) will have to be modified to include this effect by adding a random component as follows:

$$L_p = L_0 + 10\alpha \log_{10} d + X(\sigma). \tag{3.10}$$

Here, X is a random variable with a distribution that depends on the fading component. Several measurements and simulations indicate that this variation can be expressed as a log-normally distributed random variable. A log-normal absolute fading component ends up as a zero-mean Gaussian fading component in dB. Therefore, the distribution function of X is given by

$$f_{\text{SF}}(X) = \frac{1}{\sqrt{2\pi}\sigma} \exp\left(\frac{X^2}{2\sigma^2}\right), \tag{3.11}$$

where σ is the standard deviation of the received signal strength in dB caused by shadow fading.

Fade Margin:

The problem caused by shadow fading is that all locations at a given distance may not receive sufficient signal strength for correctly detecting the information. To achieve sufficient signal coverage, the technique employed is to add a *fade margin* to the path loss or received signal strength. The fade margin is usually taken to be the additional signal power that can provide a certain fraction of the locations at the edge of a cell (or near the fringe areas) with the required signal strength. For computing the coverage, we thus employ the following equation:

$$L_p = L_0 + 10\alpha \log d + F_\sigma \tag{3.12}$$

where F is the fade margin associated with the path loss to overcome the shadow fading component.

The distribution of X in Equation (3.11) is employed to determine the appropriate fade margin. Since X is a zero-mean normally distributed random variable that corresponds to lognormal shadow fading, this means that at the fringe locations, the mean value of the shadow fading is 0 dB. 50% of the locations have a positive fading component and 50% of the locations have a negative fading component. In Equation (3.10), this will mean that the locations that have the positive fading component X will suffer a larger path loss resulting in unacceptable signal strength. To overcome this, a fading margin is employed to move most of these locations to within acceptable RSSI value. This fading margin can be applied by increasing the transmit power and keeping the cell size the same or reducing the cell size by setting a higher threshold for making a handoff. In mathematical terms for $\gamma\%$ coverage, the base station should have an additional fade margin satisfying

$$1 - \gamma = 0.5erfc(F_\sigma/\sigma\sqrt{2}) = \int_{F_\sigma}^{\infty} f_{\mathrm{SF}}(x)dx\,. \tag{3.13a}$$

The function *erfc* and its inverse *erfcinv* are well known in mathematics, and they are available in MATLAB and other mathematical analysis tools. Using the *erfcinv* function, we can calculate fade margin from

$$1-\gamma = 0.5erfc(F_\sigma/\sigma\sqrt{2}) \Rightarrow F_\sigma = \sqrt{2}\times\sigma\times erfcinv\left[2\left(1-\gamma\right)\right]. \tag{3.13b}$$

After calculation of fade-margin, F_σ, for a given coverage certainty of γ, we can achieve coverage similar to Equation (3.9c) by modifying that equation to

$$d = 10^{\frac{P_t + G_A - F_\sigma - P_s - L_0}{10 \times \alpha}}. \tag{3.9d}$$

Equation (3.9c) gives the coverage for 50% certainty. For any other level of certainty, we need to calculate the associated fade margin and use Equation (3.9d). An example will further help in the understanding of this concept.

Example 3.5: Computing the Fading Margin and Coverage:

A mobile system is to provide 95% successful communication at the fringe of coverage ($\gamma = 0.95$). The variance of the shadow fading in the environment that the device is operating in is $\sigma = 8\,\mathrm{dB}$. Substituting these values in Equation (3.13), we have

$$1 - 0.95 = 0.5erfc(F_\sigma/8\sqrt{2}) \Rightarrow erfc(F_\sigma/8\sqrt{2}) = 0.1.$$

To solve this equation, we can use the *erfcinv* function in the MATLAB

MATLAB: >> erfcinv(0.1) → ans = 1.1631.

Therefore,

$$F_\sigma/8\sqrt{2} = 0.1631 \Rightarrow F_\sigma = 13.1589\,\mathrm{dB}.$$

Therefore, for this example, the fade margin to be applied is 13.16 dB. That means, if we want to calculate coverage with 95% confidence, we need to subtract 13.16 from the actual power in our coverage equation. In the last part of Example 3.4, we found that coverage was

$$d = 10^{\frac{104-40.05}{40}} = 40\,m \Rightarrow d = 10^{\frac{104-40.05-13.16}{40}} = 20\,\mathrm{m}.$$

That means if a device is at 40 m, the probability of connecting to the access point is 50% and if it is at 20 m, this probability is 95%.

So far, we have discussed achievable signal coverage in terms of the received signal strength and the path-loss model. In the following sections, we discuss how we can measure the path-loss parameters in an environment with its own multipath characteristics.

3.4.3 Empirical Measurement of Path-Loss Parameters

We can use the traditional least square (LS) method of statistical modeling to estimate the path-loss model parameters, (P_0, α, σ), given by

Equation (3.9b) from a set of measurements of the RSSI at specific known locations. Let us assume that we have measured N samples of the RSSI, P_i in dBm, from a radiating antenna at distances, d_i. To solve the problem of estimating the parameters using the LS approach, we define a cost function, which is the average of the expected received power at a distance, $\widehat{P_i} = P_0 - 10\,\alpha\log(d_i)$, and the measured power, P_i, at that location:

$$\varepsilon(P_0, \alpha) = \frac{1}{N}\sum_{i=1}^{N}(\widehat{P_i} - P_i)^2 = \frac{1}{N}\sum_{i=1}^{N}(P_0 - 10\alpha\log d_i - P_i)^2. \quad (3.14a)$$

By taking the derivatives of this cost function, we will have two sets of linear equations with two unknowns:

$$\begin{cases} \dfrac{\partial\varepsilon}{\partial\alpha} = \displaystyle\sum_{i=1}^{N} -2\times 10\log d_i(P_0 - 10\alpha\log d_i - P_i) = 0 \\ \dfrac{\partial\varepsilon}{\partial P_0} = \displaystyle\sum_{i=1}^{N} 2(P_0 - 10\alpha\log d_i - P_i) = 0 \end{cases}, \quad (3.14b)$$

and we can calculate, P_0, and α, from them. Substituting the estimated values of (P_0, α) in Equation (3.14a), we can obtain an estimate for the variance of the shadow fading:

$$\sigma^2 = \varepsilon(P_0, \alpha) = \frac{1}{N}\sum_{i=1}^{N}(P_0 - 10\alpha\log r_i - P_i)^2. \quad (3.14c)$$

The formulation in Equation (3.14) allows us to compute the necessary parameters for the RSSI linear regressive model.

Example 3.6: MATLAB Code for Calculation of Path-Loss Parameters:

In this example, we provide MATLAB code for solving the above linear regression problem to calculate the parameters for RSSI behavior. Figure 3.18(a) shows the MATLAB code to fit the data. In this code, the vector r represents the range of measurement locations and vector Pr represents the associated RSSI values in dBm. Figure 3.18(b) shows the results of curve fitting and location of a sample measurement point and its associated estimated value with the RSSI model. The program

```
%% This is sample for curve fitting your pathloss data to
find distance-power gradient
clc;clear all;close all;
% Enter your distance in meter here
r=[5.0;6.5;8.0;9.5;11.0;12.5;14.0;15.5;17.0;20.0];
% Enter your RSS in dB here
Pr=[-38.9459;-38.8443;-36.2157;-39.1163;-36.3094;-
44.6655;-40.8041;-46.7637;-42.3983;-50.2684];
% Plot your Lp vs. Distance in dB
r_dB=10*log10(r); % Careful! Change your distance in
10log(r) when plotting
% Linear Fitting
F1=fit(r_dB,Pr,'poly1');
%Plot your data with fitting
plot(F1,r_dB,Pr);
grid on
xlabel('10*log10(r)');
ylabel('Received Signal Strength[dBm]');
val=coeffvalues(F1);
disp('The Estimated Path Loss Model is:');
model1=['Pr=',num2str(val(2)),num2str(val(1)*10),'*log(r)
'];
disp(model1);
disp('Mean value of shadow fading is:');
disp(mean(Pr+20+2*r_dB));
disp('Standard Deviation of shadow fading is:');
disp(std(Pr+40+2*r_dB));
```

(a)

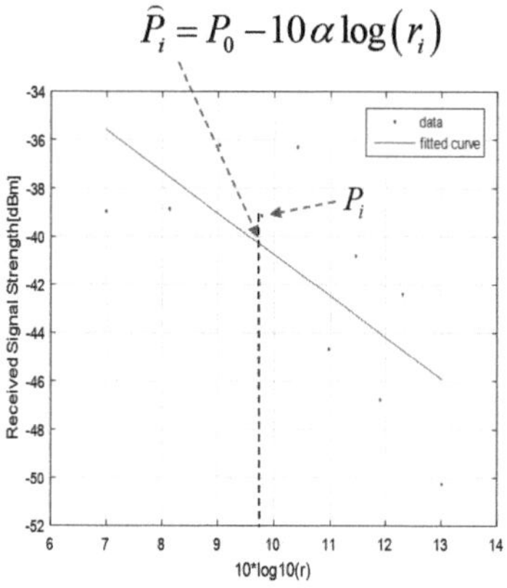

$$\hat{P}_i = P_0 - 10\alpha\log(r_i)$$

(b)

Figure 3.18 Empirical measurement of parameters of the linear regression model for behavior of the average RSSI. (a) MATLAB code to fit the data to a regression line. (b) Resulting plot with locations of a sample measurement and its estimated location using the RSSI behavior model.

also prints out the estimated model and the standard deviation of shadow fading as:

The Estimated Path Loss Model is:

```
Pr=-23.5916-17.1571*log(r)
Mean value of shadow fading is:
    -0.6353
Standard Deviation of shadow fading is:
    3.2478
```

Fitting this output to the model in Equation (2.1), we have $P_0 = -23.5916$, $\alpha = 1.71571$, and $\sigma = 3.2478$. The estimation of shadow fading has a bias of -0.6353.

The above method for modeling the average RSSI has been used by a variety of standardization organizations to model the behavior of the RSSI for different wireless communication systems. These models are used for the calculation of coverage and the expected data rate of a wireless device at a given distance from an access point or a base station. A comprehensive coverage of these models is available in various books [Pah05, Rap02, Mol12]. Here, we provide a summary of three popular models for inside the human body applications, indoor areas, and urban areas, which are useful for the analysis of RSSI-based localization applications described in this book.

3.5 Path-Loss Model Examples in Standards

Multipath arrivalMultipath arrival of the signal in indoor and urban areas, where the applications discussed in this book operate, causes extensive fluctuations of the amplitude of the received signal in time. Figure 3.19 illustrates the variation of the amplitude in dB as a function of the logarithmic distance between the transmitter and the receiver as a receiver is moved away from a transmitter. This figure also shows how we approach the modeling of these variations of the signal for different applications. The instantaneous received amplitude in a multipath environment always varies with time and with small local changes in distance. The Fourier transform of these changes is referred to as the doppler spectrum that we will explain in Section 3.3.5, and it reflects the speed of motions in the environment. The *average RSSI* in dBm is what we use for RSSI-based localization to determine the distance between an antenna and a device. The traditional method to model how the RSSI is related to the distance from the transmitter is to use linear regression and LS estimation to calculate the parameters of the model using empirical data. Path-loss modeling is based on the received signal power, P_0, or path loss from a reference point, L_0. In the path-loss model of Equation (3.9a), this reference location was 1 m from the transmitting antenna. We can generalize that to any reference location at distance d_0 from the transmitting antenna and the path loss in that reference location, L_0:

$$L_p = L_0 + 10\alpha \log_{10}(d/d_0) + X(\sigma). \tag{3.9e}$$

Intuitively, this model assumes that the RSSI in Watts or mW (not in dB) is inversely proportional to the distance raised to the power indicated by the distance-power gradient, αdistance-power gradient, α.. The reference location d_0 is an arbitrary location close to the transmitter. Shadow fading represents variations of the RSSI from the linear regression line in dB caused by objects shadowing paths between the transmitter and the receiver.

The above method for modeling the average RSSI is commonly used by a variety of standardization organizations to model the behavior of the RSSI for different wireless communication systems operating in different environments. These models are used for the calculation of coverage and the expected data rate of a wireless device at a given distance from an access point or a base station. A comprehensive coverage of these models is available in various books [Pah05, Rap02, Mol12]. Here, we provide a summary of three example models for radio propagation inside the human bodies, indoor areas, and urban areas, which are useful for the analysis of RSS-based applications described in this book, and provide different pitches on using the path-loss model in practice.

3.5.1 NIST Model for RSSI Inside the Human Body

Researchers at the National Institute of Standards and Technology (NIST) developed a path-loss model [Say09, Say10] for understanding how the signal strength attenuates from inside of the human body to the surface of the human body. This model has been adopted by the IEEE 802.15.6 standards group for body area networking. Since empirical measurements of radio propagation inside the human body are not practical or feasible, NIST researchers used the high-frequency structure simulation (HFSS) software to calculate radio signal strength attenuation inside the human body using numerical solutions to Maxwell's equations. The commercially available HFSS tool is one of the most popular software tools designed for this purpose and offers a graphical user interface for the emulation of radio wave propagation inside the human body with its organs. Using the HFSS, NIST researchers have defined two scenarios of radio propagation for deep and near-surface tissues. For each scenario, they have made multiple measurements to determine the parameters for the path-loss model given by Equation (3.9e). Table 3.2 shows the parameters of the model for the two scenarios. These simulations are made for MedRadio bands at 402–405 MHz.

3.5.2 IEEE 802.11 Model for Indoor Area

For small distances in indoor areas, the transmitter and the receiver are often in the same room where the received signal power from the direct LOS

Table 3.2 Parameters of the NIST path-loss model from implant to body surface.

Location	d_0 (mm)	L_0(dB)	α	σ(dB)
Deep inside	5	47.14	4.26	7.85
Near surface	5	49.81	4.22	6.81

paths dominates the power arriving from other paths. In these situations, the distance power gradient α is close to 2 and is associated with free space propagation (where the Friis equation applies). As the distances separating the transmitter and receiver increases such that the receiver is beyond the room where the transmitting antenna resides in, walls obstruct the direct LOS signal and reduce the received power from that path significantly. The reception of the received power from other reflected or diffracted paths increasingly contribute to the overall received signal power. The distance power gradient α in these obstructed LOS or OLOS conditions increases substantially. As a result, popular path-loss model standards designed for WLAN and wireless personal area network (WPAN) applications define *distance-partitioned* models for the behavior of the RSS. These models define different distance power gradients for different ranges of distances. We present the IEEE 802.11 recommended model as an example model for RSSI behavior in indoor areas.

Figure 3.19 shows the general two-piece distance partitioned model for RSSI behavior, recommended by the IEEE 802.11 standardization committee for indoor areas. The path loss is modeled as two piecewise linear segments separated by a break point at d_{bp} by

$$L_p = X(\sigma) + L_0 + \begin{cases} 10\alpha_1 \log_{10}(d) & ; d < d_{\mathrm{bp}} \\ 10\alpha_1 \log_{10}(d_{\mathrm{bp}}) + 10\alpha_2 \log_{10}(dr/d_{\mathrm{bp}}) & ; d > d_{\mathrm{bp}} \end{cases} . \tag{3.15}$$

The distance-power gradient in the two segments are $\alpha_1 = 2$ and $\alpha_2 = 3.5$, respectively. Note that $\alpha_2 > \alpha_1$. The top row of Equation (3.15) for distances less than the breakpoint distance is the same as Equation (3.9e) with a distance power gradient of 2 and d_0 = 1 m used as the reference distance.

The bottom row is also the same as Equation (3.9e) with the break point, d_{bp}, used as the reference distance. Note that the reference path loss at the break point is $L_0 + 10\alpha_1 \log_{10}(d_{\mathrm{bp}})$.

The IEEE 802.11 standard defines six different scenarios for path-loss models with four different break points. Table 3.3 shows the parameters associated with these models. Model A is a *flat fading* model with a single path between the transmitter and the receiver. The breakpoint distance for this model is at 5 m and the standard deviation of the shadow fading is 5 dB. Model B is recommended for a typical residential environment with LOS conditions and more than one effective path between the transmitter and the receiver. The path-loss parameters of this model are the same as those of Model A. Model C is recommended for a typical residential or small

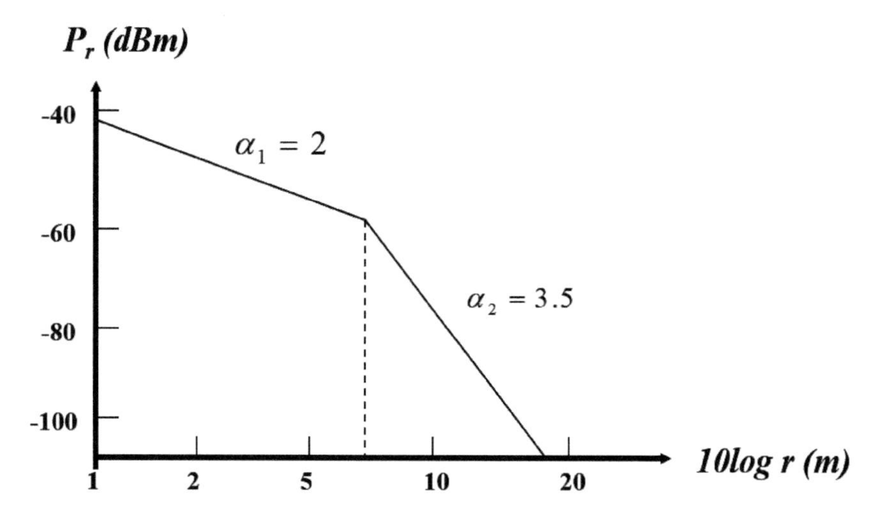

Figure 3.19 The IEEE 802.11 distance-partitioned path-loss model.

office environment with LOS and OLOS conditions between the transmitter and the receiver. The breakpoint for this model is still at 5 m, but the standard deviation of the shadow fading is increased to 8 dB. Model D is recommended for a typical office environment with OLOS conditions with a 10 m breakpoint distance and an 8 dB standard deviation of shadow fading. Model E is recommended for a typical large open space and office environment in areas with OLOS conditions and it has a breakpoint distance of 20 m and a shadow fading standard deviation of 10 dB. Model F is recommended for a large open space with indoor and outdoor environment in the areas with OLOS conditions.

Example 3.7: Coverage Using the IEEE 802.11 Distance Partitioned Model:

The transmitter power of an 802.11 device is 20 dBm and the receiver sensitivity is −90 dBm. Determine the coverage in a small office building with LOS and NLOS using Model C. Assume that antenna gains are 1 and the frequency of operation is 2.45 GHz.

Solution: We use Equation (3.15) and parameters of Model C shown in Table 3.3 for calculation of the IEEE 802.11 coverage. The maximum path loss allowed is 110 dB and the path loss at the first meter is 40.5 dB, just as it

Table 3.3 Parameters for different IEEE 802.11 recommended path-loss models for six environments.

Environments	d_{bp} (m)	α_1	α_2	σ(dB)
A	5	2	3.5	5
B	5	2	3.5	5
C	5	2	3.5	8
D	10	2	3.5	8
E	20	2	3.5	10
F	30	2	3.5	10

is in Examples 3.4. In LOS/NLOS small office areas and with a breakpoint distance of 5 m, the coverage is calculated using the following expression:

$$110 = 40.5 + 20\log 5 + 35\log\left(\frac{d}{5}\right)$$

The coverage with 50% confidence and no fade margin is

$$d = 5 \times 10^{\frac{69.5-14}{35}} = 195\,\text{m}.$$

If we increase the confidence to 95% (with an 8 dB standard deviation of shadow fading as shown in Table 3.3 for model C), we need an additional 13.2 dB fade margin (see Example 3.5) that reduces the coverage to

$$d = 5 \times 10^{\frac{69.5-14-13.2}{35}} = 82\,\text{m}.$$

3.5.3 Okumura–Hata Model for Urban Areas

Outdoor models for the RSSI behavior are mostly designed for cellular telephony applications and they are more detailed than models used for indoor areas for WLAN applications. Because the height of antennas could be on top of a hill or deep in a canyon, that affects the coverage. In addition, the range of frequencies used in the traditional cellular networks went from a hundred MHz up to a few GHz, and that has to be included as well. One of the most popular models for RSSI in outdoor areas is the Okumura–Hata model, which describes the path loss with Equation (3.9e) with distances in km. *The reference distance* is $d_0 = 1\,\text{Km}$. The other two parameters of the model are given by

$$\begin{cases} L_0 = 69.55 + 26.16\log f_c - 13.82\log h_b - a(h_m) \\ \alpha = 4.49 - 0.655\log h_b \end{cases}, \tag{3.16}$$

Table 3.4 Correction factor and the range of operation for Okumura–Hata modelOkumura-Hata model.

		Range of values	
Center frequency f_c in MHz			150–1500 MHz
h_b, h_m in meters			30–200 m, 1–10 m
$a(h_m)$ in dB	Large city	$f_c \leq 200$ MHz	8.29 [log (1.54 h_m)]^2 – 1.1
		$f_c \geq 400$ MHz	3.2 [log (11.75 h_m)]^2 – 4.97
	Medium-small city	$150 \geq f_c \geq 1500$ MHz	1.1 [log f_c – 0.7] h_m – (1.56 log f_c – 0.8)

where f_c is the frequency in MHz, and h_b, h_m are the heights of base station antenna and mobile antenna, respectively. The function $a(h_m)$ is a correction factor that depends on the frequency of operation and the environment. Table 3.4 provides functions for the calculation of the correction factor and the range of operation of this model. More elaborate channel models for a wider range of frequencies for outdoor operations are available in [Mol12]. The variance of shadow fading in urban areas varies between 3 and 10 dB [Gud91].

Example 3.8: Coverage with Okumura–Hata Model:

The receiver sensitivity of a cell phone is -126 dBm and it operates at 900 MHz. For a base station height of 100 m and a mobile height of 2 m, determine the minimum transmit power for the base station to support a 30 km coverage.

The path loss is

$$\begin{aligned} L_p &= 69.55 + 26.16 \log f_c - 13.82 \log h_b - a(h_m) + [44.9 - 6.55 \log h_b] \log d \\ &= 69.55 + 26.16 \log 900 - 13.82 \log 100 - 1.05 + [44.9 - 6.55 \log 100] \log 30 \\ &= 165.11 \text{ dB}. \end{aligned}$$

Solution: We calculate the terms in the Okumura–Hata model as follows:

$$a(h_m) = 3.2 \left[\log(11.75 h_m)\right]^2 - 4.97 = 1.05 \text{ dB}.$$

Therefore, the base station transmit power should be

$$P_r(\text{dBm}) = L_p(\text{dB}) + P_r(\text{dBm}) = 165.11 + (-126) = 39.11 \text{ dBm}.$$

In Watts, the transmit power is

$$10^{39.11/10} = 8147 \text{ mW} \approx 8 \text{ W}.10^{39.11/10} = 8147\ mW \approx 8\ W.$$

3.6 Modeling of RSSI Fluctuations and Doppler Spectrum

In the last section, we discussed the behavior of the average RSSI and its application to path-loss modeling and analysis of the effects of shadow fading. These models provide an insight into the understanding of the relation between power and distance that is essential for the deployment of wireless networks. Such a characterization of the received signal strength corresponds to a *large-scale* average value. In reality, as shown in Figure 3.19, the RSSI is rapidly fluctuating in time and locally in space due to the mobility of the mobile terminal or movement of other objects close to the transmitter and the receiving antennas causing changes in multiple signal components arriving via different paths.

Two effects contribute to the rapid fluctuations of the signal amplitude. The first is caused by the movement of the mobile terminal towards or away from the base station transmitter and is called *Doppler spectrum*. The second is caused by the addition of signals arriving via different paths, which is referred to as *short time*, *small-scale*, or *multipath fading*.

The short distance or short time variations of the RSSI, referred to as *small-scale fading*, are the rapid instantaneous changes in the received signal power caused by fast changes in the phase of the received signal from different paths due to small movements. As shown in Figure 3.19, for the analysis of the short-term variations of the channel, we are interested in the statistics of short-term multipath fading and finding the shape of the Doppler spectrum of the signal. The statistics of the short-term variations in RSSI allows us to calculate the *error rate* of different transmission techniques over the wireless medium. The statistics of the temporal multipath fading are characterized by the probability density function (PDF) of the sampled values of the fast variations of the channel. As we will see later in this chapter, the most popular distribution for this variation is the Rayleigh distribution, and for that reason, sometimes, this type of fading is referred to as Rayleigh fading.

The Doppler spectrum is the Fourier transform of the samples of the variation of the signal. It is very important to model this spectrum because if want to simulate these variations, we will need a good model. If we know the spectrum of a random signal, we can regenerate it by designing a filter with that spectrum and stimulating that filter with a noise-like random signal. The Doppler spectrum allows us to learn how to simulate variations of the channel in time to examine its impact on the transmission of *packets* over the channel before we implement the expensive hardware for a transmission technique.

The cause of all problems in the wireless channel is the multipath. In this section, we use the Friis' equation and a technique that we refer to as geometric ray tracing with simple models to show how multipath causes changes in the distance-power gradient and causes fluctuations of the RSS. Then we provide some examples of models for small-scale multipath fading used in practical applications in wireless networking.

3.6.1 Friis' Equation and Geometric Ray Tracing

To model the path loss, we started with Equation (3.7) that was representing the Friis' equation in logarithmic form, and based on that equation, we showed how we develop empirical path-loss models for different wireless systems in multipath conditions and we described how we come up with the concept of large-scale shadow fading. To demonstrate how multipath causes short-term or small-scale fading and to discover the meaning of Doppler spectrum we start with the Friis' equation in the linear form given by Equation (3.6).

As we discussed in Section 3.3.2, electromagnetic waves at the high frequencies used in wireless networks can be treated as rays and if we have the geometry of the area, we can relate the length of these rays to the geometry using principles from geometric optics, which have been in use for a couple of millennia to describe the imaging inside mirrors. Using geometric optics, we can *trace* the paths that waves between a transmitter and the receiver will travel. If for each path, we can find the amplitude and phase of the received signal, we will be in a position to analyze the received signal and explain how it causes counter-intuitive observations such as a change in the distance-power gradient or multipath fading.

We begin this discussion with an example in which we use the Friis' equation to calculate the magnitude and phase of a simple tone cosine signal when it is transmitted over a single path wireless channel.

Example 3.9: Single-Tone and Single-Path Transmission:

What are the amplitude and phase of a received signal if a single frequency cosine signal $x(t) = \sqrt{P_t} \cos 2\pi f t$ (where P_t is the transmitted power and f is the frequency of the signal) is transmitted along with a single path free-space medium?

Solution:
We denote the received signal by $y(t)$. The amplitude of the received signal is obtained by taking the square root of the received signal power. The received signal power can be determined from Friis' equation in the linear form described by Equation (3.6), i.e.,

$$P_r = \frac{P_0}{d^2} \Rightarrow \sqrt{P_r} = \frac{\sqrt{P_0}}{d}$$

Since the radio transmission environment forms a linear time-invariant system in this case, if the transmitted signal is a cosine, the received signal will also be a cosine at the same frequency with a delay of

$$\tau = \frac{d}{c}$$

That results in a phase value of

$$\phi = 2\pi f\tau = \frac{2\pi f d}{c} = \frac{2\pi d}{\lambda}$$

where $\lambda = f/c$ is the wavelength of the transmitted cosine. Then, the received signal is simply

$$y(t) = \sqrt{P_r}\cos 2\pi f(t-\tau) = \frac{\sqrt{P_0}}{d}\cos\left(2\pi f t - \phi\right).$$

Figure 3.20 illustrates the basic concept behind single-tone transmission in a free-space single-path wireless transmission medium that was described in Example 3.9. Since the channel is linear, the transmitted sinusoid is received as a sinusoid after propagation over a distance d. The *magnitude* corresponds to the amplitude $\sqrt{P_0}/d$, and the *phase* $\phi = 2\pi d/\lambda$, of the received sinusoid arriving through the unobstructed direct path between the transmitter and the receiver, are both functions of the distance. The amplitude changes slowly with the inverse of the distance and the phase rotates rapidly at the speed of one rotation which is 2π radians every d/λ. For example, with a Wi-Fi device operating at 2.4 GHz, the wavelength is 12.5 cm (see Example 3.3) and the phasor representing a path has one full rotation every 12.5 cm. For the same motion, the amplitude of the signal does not change significantly. As we will see later, this observation helps us to understand the cause of multipath fading.

To analyze the effects of multipath on single-tone transmission, we extend the results of single-tone transmission provided in Example 3.9 to a scenario

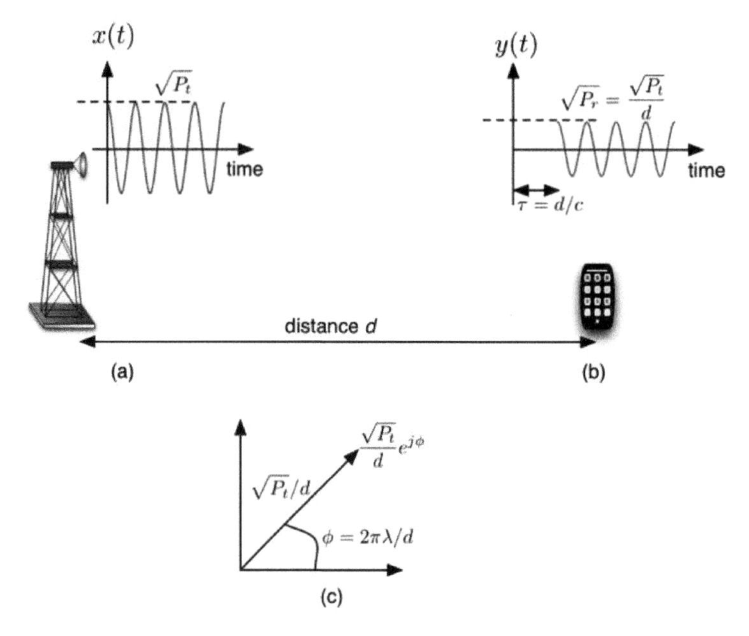

Figure 3.20 Visualization of the basic concept behind single-tone transmission in a free space. (a) Transmitted signal. (b) Received signal. (c) Phasor diagram representation of the signal.

where we have a single path arriving after reflection on a wall. For a reflected path from a wall, with minor modifications, we can use the same equations for calculation of the received signal amplitude and phase. As shown in Figure 3.21, the difference between a direct LOS and a reflected path is that, after reflection, the incident propagated wave from the wall has a loss in its amplitude according to the reflection coefficient and it changes the polarity of its phase. Therefore, if a signal arriving along a path reflected from a wall with a length of d_i and a reflection coefficient of a_i, the amplitude and phase of the received signal are $a_i\sqrt{P_0}/d$ and $\phi_i = 2\pi d_i/\lambda + \pi$. If we represent the reflection coefficient by a negative number, then the addition of π to the phase shift is not necessary. This simple observation allows us to explain the effects of multipath, when, in addition to the direct LOS path, we have reflected paths as well.

We begin this discussion with an example scenario with two mobile devices in a large open area, shown in Figure 3.22(a), in which we have three paths, the direct LOS path and the paths reflected from the ceiling and the floor of the building. In this situation, we can make a reasonable assumption that the effects of other paths are negligible.

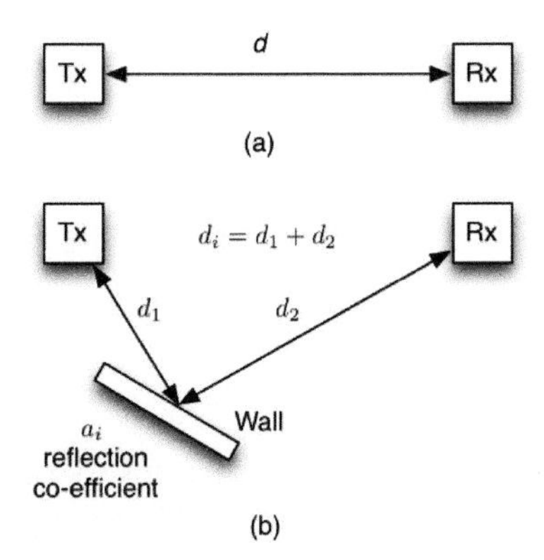

Figure 3.21 Comparison of direct and reflected paths. (a) Direct LOS path. (b) A reflected path with reflection coefficient.

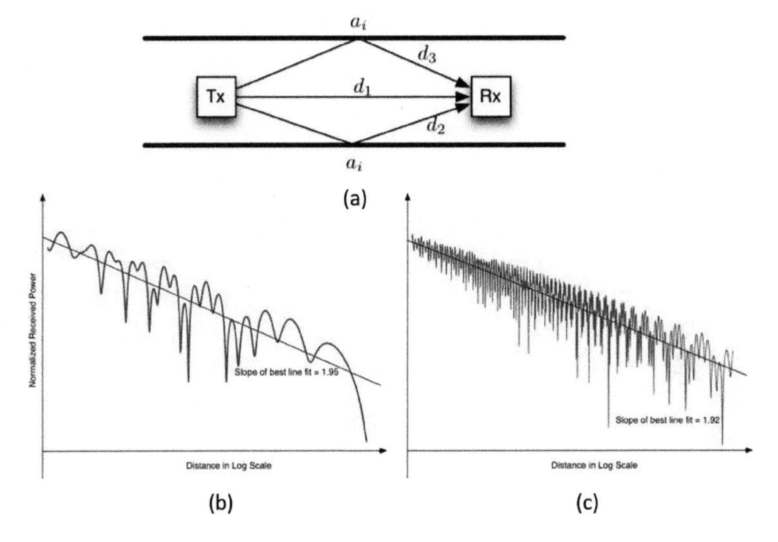

Figure 3.22 Ray tracing in an open area. (a) The scenario of operation. (b) Normalized received power versus distance at 1 GHz and (c) at 10 GHz.

Example 3.10: Multipath Fading in an Open Indoor Area:

Consider the three-path open indoor area scenario of operation shown in Figure 3.22(a), assuming the height of the ceiling is 5m, the antennas are located at 1.5 m above the floor, and the reflection coefficient is $a_i = -0.7$.

For this scenario:

a) Give equations for calculation of the RSSI from all paths if the transmitted power and the antenna gains are normalized to 1.
b) Use MATLAB to plot the RSSI in dB for $1 < d < 100$ m to demonstrate the formation of multipath fading. Assume the frequency of operation is 1 GHz.
c) Repeat (b) for a frequency of 10 GHz.

Solution:

a) The amplitude and phase of the ith path is given by $a_i\sqrt{P_0}e^{j\phi_i}/d_i$ where for the LOS path, $a_i = 1$ and for the other two paths, $a_i = -0.7$. Therefore, the amplitude and phase of the received signal are calculated from $\sqrt{P_0}\sum_{i=1}^{3}\frac{a_i}{d_i}e^{j\phi_i}$, the RSSI is the magnitude square of the complex received signal:

$$P_r = \left|\sqrt{P_0}\sum_{i=1}^{3}\frac{a_i}{d_i}e^{j\phi_i}\right|^2 = P_0\left|\sum_{i=1}^{3}\frac{a_i}{d_i}e^{j\phi_i}\right|^2,$$

where with normalized transmitted power and antenna gains, $P_0 = (\lambda/4\pi)^2$.

In this example, the ceiling height is assumed to be 5 m and the antennas are 1.5 m above the floor. The reflection coefficients are assumed to be $a_1 = +1$ (the LOS path) and $a_2 = a_3 = -0.7$ for the other two paths. The distance of the direct path is the actual distance between the transmitter and the receiver, and the distances for the path from the ground and the ceiling is given by

$$d_2 = 2 \times \sqrt{\frac{d_1^2}{4} + (1.5)^2} \quad \text{and} \quad d_3 = 2 \times \sqrt{\frac{d_1^2}{4} + (3.5)^2}.$$

a) The following MATLAB code can be used to determine the normalized amplitude and phase of the received signal for a variety of parameters. Figure 3.22(b) shows the normalized received power versus distance calculated for distances ranging from 1 to 100 m from the MATLAB simulation. The plot shows power in decibels and distance on a logarithmic scale. This figure demonstrates the power fluctuations due to multipath fading for a frequency of 1 GHz.
b) Figure 3.22(c) shows the results for the frequency of 10 GHz.

```
%Define parameters
c = 3e8;
Pt = 1; Gr = 1; Gt = 1;
a = [1, -0.7, -0.7];
fc = [100e6 1e9 10e9];
lambda = c./fc;
d = logspace(0,2,1000);

%Define NLOS distance vectors d1 and d2
d1 = 2*sqrt(0.25*(d.^2)+(1.5^2));
d2 = 2*sqrt(0.25*(d.^2)+(3.5^2));

%Part 1abcd:
for i=1:length(fc)

  %Calculate P0 for fc(i)
  P0 = 1e-3;
  %P0 = Pt*Gr*Gt*((lambda(i)/(4*pi))^2);

  %Calculate phases for fc(i)
  phi1 = -(2*pi*fc(i)*d)/c;
  phi2 = -(2*pi*fc(i)*d1)/c;
  phi3 = -(2*pi*fc(i)*d2)/c;

  %Calculate received power for fc(i)
  Vr = (a(1)*(exp(j*phi1)./d)+a(2)*(exp(j*phi2)./d1)+a(3)*(exp(j*phi3)./d2));
  Pr_dB = 10*log10(P0*abs(Vr.^2));

  %Find the best-fit curve for the received power plot
  bf = polyfit(10*log10(d),Pr_dB,1);
  bf_val = polyval(bf,10*log10(d));

  %Plot power vs. distance
  figure(i)
  semilogx(d,bf_val,'r:',d,Pr_dB,'b-');

  %Labels
  xlabel('Distance [m]'); ylabel('Received Power [dB]');
  title('Received Power vs. Log Distance');
end
```

Figure 3.22 illustrates the calculations in the above example. This figure indicates that while the average power decreases with distance, the power also fluctuates by as much as 20–30 dB at a rate proportional to the frequency of operation. To explain what causes the fluctuations, we resort to a phasor diagram in the complex plane. Figure 3.23 shows the three paths in the complex plane and the resulting vector sum of the three path vectors each represented by a magnitude and a phase. As a mobile user moves away from the transmitter, these three vectors constantly change their amplitude and phase. However, the amplitude of the paths changes slowly (proportional to the inverse of the distance), but the phase changes rapidly at a rate of $2\pi/\lambda$ radians per meter. This means that for a mobile with a carrier frequency of 1 GHz, we have a 360° change in the phase every $\lambda = 33\,\mathrm{cm}$. Therefore, to visualize the received signal strength in a multipath environment, one should consider Figure 3.23 when all the amplitudes and phases are changing – amplitudes very slowly and phases very rapidly. Each vector is shrinking in its amplitude slowly while it rotates like a road runner. The received signal amplitude is the vector sum of all path amplitudes and phases. When all paths are in line, they add up and result in a strong amplitude. When they are aligned against one another, the result is very small amplitudes registering the fading in the RSS. Therefore, as mobile moves, it observes extensive fluctuations in its amplitude caused by different combinations of the phases that add up or degrade the overall amplitudes. The rate of these variations and occurrence of fading is proportional to the speed of rotation and consequently the wavelength of the signal. When we increase the frequency to 10 GHz as in Figure 3.22(c), the rate of occurrence of the fades increases by ten times. The

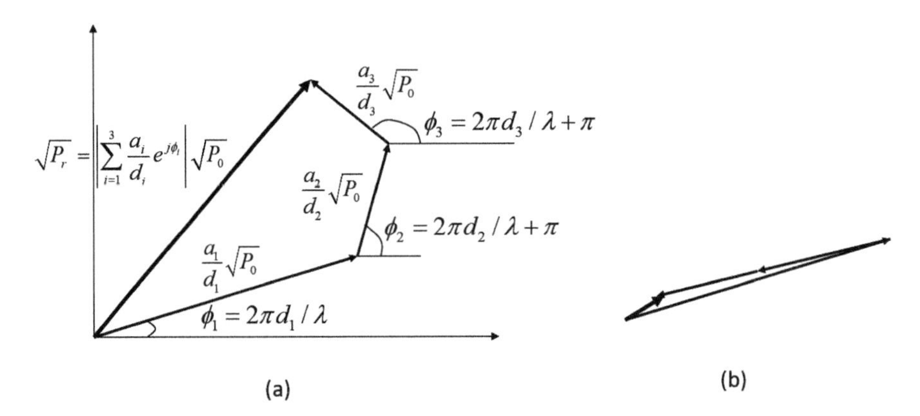

Figure 3.23 Visualization on NB communication over multipath. (a) Phasor diagram representing how multipath components adds up together. (b) Occurrence of a multipath fading.

rate of variations of the phase of the multipath components is proportional to the frequency.

To connect this example to path-loss modeling, the reader should note that the slope of the best fit line to the RSSI is the distance-power gradient that is shown in Figure 3.22(b) and (c). This slope remains very close to 2 (1.95 and 1.92), which is consistent with the free space and IEEE 802.11 model for the path loss. The IEEE 802.11 model assumes a distance-power gradient of 2 for open indoor areas. The average power, however, does not show any shadow fading because the details of furniture or other objects contributing to the shadowing are not included in the scenario and there is no wall between the transmitter and the receiver.

In Example 3.10, we showed that as the distance between the transmitter and the receiver increases, the rapid changes in the phase of the multipath components cause rapid multipath fading. In a typical environment in indoor or urban areas, even when we move along a circle and keep the distance constant, the multipath components change their phases and cause rapid multipath fading. Even if we keep the distances constant and people or vehicles move close to the transmitter and the receiver, the phase of the multipath components changes, and multipath fading is observed. Figure 3.24 shows the measured average RSSI values in a laptop of a signal transmitted by an IEEE 802.11 access point. Note the large variations in the RSSI samples caused by multipath fading, even though the RSSI samples are averaged values.

In this section, we explained how multipath causes fluctuations in the RSS. As we explained in this section, these fluctuations in RSSI of the

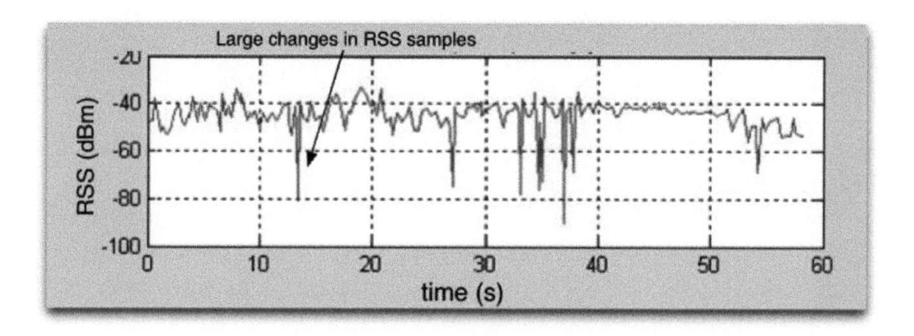

Figure 3.24 Measurement of the received signal strength from an IEEE 802.11b/g access point for a fixed location of a laptop.

received signal in wireless links are modeled by the statistics of multipath fading and the Doppler spectrum. Next, we provide some examples of popular models used for these purposes.

3.6.2 Modeling of Small-Scale Fading

Multipath fading results in fluctuations of the signal amplitude because of the addition of signals arriving with different *phases*. This phase difference is caused since signals have traveled different distances by traveling along different paths. Since the phase of the arriving paths is changing rapidly, the received signal amplitude undergoes rapid fluctuation that is often modeled as a random variable with a distribution.

To model these fluctuations, we can generate a histogram of the amplitude of the received signal in time. The density function formed by this histogram represents the distribution of the fluctuating values of the amplitude of the received signal. The most used distribution for multipath fading is the Rayleigh distribution whose PDF is given by

$$f_{ray}(r) = \frac{r}{\sigma^2} \exp(-\frac{r^2}{2\sigma^2}), \quad r \geq 0. \tag{3.17a}$$

Here, it is assumed that all signals suffer nearly the same attenuation but arrive with different phases. The random variable corresponding to the signal amplitude is *r*. Theoretical considerations indicate that the sum of such signals will result in the amplitude having the Rayleigh distribution. This is also supported by measurements at various frequencies [Pah05]. When a strong LOS signal component also exists, the distribution is found to be Rician, and the PDF of such a distribution is given by

$$f_{\text{ric}}(r) = \frac{r}{\sigma^2} \exp(\frac{-(r^2 + K^2)}{2\sigma^2}) I_0(\frac{Kr}{\sigma^2}), \quad r \geq 0, K \geq 0. \tag{3.17b}$$

Here, *K* is a factor that determines how strong the LOS component is relative to the rest of the multipath signals.

These equations are used to determine what fraction of time a signal is received such that the information it contains can be decoded or what fraction of area receives signals with the requisite strength. The remainder of the fraction is often referred to as outage.

Small-scale fading results in very high bit error rates. To overcome the effects of small-scale fading, it is not possible to simply increase the transmit power because this will require a humungous increase in the transmit power.

A variety of techniques are used to mitigate the effects of small-scale fading including error control coding with interleaving, diversity schemes, and using directional antennas. These techniques will be discussed in Chapter 3.

3.6.3 Modeling of Doppler Spectrum

Distributions of the amplitude of a radio signal presented in the previous section demonstrate how a signal is undergoing small-scale fading. In general, it is also important to know for what time a signal strength will be below a value (duration of fade) and how often it crosses a threshold value (frequency of transitions or fading rate). This is particularly important to design the coding schemes and interleaving sizes for efficient performance. We see that this is a second-order statistic, and it is obtained by what is known as the *Doppler spectrum* of the signal.

Doppler spectrum is the Fourier transform of the fluctuations of the received signal strength. Figure 3.25 [How90] demonstrates the results of measurements of amplitude fluctuations in a signal and its Doppler spectrum

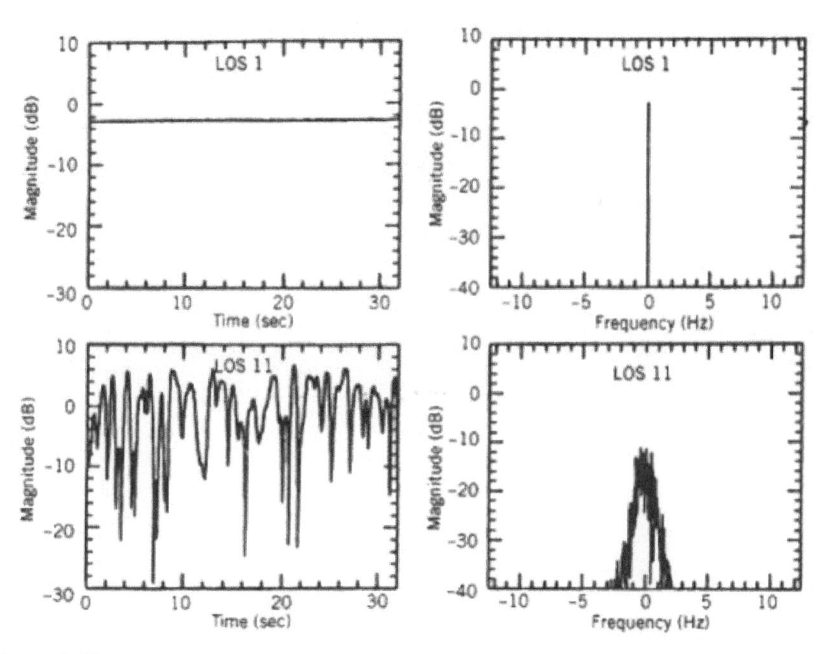

Figure 3.25 Measured values of the Doppler (a) with no motion and (b) receiver moving randomly.

under different conditions. In Figure 3.25(a), the transmitter and receiver are kept constant and nothing is moving in close to them. The received signal has a constant envelope and its spectrum is only an impulse. In Figure 3.25(b), the transmitter is randomly moved to result in fluctuation of the received signal. The spectrum of this signal is now expanded over a spectrum of around 6 Hz reflecting the rate of variations of the received signal strength. This spectrum is referred to as the Doppler spectrum.

In the mobile radio applications, the Doppler spectrum for a Rayleigh fading channel is usually modeled by

$$D(\lambda) = \frac{1}{2\pi f_m} \times [1 - (\lambda/f_m)^2]^{-1/2} \qquad \text{for } -f_m \leq \lambda \leq f_m. \quad (3.17)$$

Here, f_m is the maximum Doppler frequency possible and is related to the velocity of the mobile terminal via the expression $f_m = v_m/\lambda$ where v_m is the mobile velocity and λ is the wavelength of the radio signal. This spectrum, commonly used in mobile radio modeling, is also called the classical Doppler spectrum and is shown in Figure 3.26. Another popular model for the Doppler spectrum is the uniform distribution that is sometimes used for indoor applications [Pah05]. From the rms Doppler spread, it is possible to obtain the fade rate and the fade duration for a given mobile velocity [Pah05]. These

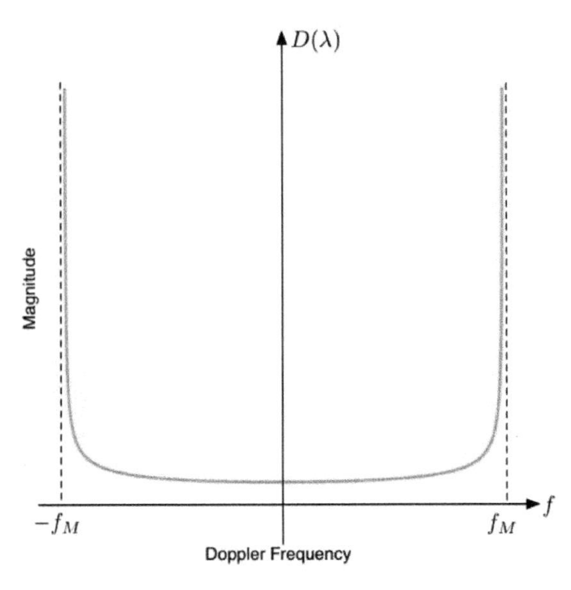

Figure 3.26 Classical Doppler spectrum for mobile radio.

values can then be used in the design of appropriate coding and interleaving techniques for mitigating the effects of fading. Diversity techniques are useful to overcome the effects of fast fading by providing multiple copies of the signal at the receiver. Since the probability that all these copies are in fade is small, the receiver can correctly decode the received data. Frequency hopping is another technique that can be used to combat fast fading. Because all frequencies are not simultaneously under fade, transmitting data by hopping to different frequencies is an approach to combat fading. This is discussed in Chapter 4.

3.7 Wideband Modeling of Multipath Characteristics

The phasor analysis based on the Friis equation that we presented in Section 3.5.1 for a geometric optical explanation of the cause of multipath fading for the RSSI assumed that we transmit a sinusoid signal at a given frequency. In circuits and systems, this analysis is sometimes referred to as frequency domain analysis because as shown in Figure 3.27(a), in the frequency domain, the spectrum of a sinusoid has only a single impulse representing a frequency tone.

A sinusoid is an ideal signal in the frequency domain. The equivalent of the ideal signal in the time domain is an ideal impulse, shown in Figure 3.27(b), that we use to find the time domain characteristics of a system. The impulse in time has a constant spectrum in frequency (i.e., it has all the frequency components with the same power). Using an ideal impulse, one can determine the channel impulse response (CIR) that can be used to find out what happens to a waveform if it is transmitted over a multipath channel. Waveforms are used in communications to carry bits of information and in localization to determine the distance between the transmitter and the receiver. Knowing what happens to a transmitted waveform is essential for the design of efficient wireless access and localization systems. The behavior of paths in forming the CIR fluctuates as objects or devices move in the environment. We can benefit from these fluctuations to design other cyberspace applications for motion and gesture detection or authentication and security [Pah20].

3.7.1 Channel Impulse Response and Bandwidth

If we consider the geometric optics-based ray tracing technique that we have developed so far, we realize that in ideal conditions, if we can transmit an impulse, at the receiver, we receive several impulses each arriving along a

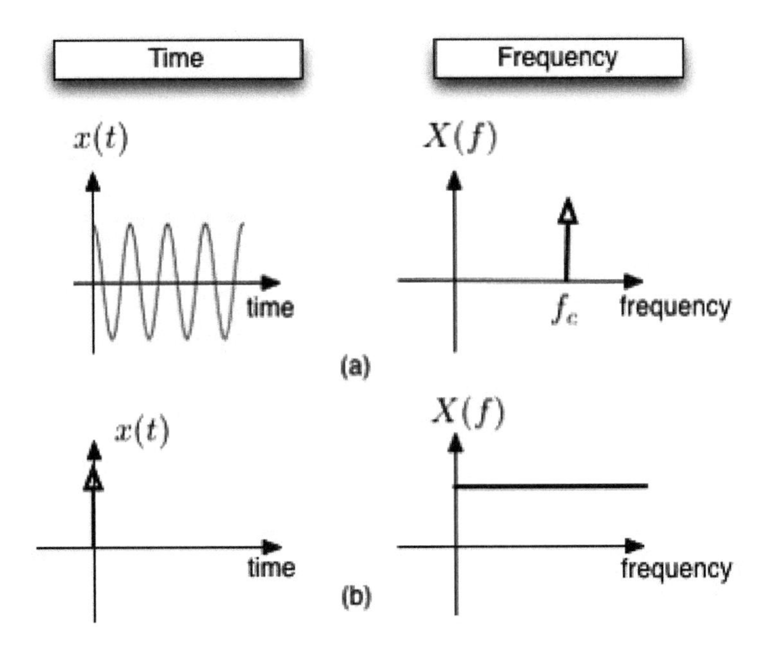

Figure 3.27 Time-frequency characteristics of signals used for channel modeling (a) a sinusoid as an ideal narrow band signal and (b) an impulse, the ideal wideband signal.

different path. In other words, the CIR of a multipath channel is a discrete function of the form

$$h(\tau) = \sum_{i=1}^{L} A_i \delta(\tau - \tau_i) e^{j\varphi_l} \tag{3.19}$$

where$A_i = a_i\sqrt{P_0}/d_i$ and $\phi_i = 2\pi d_i/\lambda$ are the amplitude and phase of ith path and $\tau_i = d_i/c$ is the time of arrival of the path. The complex value CIR of a three-path channel is shown in Figure 3.28(a). The magnitude square of the complex CIR is called the delay power spectrum, shown in Figure 3.28(b):

$$Q(\tau) = |h(\tau)|^2 = \sum_{i=1}^{L} P_i\, \delta(\tau - \tau_i) \tag{3.20}$$

where$P_i = |A_i|^2$ is the power of the signal arriving along the ith path. The physical meaning of the delay power spectrum is that it represents the received power arriving along different paths as a function of the delay of

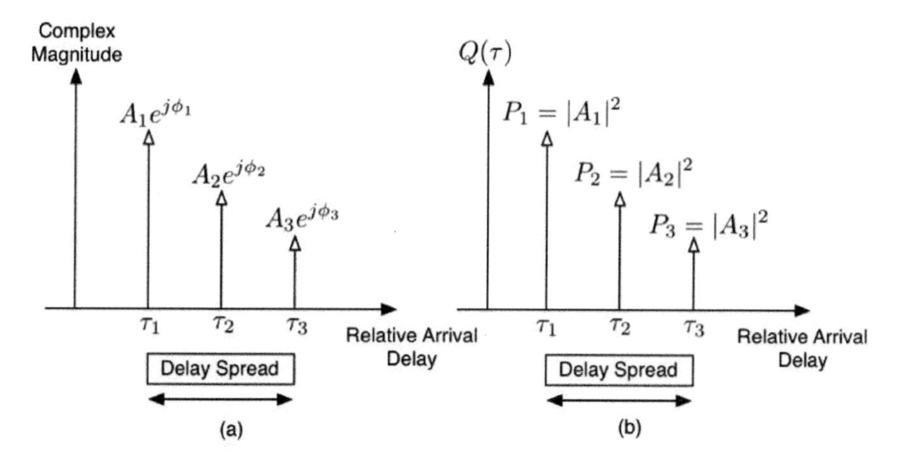

Figure 3.28 Multipath characteristics. (a) Complex CIR between a transmitter and a receiver. (b) The delay power spectrum.

arrival of these paths. Therefore, the horizontal axis is not real-time; it is the delay or lag between the various multipath components.

To measure a typical CIR, we need to transmit a narrow pulse that resembles an impulse. The narrower the pulse, the wider is the required bandwidth for transmission of the pulse. To measure (or *resolve*) all multipath components, the bandwidth should be wide enough that its inverse is proportional to the difference in the delay of arrival of different paths $\Delta\tau$. This delay reflects the intensity of the multipath arrivals. Since delay is a function of distance, the difference between arrivals or intensity of the paths is related to the difference in the path lengths Δd. The system bandwidth requirement to isolate the paths is approximately $W = 1/\Delta\tau = c/\Delta d$.

Example 3.11. Resolving Multipath Components in Indoor and Outdoor Areas:

In indoor areas used for WLAN applications, distances on the order of meters separate walls and other objects. Therefore, it is reasonable to assume that a measurement system used for measuring the indoor multipath should be able to resolve multipath components that are up to 1 m apart. For separating that has 1 m difference in their lengths, we need a system with a bandwidth of $W = 3 \times 10^8 = 300\,\text{MHz}$. If we consider outdoor areas where buildings are around tens of meters apart, our measurement system may require bandwidth on the order of 30 MHz to separate paths with 10 meter difference in their lengths. If we need a measurement system for WPANs in which devices may be in fractions of meters, we may need a measurement system with a bandwidth around 1 GHz.

Figure 3.29 shows a sample measured time and frequency response of a typical radio channel. In the time domain, shown in Figure 3.29(a), a transmitted narrow pulse arrives as multiple paths with different strengths and arriving delays. In the frequency domain, shown in Figure 3.29(b), the response is not flat, and it suffers from deep frequency selective fades. This measurement is taken in a typical indoor office area. The bandwidth of the measurement system is 200 MHz (from 900 to 1100 MHz) and the center frequency is 1 GHz.

In practical applications, when we transmit a waveform for communication or localization, knowing the required bandwidth for operation in an environment is very important. In localization applications, if we want to measure the distance based on the time-of-flight of the signal, we have to synchronize the transmitter and the receiver and determine the time difference between the peak of the transmitted pulse and the first arriving path at the receiver. In this application, the bandwidth must be large enough so that the first path is isolated from the other paths. The bandwidth requirement is then very similar to bandwidth requirements for the multipath channel measurement systems described earlier in this section. The bandwidth requirement for wireless communications is different because, for reliable communications, the emphasis is on the *symbol transmission rate*. For a given multipath scenario, we want to increase the rate of transmission of our symbols, but increasing the symbol transmission rate or bandwidth beyond certain values will cause ISI.

3.7.2 Multipath Spread, ISI, and Bandwidth

In Figure 3.29, the delay between the arrival of the first and the last path is referred to as the excess multipath delay spread or simply the *delay spread* of the channel. In wireless communications, the inverse of the width of the transmitted symbols approximately represents the required bandwidth for data transmission. One of the significant problems caused by this phenomenon along with fading is ISI. If the multipath delay spread is comparable to or larger than the symbol duration, the received waveform representing one symbol spreads into the waveform representing adjacent symbols and produces ISI. The ISI distorts the symbols so that the

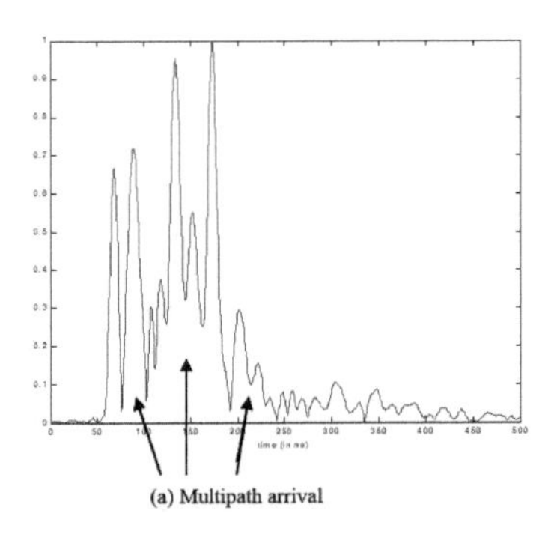
(a) Multipath arrival

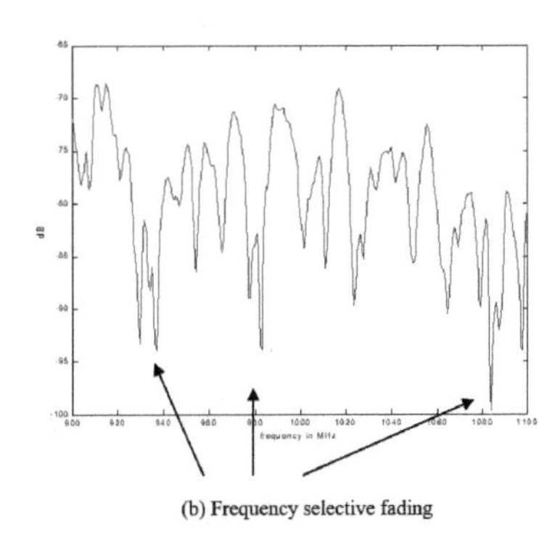
(b) Frequency selective fading

Figure 3.29 Typical time and frequency response of a radio channel.

receiver cannot distinguish between the possible transmitted symbols, thereby resulting in irreducible errors in the transmitted information.

To explain this phenomenon, consider Figure 3.30, in which we want to transmit a rectangular waveform for communication purposes over a three-path channel. The information is coded in the amplitude of this waveform, and every T_s seconds, we transmit a symbol. The data rate or the required bandwidth for implementation of this transmission system is $W = 1/T_s$. Because of the multipath, at the receiver, we have three waveforms that add together, and the resulting waveform stretches beyond T_s, the duration allocated to the symbol, and interferes with the transmission of the next symbol causing ISI. As we increase the transmission data rate, the duration of the waveform T_s is reduced and the amount of ISI is increased. The increase in the ISI causes degradation in the performance of the transmission system. In data communications over single path channels, degradation of transmission performance is due to the background noise. When we encounter such a situation, we can increase the transmission power so that the received signal is stronger than the fixed background noise. In an ISI environment, increasing the transmitter power does not solve the problem because it increases the amount of ISI as well.

The amount of ISI depends on the delay spread and the strength of the individual paths with respect to the first path. The second central moment of the multipath intensity profile is called the *rms delay spread* and it is used as a measure of the ISI. The basic definition of the second central moment is given by

$$\tau_{\mathrm{rms}} = \sqrt{\overline{\tau^2} - (\overline{\tau})^2}$$

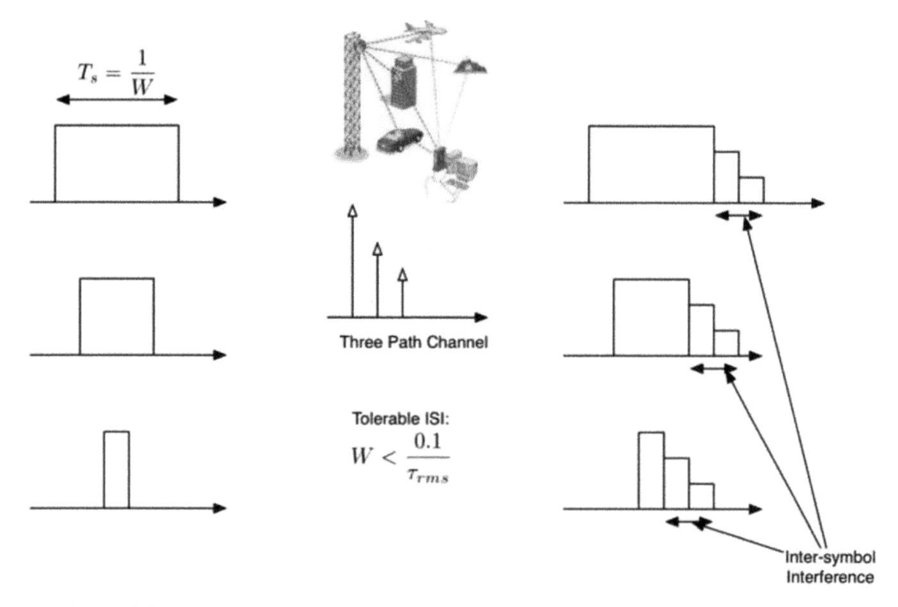

Figure 3.30 Relation among CIR, ISI, and symbol transmission rate or bandwidth.

where the *n*th moment is defined by

$$\overline{\tau^n} \equiv \frac{\sum\limits_{k=1}^{N} \tau_k^n P_k}{\sum\limits_{i=1}^{L} P_k}, \qquad n = 1, 2$$

from which we have

$$\tau_{\text{rms}} = \sqrt{\frac{\sum_{k=1}^{N} \tau_k^2 P_k}{\sum_{k=1}^{N} P_k} - \left(\frac{\sum_{k=1}^{N} \tau_k P_k}{\sum_{k=1}^{N} P_k}\right)^2}. \tag{3.21}$$

Example 3.12: Calculation of the rms Delay Spread:

Consider a two-path channel, with an impulse response comprising multipaths with arrival delays of$\tau_1 = 0\,(n\,\text{sec})$ and$\tau_2 = 50\,(n\,\text{sec})$ and path powers of $P_1 = 1\,(0\,\text{dBm})$ and $P_2 = 0.1\,(-10\,\text{dBm})$, shown in Figure 3.31. Determine the rms delay spread from the direct calculation of moments and using Equation (3.21).

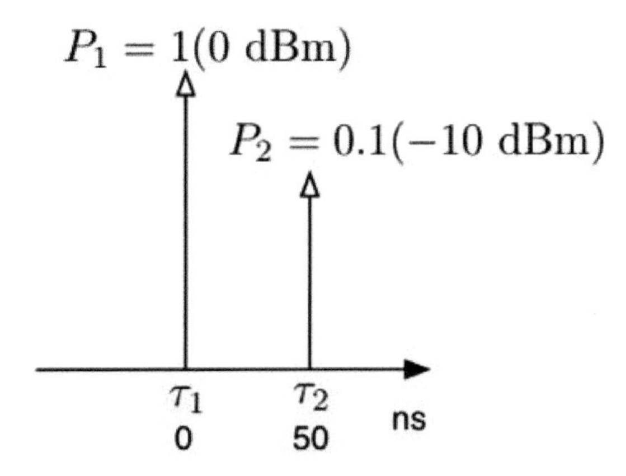

Figure 3.31 The delay power spectrum of the channel used in Example 3.12.

Solution: For the direct calculation of the rms delay spread, using the expression for the moments, we can calculate the first and second moments as

$$\overline{\tau} = \frac{0 \times 1 + 50 \times 0.1}{1 + 0.1} = 4.55(n\,\text{sec})$$

$$\overline{\tau^2} = \frac{0 \times 1 + 2500 \times 0.1}{1 + 0.1} = 227.27(n\,\text{sec}).$$

Then we can determine the rms delay spread

$$\tau_{\text{rms}} = \sqrt{\overline{\tau^2} - (\overline{\tau})^2} = \sqrt{227.27 - (4.55)^2} = 14.37(\text{n sec}).$$

We could also determine the same value using Equation (2.19) in one step

$$\tau_{\text{rms}} = \sqrt{\frac{0 \times 1 + 2500 \times 0.1}{1 + 0.1} - \left(\frac{0 \times 1 + 50 \times 0.1}{1 + 0.1}\right)^2} = 14.37(\text{n sec}).$$

Considering Figure 3.20, if we want to have an intuition on the SNR in the presence of a multipath, the strength of the signal relative to ISI noise is related to the duration of the transmitted pulse. The longer the length of the pulse, T, the smaller will the effects of the ISI be. Therefore, $T/\tau_{\text{rms}} = 1/W \times \tau_{\text{rms}}$ is a measure of the signal to ISI noise. As we will see in Chapter 4, most basic digital transmission systems operate at a reasonable error rate when their SNR is above 10 dB. Considering this value, one can conclude that reliable operations of a simple digital communication link over a multipath channel are possible if the symbol transmission rate or the bandwidth of the system is less than 10% of the inverse of the rms delay spread or

$$W < \frac{0.1}{\tau_{\mathrm{rms}}} \tag{3.22}$$

If the bandwidth of the transmission system over a multipath channel follows the above rule, the shape of the transmitted symbol is preserved without significant distortions in the shape of the waveform. As the inverse of the rms delay spread plays an important role in the calculation of the data transmission rate over multipath channels, in the literature, this inverse is sometimes referred to as the *coherence bandwidth* B_c of the channel.

The rms delay spread varies depending on the type of environment. In indoor areas, it could be as small as 30 ns in residential areas or as large as 300 ns in factories [Pah05]. In urban macro-cellular areas, the rms delay spread is on the order of a few microseconds. This means that the maximum data rates that can be supported by a simple binary modem in indoor areas can be as high as 6.7 Mbps (at 30 ns) and in wide-area cellular networks, it can be as low as 50 kbps (at 4 μs). This observation indicates that as we increase the range of coverage of a wireless network, the distances of objects causing multipath to become larger and the rms delay spread increases resulting in a lower supportable data rate with simple schemes using a single frequency carrier.

To support higher data rates, different receiver techniques are necessary. Equalization is a method that tries to cancel the effects of multipath delay spread in the receiver. Direct sequence spread spectrum enables resolving the multipath components and exploiting them to improve performance. OFDM uses multiple carriers, spaced closely in frequency, each carrying low data rates to avoid ISI. Beamforming with MIMO antenna systems reduces the number of multipath components, thereby reducing the total delay spread itself. We discuss these topics in Chapter 4. Here, we have discussed a simple method to relate the bandwidth to the ISI.

3.7.3 Wideband Channel Models in Standardization Organizations

In reality, $Q(\tau)$ is a two-dimensional function,$Q(\tau, t)$ of *time* and *delay* of arrival. However, for larger data rates, it is a slow-time varying function or channel, and for practical purposes, we can represent it as a function of only delay [Pah05]. The physical meaning of slow-time varying channel is that when we are transmitting a waveform for communication or localization purposes, during the transmission of the symbol, the channel is stable or invariant. We have already modeled the rate of variations of the channel by the Doppler spectrum function$D(\lambda)$ that reflects the effects of movement in the area of operation. In a typical wireless application, the rate of variations of the channel is at most several hundred Hz, while typical transmission rates are on the order of mega symbols per second. In this situation, a slow-time varying model is very reasonable. Under this assumption, we divide the channel behavior into a static and a dynamic component. The static behavior is represented by the delay power spectrum, $Q(\tau)$, and the dynamic behavior by the Doppler spectrum$D(\lambda)$. In classical radio channel modeling [Pah05], a combination of these two functions is called the *scattering function* and it is defined by

$$S(\tau, \lambda) = Q(\tau) \times D(\lambda). \tag{3.23}$$

If we complement specifications for Equation (3.23) with a path-loss model, we have a complete set of analytical models to simulate a channel for both coverage and communication performance evaluation purposes.

Standardization organizations usually provide channel models that identify the scattering function and path-loss models for different scenarios. One of the major challenges for wireless standardization organizations is to compare and select the best modem design for the physical layer implementation among multiple proposed systems. To have a fair comparison among these proposed alternatives, a commonly accepted channel model is needed. After the completion of the standard, these models are used by manufacturers for the design and performance evaluation of their products. Since the bandwidth and environments in which these channel models are used are different, most standardization groups come up with their own standard model(s). Since we have different channel models for the wide-area and LANs, we treat their details in separate sub-sections in the relevant chapters.

We proceed to explain the wideband model recommended by GSM as a simple example to understand the channel models recommended by standardization organizations.

Example 3.13: A Wideband Model for the GSM Cellular Networks:

The GSM standardization group defines a set of channel profiles with discrete delay power spectra of different types for rural areas, urban areas, and hilly terrains [GSM91]. The basic difference between these models is the value of

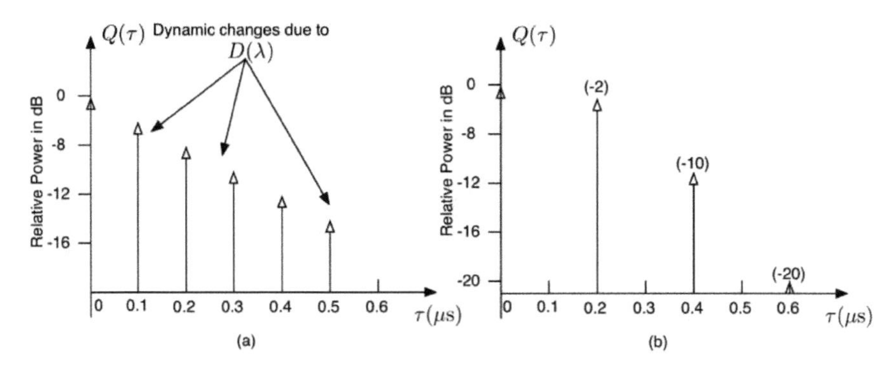

Figure 3.32 Two options for delay power spectrums recommended by the GSM committee. (a) 6-paths. (b) 4-paths.

the rms delay spread and the number of multipath components that are used to represent the channel profile.

In this example, we only describe the model used for rural areas. This model defines a delay power spectrum with two options for implementation with six or four multipath. Figure 3.32 shows the two delay power spectra recommended for rural areas. In the six-path model, the multipath are 0.1 μs apart and they cover a delay spread of 0.5 μs. The power of the received signal at each path starts at 0 dB and decays by 0.4 dB with each successive path. In the four-path model, the delay spacing is 0.2 μs and it covers a delay spread of 0.6 μs. The relative powers of the paths are 0, −2, −10, and −20 dB, respectively. Both the six-path and four-path models roughly provide the same rms delay spread. If we want to evaluate the effects of multipath on the design of different modems, both models should provide similar results. The six-path option provides for a more refined model at the expense of additional hardware for the implementation of two more paths. The bandwidth of the GSM channels is 200 kHz resulting in pulses with approximately 5 μs width. The delay spread of the channel is around 10% of this value that follows our bandwidth constraint for manageable ISI defined by Equation (3.22).

The first path in both profiles is assumed to have a Rician distribution given by Equation (3.17b) because they are assumed to be along with a direct LOS path. The rest of the paths are assumed to have classical Rayleigh distributions given by Equation (3.17a). The Doppler spectrum choices for each path or tap of the model are either Rician or the classical Rayleigh. In a manner like the simulation of narrowband signals, the Doppler power spectrum for the classical Rayleigh model is the one given by Equation (3.17) and illustrated in Figure 3.26. The Rician spectrum is the sum of the classical Doppler spectrum in Equation (3.17), and one direct path, weighted so that the total multipath power is equal to that of a direct path alone. It is given by

$$D(\lambda) = \frac{0.41}{2\pi f_m}\left[1-(\lambda/f_m)^2\right]^{-1/2} + 0.91\delta(\lambda - 0.7f_m), \quad -f_m < \lambda < f_m. \tag{3.24}$$

If we complement this multipath model with the Okumura–Hata path-loss model, we have a complete model for the behavior of the channel in the specified area. Using these models, we can predict the coverage of the system and simulate the effects of fluctuations of the channel as well as the effects of multipath on the waveforms transmission.

3.7.4 Simulation of Channel Behavior

In terms of hardware or software simulations, the separation of the static and dynamic behavior allows us to implement the behavior in the delay variable with a tapped delay line spaced at the value given by delay power spectrum, $Q(\tau)$, and use a filtered complex Gaussian noise with the shape of Doppler spectrum, $D(\lambda)$, to simulate the time variations of each tap. This concept is illustrated in Figure 3.33. In Figure 3.33(b), we have paths with different arrival delays, identified by $Q(\tau)$, implemented in parallel branches. Figure 3.33(a) shows the implementation of the amplitude and phase of each path using a filtered Gaussian noise with a spectrum shape prescribed by $D(\lambda)$. The simulated complex channel fluctuations in Figure 3.33(a) are scaled by the strength of the path so that the overall channel response in Figure 3.33(b) provides for the delay power spectrum defined in Equation (3.20). In general, the delayτ_i is a random variable as well, but for simplicity of implementation traditional standardization organizations, assume fixed values for the delay and try to fit the rms delay spread of the multipath profile with the typical measurements in the environment that model is designed for.

The main objective is the development of a model for wideband characteristics of the channel is to develop a foundation for the design and comparative performance evaluation of wireless modems. The analysis of the performance was traditionally performed using analytical equations and the calculation of the analytical equations using digital computers. As the speed of computers and digital hardware in general increased, the models were also used for real-time hardware and computer software simulations of the channel behavior.

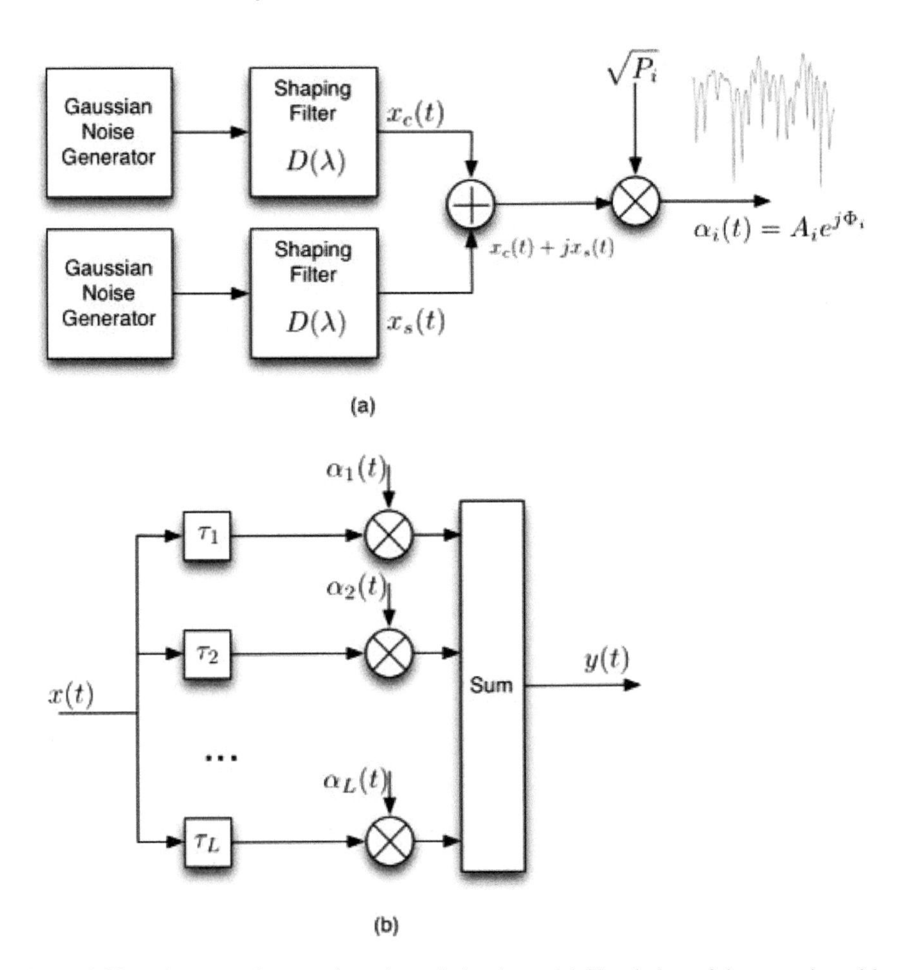

Figure 3.33 Elements of a complete channel simulator. (a) Simulation of the tap gains with dynamic behavior of $D(\lambda)$. (b) Simulation of static multipath using $Q(\tau)$.

3.7.5 Channel State Information and MIMO Channels

MIMO systems need models for the spatial channel characteristics. The *spatial wideband channel models* not only provide the delay-power spectrum presented by Equation (3.20) but also the angle of arrival of the multipath components. The advent of antenna array systems that are used for interference cancelation and position location applications has made it necessary to understand the spatial properties of the wireless communications channel.

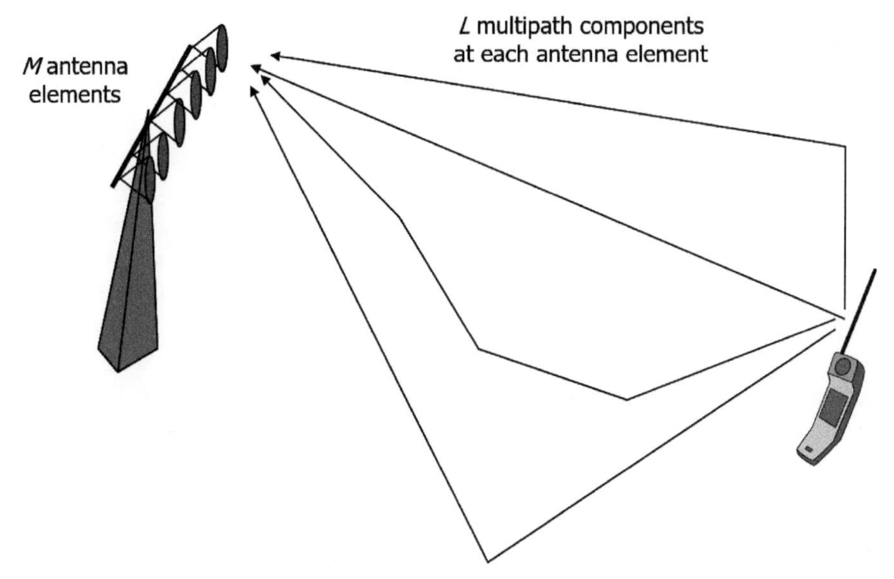

Figure 3.34 Single-input multiple-output model.

We start with *single-input multiple-output* (SIMO) radio channel models [ERT98]. In these models, a typical cellular environment is considered where it is assumed that the mobile transmitters are relatively simple, and the base station can have a complex receiver with adaptive smart antennas with *M* antenna elements. As shown in Figure 3.34, the multipath environment is such that up to *L* signals arrive at the base station from different mobile terminals (l) with different amplitudes (α) and phases (φ) at different delays (τ) from different directions (θ). These are in general time-invariant, and, as a result, the CIR is usually represented by

$$\vec{h}(t) = \sum_{l=1}^{L(t)} \alpha_l(t) e^{j\varphi_l(t)} \delta\left(t - \tau_l(t)\right) \vec{a}\left(\theta_l(t)\right). \tag{3.25}$$

Note that the CIR is now a *vector* rather than a scalar function of time. The quantity $\vec{a}(\theta(t))$ is called the array response vector and will have *M* components if there are *M* antenna array elements. Thus, there are *M* CIRs each with *L* multipath components. A variety of models are available in [ERT98]. The amplitudes are usually assumed to be Rayleigh distributed although they are now dependent on the array response vector $\vec{a}(\theta(t))$ as well.

An extension of this model to the scenario where there are N mobile antenna elements and M base station antenna elements is called a MIMO channel [PED00]. In this case, the CIR is an $M \times N$ matrix that associates a *transmission coefficient* between each pair of antennas for each multipath component. Experimental results and models are considered in [PED00, KER00].

There appears to be tremendous potential for improving capacity using MIMO antenna systems. Capacity increases between 300% and 500% are possible in cellular environments. Spectrum efficiency is also increased. For example, a 4 × 4 antenna array system over the MIMO channel can provide a spectral efficiency of 27.9 b/s/Hz [PED00] compared with spectral efficiencies of 2 b/s/Hz in traditional single-input single-output (SISO) radio systems.

3.8 APPENDIX A3: What is dB?

Decibel or dB is usually the unit employed to compute the logarithmic measure of power and power ratios. The reason for using dB is that all computation reduces to addition and subtraction rather than multiplication and division. Every link, node, repeater, or channel can be treated as a *black box* (see Figure A3.1) with a decibel gain. The decibel gain of such a black box is given by

$$\text{dB gain} = 10\log\left(\frac{\text{power of output signal}}{\text{power of input signal}}\right) = 10\log\left(\frac{P_{out}}{P_{in}}\right) \quad \text{(A3.1)}$$

This corresponds to the *relative* output power with respect to the input power. The logarithm is always to the base 10. If the ratio in (B.1) is negative, it is a decibel loss. The decibel gain relative to absolute power of 1 mW is denoted by dBm and that relative to 1 W is denoted by dBW. For example, if the input power is 50 mW, relative to 1 mW, the input power is 10 log (50 mW/1 mW) = 16.98 dBm. If this is followed by a link having a loss of 10 dB, the absolute power at the output of the link will be 16.98 – 10 = 6.98 dBm. Relative to 1 W, these values will be 10 log (50×10^{-3}/1) = -13 and –23 dBW, respectively. Antenna gains are represented similarly with respect to an *isotropic* antenna (that radiates with a gain of unity in all directions) or a *dipole* antenna. The former gain is in units of dBi and the latter in units of dBd. The units in dBi are 2.15 dB larger than the units in dBd.

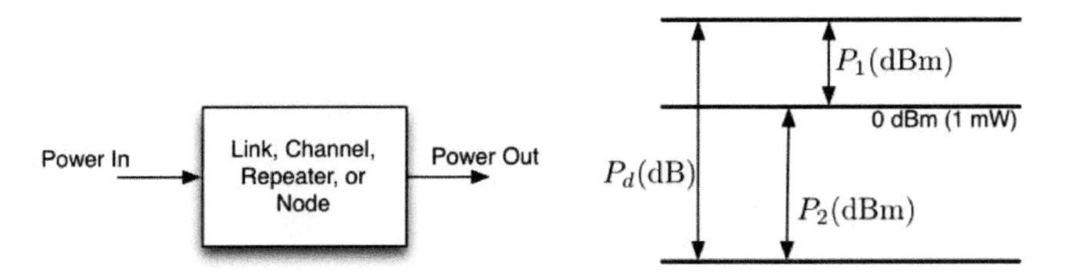

Figure 3.35 Decibel. (a) Overall concept. (b) Relation between dB and dBm.

Assignments

1. What is a voiceband modem and how did it play a major role in the evolution of PHY technologies in Ethernet and Wi-Fi?
2. What is the meaning of the medium bandwidth?
3. What is the difference between the user data rate and symbol transmission rate in a digital communication link?
4. What is the difference between thermal noise and ISI?
5. How does ISI caused by limitations of the transmission medium bandwidth differs from ISI from multipath arrival of RF signals?
6. What are co-channel and cross-channel interferences in wireless networks?
7. What are RSSI and SNR in communication devices and how do they relate to one another?
8. What are the practical values of medium bandwidth, user data rate, and SNR and how does Shannon–Hartley bound relates them to each other?
9. What are the typical bandwidths of Cat-3, Cat-5, and Cat-7 TP wirings? Which one is used for telephone and which one for Ethernet?
10. What are the NEXT and FEXT and how do they affect the bandwidth of a TP wiring?
11. Compare cable, TP, and fiber optic lines in terms of length of coverage, availability in existing buildings, cost of installation, and radiated interference.
12. What are the different grades of optical fiber lines and how do they differ in the bandwidth and path loss?
13. How does the coverage in guided and wireless media differ from each other?
14. Explain what path-loss gradient means. Give some typical values of the path-loss gradient in different environments.
15. What is the Doppler spectrum and how can one measure it?

16. Differentiate between shadow fading and fast fading.
17. What distributions are used to model fast fading in LOS situation and in OLOS situations?
18. What are the differences between multipath, shadow, and frequency-selective fading?
19. For position location applications, how are wideband radio channels classified? How is this classification useful?
20. What is the difference between a SIMO and a MIMO radio channel?

Instructor's solution available on River Publishers' website:
https://www.riverpublishers.com/book_details.php?book_id=919

Problem 1:
Assuming in Figure 3.1(b) amplitude of the signal is in millivolts, and T_s = 0.1 nano-seconds:

(a) What are the symbol transmission rate, R_s, in giga symbols per-second (GSps), needed medium bandwidth, W, in GHz, and application data rate, R_b, in Gbps?
(b) What is the number of bits per symbol, m, and the number of symbols, M?
(c) What is the energy of each symbol, E_{si}, and the average received energy, E_s?
(d) What is the received signal strength in micro-W and in dBm?
(e) If the transmitted power was 40dBm, what was the path loss in dBm?

Problem 2:
Use Shannon–Hartley equation to calculate the minimum SNR requirement to carry a Gigabit Ethernet signal over one pair of 100 meters Cat-3 TP wiring with a bandwidth of 25MHz.

Problem 3:

(a) The bandwidth of the Bluetooth is W = 1 MHz; what is the background noise for the system in dBm?
(b) If the minimum RSSI from the Bluetooth for a device operating in a room is −96 dBm, what is the minimum SNR of the Bluetooth devices?
(c) Use Shannon–Hartley equation to calculate the maxim data rate for communications over a Bluetooth link.

Problem 4:
Determine the maximum length of a Gigabit Ethernet Fiber Optic cable using multimode 62.5/125 fiber at 1300 nm wavelength. Assume the link budget of

the LED and the photo sensitive diodes used for transmission is 34 dB and the loss per km of the fiber optics cable is 1.5 dB. Leave a 3 dB safety margin in your calculations with an allowance of 2 splices each with 0.1 dB loss.

Problem 5:
Assuming that wireless devices use an antenna length of one-fourth of the wavelength for the transmitted frequency, what are the typical antenna lengths for the cellular phones operating at 900 and 1800 MHz bands and WLANs operating at 2.4 and 5.2 GHz.

Problem 6:

(a) Using MATLAB or any computational software, plot the path gain of the Cat-3 and Cat-5 cables as a function of distance for a frequency of 10 MHz.
(b) Use the plot to find the path loss for a 100 m Cat-3 cable.
(c) What would be the length of a Cat-5 for the same path loss?
(d) What would be the maximum length of a Cat-5 cable if it carries legacy Ethernet's 10 Mbps signal? Explain.

Problem 7:
What is the received power (in dBm) in the free space of a signal whose transmit power is 1 W and carrier frequency is 2.4 GHz if the receiver is at 1 mile (1.6 km) from the transmitter? What is the path loss in dB?

Problem 8:
Use the Okumura–Hata and COST-231 models to determine the maximum radii of cells at 900 and 1900 MHz, respectively, having a maximum acceptable path loss of 130 dB. Use $a(h_m)$ = 3.2 [log $(11.75\ h_m)^2$ – 4.97 for both cases.

Problem 9:
Table P3.1 provides the minimum required RSSI for an IEEE 802.11b device to operate at different rates.

(a) Calculate the coverage associated with each data rate in the table.
(b) Plot the staircase function of the Data Rate vs. RSS.
(c) Plot the staircase function of the Data Rate vs. Distance.
(d) If a mobile terminal moves away from an 802.11 AP and it goes out of the coverage area, calculate the average data rate that the terminal observes during the movement in the coverage area of the AP.

Table P3.1 Data rate and minimum power requirement for IEEE802.11b.

Data rate (Mbps)	RSSI (**dBm**)	Coverage (m) using IEEE 802.11 path-loss model *D*
11	−82	
5.5	−87	
2	−91	
1	−94	

Problem 10:

The transmitted power in the IEEE 802.11g is 100 mW. When the terminal is close to the access point (AP), the maximum data rate is 54 Mbps that requires a −72 dBm received signal strength (RSS). The minimum supported data rate is 6 Mbps that requires a minimum of −90 dBm of RSS.

(a) Determine the coverage of the AP for 54 and 6 Mbps in a small office using the IEEE 802.11 channel model.
(b) Repeat (a) assuming a single distance-power gradient of a = 2.5 and compare your results with the results of part (a).

Problem 11:

In a mobile communications network, the minimum required SNR is 12 dB. The background noise at the frequency of operation is −115 dBm. If the transmit power is 10 W, transmitter antenna gain is 3 dBi, the receiver antenna gain is 2 dBi, the frequency of operation is 800 MHz, and the base station and mobile antenna heights are 100 and 1.4 m, respectively, determine the maximum in-building penetration loss that is acceptable for a base station with a coverage of 5 km if the following path-loss models are used:

(a) Free space path-loss model with $\alpha = 2$.
(b) Two-ray path-loss model with $\alpha = 4$.
(c) Okumura–Hata model for a small city.

Problem 12:

Signal strength measurements for urban microcells in the San Francisco Bay area in a mixture of low-rise and high-rise buildings indicate that the path loss *Lp* in dB as a function of distance *d* is given by the following linear fits:

$$L_P = \begin{cases} 81.14 + 39.40 \log f_c - 0.09 \log h_b \\ \quad + [15.80 - 5.73 \log h_b] \log d, \ for\ d < d_{bk} \\ \quad [48.38 - 32.1 \log d_{bk}] + 45.7 \log f_c \\ \quad + (25.34 - 13.9 \log d_{bk}) \\ \quad \log h_b + [32.10 + 13.90 \log h_b] \log d \\ \quad + 20 \log (1.6/h_m), \ for\ d > d_{bk} \end{cases}.$$

Here, d is in kilometers, the carrier frequency f_c is in GHz (that can range between 0.9 and 2 GHz), h_b is the height of the base station antenna in meters, and h_m is the height of the mobile terminal antenna from the ground in meters. The *breakpoint* distance d_{bk} is the distance at which two piecewise linear fits to the path-loss model have been developed and it is given by $d_{bk} = 4\ h_b h_m/1000\ \lambda$, where λ is the wavelength in meters (when you calculate d_{bk} using units of meters for h_b, h_m, and λ, the scaling factor of 1000 makes the units of d_{bk} as km). What would be the radius of a cell covered by a base station (height 15 m) operating at 1.9 GHz and transmitting power of 10 mW that employs a directional antenna with a gain 5 dBi? The sensitivity of the mobile receiver is -110 dBm. Assume that $h_m = 1.2$ m. How would you increase the size of the cell?

Problem 13:

A mobile system is to provide 95% successful communication at the fringe of coverage with a location variability having a zero mean Gaussian distribution with a standard deviation of 8 dB. What fade margin is required?

Problem 14:

Sketch the power-delay profile of the following wideband channel. Calculate the excess delay spread, the mean delay, and the rms delay spread of the multipath channel described in Table P3.2. A channel is considered a "wideband" if the inverse of the rms multipath spread is smaller than the data rate of the system. Would the channel be considered a wideband channel for a binary data system at 25 kbps? Why?

Table P3.2 Delay-power profile for Problem 14.

Relative delay in microseconds	Average relative power in dB
0.0	−1.0
0.5	0.0
0.7	−3.0
1.5	−6.0
2.1	−7.0
4.7	−11.0

Problem 15:

The modulation technique used in the existing advanced mobile phone system (AMPS) is analog FM. The transmission bandwidth is 30 kHz per channel and the maximum transmitted power from a mobile user is 3 W.

The acceptable quality of the received SNR is 18 dB and the power of the background noise in the system is –120 dBm. Assuming that the height of the base and mobile station antennas are h_b=100 m and h_m = 3 m, respectively, and the frequency operation of is f = 900 MHz, what is the maximum distance between the mobile station and the base station for an acceptable quality of communication?

(a) Assume free space propagation with transmitter and receiver antenna gains of 2.
(b) Use Hata's equations for Okumura's model in a large city.

Problem 16:

The IEEE 802.11 WLANs operate at a maximum transmission power of 100 mW (20 dBm) using multiple channels with different carrier frequencies. The IEEE 802.11g uses 2.402-2.480 GHz bands and the IEEE 802.11a uses 5.150–5.825 GHz bands. Both standards use OFDM modulation with a bandwidth of 20 MHz.

(a) Calculate received signal strength in dBm at 1 m distance of an IEEE 802.11g access point for the smallest and the largest possible carrier frequencies in the band. Assume that transmitter and receiver antenna gains are 1, and in 1 m distance, signal propagation follows the free-space propagation rules.
(b) Repeat (a) for the IEEE 802.11a WLANs.
(c) Compare the received signal strengths at 1 m distance of the IEEE 802.11g and IEEE 802.11a devices. Use the middle of the allocated band for each standard as the carrier frequency in your calculations.
(d) Compare the rate of the received signal fluctuations (Doppler shift), due to the change in frequency of operation, for the IEEE 802.11g and the IEEE 802.11a. Use the middle of the allocated for each standard as the carrier frequency in your calculations.

Problem 17:

A multipath channel has three paths at 0, 50, and 100 nsec with the relative strengths of 0, –10, and –15 dBm, respectively.

(a) What is the multipath spread of the channel?
(b) Calculate the rms multipath spread of the channel.
(c) What would be the difference between multipath spreads and rms multipath spreads of this three-path channel and a two-path channel formed by the first and the third path of this profile.

Problem 18:

In the 1900 MHz bands, measurements [Bla92] show that the rms delay spread increases with distance. An upper bound on the rms delay spread is given by the equation$\tau = e^{0.065L(d)}$ in ns where $L(d)$ is the mean path loss in dB as a function of distance d between the transmitter and receiver. The path loss itself is given by the equation $L(d) = L_0 + 10\alpha \log_{10}(d/d_0)$ where $L_0 = 38$ dB and $\alpha = 2.2$ for $d < 884$ m and $\alpha = 9.36$ for $d > 884$ m. The standard deviation of shadow fading is 8.6 dB. Assume that you are using a transmission scheme that has a symbol rate of 135 KSps without equalization. If the maximum allowable path loss is 135 dB, what limits the size of the cell – the rms delay spread or the outage at 90% coverage at the cell edge? Explain clearly all of your steps.

Project 1: BLE Signal for COVID-19 Social Distancing

In this project, we use the Bluetooth low energy (BLE) signal for estimating the observation of the social distance protocol of 6-feet for COVID-19. The BLE devices broadcast periodic messages for pairing with other proximity devices with BLE. We can measure the RSSI of the broadcast BLE of a smartphone to measure the distance of the person carrying the smartphone to make sure the distance protocol is observed. In this project, you will design a path-loss model for the BLE and use your model to estimate the distance from another device with BLE.

Data Collection and Modeling of BLE Signal:

(a) Download LightBlue or an alternative APP on your smartphone phone.

(b) Test the APP to see if you can measure the RSSI from BLE devices in your proximity.

(c) Select a fixed BLE enabled device, e.g., a printer, and go to several distances (e.g., seven) and make five measurements in each location and record them in a table (Figure P3.1(a)).

(d) Using LS estimation MATLAB code provided in the class for RSSI linear regressive behavior model, $P_{r-\mathrm{dBm}} = P_{0-\mathrm{dBm}} - 10\alpha \log_{10} d + X(\sigma)$, find the estimate of the RSSI at 1 m, $P_{0-\mathrm{dBm}}$, the distance power gradient, α, and standard deviation of the shadow fading, σ, for the environment the device is operating in.

(e) The IEEE 802.11 model for path loss for open area and short distances recommends ($\alpha = 2\,,\ \sigma = 5$); how do your results compare with that of the IEEE 802.11 model? Discuss reasons for the differences, if any.

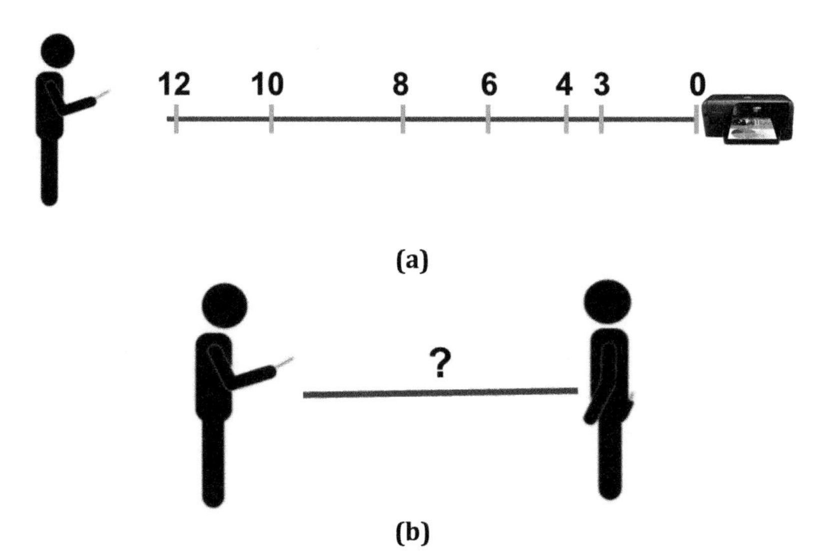

Figure P3.1 COVID-19 test with BLE. (a) Measurement scenario for RSSI behavior modeling. (b) Test scenario for performance analysis.

Testing the Results:

(a) Using the model, measure your distance from someone (Figure P3.1(b)) from $\hat{d} = 10^{\frac{\mathrm{RSSI}-P_{0-\mathrm{dBm}}}{10\alpha}}$ and compare it with the exact distance, d, to calculate the distance measurement error, $\mathrm{DME} = d - \hat{d}$, for your model.

(b) Repeat the experiment for five locations with less than 6-feet and five locations with more than 6-feet distance.

(c) Find the mean and variance of the DME when the distance rule was not observed and when it was observed.

Project 2: Simulation of Multipath Fading

Figure P3.2 shows two mobile robots communicating in a large open indoor area with a ceiling with a height of 5 m and two antennas that are 1.5 m above the ground. Communication between the terminals is taking place through three paths: the direct path, the path reflected from the ground, and the path reflected from the ceiling. The reflection coefficients from the ground and the ceiling are 0.7, and each reflection causes an additional 180° phase shift.

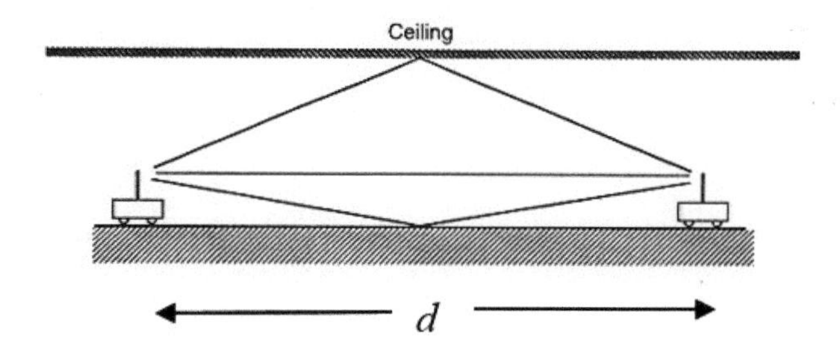

Figure P3.2 An indoor scenario for path-loss modeling using Ray tracing.

(a) If the transmitted power is 1 mW, derive an expression for calculation of P_o, the free space received signal strength in 1 m distance from the transmitter, as a function of the frequency of operation, f.
(b) Derive an expression for calculation of the amplitude, delay, and phase of each of the arriving paths as a function of distance, d, and the frequency of operation, f.
(c) Derive an expression for calculation of the RSS as a function of d and f.
(d) Use MATLAB to plot the RSSI versus distance for $1 \text{ m} < d < 100 \text{ m}$ for center frequencies of 900 MHz, 2.4 GHz, and 5.2 GHz (similar to the plot provided in Figure 3.3).
(e) Discuss the relation between the received signal strength, rate of fluctuations, and frequency of operation.
(f) Use the results for 2.4 GHz to design a two-piece path-loss model for the RSSI by determining a suitable break-point and two distance power gradients in the two regions. Compare your model with the IEEE 802.11 models and discuss your observations.

Project 3: The RSSI in IEEE 802.11

There are several software tools (e.g., WirelessMon by PassMark) that can be used to gather information about access points in close proximity. These tools provide multiple features, but we are going to use it to log the RSS from chosen APs at different locations to check with 802.11 models. The following steps can be used to make an RSSI measurement using these tools.

(a) Install a software tool for measurement of RSSI (e.g., you can download wirelessmon.exe from http://www.passmark.com/products/wirelessmonitor.htm) in your laptop.

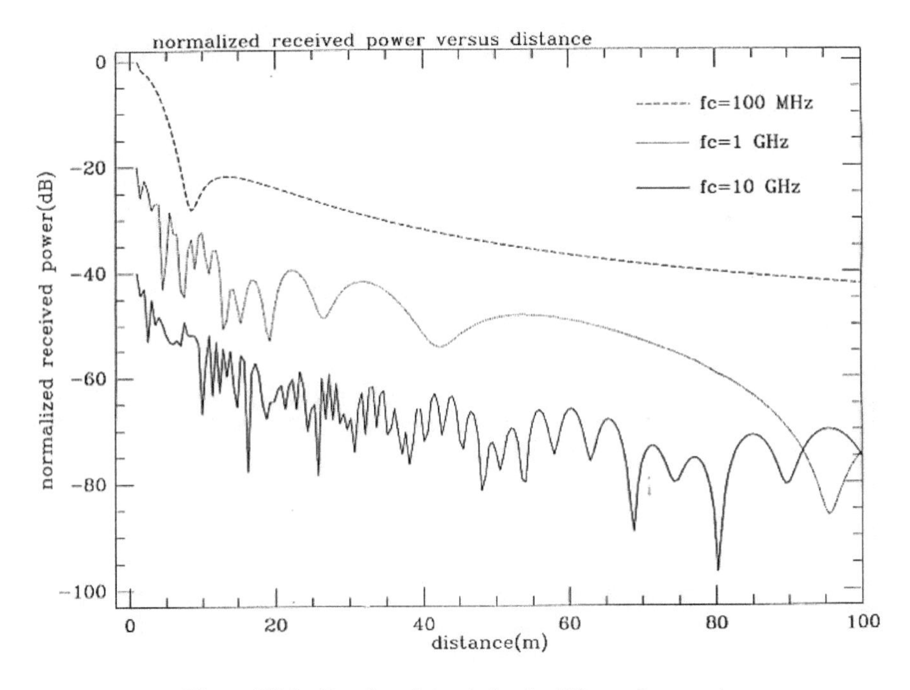

Figure P3.3 Results of simulation in different frequencies.

(b) Set the software to monitor the access point of your choice; the access points can be distinguished from each other by their MAC addresses.

(c) Modify the logging options of the software for recording the characteristics of an AP.

(d) Record the RSSI readings from a specific AP.

(e) Do war driving on a specific floor of your building, for which you have a schematic available, to find the exact location of AP on that floor. Show the locations in the schematics of the building.

(f) Select at five different locations on the floor of your choice which are approximately 1, 5, 10, 20, and 30 m away from your AP of choice. Spread the points over the entire floor and mark them on your schematic floor plan. Determine the distance from the selected points to each of the AP locations on that floor.

(g) Measure RSSI at each location for at least 1 minute. Calculate the average power received from each AP in each location and record them in a table that relates the distance to RSSI from your target AP.

Figure P3.4 Location of the transmitter and first five locations of the receiver used for calculation of the RSSI and path loss.

(h) Use the table to generate the scattered plot of the average RSSI (in dBm) vs. distance in logarithmic form for all APs in your target floor.
(i) Find the best fit 802.11 model for your data in the scattered plot.
(j) Use www.speakeasy.net/speedtest to record the measured data rate in each of the five locations.
(k) Explain the correlation between the throughput from speakeasy and the power and distance in each location.

Project 4: Coverage and Data Rate Performance of the IEEE 802.11b/g WLANs

I. Modeling of the RSS

To develop a model for the coverage of the IEEE 802.11b/g WLANs, a group of undergraduate students at the Worcester Polytechnic Institute (WPI) measured the RSS in six locations in the third floor of the Atwater Kent Laboratory (AKL) at WPI, shown in Figure P3.4. After subtracting the RSSI from the transmitted power recommended by the manufacturers they calculated the path loss for all the points that are shown in Table P3.3.

To develop a model for the coverage of the WLANs, they used the simple distance-power gradient model:

$$L_p = L_0 + 10\alpha \log_{10}(d)$$

in which d is the distance between the transmitter and the receiver, L_p is the path loss between the transmitter and the receiver, L_o is the path loss

Table P3.3 The distance between the transmitter and the receiver and the associated path loss for the experiment.

Distance (m)	Number of Walls	L_p (dB)
3	1	62.7
6.6	2	70
9.5	3	72.75
15	4	82.75
22.5	5	90
28.8	6	93

in 1 m distance from the transmitter, and α is the distance-power gradient. One way to determine L_o and α from the results of measurements is to plot the measured L_p and $\log_{10}(d)$ and find the best fit line to the results of measurements.

(a) Use the results of measurements by students to determine the distance-power gradient, α, and path loss in the 1 m distance from the transmitter, L_o. In your report, provide the MATLAB code and the plot of the results and the best fit curve.

(b) Manufacturers often provide similar measurement tables for typical indoor environments. Table P3.4 shows the RSSI at different distances for open areas (an area without a wall), semi-open areas (typical office areas), and closed areas (harsher indoor environments) provided by Proxim, one of the manufacturers of WLAN products. Use the results of measurements from the manufacturer and repeat part (a) for the three areas used by the manufacturer. Which of the measurement areas used by the manufacturer resembles the third floor of the AKL used by the students? Assume that the transmitted power used for these measurements was 20 dBm. In your report, give the curves used for calculations of the distance-power gradient in different locations.

Table P3.4 Data rate, distance in different areas, and the RSSI for IEEE 802.11b (Source: Proxim).

Data rate (Mbps)	Closed area	Semi-open area (m)	Open area (m)	Signal level (dBm)
11	25	50	160	−82
5.5	35	70	270	−87
2	40	90	400	−91
1	50	115	550	−94

II. Coverage Study

IEEE 802.11b/g WLANs support multiple data rates. As the distance between the transmitter and the receiver increases, the WLAN reduces its data rate to expand its coverage. The IEEE 802.11b/g standards recommend a set of data rates for the WLAN. The first column of Table P2.5 shows the four data rates supported by the IEEE 802.11b standard and the last column represents the required RSSI to support these data rates. Table P3.5 shows the data rates and the RSSI for the IEEE 802.11g provided by Cisco.

Table P3.5 Data rates and the RSSI for the IEEE 802.11g (Source: Cisco).

Data rate (Mbps)	RSSI (dBm)
54	−72
48	−72
36	−73
24	−77
18	−80
12	−82
9	−84
6	−90

(a) Plot the data rate versus coverage (staircase functions) for the IEEE 802.11b WLANs for closed, open, and semi-open areas using Table P2.4 provided by Proxim. Discuss the coverage vs. data rate performance in different areas and relate them to the value of a of different areas, calculated in part I of the project.

(b) Using a and L_0 found for the third floor of the AKL, plot the data rate versus coverage (staircase functions) for IEEE 802.11b and g WLANs operating in that area. Discuss the differences in data rate vs. coverage performance of the 802.11b and g in the third floor of the AKL.

4

Physical Layer Communications

4.1 Information Transmission

In Chapter 3, we described the behavior of different wired and wireless media. To transfer the information over a medium, we need information transmission techniques. Several analog and digital information transmission techniques have evolved in the past century. Fundamentally, these techniques are either used for transmitting a waveform over the transmission medium or to process the information so that it uses the transmission facilities efficiently. In this chapter, we provide an overview of transmission techniques that gradually moved into implementations of popular physical (PHY) layers in modern information networks. In Chapter 4, we present a summary of coding techniques and protocols used for reliable packet transmission. The material presented in these chapters prepares the reader for an understanding of the reasons behind the development of wide-area network (WAN), local area network (LAN), and personal area network (PAN) information networking alternatives in modern times. The presented material is carefully selected to avoid complicated signal processing details which need training in electrical engineering. The audience of the book is assumed to be computer engineering and science students. However, electrical engineering students may have an easier time grasping this material because they may have studied similar material in other courses.

Figure 4.1 shows a general block diagram for different components involved in information transmission. The information source is first encoded into digital information or a bitstream which then passes through a channel coding process that increases its integrity and protects the data against disturbances caused by the transmission channel. The encoded data stream is then mapped into a set of analog waveforms $s_i(t)$ each representing a set of transmitted information bits. Symbols are transmitted every T_s seconds referred to as the symbol duration or symbol interval. The waveform

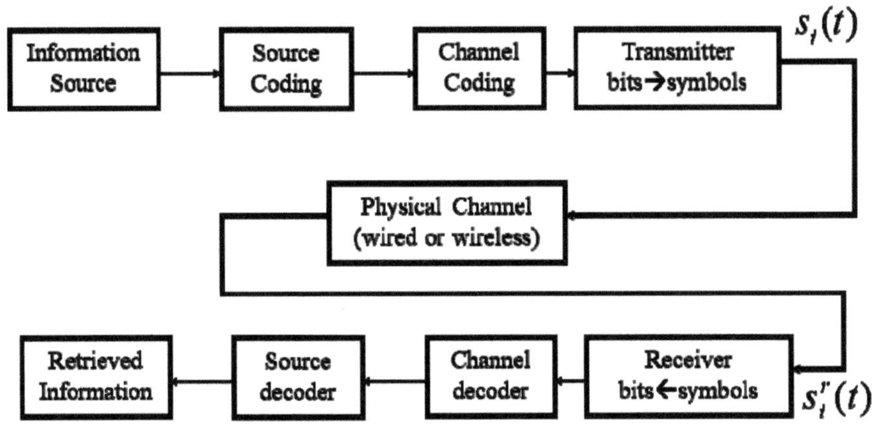

Figure 4.1 Overview of transmission and reception of digital data in the PHY layer.

representing the symbols passes through the channel which may make changes to the shape of the signal and disturb it with additive noise. The received symbol is then processed at the receiver which extracts the best-processed estimate of the transmitted bits associated with the received symbol which are then decoded to retrieve the transmitted information. We have structured this chapter to address bit transmission techniques. In networking, we use packets or frames for transmission.

In the rest of this section, we discuss issues related to wired and wireless transmission followed by introducing the simplest implementation of the physical layer using baseband transmissions. Section 4.2 describes multisymbol transmission and signal constellations used to express the details of more complex transmission techniques. Section 4.3 is devoted to the performance analysis of the PHY layer. In this section, we describe the effects of noise and fading on the performance of modems and we introduce the concept of diversity techniques and their impact on the performance of wireless networks. Section 4.4 describes the effects of multipath fading and introduces the diversity techniques fundamentals to remedy the harmful effects of multipath in restricting the data rate. Section 4.5 provides an overview of the modern techniques used for wideband or high data rate wireless networks. Direct sequence and frequency-hopping spread spectrum (DSSS/FHSS), orthogonal frequency division multiplexing (OFDM), space–time coding (STC), and multiple-input multiple-output (MIMO) transmission techniques are described in this section.

4.2 Fundamentals of Transmission and Signal Constellation

In principle, transmission techniques used in information networks are applicable to all wired and wireless modems because the basic design issues are common to both systems. In general, we would like to transmit data with the highest achievable data rate with a minimum expenditure of signal power, channel bandwidth, and transmitter and receiver complexity. In other words, we usually want to maximize both bandwidth efficiency and power efficiency and minimize the transmission system complexity. However, the emphasis on these three objectives varies according to the application requirement and medium for transmission, and certain details are specific to applications and transmission media. Also, these design objectives are often conflicting, and the tradeoffs decide what factors are considered more important than others.

We start our technical discussions on transmission techniques with *baseband transmission* using line coding techniques. In baseband transmission, the digital signal is transmitted without modulating it over a carrier at a higher frequency. In line-coded transmission, the digital data stream is line-coded to facilitate synchronization at the receiver and avoid the DC offset during transmission. Baseband line coded signaling is commonly used in short distance wired as well as wireless applications. In wired applications, baseband signaling using differential Manchester line coding is used in the IEEE 802.3 Ethernet, the dominant standard for LANs, as well as IEEE 802.5 Token Ring, the competitor of Ethernet in the early days of the LAN industry. Baseband transmission is also used in long-haul optical communications in the synchronous optical network (SONET) and Ethernet optical hierarchies. In wireless communications, baseband transmission with line coding is popular in high speed diffused and directed beam infrared (IR) wireless LANs as well as ultra-wideband (UWB) and millimeter wave (mmWave) pulse transmission communications systems. We discuss the binary baseband transmission first to provide a clear understanding of the transmission issues through simple examples.

4.2.1 Line Coding and Binary Baseband Transmission

If the data stream produced by a computer is applied directly to the wires, the receivers will have difficulty in synchronizing with the transmitted symbols. To provide better synchronization at the receiver, the format of the incoming data stream is modified before transmission. This modification process is often referred to as line coding. We now provide more details of several popular line coding techniques with a discussion on why they have evolved and examples of how they work.

Figure 4.2 shows examples of popular line coding techniques. The non-return to zero (NRZ) line coding, shown in Figure 4.2(a), just uses two different amplitudes for the two different binary digits. On wired links, these two amplitudes are selected at the same level with opposite polarities. With this setup, the average transmitted amplitude of the waveform is zero providing a zero DC component for the transmission. It can be shown that this form of transmitted symbols provides the best performance with fixed transmission power. In optical communications, however, we have light emitting diodes (LEDs) which cannot implement polarity. Usually, the light is either on or off. Hence, the transmission does not have a zero average or DC value in this case. In the case of NRZ-I, shown in Figure 4.2(b), in the computer communication community, the letter "I" stands for inverted. Here the term "inverted" does not mean that the amplitude is inverted, it means that the information bit is encoded at the edge of the bit. In Figure 4.2(b), whenever we have a "0," we have no transition at the edge, and when we have a "1," we have a transition at the edge. This arrangement in the telecommunication literature is referred to as differential coding. The advantage of this type of coding is that we can afford 180° phase ambiguity, which means if the polarization of the received data is the reverse of the transmitted data, still coding works. In baseband communications, receiver digital circuitry is often edge triggered which better suits this type of coding.

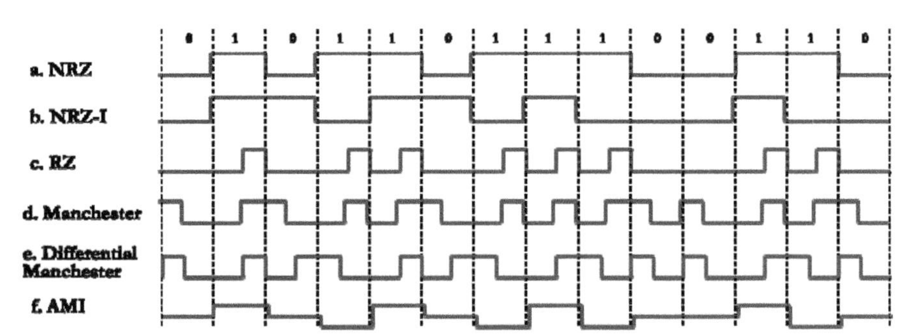

Figure 4.2 Examples of popular line coding techniques. (a) Non-return to zero (NRZ). (b) NRZ inverse. (c) Return to zero (RZ). (d) Manchester coding. (e) Differential Manchester coding used in 802.3 Ethernet. (f) Three-level alternate mark inversion (AMI).

NRZ in Figure 4.2(a) means that the signal amplitude does stay at the same level throughout the transmission of a bit. In contrast, with a return to zero (RZ) in Figure 4.2(c), during transmission of the symbol "1," in the middle of the transmission of a bit, the amplitude is changed. The transition in the middle of the bit in RZ coding assures that if we have a string of "1"s, even then, for every symbol, we have a transition enabling clock recovery. In general, the receiver synchronizes based on the transitions in the received signal; more transitions provide better synchronization at the receiver. The advantage of RZ over NRZ is these additional transitions. The disadvantage of the RZ approach is that the transmitted pulses are two times narrower consuming twice the bandwidth of NRZ coding.

Figure 4.2(d) shows Manchester coded information bits. In this coding technique, the digit "0" is represented by a pulse starting at a high level that switches to the low level in the middle of the transmission of the bit. The symbol for "1" is the inverse of "0" which starts low and ends at the high amplitude. This code has the basic concepts of the RZ transmissions with the additional advantage that it has a transition for both symbols, while with RZ, the transitions were only used on the digit "1." In optical communications, where we turn an LED on and off to transfer two digits for binary communications, RZ can be easily implemented by reducing the on-time of the LED to a half. This will also reduce the power consumption of the signal to a half, which is very useful for applications such as remote control where we want to have very long battery life. In remote controls for devices like TVs, to save even more in terms of the transmission power, rather than keeping the LED on for half of the time, we may turn it on and off several times [Pah95]. The edge triggered variation of the Manchester code, called differential Manchester coding, is shown in Figure 4.2(e). In this code, we always have a transition in the middle of the bit. However, if we have a "0," we have a transition at the start of the bit, and if we have "1," there is no transition at the beginning of the bit. Differential Manchester coding was the first physical layer transmission techniques recommended by the IEEE 802.3 Ethernet standardization committee.

4.2.2 Multi-Level Transmission and Signal Constellation

Figure 4.2 also shows a three-level line coding technique (Figure 4.2(f)) called alternate mark inversion (AMI), in which we transmit the "1" symbols as a pulse with alternating polarities. In a manner similar to RZ, this approach provides for better synchronization, but unlike RZ, it does not double the bandwidth of the system. In this section, we only referred to some basic example and the issues which are involved in the design of line coding techniques. AMI was used for transmission in first-generation pulse code modulation (PCM) signals in core public switched telephone network (PSTN) and the reason was that it was not producing much of DC offset at the receiver helping signal reach longer distances. The difference between AMI and other line coding techniques in Figure 4.2 is that all other methods use two voltage levels, but AMI has three. Three levels do not suit optical transmission because optical lines simply switch the transmitter light on and off for signaling. In copper wires, we can use any binary voltage level for communications; besides, we can increase the number of levels and so send more bits per symbol to increase the data rate. We refer to these transmission methods as multi-level transmission systems.

Figure 4.3 (the same as Figure 3.1b) shows an example of multi-level transmission line coding method with four-level or symbol transmission. Symbols are transmitted every T_s second which we may refer to as the symbol time. Each symbol carries 2 bits of encoded data. The stream of data is packed into blocks of length 2 bits to transmit with one of the four symbols. Since the shape of the waveforms is the same and only their amplitude is different in digital transmission literature, we refer to this communication technique as *pulse amplitude modulation* (PAM). The four possible pulses in Figure 4.3, referred to as 4-PAM, take$(\pm 1\,,\,\pm 3)$ volts amplitudes with the same basic pulse shape as $s_1(t)$. The energy of a symbol is proportional to the square of its amplitude since amplitude levels for the transmitted pulses are different, and the energy that each symbol carries is different. The average energy of transmitted symbols divided by the duration of the pulse is the average transmitted power of the transmitter. If we use voltage-level zero as a level with four pulses$(0, \pm 1\,,\,\pm 2)$, we have five symbols called 5-PAM. 5-PAM is more efficient than 4-PAM because it sends one extra symbol that reduces the average transmission power. Now two questions come to mind; first, how much power do we save if we use 5-PAM and how should we map the data stream into five levels?

A popular graphical method to illustrate the relative energy of the symbols used in digital communication standards is to employ a "signal constellation." The signal constellation is a way of showing the signals as points in space rather than referring to the actual time-dependent waveform representing symbols, shown in Figure 4.3 for 4-PAM. In the signal constellation representation of a transmission method, we use the square root of the energy of the signal as a representation of its amplitude, as shown in Figure 4.4.

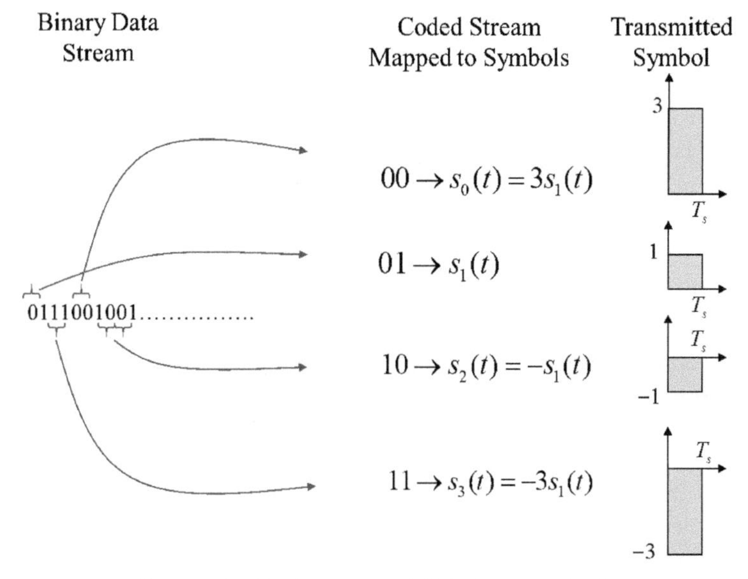

Figure 4.3 Basic concept of digital communication.

Figure 4.5 illustrates the signal constellation for 5-PAM. This way, the transmitted symbols in the case of a PAM-based digital communication system are illustrated in a one-dimensional graph. The average energy per symbol is the average energy of the constellation which is simply the average of square of the distance of the points from the center of the coordinate system. The signal constellation also illustrates the line coding approach to map the data stream to symbols of the constellation and the probability of transmission of each symbol. A signal constellation reflects all important parameters associated with a digital transmission technique and a standard organization uses it to define characteristics of complex transmission systems.

4.2.3 Transmission Parameters and Signal Constellation

The signal constellation defines the characteristics of the transmitted information symbols. At the receiver, these symbols arrive in additive noise. Table 4.1 provides a summary of important parameters that are used in digital communications to relate the transmission to signal-to-noise ratio (SNR) of a transmission technique. The first set of parameters is related to the transmission rates and their relation to the number of bits and the number of symbols. The data transmission rate, R_b, is the coded information bit rate of the link in bits per second (bps). The symbol transmission rate, $R_s = 1/T_s$, is the symbol transmission rate in symbols per second (sps). If we have M symbols for digital communications in the constellation and each symbol is transmitted with the probability of p_i and it carries m_i bits, then the average number of bits per symbol is given by

$$m = \sum_{i=0}^{M-1} p_i \times m_i. \tag{4.1}$$

With an average of $m$bits per transmitted symbol, the bit rate$R_b = m \times R_s$.

The second set of parameters is related to the energy per symbol and the transmission power. The transmitted energy per symbol is given by

$$E_{s_i} = \int |s_i(t)|^2 \, dt \tag{4.2}$$

and the average transmitted energy in Joules is

$$E_s = \sum_{i=0}^{M-1} p_i E_{s_i}, \tag{4.3}$$

where p_i is the probability of transmission of a symbol, specified in the constellation. For equally likely symbols (symbols transmitted with equal probability) of $p_i = 1/M$ and $M = 2^m$. The transmitted power is related to average to the average symbol transmission and symbol transmission time by $P = E_s/T_s$ in Watts.

In digital communication literature, we also define the average energy per bit is given by

$$E_b = \frac{E_s}{m}, \tag{4.4}$$

which is an abstract value allowing comparison of energy allocated to every transmitted bit and make it easy to compare two constellations. If the variance of the received background noise is denoted by N_0, the received SNR per symbol and per bit will be given by

$$\frac{E_s}{N_0} \text{ and } \frac{E_b}{N_0}, \tag{4.5}$$

respectively. The received SNR per symbol is representative of the physical received signal power compared with the background noise power.

Table 4.1 Summary of important parameters used in digital communications.

User bit rate	R_u
Coding rate	$R<1$
Coded user bit rate	$R_b = R_u / R$
Symbol transmission rate	R_s
Number of symbols	M
Average number of bits per symbol	$m = \sum_{i=0}^{M} p_i \times m_i$
Energy per symbol	$E_{s_i} = \int \lvert s_i(t) \rvert^2 \, dt$
Average energyper symbol	$E_s = \sum_{i=0}^{M} p_i E_{s_i}$
Energy per bit	$E_b = E_s / m$
Variance of the noise	N_0
Signal to noise ratio	E_s / N_0
Signal to noise ratio per bit	E_b / N_0

The received symbols are corrupted by noise and if we map them to the constellation, they will be typically in a location close to the transmitted symbol. On rare occasions, the received point in the constellation gets closer to another symbol causing the receiver to make an error in detection. The probability of the symbol error with Gaussian distributed noise can be approximated by [Pah05]

$$P_s \approx \frac{1}{2}\text{erfc}\left(\frac{d}{2\sqrt{N_0}}\right) \geq \frac{1}{2}e^{-\frac{d}{2\sqrt{N_0}}}, \tag{4.6}$$

where d is the distance between neighboring symbols in the constellation.

We have defined several parameters and relationships among them; examples will make these concepts much clearer.

Example 4.1: Signal Constellation and Energy Per Symbol for 4-PAM:

Figure 4.4 shows the 4-PAM signal constellation for the digital communication system described in Figure 4.3 for which we have

$$\begin{cases} M = 4, \; m = 2, \\ E_s = 2E_b \end{cases}$$
$$p_0 = p_1 = p_2 = p_3 = \tfrac{1}{4}.$$

Then, the relation between symbol rate and user data rate is $R_b(\text{bps}) = 2R_s(\text{sps})$.

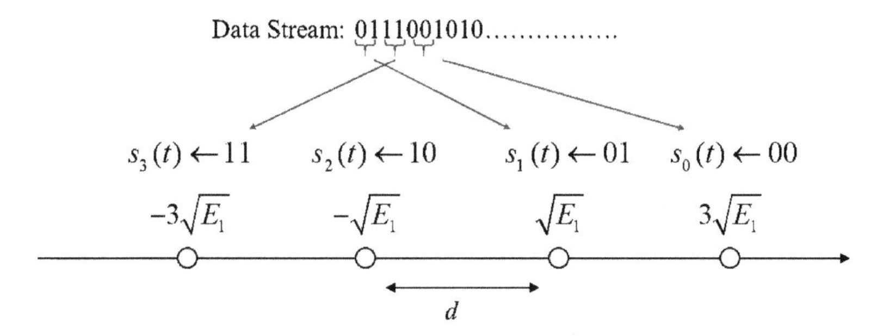

Figure 4.4 The signal constellation for four symbols (PAM).

The average energy per symbol is

$$E_s(\text{joules}) = \sum_{i=0}^{3} \frac{1}{4} E_i = \frac{9+1+1+9}{4} E_1 = 5E_1$$

since, for this constellation,

$$d = 2\sqrt{E_1} = \sqrt{\frac{4E_s}{5}},$$

the average energy pe symbol is

$$E_s = 2E_b = 1.25 \times d^2.$$

Substituting in Equation (4.6), the symbol error rate is

$$P_s \approx \frac{1}{2}\text{erfc}\sqrt{\frac{4E_s}{5N_0}} = \frac{1}{2}\text{erfc}\sqrt{\frac{4 \times 2E_b}{5N_0}}.$$

Example 4.2: Energy in a 5-PAM System:
Figure 4.5 shows a signal constellation for the 5-PAM digital communication system. In this system, the first 2 bits of the stream of data are checked and if we have 00, 01, or 10, we map them to their associated symbols in the middle of the constellation. If we have a 11, however, we wait for the next bit to form a block of 3 bits. The two possible 3 bits patterns of 110 and 111 are then mapped to the corner points of the constellation. For this constellation, $M = 5$, and

$$p_0 = p_4 = \tfrac{1}{8};\;\; m_0 = m_4 = 3\,(\text{bits})$$
$$p_1 = p_2 = p_3 = \tfrac{1}{4};\;\; m_1 = m_2 = m_3 = 2\,(\text{bits}).$$

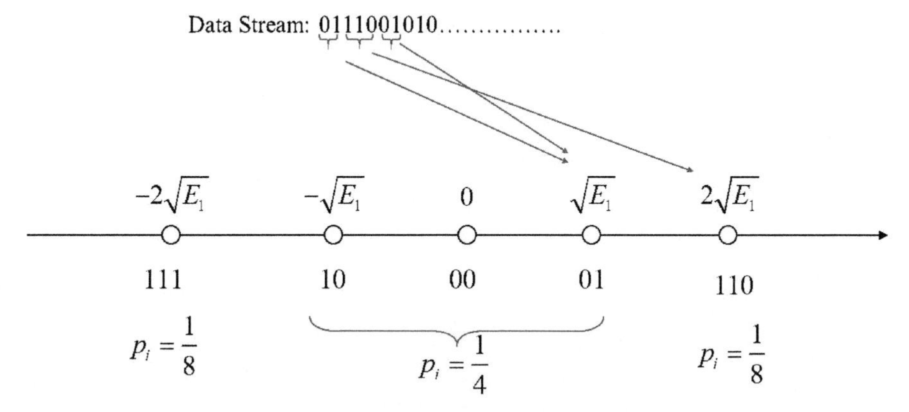

Figure 4.5 The signal constellation for 5-PAM.

Therefore,

$$\begin{cases} m = \sum\limits_{i=0}^{4} p_i \times m_i = 2 \times \frac{1}{8} \times 3\,(\text{bits}) + 3 \times \frac{1}{4} \times 2\,(\text{bits}) = 2.25\,(\text{bits}) \\ E_s = 2.25 E_b \end{cases}$$

and $R_b = 2.25 R_s$ which indicates that, with the same duration of pulses (the same bandwidth), we have a system that provides a higher data rate than the 4-PAM system. Following the same pattern as Example 4.1, we can find the average energy per symbol as well as the probability of symbol error

$$E_s = \sum_{i=0}^{5} p_i E_i = \tfrac{1}{8} \times 4E_1 + \tfrac{1}{4} \times E_1 + \tfrac{1}{4} \times 0 + \tfrac{1}{4} \times E_1 + \tfrac{1}{8} \times 4E_1 = \tfrac{3}{2} E_1$$
$$d = \sqrt{E_1} = \sqrt{\tfrac{2E_s}{3}} \Rightarrow E_s = 1.5d^2$$
$$P_s \approx \tfrac{1}{2}\text{erfc}\sqrt{\tfrac{2E_s}{3N_0}} = \tfrac{1}{2}\text{erfc}\sqrt{\tfrac{2\times 2.25E_b}{3N_0}}.$$

Therefore, for the same distance (the same symbol error rate), 5-PAM needs $1.5/1.25 = 1.2$ times transmitted energy per symbol (the same as the power). Power is usually measured in dB in which we need $10\log(1.2) \simeq 0.8\,(\text{dB})$ $10\log(1.2) \simeq 0.8\,(\text{dB})$ more power to support 2.25 times higher data rate.

The 5-PAM constellation can be implemented for wired baseband communications, but it is not suitable for wireless communications because the symbol represented by zero energy cannot be easily created in unreliable noisy wireless medium.

4.2.4 Multi-Dimensional Signal Constellations

Multi-dimensional constellations are widely used in transmission techniques defined by the standard organizations for access to the PSTN and the Internet. Historically, they were first employed in the early 1970s in voiceband data communications to achieve 9600 bps transmission over 4-wire telephone lines [Pah88], and, later, they became popular in high-speed Ethernet, DSL modems, cable modems, wireless LANs, and high data rate cellular networks to become the dominant data transmission method for wired and wireless communications. Multi-dimensional signal transmission basically defines symbols in a multi-dimensional space. Each dimension carries a stream of symbols, each containing a few bits of information. These streams of data bits are then carried over different communication lines.

In this section, we study the two-dimensional signal constellations for digital communications, the most popular and widely used constellations. Two-dimensional communication needs two lines. These lines can be wires or wireless, and the data can be transmitted as baseband or carrier modulated signals. Carrier modulated signals can carry the information in their magnitude and phase, creating a natural two-dimensional system over one carrier. In 2D constellations, we define symbols in two dimensions and all the practical parameters for the constellation remain the same as those defined in Equations (4.1)–(4.6) defined in the previous section. In a 2D constellation, the energy of each symbol remains as the square of its distance from the origin. Again, we start our discussions with an example to clarify the concept.

Example 4.3: Two-Dimensional 16-QAM Constellation:

Assume that we have two independent 4-PAM transmission systems, each having its own four symbols and associated bits. We can create blocks of 4 bits and map each of 2 bits to one of the two 4-PAM systems. If we define a two-dimensional signal constellation for this system, we can represent the constellation as shown in Figure 4.6. In the literature square, 2D constellations are referred to as quadrature amplitude modulation (QAM), and the name of this specific QAM is 16-QAM. The transmission parameters of 16-QAM constellation are

$$M = 16,\ m = 4, \quad \Rightarrow E_s = 4E_b \text{ and } R_b = 4R_s$$
$$p_i = \tfrac{1}{16}$$

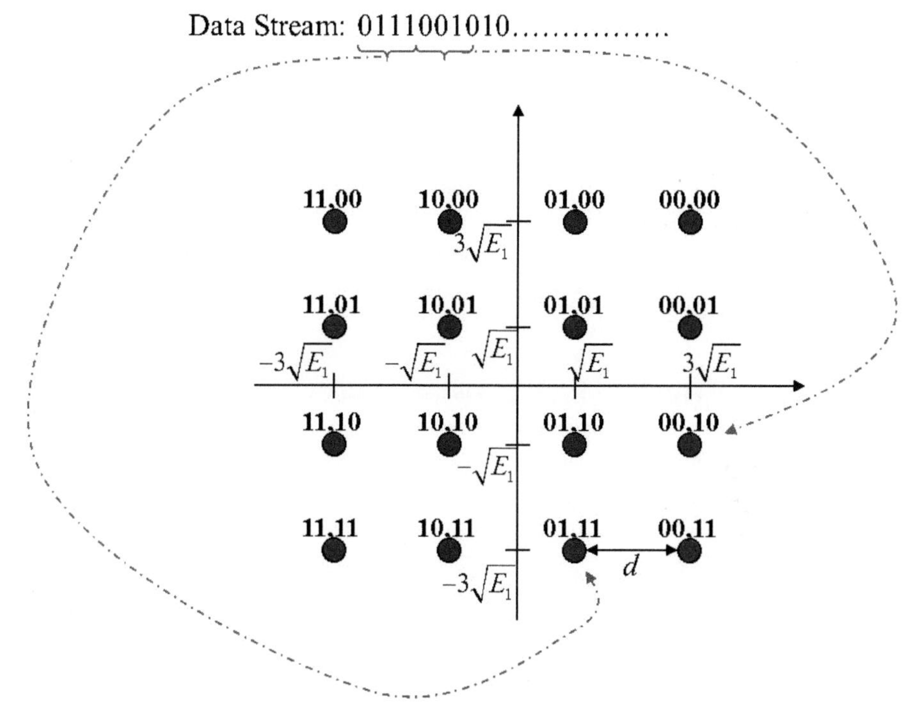

Figure 4.6 The two-dimensional signal constellation associated with the one-dimensional constellation shown in Figure 4.4.

where

$$E_s = \sum_{i=0}^{15} \frac{1}{16} E_i = \frac{4 \times 18 + 8 \times 10 + 4 \times 2}{16} E_1 = 10E_1.$$

Since

$$d = 2\sqrt{E_1} = \sqrt{\frac{2E_s}{5}} \Rightarrow E_s = 2.5 \times d^2,$$

the probability of error is calculated from

$$P_s \approx \frac{1}{2} \text{erfc} \sqrt{\frac{2E_s}{5N_0}}.$$

This constellation is called a QAM constellation because it uses two independent or orthogonal PAM systems. In this case, it is 16-QAM with 16 signal points. QAM is the choice of IEEE 802.11 standard and 4G/5G cellular networks. The transmission systems in these standards are multi-rate using different sizes of QAM constellations carrying 1–8 bits per symbol. We discuss these methods when we introduce these standards in Chapter 7.

Another important constellation very popular in the wired transmission standards is the 5×5PAM used in several Ethernet transmission standards. Figure 4.7 shows the details of this signal constellation. 5×5PAM is indeed the 2D version of the 1D 5-PAM (Figure 4.5). 5×5PAM, shown in Figure 4.7, combines two 5-PAM constellations with the same coding method described shown in Figure 4.5. Following our tradition in this section, we continue our explanation with an example.

Example 4.4: The 5×5 PAM Constellation in Ethernet:
In the calculation of transmission parameters of 5×5PAM constellations, we should remember that, in this constellation, similar to 5-PAM, transmitted symbols are not sent with equal probability. The $M = 25$ symbols are divided into three classes, 9-inner, 12-side, and 4 in the corners, each carrying a

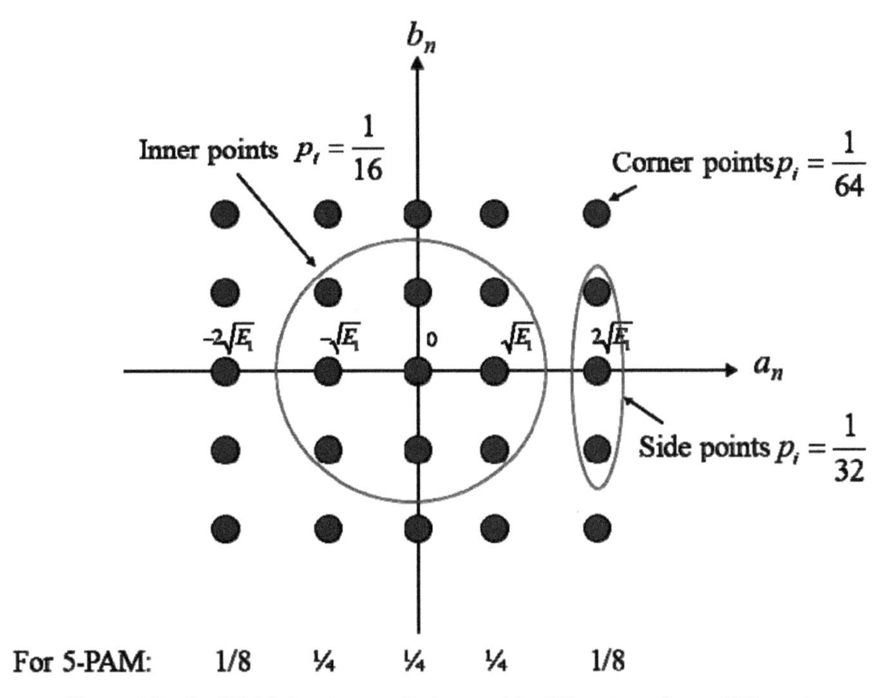

Figure 4.7 5×5PAM signal constellation used in different versions of Ethernet.

different number of bits with a different probability of transmission.

$$\begin{array}{l}\text{For inner points}: \ p_i = 1/16, \ m_i = 4(\text{bits}) \\ \text{For side points}: \ p_i = 1/32, \ m_i = 5(\text{bits}) \\ \text{For corner points}: \ p_i = 1/54\,, \ m_i = 6(\text{bits}).\end{array}$$

Therefore,

$$\begin{array}{l}m = \sum_{i=0}^{24} p_i m_i = 9 \times \frac{1}{16} \times 4 + 12 \times \frac{1}{32} \times 5 + 4 \times \frac{1}{64} \times 6 = 4.5(\text{bits}) \\ R_b = 4.5R_s, \ E_b = \frac{E_s}{4.5}\end{array}$$

$$\begin{aligned}E_s = \sum_{i=0}^{24} p_i E_i &= 0 + 4 \times \tfrac{1}{16} \times E_1 + 4 \times \tfrac{1}{16} \times 2E_1 + \\ &\quad 4 \times \tfrac{1}{32} \times 4E_1 + 8 \times \tfrac{1}{32} \times 5E_1 + 4 \times \tfrac{1}{64} \times 8E_1 = 3E_1\end{aligned}$$

$$\begin{array}{l}d = \sqrt{E_1} = \sqrt{\frac{E_s}{3}} \\ E_s = 3 \times d^2\end{array}$$

Then,

$$P_s \approx \frac{1}{2}\text{erfc}\sqrt{\frac{E_s}{3N_0}} = \frac{1}{2}\text{erfc}\sqrt{\frac{2.25E_b}{3N_0}}.$$

To map the binary data stream to this constellation, one may treat the stream as two sequential streams of 5-PAM encoded data using the encoding technique described in Example 3.4. Similar to that example, if the first 2 bits in the stream are 00, 01, or 10, they are assigned to a_n values representing the three middle horizontal coordinates, and if it is 11, the next bit is read to make the a_n assignment. After completion of the horizontal bit assignment, the next 2 or 3 bits in the stream are read and mapped with the same rules to determine the appropriate b_n value to identify the point in the constellation. Comparing this with just the 5-PAM transmission scheme, this constellation has two times higher data rates at the expense of $10\log(3/2) = 1.76\,\text{dB}$ more power. Like the 5-PAM scheme, this constellation is only good for baseband data transmission that is restricted to wired transmission and we need two sets of wires to implement it. If we were using two independent 5-PAMs over the two sets of wires, then we would have to double the transmission power (3 dB more power). In that regard, with the same data rate and the same number of wires, 5×5 PAM provides a $3 - 1.76 = 1.24(\text{dB})$ edge that results from correlating the two streams of symbols over the two sets of wires.

4.2.5 Effects of Coding on Data Rate

Digital communications rely on modulation and coding techniques. We described popular modulation techniques, PAM and QAM. In Sections 4.2.2 and 4.2.4 and in Section 4.2.1, we described line coding techniques applied to baseband transmission. Another class of popular coding techniques is error correction coding methods applied to the packets arriving at the physical layer of the communication systems. When the data stream from higher layers arrives at the PHY, these error correcting codes add a certain number of parity bits to the transmitted short blocks of information to create a coded stream. The codes usually are identified with their type and rate. We have error correcting block codes such as Hamming codes and we have convolutional codes at a different rates. The rate of a code is the ratio of the data bit over the coded data bits with parities. For example, a Hamming code may add three parities to every 4 bit to create a 7-bit block code. The code is then referred to as a 4/7 bit, meaning every 7 bit of the coded stream carries 4 bits of data. This code is referred to as (7,4) hamming code and it can detect 1-bit error in the 7-bit coded received signal. Convolutional codes are state machines converting a multiple stream of data to a higher level stream of data. For example, a 2/3 convolutional code is a state machine with two inputs and three output streams. The coded data stream includes the parity as well as data, and it needs a higher rate medium for transmission. If we represent the coded rate of the data by R_b and the coding rate by R, the user data rate before coding is $R_u = R_b/R$. An example will further classify this concept.

Example 4.5: IEEE 802.11 54 Mbps Data Stream with 64-QAM: Figure 4.8 shows the 64-QAM signal constellation used in IEEE 802.11a,g,n,ac for 54 Mbps data transmission. This constellation has 6 bits per symbol and the symbol transmission rate is $R_s = 12\,\mathrm{Msps}$. Therefore, the coded bit rate is $R_b = 6(\mathrm{bps}) \times 12(\mathrm{KSps}) = 72(\mathrm{Mbps})$. However, the transmitted data is encoded by a rate $R = 3/4$ convolution codes which turn the effective user data rate to $R_u = 72(\mathrm{Mbps}) \times \frac{3}{4} = 54(\mathrm{Mbps})$. These convolution codes add an equivalent of 1 bit to every 3 bits of a data stream to increase its integrity and provide for error corrections. These codes are very popular in all wireless networks.

Another class of popular code in communication networks is error detecting codes. These are a set of parity bits that we add to a block of information based on certain rules. At the receiver, the detect block of data passes through the same roles and generates the parities again. If the parity

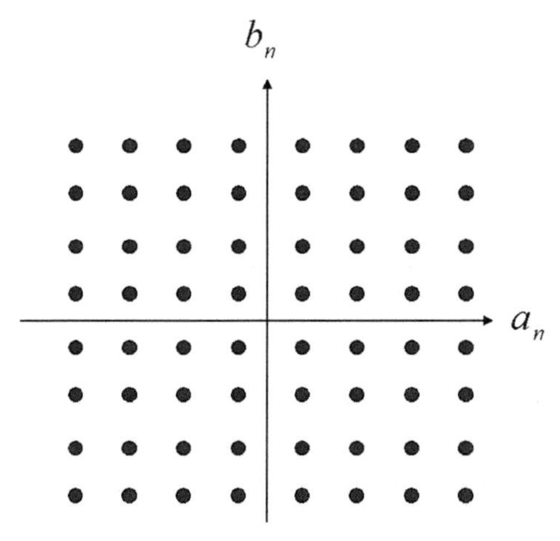

Figure 4.8 QAM constellations for 4 and 6 bits per symbol.

codes generated at the receiver are not the same as the transmitted parities, the receiver knows that there is an error in packet and can ask for retransmission of the packet. The most popular parity codes in IEEE 802 community are the cyclic redundancy check (CRC) codes. These codes are commonly used to create the checksum for variable length and long blocks of data. These codes can be implemented with small overhead and simple hardware. The basic idea of a CRC code is that it interprets the long block of data as a binary number, divides it by a prime number of length $n + 1$ bits, and uses the n-bit remainder as the checksum for the coding. At the receiver, the main data is divided by the same prime number, and the remainder is checked to detect the occurrence of an error in the transmission of the block. As an example, the IEEE 802.3 Ethernet uses CRC codes of length 33 bits with a 32-bit remainder as a checksum for its long and variable length packets. The Ethernet packets carry variable length data of up to 1500 bytes and overhead of up to 22 bytes. This long binary number is divided by a 33-bit prime number to form a CRC-32 and the remainder is used as the checksum. There are simple hardware implementations that can efficiently implement this division. Here, to show the basic concept of CRC codes, we give a simple example with a few bits of data.

Example 4.6: CRC Parity Code Generation:

In this example, we describe CRC codes using the prime number 11, represented by binary code [1011]. The checksum is the remainder after dividing the original block of bits by this prime number. Assume that our information bit stream is

$$I = \begin{bmatrix} 1 & 0 & 0 & 1 & 1 & 0 & 1 \end{bmatrix}.$$

The parity for this information is generated by division of [1001101000] by the prime number [1011] as shown in Figure 4.9(a). The resulting remainder checksum is [101] and is then added to the transmitted information to form the coded transmitted symbol:

$$C = [1001101101].$$

As shown in Figure 4.9(b), at the receiver, the received coded word is divided by [1011] and the remainder should be zero. If the remainder is not zero, it shows that packet is erroneous.

Depending on the length of the packet, the parity codes will affect the user application data rate. For the IEEE 802.3, this overhead is very small because 4-byte CRC code adds to a packet with another 22 bytes of overhead plus the packet length that can go up to 1500 bytes. In this section, we introduced examples of applied codes to communications and networking and how they impact the user data rate. In the following chapter, when we introduce details of standards, we will have more discussions on details of practical codes.

```
           1010011                           1010011
1011)1001101000                    1011)1001101000
     1011                               1011
      1010                               1010
      1011                               1011
        1100                               1100
        1011                               1011
         1110                               1110
         1011                               1011
          101                                101
          (a)                                (b)
```

Figure 4.9 Example of CRC code parity. (a) Generation at transmitter. (b) Check at the receiver.

4.3 Performance Analysis Using Signal Constellation

In the digital transmission of binary information over a medium, we have four fundamental characteristics, bandwidth of the medium, R_s, the received SNR, E_s/N_0, the error rate, P_s, and the speed of the data, R_b. The bandwidth in the wired medium is governed by the type of the media and in radio frequency (RF) signals with FCC regulations. The received SNR is a measure of the transmitted power, which is a design parameter, against the fixed background noise that is produced by the device or the environment. Error rate demonstrates the quality or performance of the transmission system. Digital communication literature provides detailed methods for calculation and plotting relation between error rate versus SNR, like those given by Equation (4.6). These plots are used for comparative performance evaluation of different transmission techniques. Another approach to analyzing the performance of the transmission systems is to resort to the calculation of the Shannon–Hartley bounds on performance relating the data rate to the SNR and the bandwidth. These bounds show the maximum number of bits that a symbol can carry over a given bandwidth with a given SNR. In information theory, these bounds are referred to as channel capacity. In the remainder of this section, we demonstrate how we use these tools in practice to analyze the performance of PHY of communications networks and how we can calculate the channel capacity.

4.3.1 Plots of Error Rate Versus Signal-to-Noise Ratio

In a wired or wireless transmission scheme, the transmitted symbols are corrupted by additive background noise. In digital communications, the additive noise causes erroneous decisions in detecting the transmitted symbols at the receiver. In digital communication literature, to demonstrate the performance of these transmission techniques, we often plot the symbol transmission error rate P_s in logarithmic form against the SNR per bit $E_b/N_0 = (E_s/N_0)/m$ in dB. These plots allow us to determine the required level of the transmitted power to achieve a certain error rate, which is a subjective criterion imposed by the application. For example, in digital voice transmission, we may accept error rates on the order of 10^{-2} (one out of hundred bits), while for voiceband data communications, we expect error rates on the order of 10^{-5}, and for applications in wired LANs, one may think of error rates lower than10^{-8}. SNR is a measure of how much received power is required to detect information in a signal correctly with a given probability.

In addition, during the design process, we use these plots to compare different alternatives for data transmission methods. We resort to an example to clarify this discussion.

Example 4.7: Performance of IEEE802.11 Transmission Methods: Most recent IEEE 802.11 popular standards, such as a, g, n, and ac, use a variety of QAM constellations. In this example, we compare the performance of 4-, 16-, 64-, and 256-QAM. A constellation of these modems are square, and they carry 2, 4, 6, and 8 bits, respectively. To calculate the error rate of these modems versus SNR, similar to all examples in Section 4.2.4, we need to calculate the energy per symbol, E_s, as a function of minimum distance, d, and then plug the results in Equation (4.6). It can be shown that

$$E_{s,m} = \frac{M-1}{6}d^2 \Rightarrow E_{b,m} = \frac{M-1}{6 \times m}d^2.$$

Then, the error rate for all of these constellations is given by

$$P_s \approx \frac{1}{2}\text{erfc}\left(\frac{6m}{M-1} \times \frac{E_b}{N_0}\right); \quad m = 2, 4, 6, 8; \; M = 2^m.$$

The following MATLAB code has produced the plot for all of these four QAM transmission schemes, which is shown in Figure 4.10:

```
EbNo_db = linspace(0, 25, 100);
EbNo  = 10.^(EbNo_db/10);
pe_0 = 1e-5;
 for m=1:4
  ms=2*m;
  M = 2^ms;
  a = 1/2;
  b = 3*ms/(2*(M-1));
  pe(ms-1,:) = a * erfc (sqrt(b*EbNo));
  EbNo_0(ms-1) = (1/b) * (erfinv(1-(pe_0/a)))^2;
  EbNo_0(ms-1) = 10*log10(EbNo_0(ms-1));
  s=[int2str(M),' QAM: required EbNo = ',num2str(EbNo_0(ms-1)),' dB for a Pe of
',num2str(pe_0)];
  disp(s);
end;
figure(1)
for m=1:4
  ms=2*m;
  semilogy(EbNo_db, pe(ms-1,:))
  hold on
end;
```

```
semilogy([0 25], [pe_0 pe_0],'r')
axis([0 25 1e-6 1])
set(gca, 'XTick', [-10:2:25])
title('Symbol Error Rate versus SNR/bit')
xlabel('Eb/N0 (dB)')
ylabel('Ps')
```

This code also prints the SNR requirement for error rate of 10^{-5}

4 QAM: required EbNo = 9.5879 dB for a Pe of 1e-05

16 QAM: required EbNo = 13.5673 dB for a Pe of 1e-05

64 QAM: required EbNo = 18.0388 dB for a Pe of 1e-05

256 QAM: required EbNo = 22.8614 dB for a Pe of 1e-05

Each constellation needs approximately 5 dB more power to maintain the same error rate. Another observation is that for each 3 dB of additional SNR,

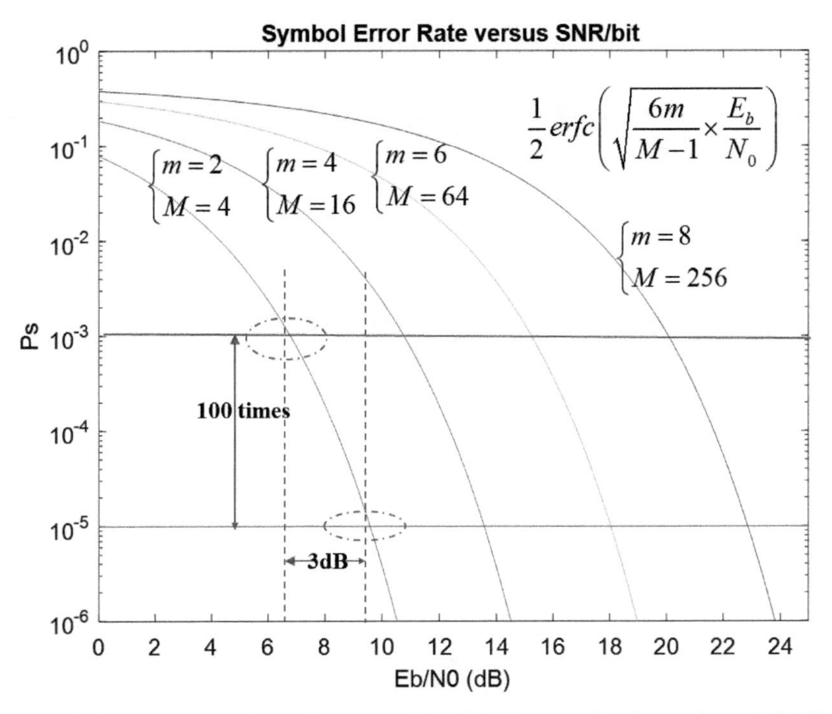

Figure 4.10 Error rates of 4-, 16-, 64-, and 256-QAM versus signal-to-noise ratio in dB.

the error rate drops 100 times. For example, if we want to improve the error rate from $10^{-3}$to 10^{-5}, we need 3 dB or two times more transmission power with the same background noise.

Another interesting observation from Figure 4.10 is that the difference between the error rate of any of the water-fall plots for 100 times less error is approximately 3 dB which was what we expect because they are different by a factor of 2.

4.3.2 Shannon–Hartley Bounds on Achievable Data Rate

In our discussions thus far, we have shown how the design of the signal constellation relates to the achievable data rate and how we can calculate the error rate for a given received SNR. At this point, it is useful to consider the ultimate limits on data rate and efficiency that are theoretically achievable. This is best done by examining Shannon–Hartley's well-known formula:

$$R_b = W \log_2 \left(1 + \frac{E_s}{N_0}\right) \quad \text{bps} \tag{4.7a}$$

where R_b is the maximum achievable information transfer rate in bps in a bandwidth of W Hertz, and the signal-to-noise power ratio of E_s/N_0 [Sha48]. We can reasonably replace the bandwidth with the symbol transmission rate, R_s. This simple and powerful theorem relates the three most important transmission parameters, power, bandwidth, and maximum achievable data rate, and the ultimate guideline for understanding the limits to the performance of a transmission technique is considered. For example, if we apply the above equation to the voiceband telephone channel which has a bandwidth of 4 kHz and a typical SNR of 30 dB, we obtain a theoretical channel capacity or maximum achievable data rate of about 40 Kbps. If we want to increase the data rate since bandwidth and the maximum transmitted power are fixed, we need to improve the line conditions to reduce the noise at the receiver. In V.90 modems, they do that in the downstream channel by eliminating the analog circuitry in the connection between the user and the PSTN to achieve 56 Kbps.

This simple bound provides us an idea to relate the bandwidth, data rate, and the SNR to find achievable transmission rates over a specific channel. If we consider Equation (4.6) for the calculation of symbol error rate and its relation to different transmission signal constellation, in Example 4.7 and Figure 4.10, we observe that for a fixed SNR, if we increase the constellation size, both data rate and error rate increase. Increase in data rate is desirable, but increase of error rate is undesirable. The maximum error rate is 0.5 in which half of the information is in error, meaning what we receive does not contain any information because we have 50% assurance on the accuracy of the received bit. The same assurance that we have from guessing what the result of flipping a coin is! The Shannon–Hartley bound calculates maximum achievable data rate when$P_s = 0.5$. This bound provides the Utopic value of achievable data rate; we can use it to assess the quality of a modulation and coding technique to see how close it can get to the ideal achievable data rate.

To relate signal constellation design to Shannon–Hartley bounds, we assume $W = R_s = mR_b$ and rewrite Equation (3.8) in the following format:

$$m = \frac{R_b}{W} = \log_2\left(1 + \frac{E_s}{N_0}\right) = \log_2\left(1 + m \times \frac{E_b}{N_0}\right).$$

Then, we will have

$$\gamma_b = \frac{E_b}{N_0} = \frac{W}{R_b}\left(2^{\frac{R_b}{W}} - 1\right) = \frac{1}{m}\left(2^m - 1\right). \tag{4.7b}$$

This equation describes channel capacity in terms of two convenient normalized parameters, $\gamma_b = \frac{E_b}{N_0}$ and $m = \frac{R_{b-\max}}{W}$. The first parameter is the minimum value of *SNR per bit* required for reliable transmission of data at maximum data rate over a channel with bandwidth *W*. The second parameter, $m = \frac{R_b}{W}$, simply normalizes the bit rate by the bandwidth to show a maximum number of bits per symbol. Therefore, this equation relates the achievable data rate to the bandwidth of the system and the SNR which provides us with a convenient framework for assessing the efficiency of the communications of any chosen modulation scheme.

In Figure 4.11, we show the plot of Equation (4.8b) illustrating relation between $m = R_b/W$ versus $\gamma_b = E_b/N_0$. Note that the lower portion of the scale is expanded for convenience in drawing the figure. This figure essentially represents a bandwidth-versus-efficiency plane, and the capacity curve divides the plane into two regions. The area on the left of the curve defines the region in which reliable communication cannot be achieved; that is, no modulation or coding scheme can be devised to operate in that region. In the right-hand area of the figure, which defines the region of achievable signal designs, design points are shown for several modulation methods, which we have discussed earlier in Example 4.7 and Figure 4.10. For all the cases shown, the delivered error rate is 10^{-5}. The displacement of each design point from the capacity boundary indicates how close the communication efficiency of the corresponding modulation scheme comes to the capacity

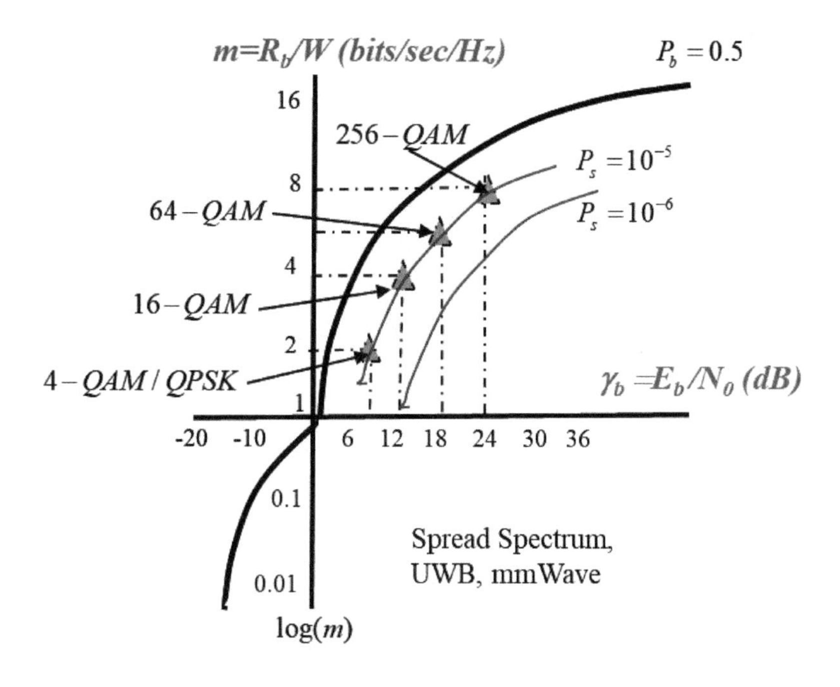

4 QAM: required EbNo = 9.5879 dB for a Pe of 1e-05
16 QAM: required EbNo = 13.4345 dB for a Pe of 1e-05
64 QAM: required EbNo = 17.7869 dB for a Pe of 1e-05
256 QAM: required EbNo = 22.5032 dB for a Pe of 1e-05

Figure 4.11 Channel capacity and comparison of several modulation methods at BER of 10^{-5} and 10^{-6}.

limit. The horizontal displacement measures the shortfall in terms of SNR per bit, while the vertical displacement measures the shortfall in terms of bandwidth utilization. Note that if we were to plot the modem design points for a lower level of delivered bit error rate (BER), the points would all move to the right (i.e., further away from the capacity boundary), whereas if we used a higher BER, they would move closer to the capacity boundary.

It is conventional to call the region of $R/W > 1$ the *bandwidth-limited region* of operation and to call the region of $R/W < 1$ the *power-limited region* of operation. The bandwidth-limited region includes all the modulation schemes we have described in Section 4.2, where rigid channel bandwidth limitations are imposed by the medium or FCC regulations. There we see that the *QAM* modem signal constellations provide steadily increasing bandwidth utilization as *M* is increased.

By inspecting the figure, we can conclude that if the bandwidth is much more than the data rate (e.g., original direct-sequence SS 802.11 with a 2 Mbps data rate and 26 MHz bandwidth), we can operate at a smaller SNR per bit (which results in larger coverage for the radio).

4.4 Performance in Multipath RF Channels

Ideally, digital communications over radio channels are the same as communication over guided media. Except that the attenuation model is different, the bandwidth for communication is allocated by FCC, and we have inference noise. This remains that way if the communication is through a single path, for example, between aerial devices in space, for example, between satellites and airplanes. In indoor and urban areas, where all popular wireless networking systems work, the transmitted signals bounce from objects surrounding the transmitter and the receiver, and, as we explained in Section 4.3, it causes fading in the received signal. Fading is a heavy fluctuation of received signal strength on the orders of 20–30 dB. A system designed with a very low error rate of 10^{-2} to 10^{-8}, then goes to fade and have an error rate close to 0.5. When fade occurs, we have many errors bringing the average error rate substantially lower. In this section, we first show the effects of fading on the performance of digital communications, then we introduce diversity techniques as methods to mitigate the effects of fading.

4.4.1 Effects of Fading on Performance Over Wireless Channels

One of the main characteristics of the wireless medium affecting the performance of a transmission technique is large fluctuations of the received power level, referred to as *fading*. As opposed to wired channels, the received signal from wireless channels suffers from strong amplitude fluctuations (on the order of 30–40 dB) that cause fading in the received signal. During periods of signal fading, the error rate of the transmission system increases substantially and when the system is out of fade, the error rate becomes negligible.

Figure 4.12 shows the basic concept behind the strange and complex behavior of transmission techniques over fading wireless channels. Due to the fading effect, the SNR per bit, γ_b, randomly fluctuates in time. The average SNR per bit, $\overline{\gamma_b}$, represents the transmitted power that can be regulated by

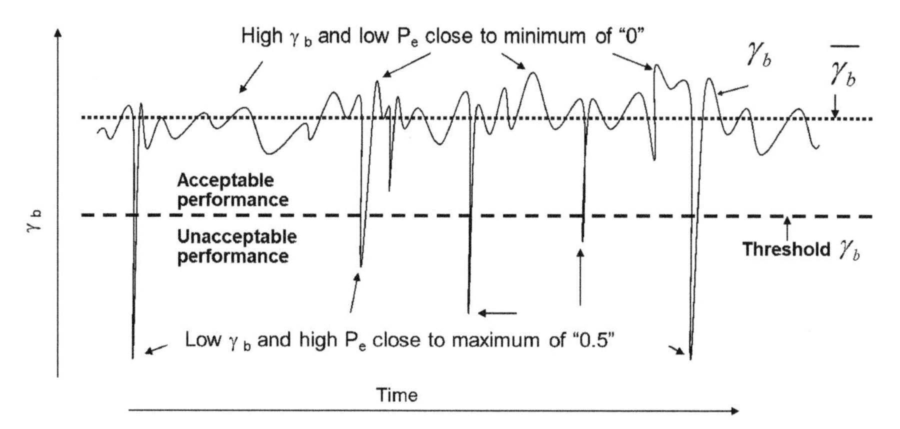

Figure 4.12 Relation between error rate, outage rate, and fading characteristics.

the designer. Any application has an acceptable error rate, when γ_b crosses a specified "threshold" and the error rate drops below the acceptable error rate. The fraction of time in which the error rate is unacceptable is called the *outage rate* or outage probability for those applications. The error rate when the signal is above the threshold is always very small (close to zero), and when it is below the threshold, it is very high (close to 0.5). Therefore, most errors occur during deep fades when the signal level crosses the threshold. This is a very important observation – to remedy the effects of fading, we need to find methods that can recover bits corrupted during the occurrence of deep fading.

To evaluate the performance over a fading radio channel, either the average bit error rate, $\overline{P_e}$, or the probability of outage versus average received SNR per bit, $\overline{\gamma_b}$, is used. The average bit error rate and the outage rate may look different. However, they represent the same phenomenon, and, in many cases, they look similar. To calculate any of these performance measures, we need a model for the variations of $\overline{\gamma_b}$. The most common model used for variations of signal amplitude in fading channels is the Rayleigh distribution for which $\overline{\gamma_b}$ follows an exponential distribution

$$f_\Gamma\left(\gamma_b\right) = \frac{1}{\gamma_b} e^{-\gamma_b/\bar{\gamma}_b}$$

in which γ_b is the instantaneous SNR per bit and $\overline{\gamma_b}$ is the average SNR per bit over a long time. As we showed in Example 4.7, the error rate of different QAM constellations follows the same pattern using the erfc function. If we consider the exponential asymptotic bound for errors described in Equation (4.6), this error rate becomes

$$\begin{cases} P_s \approx \frac{1}{2}\text{erfc}\left(\frac{6m}{M-1} \times \frac{E_b}{N_0}\right) = \frac{1}{2}\text{erfc}\left(\alpha_m \times \frac{E_b}{N_0}\right) \leq \frac{1}{2}e^{-\alpha_m \times \gamma_b} \\ \alpha_m = \frac{6m}{M-1} \end{cases},.$$

which means the error rate in any constellation can be represented by an exponential function of a parameter multiplied with the SNR per bit. Then, the average error rate, $\overline{P_s}$, over all possible values of the SNR per bit, γ_b, is given by

$$\bar{P}_s = \frac{1}{2\bar{\gamma}} \int_0^{\infty} e^{-\frac{\gamma_b}{\bar{\gamma}_b}} e^{-\alpha_m \times \gamma_b} d\gamma_b = \frac{1/2}{1+\alpha_m \bar{\gamma}_b} \simeq \frac{1/2}{\alpha_m \bar{\gamma}_b} \tag{4.8}$$

The average BER, $\overline{P_e}$, is an inverse function of the average SNR per bit, $\overline{\gamma_b}$. To compare this relationship with the relation between the error rate, P_e, and SNR per bit, γ_b, for non-fading wired channels, remember that in non-fading channels, a 3 dB change in transmission power decreased the bit error rate by two orders of magnitude (hundred times). Equation (4.8) reveals that, for fading channels, we need 20 dB (100 times) more average power to increase the average bit error rate by two orders of magnitude. Figure 4.13(a) shows the average probability of error in logarithmic form versus average SNR per bit in dB for $\alpha_m = 2$. This relation is a line with a slope of one. Comparing this curve with the waterfall-like plots of bit error rate versus signal to noise per bit, shown in Figure 4.13(b), we need much more power to overcome the effects of fading to achieve the same error rates. For example, to achieve an error rate of 10^{-5}, we need a γ_b of less than 10 dB on a the non fading wired channel compared with an average signal to noise per bit, $\overline{\gamma_b}$, or more than 45 dB needed to achieve the same average error rate on a fading wireless channel. This difference means that achievement of the same error rate for the radio channel requires more than three thousand times (35 dB) more power. Designers of the radio modems have worked hard in the past half a century to close this gap between the performance over non-fading wire and fading wireless channels, and they have come up with a number of innovative solutions such as STC and MIMO antenna systems which are used as the latest advancement in the design of the physical layer of wireless networks. All these solutions take advantage of the so-called diversity techniques that we will discuss next.

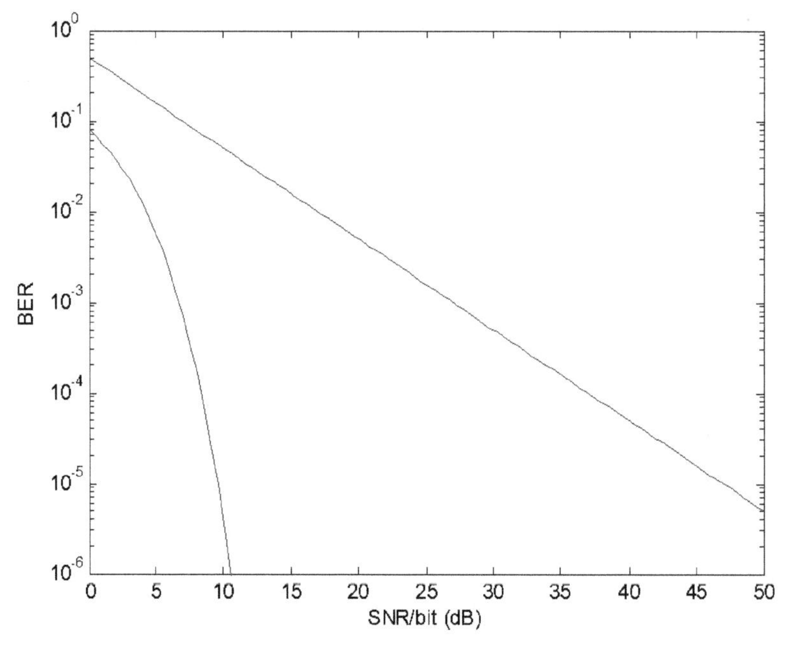

Figure 4.13 Average BER in logarithmic form versus average SNR/bit in dB for $m/a = 2$ (a) over a Rayleigh fading wireless channel and (b) over a non-fading wired channel.

4.4.2 Diversity Techniques to Remedy Fading

As we observed in the previous section, fading is manifested as signal amplitude fluctuations over a wide dynamic range. During short periods of time, the channel goes into deep fades causing significant numbers of errors that virtually dominate the overall average error rate of the system. To compensate for the effects of fading when operating with a fixed-power transmitter, the power must typically be increased by several orders of magnitude relative to the non fading operation. This increase of power protects the system during the short intervals of time when the channel is deeply faded. A more effective method of counteracting the effects of fading is to use diversity techniques in the transmission and reception of the signal. The concept here is to provide multiple copies of the received signals whose fading patterns are different (and hopefully independent as we will see later). With the use of diversity, the probability that all the received signals are in a fade at the same time reduces significantly, which, in turn, can yield a large reduction in the average error rate of the system.

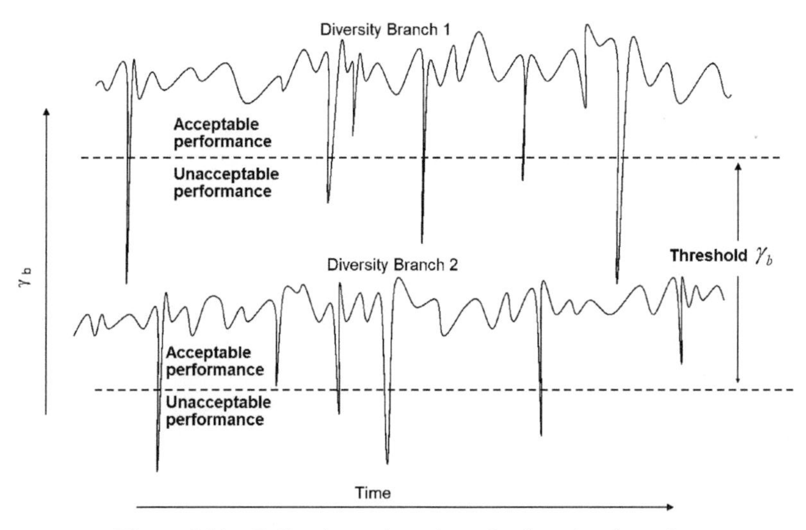

Figure 4.14 Fading in two branches of a diversity channel.

Figure 4.14 shows fluctuations in two branches of a diversity channel and how they help in the reduction of overall error rates. When one of the branches is in a deep fade, causing a large number of errors, the correct data can be retrieved from the other branch. In a diversity channel, large numbers of errors can occur when all branches are in deep fade at the same time. Since the probability of a deep fade occurring in all branches is much lower than that in only one branch, the error rate on a diversity channel is much less than that on a single-branch fading channel. The occurrence of deep fading on all branches is a function of correlation among different branches and of the number of diversity channels. As the correlation among the diversity branches decreases and they become independent and the number of branches increases, the error rate decreases.

Diversity can be provided spatially by using multiple antennas, in frequency by providing signal replicas at different carrier frequencies, or in time by providing signal replicas with different arrival times. It is conventional to refer to the diversity components as *diversity branches*. We assume that the same symbol is received along different branches, with each branch exposed to a *separate* random fluctuation. This has the effect of reducing the probability that the received signal will be faded simultaneously on all the branches; this, in turn, reduces the overall outage probability as well as the average BER.

A variety of techniques are available for the reception of the diversity signals. In the most popular and the optimum method of combining, called *maximal-ratio combining*, the diversity branches are weighted prior to summing them, each weight is proportional to the received branch signal amplitude. Let us assume that the amplitudes of the signals received on different branches are all uncorrelated Rayleigh-distributed random variables and all branches of diversity have the same average received signal power and the average SNR on each branch is denoted by $\overline{\gamma_b}$. The probability distribution function of the post-combining of a maximum ratio combiner SNR is then given by the gamma function

$$f_\Gamma\left(\gamma_b\right) = \frac{1}{(D-1)!\overline{\gamma_b}^D}\gamma_b^{D-1}e^{-\gamma_b/\overline{\gamma_b}}, \tag{4.9}$$

where D is the order of diversity. It can be shown that [Pro00, Pah05] using Equations (4.8) and (4.9), the average probability of error for the maximal-ratio combiner output is given by

$$\overline{P_e} = \int_0^\infty f_\Gamma(\gamma_b)P_b(\gamma_b)d\gamma_b \approx \left(\frac{1}{8\alpha_m\overline{\gamma_b}}\right)^D \begin{pmatrix} 2D-1 \\ D \end{pmatrix}, \tag{4.10}$$

where we use the standard notation for a binomial coefficient:

$$\begin{pmatrix} N \\ k \end{pmatrix} = \frac{N!}{(N-k)!k!}.$$

The expression in Equation (4.10) shows that average BER performance at the maximal-ratio combiner output improves exponentially with increasing D, the order or number of branches of diversity. Figure 4.15 shows the average probability of error $\overline{P_e}$ versus average SNR per bit $\overline{\gamma_b}$, for different orders of diversity. Included in the figure is the error rate curve for steady-signal reception. As we saw earlier, with a single antenna, we lose $30-35$ dBin performance relative to steady-signal reception at reasonable levels of error rate. With two independent diversity branches, the performance loss is reduced to about 25 dB, and with four orders of diversity, the signal-to-noise penalty is reduced to around 10 dB. With additional orders of diversity, the penalty relative to non-fading can be further reduced. There will, of course, be a practical limit to the order of diversity implemented because, for example, one cannot put an arbitrarily large number of antennas into a communications terminal.

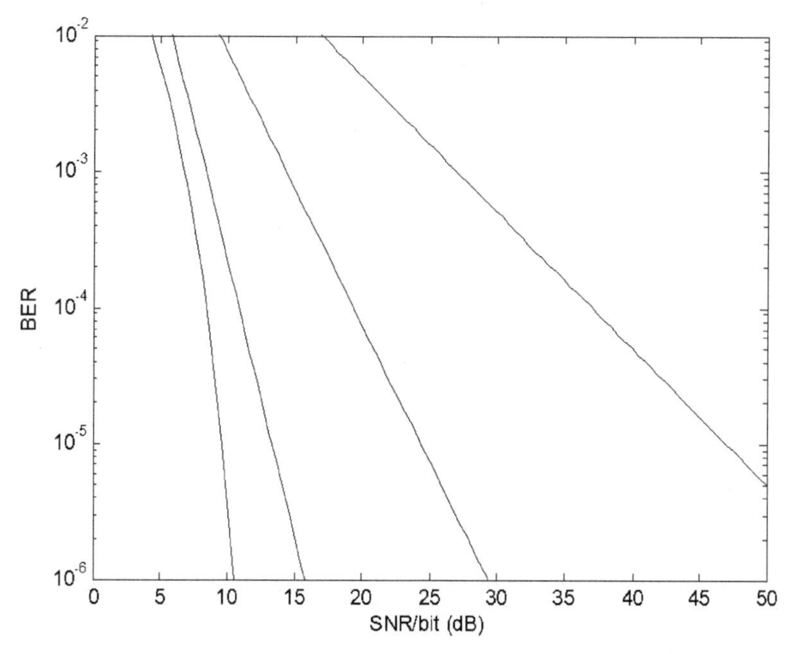

Figure 4.15 Average BER versus average SNR per bit for different orders of diversity.

4.5 Wireless Modems Technologies

Another major difference between wireless and wired channels is that the wireless channel is a multipath channel. In a multipath channel, the shape of the received pulse and the time duration of the signaling pulse are both changed due to multipath arrivals. The difference between the first and the last arriving pulses at the receiver due to the same transmitted pulse is the *delay spread* of the channel. If the symbol duration is much larger than the multipath spread of the channel, all pulses received via different paths arrive roughly on top of one another causing only amplitude fluctuations and fading that was discussed in the previous section. If the ratio of the delay spread to the pulse duration becomes considerable, the received pulse shape is severely distorted, and it also interferes with neighboring symbols causing inter-symbol interference (ISI). In addition to SNR fluctuations due to fading effects, the interference power also degrades the performance. However, the ISI effect of multipath degrades the performance in a different manner than fading. The effects of fading can be compensated via an increase in the transmit power by a fading margin. An increase in the transmit power cannot compensate for the effects of ISI. This is because an increase in the transmit power increases the signal as well as the ISI interference power, keeping the signal to interference ratio at the same level.

The effects of ISI caused by multipath are the main obstacle for high-speed communications over wireless channels. As we increase the data rate, the duration of the transmitted symbols decreases, and the ISI caused by multipath increases. As a result, handling the ISI effects of multipath in hope of using the diversity of the received signal from different paths has been another major area of research for the past several decades. As a result of this research, a number of signal processing techniques such as OFDM, adaptive equalization, SS, and UWB communications have emerged to take advantage of the multipath arrival and to achieve higher data rates, referred to as wideband modems, for wireless RF transmissions. The most popular of these techniques is OFDM, adopted by the IEEE 802.11 community. OFDM combined with MIMO technologies has enabled wireless LANs (WLANs) to increase their first-generation data rates of 2 Mbps using SS technology to over 100 Mbps in IEEE 802.11n. In the rest of this section, we provide an overview of the most important of these technologies, SS transmission, OFDM, and MIMO.

4.5.1 Spread Spectrum Transmissions

SS technology was first invented during the Second World War, and it has dominated military communication applications, where it is attractive because of its resistance to interference and interception, as well as its amenability to high resolution ranging. In the later part of the 1980s, commercial applications of SS technology were investigated, and, today, it is the transmission technique used in 3G cellular, a number of proprietary cordless telephones, original IEEE 802.11 WLAN, IEEE 802.15 Bluetooth, ZigBee, and UWB wireless PANs (WPANs). The voice-oriented digital cellular and PCS industries have selected SS technology to support code division multiple access (CDMA) networks as an alternative to time division multiple access (TDMA)/frequency division multiple access (FDMA) networks in order to increase system capacity, provide a more reliable service, and provide soft handoff of cellular connections that will be discussed in more detail in subsequent chapters. In the WLAN industry, SS technology was adopted primarily because the first unlicensed frequency bands suitable for high-speed radio communication were industrial, scientific, and medical (ISM) bands, which were initially released by the FCC under the condition that the technologies use spread spectrum. In Bluetooth, ZigBee, and UWB WPANs, spread spectrum is adopted because of the simplicity of implementation and low power consumption which is necessary for *ad hoc* and sensor network implementations.

The main difference between spread spectrum transmission and traditional radio modem technologies are that the transmitted signal in spread spectrum systems occupies a much larger bandwidth than the traditional radio modems. Compared to baseband impulse transmission techniques, the occupied bandwidth by spread spectrum is still restricted enough so that the spread spectrum radio can share the medium with other spread spectrum and traditional radios in a frequency division multiplexed format. There are two basic methods for spread spectrum transmission: direct sequence spread spectrum (DSSS) and frequency hopping spread spectrum (FHSS).

Frequency Hopping Spread Spectrum:
The FHSS technique was first invented to protect guided torpedoes from jamming by the German movie star *Hedy Lamarre*, who had no technical training – therefore, it must be a relatively simple technology. To avoid a jammer, the FHSS transmitter shifts the center frequency of the transmitted signal. The shifts in frequency or *frequency hops* occur according to a random pattern that is only known to the transmitter and the receiver. If we move the center frequency randomly among one hundred different frequencies, then the required transmission bandwidth is a hundred times more than the original transmission bandwidth. We call this new technique ***an spread spectrum technique*** because the spectrum is spread over a band that is a hundred times larger than the original traditional radio. FHSS can be applied to both analog and digital communications, but it has been applied primarily for digital transmissions.

FHSS modulation technique can be thought of as a two-stage modulation technique. In the first stage, the input data stream is modulated with a traditional modem, and, in the second stage, the center frequency is changed according to a random hopping pattern generated by a random number generator. Ideally, the random pattern or spreading code is designed so that the occurrence of frequencies is statistically independent of one another. At the receiver first a de-hopper synchronized to the transmitter repeats the hopping pattern of the transmitted signal, then a traditional demodulator detects the received data. In a digital implementation of this system, the sampling rate is the same as the sampling rate of the traditional system leaving the complexity of the implementation in the same range as traditional modems. As we will see later, DSSS needs much higher sampling rates and, consequently, a more complex hardware implementation.

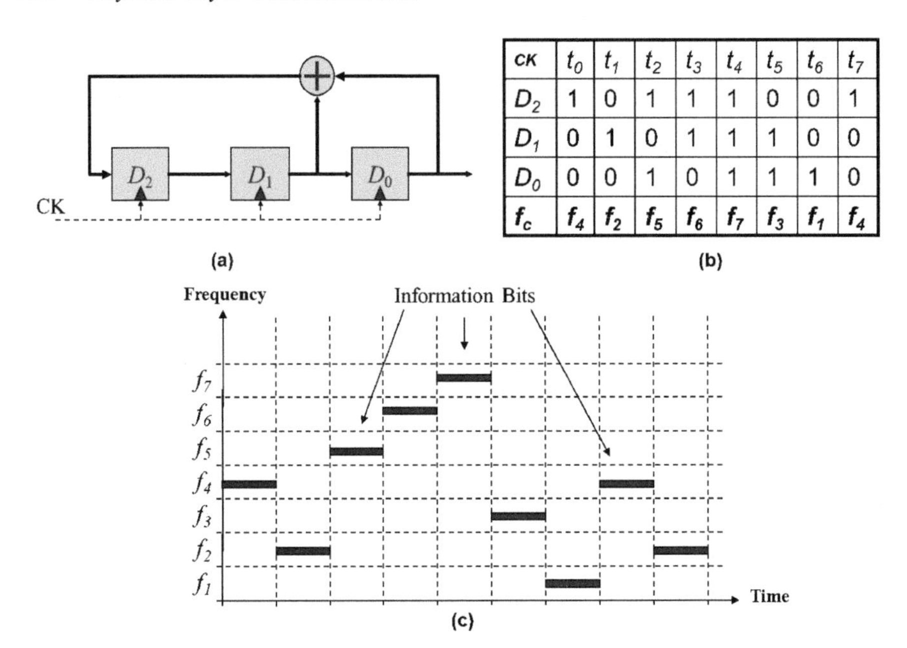

Figure 4.16 FHSS using a length seven PN sequence. (a) Three-stage state machine for implementation of the code. (b) States of the memory in time and frequency. (c) Transmission of information bits in time and frequency in FHSS.

Figure 4.16 shows the hopping pattern and associated frequencies for a frequency hopping system transferring data packets over the air. A three-state recursive machine is used for the generation of the code that determines the frequency to be hopped to. This machine, shown in Figure 4.16(a), generates a sequence of seven states representing all seven non-zero values which can be formed by three binary digits periodically. Figure 4.16(b) shows all seven states of the machine and their associated number which is used as the frequency index for FHSS. Each packet is transmitted using one of these frequencies. The sequence of frequencies is f_3, f_5, f_6, f_1, f_4, f_8, f_2, f_7 before returning to the first frequency f_3. Figure 4.16(c) shows the transmission of the information bits in time and frequency in this FHSS system. The random sequence used in these situations is referred to as pseudo-noise (PN) sequences.

In FHSS, the hopping of the carrier frequency does not affect the performance in the presence of additive noise because the noise level in each hop remains the same as the noise level of the traditional modems. Therefore, the performance of the FHSS systems in non-interfering environments remains the same as the performance of the traditional systems without frequency hopping. In presence of narrowband interference, the signal-to-interference ratio of a traditional modem operating at the frequency of the interferer becomes very low, corrupting the integrity of the received digital information. The same situation happens in frequency selective fading channels when the center frequency of a traditional system coincides with a deep frequency selective fade. In a FHSS system, since the carrier frequency is constantly changing, the interference or frequency selective fading only corrupts a fraction of the transmitted information and transmission in the rest of the center frequencies remains unaffected. This feature of FHSS is exploited in the design of wireless networks to provide a reliable transmission in the presence of interfering signals or when a system works over a frequency selective fading channel.

Multipath conditions in wireless channels cause frequency selective fading which results in very poor performance in certain frequency regions, while performance in other frequencies is acceptable. When a frequency selective fading occurs, traditional systems operating over center frequencies coinciding with the faded frequencies cannot operate properly. As shown in Figure 4.17, an FHSS system can be designed so that the deep fades in the environment only corrupt a small fraction of hops leaving the rest of the hops for successful retransmissions. In an indoor environment, the width of the fade is around several MHz, and the FHSS system used in IEEE 802.11 or Bluetooth uses hops that are 1 MHz apart from one another. Therefore, if a hop occurs in a deep fade and the data transmitted in that hop is not reliable, the retransmitted data packet in the next hop will be successful.

Frequency hopping SS allows the coexistence of several transmissions in the same frequency band using different hoping codes. Different users could be a member of the same network following a coordinate hop pattern or two different networks each using their own pattern. For example, IEEE 802.11 FHSS and Bluetooth both use the same bandwidth of 1 MHz and the same 78 channels in the 2.4 GHz ISM bands. Members of each network coordinate their transmissions while both coexist in the same band. Then multi-user interference occurs when two different users transmit on the same hop frequency. If the codes are random and independent from one another, the "hits" will occur with some calculable probability. If the codes are synchronized and the hopping patterns are selected so that two users never hop to the same frequency at the same time, multiple-user interference is eliminated.

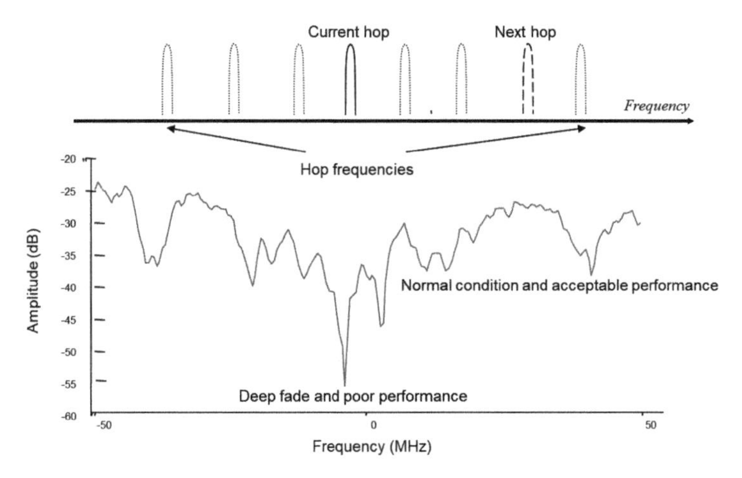

Figure 4.17 Frequency selective fading and FHSS.

Direct Sequence Spread Spectrum:

In a manner like FHSS, DSSS can be thought of as a two-stage transmission technique. In the first stage, each transmitted information bit is coded into *N* smaller pulses referred to as chips using a PN sequence. In the second stage, the chips are transmitted over a traditional digital modulator. At the receiver, the transmitted chips are first demodulated and then passed through a correlator to calculate their autocorrelation function (ACF). We define the ACF of a sequence $\{b_i\}$ as

$$R(k) = \sum_{i=0}^{N-1} b_i b_{i-k} = \begin{cases} N & ; k = mN \\ -1 & ; \text{otherwise} \end{cases}.$$

The PN-sequence is a periodic sequence of length *N*, and, therefore, the correlation is calculated over one period of the sequence. The ACF of a good random code has a very high peak of height *N* at $k = 0$, which is usually referred to as the *processing gain* of the receiver. The value of this ACF for $k \neq 0$ is far below the peak value. Therefore, DSSS systems use the peak of the ACF to detect the transmitted bit.

Example 4.7: DSSS Using Barker Code in IEEE 802.11:

A Barker code of length 11, used in the IEEE 802.11 as the spreading signal for the DSSS physical layer, is given by $\{1\ 1\ 1\ -1\ -1\ -1\ 1\ -1\ -1\ 1\ -1\}$[1].

[1]Note that we used "-1" instead of "0" digit. This way, the exclusive OR operation used in digital addition changes to standard multiplication which is easier for calculations.

Figure 4.18 shows a data bit “1” in a binary communication DSSS system, the transmitted Barker code for the data bit, and the ACF at the receiver with its high peak and low sidelobes.

Barker code has the same autocorrelation properties as other maximum length PN-sequences, but the length does not need to follow $2^m - 1$ values. The Barker code was adopted by IEEE 802.11 for the original DSSS system at 1 and 2 Mbps because FCC has allowed a minimum processing gain of 10 for the systems operating in unlicensed ISM bands. The closest maximum length codes would be of length 15 which has 15/11 times lower bandwidth efficiency than the Barker code. The maximum length PN-sequences are widely used in CDMA cellular networks and the time-of-arrival based geolocation techniques.

In cellular networks, DSSS is used for code division multiple access. In a multi-user DS-CDMA environment, different codes are assigned to different users. In other words, each user has their own unique “key” code that is used to spread and despread only his messages. The codes assigned to other users are selected so that during the despreading process at the receiver, they produce very small signal levels (like noise) that are on the order of the sidelobes of the ACF. Consequently, they do not interfere with the detection of the peak of the ACF of the target receiver. In this manner, each user is a source of noise for the detection of other users’ signals. As the number of users increases, the multi-user interference increases for all of the users. This phenomenon continues up to a point when the mutual interference among all terminals stops the proper operation for all of them.

In time-of-arrival based geolocation, the peak of the ACF is used to determine the relative time of arrival of the signal to calculate the time of flight of the signal between the transmitter and the receiver. Since radio wave propagates at the speed of light, the time of flight is used to determine the distance between a transmitter and a receiver. The global positioning system (GPS) uses DSSS transmission to determine the distance between different satellites and a mobile user which is further processed for localization of a terminal. The accuracy of the estimate of the flight time is a function of the SNR. In DSSS, we can increase the SNR by increasing the processing gain of the codes which is related to the averaging time for calculation of the ACF.

The bandwidth of any digital system is inversely proportional to the duration of the transmitted pulse or symbol. Since the transmitted chips are N times narrower than data bits, the bandwidth of the transmitted DSSS signal is N times larger than a traditional system without spreading. As a result, N is also referred to as the *bandwidth expansion* factor. In a manner like FHSS,

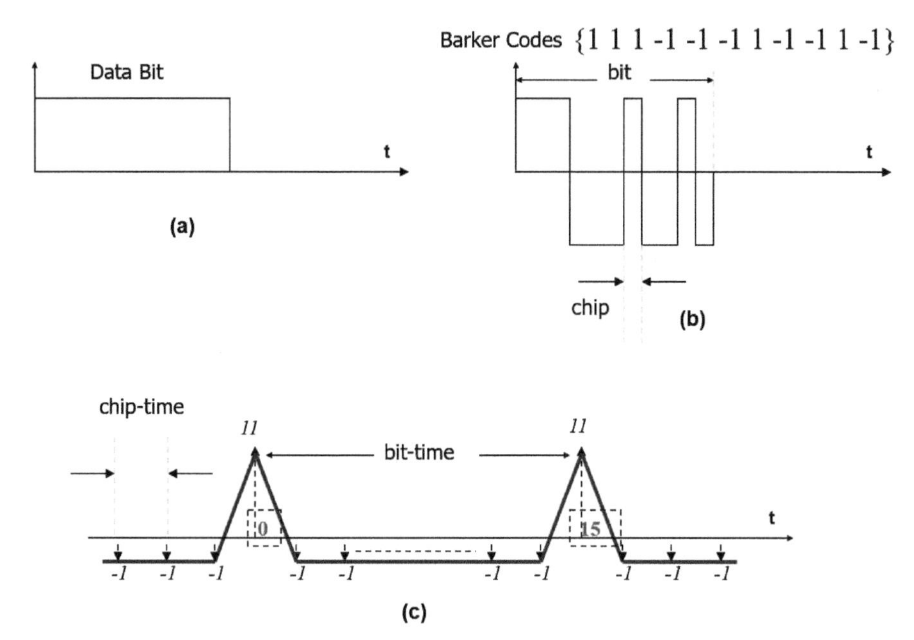

Figure 4.18 Barker code used in DSSS signal in IEEE 802.11. (a) A transmitted bit. (b) The 11-chip Barker code. (c) Circular autocorrelation function of the code.

DSSS is also anti-interference and resistant to frequency selective fading. The transmission bandwidth of the DSSS is always wide, while the FHSS is a narrowband system hopping over several frequencies in a wide spectrum. As a result, there is some distinction between the two methods. The DSSS systems provide a robust signal with a better coverage area than FHSS. The FHSS system can be implemented with much slower sampling rates saving in the implementation costs and power consumption of the mobile units.

As shown in Figure 4.18, for any transmitted bit, the output of the correlator function at the receiver produces a narrow pulse with a height of N and a base that is twice the chip duration. Therefore, a DSSS system can be thought of as a pseudo-pulse transmission technique receiving a narrow pulse designating each transmitted bit. In a multipath radio channel environment, different paths will bring different pulses at the receiver at different times and an intelligent receiver can use these pulses as a source of *time diversity* to improve its performance. These receivers are referred to as Rake receivers which are commonly implemented in DSSS wireless transmission systems to achieve reliable communications.

4.5.2 Orthogonal Frequency Division Multiplexing (OFDM)

What is known as OFDM today is a method of implementation of multi-carrier modulation (MCM) using orthogonality of the adjacent carriers. An MCM system is indeed a frequency division multiplexing (FDM) system for which a single user uses all the FDM channels together. OFDM is an implementation of MCM that takes advantage of the orthogonality of the channels and develops a computationally efficient implementation based on the fast Fourier transform (FFT) algorithm.

MCM was first evaluated for high-speed voiceband modems in the early 1960s and it was augmented with an FFT implementation for the same application in the early 1980s. It found its way into wireless local area networks (IEEE 802.11), digital subscriber line (DSL) modems, cable modems (IEEE 802.14), wireless metropolitan area networks (IEEE 802.16), wireless personal area networks (IEEE 802.15), and many other wired and wireless applications in the 1990s. The concept here is very simple. Instead of transmitting a single stream at a rate of R_S symbols/s, we use N streams over N-carriers spaced by about R_S/N Hz, each carrying a stream at the rate R_S/N symbols/s. The primary advantage of MCM is its ability to cope with severe channel conditions such as high attenuations at higher frequencies in long copper lines or the effects of frequency-selective fading in wireless channels. Sub-channels provide a form of frequency diversity, which can be exploited by applying error-control coding *across symbols* in different sub-channels. In the OFDM implementation, this latter technique is referred to as coded OFDM or COFDM. To further improve the performance of an OFDM transmission, one may measure the received signal power in different sub-channels and, using a feedback channel, may adjust the specification of the transmitted sub-carriers to optimize performance. With these features of OFDM, it has become an ideal solution for broadband transmissions. In OFDM increasing, the data rate is simply a matter of increasing the number of carriers. The limitations are complexity of implementation and the limitation on the transmitted power.

Figure 4.19 represents an MCM system with N-carriers and a channel frequency response with frequency selective fading. As shown in this figure, carriers operating at different frequencies are exposed to different channel gains. Therefore, the received signals in individual carriers have different SNRs and bit error rate qualities. If some redundancy in the carriers is imposed on the system, the errors caused by low SNR in poor channels can be recovered. This redundancy can be easily achieved by scrambling the data

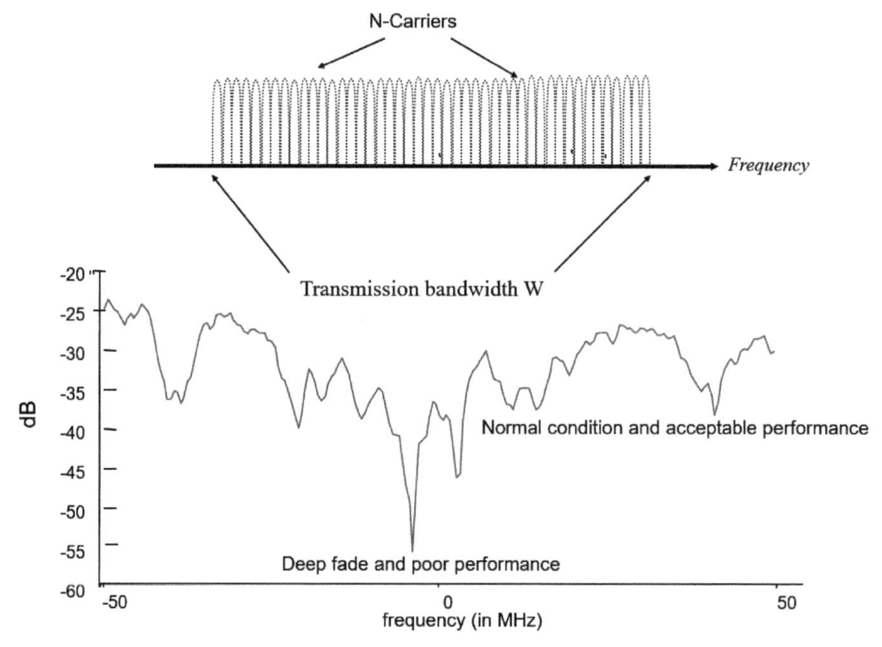

Figure 4.19 Frequency selective fading and multi-carrier transmission.

before transmission. Although the entire bandwidth used by the system is exposed to frequency selective fading, individual channels are only exposed to flat fading that does not cause ISI which needs computationally intensive receivers making use of adaptive equalization techniques. In practice, to avoid overlap between consecutively transmitted symbols, a time guard is enforced between transmissions of two OFDM pulses that will reduce the effective data rate. Also, some of the carriers are dedicated to the synchronization signal and some are reserved for redundancy. An example will bring together all details of implementation.

Example 4.8: Implementation of OFDM in IEEE 802.11a:
Figure 4.20 shows the 64 sub-channel implementation of OFDM for the IEEE 802.11a,g physical layer specifications. Each channel carries a symbol rate of 250 KSps. We have 48 sub-carriers devoted to information transmission, 4 sub-carriers for pilot tones used for synchronization, and 12 reserved for other purposes. The user symbol transmission rate is 48 $\times$ 250 KSps = 12 MSps. The bit transmission rate depends on the number of bits per symbol (constellation used for transmission) and the rate of the convolutional code

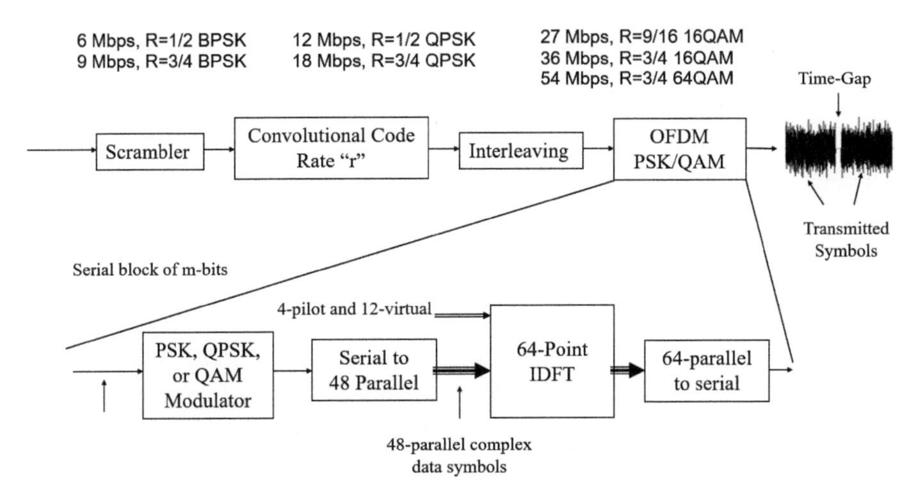

Figure 4.20 Block diagram of the OFDM transmission system of the IEEE 802.11 standard with details of various coding techniques involved in transmission.

used for transmission (see Chapter 4). For binary phase shift keying (BPSK), an $R = 1/2$ coding rate, and 1-bpS, the data rate is 12 (MSps) $\times$ 1/2 $\times$ 1 (bps) = 6 (Mbps) and for 64-QAM with 6-bps and convolutional coding of rate R = 3/4, we have 12 (MSps) $\times$ 3/4 (bps) $\times$ 6 (bps) = 54 (Mbps). As the distance between the transmitter and the receiver is increased, the data rate is reduced by adjusting the coding rate and the symbol transmission rate (size of the constellation). The fall-back data rates are 54 Mbps to 36, 27, 18, 12, 9, and, finally, 6 Mbps to cover distances of up to around 100 m. The guard time between two transmitted symbols is 800 ns compared to the symbol duration of 1/250 KSps = 4000 ns with a time utilization efficiency of 4000/4800 = 83%. The occupied bandwidth is 20 MHz providing a channel occupancy of 20 MHz/64 = 312.5 kHz per sub-channel. Therefore, the bandwidth efficiency is 250 KSps/312.5 kHz = 0.8 Symbols/s/Hz.

4.5.3 Space–Time Coding

STC techniques are used for wireless communication systems with multiple transmit antennas and single or multiple receive antennas. STC techniques are realized by introducing temporal and spatial correlation into the signals transmitted from different antennas. Using STC does not require increasing the total transmitted power or transmission bandwidth. The overall diversity gain of the STC technique results from combining the time diversity obtained from coding with the space diversity obtained from using multiple antennas. In wireless networks, a number of antennas can be deployed in the access point or base station while the mobile terminal's receiver usually has one main antenna with some possible support from other antennas depending on the size of the terminal. In traditional multiple access point or base station antenna systems, all transmit antennas carry the same signal and the signal received at each receiver antenna is the summation of all received signals from different transmit antennas. The mobile station combines the diversified signal from different antennas to optimize the performance. The basic principle of STC is to encode the transmitted symbols from different antennas at the base station and modify the receiver to take advantage of space and time diversity of the arriving signal from multiple transmitter antennas. Using STC at the base station, we can improve the performance of the downlink (base to mobile) channel significantly to support asymmetric applications such as Internet access, where the downlink data stream operates at a much higher rate than does the uplink data stream.

Using STC, significant increases in throughput over a single antenna system is possible with only two antennas at the base station and one or two antennas at the mobile terminal. It can be implemented for block [Ala98] or convolutional codes [Tar98, Nag98] with simple receiver structures. To show the basic concept of STC, we describe the simple two transmit and one received antenna block coding system known as Alamouti coded STC system [Ala98] in Appendix A. Also, Alamouti in [Ala98] introduced a simple MIMO scheme for two transmit and two receiving antennas, and the simulation results show that the performance is identical to a system that uses maximal ratio combining with one transmit and four receiver antennas. Therefore, Alamouti has shown that with his simple block coded STC for two transmit and one and two receiver antennas, one can obtain the same diversity performance as achieved with optimum maximum ratio combiner (MRC)with two and four received antennas. More elegant approaches that combine transmit diversity with channel coding, similar to Trellis-coded modulation (TCM), is also available in [Tar98], [Nag98]. A good overall overview of STC and its applications is provided in [Dha02].

4.5.4 Capacity MIMO Antenna Systems

MIMO antenna systems have recently emerged as one of the promising technologies for next generation wireless networks. In general, MIMO systems combine the transmitting scheme and the detection process in a way that the overall performance of the system is improved. In Section 4.4.2, we demonstrated the general diversity concept (e.g., by using a single transmit antenna and multiple receiver antennas, we can take advantage of space diversity) to substantially improve the performance of transmission techniques over wireless fading channels. In modern literature, this traditional approach to providing space diversity is referred to as single-input multiple-output (SIMO) antenna systems. In Section 4.4.3, we introduced STC as a method for implementation of MIMO systems where we have a number of transmitting antennas at the access point or the base station with one or a few receiver antennas at the mobile terminal. In this section, we discuss the bounds on the performance of a general MIMO system to show why, in recent years, this area has gained so much attention for the design of next generation wireless networks.

As we showed in Section 4.3.2, the normalized channel capacity in bits/s/Hz using Shannon–Hartley formula is given by Equation (4.7):

$$m = \frac{R_b}{W} = \log_2(1 + \frac{E_S}{N_0}).$$

Following the original derivations for the capacity of MIMO provided in [Tel95] and [Fos98b], the capacity of a MIMO channel is given by

$$m = \frac{R_b}{W} = \sum_{i=1}^{M} \log_2(1 + \frac{E_S}{N_0}\frac{1}{N}\lambda_i) = \sum_{i=1}^{M} \log_2(1 + \frac{mE_b}{N_0}\frac{1}{N}\lambda_i), \quad (4.11)$$

where λ_i's are the eigenvalues of the $N \times M$ cross-correlation matrix of the *channel gains* between the elements of the transmitter and the receiver antennas. For $M = N = 1$ and $\lambda_i = 1$, this equation is reduced to Equation (4.7) providing the bounds on the capacity for single transmitter and single receiver antenna systems.

To illustrate the bounds on performance improvement using MIMO systems, we assume the same number of transmitter and receiver antennas, $N = M$ and no interference among the signals from different receiving antennas so that the received signals are equal and uncorrelated, $\lambda_i = 1$. With these ideal assumptions, Equation (4.11) reduces to

$$m = \frac{R_b}{W} = N \log_2(1 + \frac{mE_b}{N_0}\frac{1}{N}) \quad . \quad (4.12)$$

Following the same algebraic manipulations used in Section 4.4.2, we can write this equation as

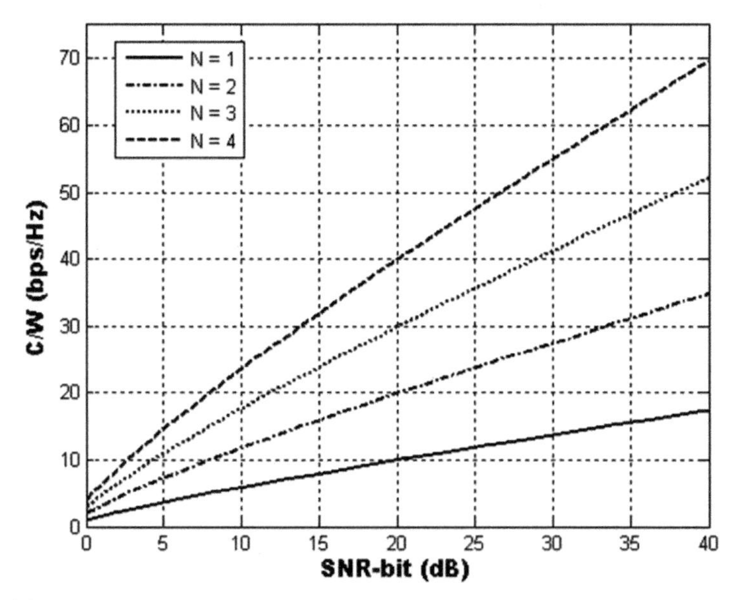

Figure 4.21 Comparison of the capacity of the SISO $N \times N$ MIMOs under ideal conditions.

$$\gamma_b = \frac{E_b}{N_0} = \frac{N}{m}(2^{\frac{1}{N} \times m} - 1). \tag{4.13}$$

Figure 4.21 shows the capacity in bps/Hz versus SNR per bit in dB for different values of *N*. As we increase the number of antennas from 1 × 1 SISO to 2 × 2, 3 × 3, and 4 × 4 MIMO antenna systems, we observe substantial increase in the channel capacity. This substantial increase of the capacity obtained by using MIMO antenna systems has motivated extensive research resulting in the emergence of new commercial technologies such as IEEE 802.11n and ac capable in multiple streaming and a huge increase in data rate.

Appendix 4A: STC for Two Transmitter and One Receiver

Figure A4.1 shows the basic operation of the traditional MRC with one transmitter and two receiver antennas. The transmitted symbol in the constellation, s_0, arrives at time T at two antennas with different random gains, $\{h_0, h_1\}$. The received sampled signal from two antennas is given by

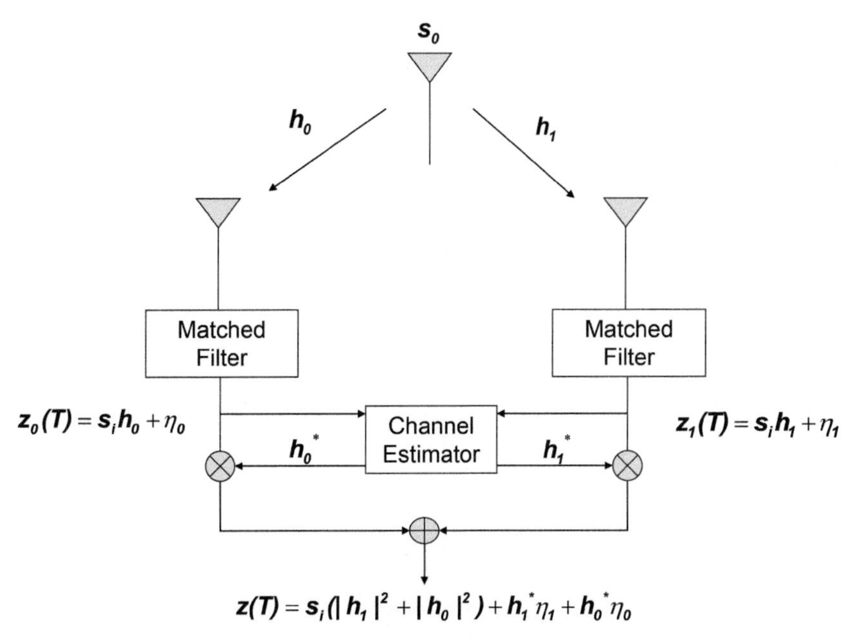

Figure A4.1 Implementation of the MRC for two transmit and one received antenna.

$$z_0(T) = s_0 h_0 + \eta_0$$
$$z_1(T) = s_0 h_1 + \eta_1,$$

where $\{\eta_0, \eta_1\}$ is the noise added by each channel. In MRC, each received signal is scaled with the signal strength of that antenna which is accomplished by multiplying the first branch by the estimated value of $h_o{}^*$ and second branch by estimated value of $h_1{}^*$. As shown in Figure A4.1, when we add the scaled branches of the diversified branches, we will have the received signal

$$z(T) = s_0(|h_1|^2 + |h_0|^2) + h_1{}^*\eta_1 + h_0{}^*\eta_0. \tag{A.1}$$

The received points in the constellation are scaled by $|h_1|^2 + |h_0|^2$and distorted by $h_1{}^*\eta_1 + h_0{}^*\eta_0$[2].

To show the basic concept of STC, we now describe the simple two transmit and one received antenna block coding system known as Alamouti coded STC system [Ala98]. Figure A4.2 illustrates the operation of the two transmit and one receive antenna system. Alamouti showed a simple transmission and detection scheme that reproduces Equation (A.1) with one-transmit and two-received antenna systems. Alamouti' s transmitted block

[2]The variance of this noise is $\left[|h_1|^2 + |h_0|^2\right] \sigma_\eta{}^2$

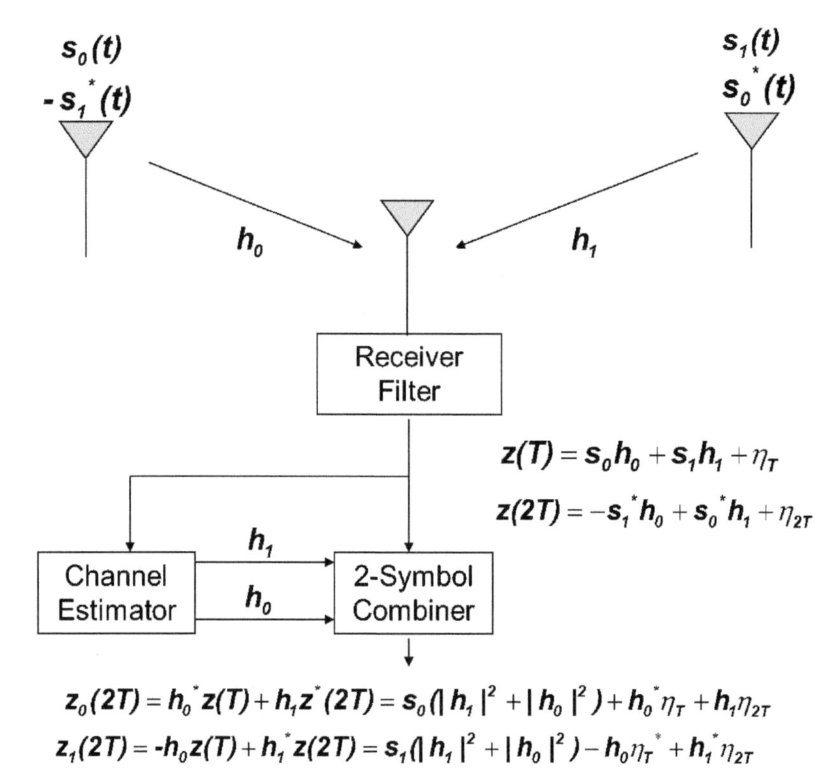

Figure A4.2 Simple Alamouti code for two transmit and one receive antennas.

code operates on a sequence of two symbols $\{s_0, s_1\}$, as shown in Figure A4.2, this sequence being time-coded into two sequences $\{s_0, -s_1{*}\}$ and $\{s_1, s_0{*}\}$ to be then space-coded by the first and second antennas, respectively. Each of these two sequences has two symbols that are transmitted in two consecutive time slots of one of the antennas in parallel with transmission of the other sequence in the other antenna. Between the two transmitter antennas and one receiver antenna, we have two channel gain factors h_0 and h_1 that are samples of two independent random processes representing the fading characteristics of the channel. Since all mobile channels are slow fading channels, the value of the channel gains during the transmission of two symbols remains the same. Therefore, the received signal after sampling at the receiver for the first and the second symbols are given by

$$z(T) = s_0h_0 + s_1h_1 + \eta_T$$

To form the decision variables for the two transmitted symbols, Alamouti suggests the following transformation on the received two samples to form two new variables observed at the end of the two intervals:

$$z_1(2T) = -h_0 z(T) + {h_1}^* z(2T) = s_1(|h_1|^2 + |h_0|^2) - h_0 {\eta_T}^* + {h_1}^* \eta_{2T}.$$

The statistical behavior of these decision variables is identical, and they are the same as the statistical behavior of the decision variable of the traditional MRC receiver shown by Equation (A.1). Consequently and not surprisingly, results of simulations for two transmit and one received antennas provided in [Ala98] are identical to the results of the analysis of MRC for one transmit antenna and two receiver antenna we presented earlier in Equation (4.12).

Assignments

Questions

a) In computer networking literature, sometimes they use the word bandwidth instead of data rate. Is that correct all the time? Explain why.
b) Discuss the advantages and disadvantages of differential Manchester coding, used in legacy Ethernet, in terms of usefulness to establish synchronization between the transmitter and the receiver and the bandwidth efficiency.
c) The word non-returns to zero used in line-coding techniques refers to what feature of the transmitted symbols?
d) Why are line transmission techniques not used in radio communications but used in wired and IR communications?
e) How does the number of symbols, symbol transmission rate, bandwidth, and data rate relate to one another?
f) How do we implement a system to carry the symbols of a 2D signal constellation over guided or wireless media?
g) Why is 5×5 PAM constellation not implemented for communication over wireless medium?
h) Why are bandwidth and power efficiency so important for wireless networking?
i) To maintain approximately the same error rate with the addition of 1 bit per symbol to a PAM system, how much increase in transmission power is needed?

j) What is the additional power for transmission of one additional bit per symbol in a QAM system?
k) Use Shannon–Hartley bounds to determine the maximum data rate achievable over a voiceband telephone channel with a bandwidth of 4 kHz and minimum SNR of 30 dB.
l) Use Figure 4.10 to give an estimate for the SNR requirement for 16-QAM modulation and compare that with the Shannon bound for 16-symbol transmission.
m) Figure 4.11 shows classification transmission techniques into band-limited and power-limited regions. What is the meaning of these terminologies?
n) Use Figure 4.13 to explain the difference between error patterns in fading wireless and a non-fading guided medium.
o) Explain why, in Figure 4.13 for the same bit error rate, we need a much higher SNR for the fading channel.
p) Use Figure 4.15 to explain why diversity techniques are so effective in improving the performance over fading channels.
q) Differentiate between frequency hopping and DSSS.
r) What is the difference between a MIMO system and a traditional system using multiple antennas to obtain space diversity?
s) What is the difference between OFDM and COFDM and why does COFDM has a better performance?

Instructor's solution available on River Publishers' website:
https://www.riverpublishers.com/book_details.php?book_id=919

Problems

Problem 1:

We want to transmit a 620 × 620 pixel image with 3 bytes per pixel using a 56 Kbps voiceband modem connecting Boston to San Francisco.

(a) How long would it take to transmit the picture over the channel?
(b) What is the propagation delay for communication between two mobile terminals?

Note: You can use Google map or other direction finding software to find the approximated distance between two cities.

Problem 2:

The standard pulse shape used in most short-distance cable communication applications such as RS232 is a rectangular pulse. If the voltage used for the amplitude of the pulses is $\pm A$ volt, the data rate is R, and the variance of the received noise is N_0, determine the SNR per bit.

Problem 3:

In the following differential encoded Manchester coded signal:

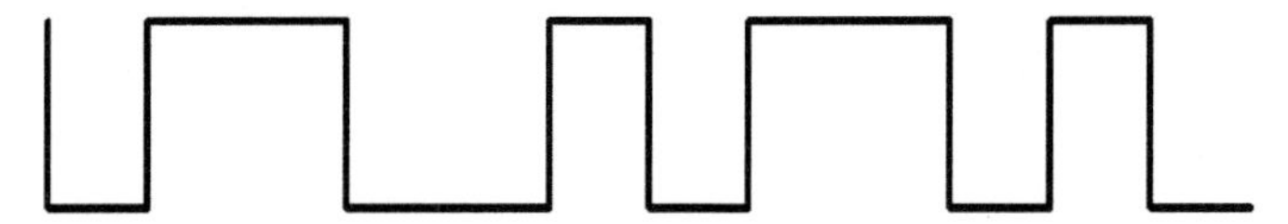

(a) Show the beginning and the end of each bit.
(b) Identify all the bits in the data sequence.
(c) Identify the bits if it was non-differential Manchester coded.

Problem 4:

For the given sequence of data

Data sequence: | 0 | 0 | 1 | 1 | 0 | 0 | 1 | 0 | 1 |

(a) Draw the waveform if the data is NRZ-I encoded.
(b) Draw the waveform if the data is RZ encoded.
(c) Draw the waveform if the data is Manchester encoded.
(d) Draw the waveform if the data is differential Manchester encoded.
(e) Draw the waveform if the data is AMI coded.

Problem 5:

Show how two binary phase shift keying transmissions can operate simultaneously in the same radio channel by using two carriers at the same frequency in phase quadrature. Draw a block diagram for the transmitters and receivers. Give the overall data rate for this quadrature carrier system as a function of available channel bandwidth.

Problem 6:

Assume that we have a BPSK modem operating on a voiceband channel with a symbol transmission rate of 2400 sps. We want to increase the data rate to 19.2 Kbps.

(a) If we increase the number of points in the constellation until the data rate becomes 19.2 Kbps while the baud rate remains at 2400 sps, what is the number of points in the constellation?

(b) What is the bandwidth efficiency of the modulation technique?
(c) What is the additional power required for the transmitter to keep the quality the same as before?

Problem 7:

(a) Give the 5×5PAM signal constellation and show the probability of transmission of each symbol.
(b) How can we implement this 2D constellation on the wires?
(c) Calculate the average number of bits per symbol for this constellation.
(d) If we use this constellation over a wireline with a bandwidth of 20 MHz, what would be the effective bit rate of the modem?

Problem 8:

(a) Derive the relation between the minimum distance and average energy in the constellation for the 16-QAM
(b) Repeat (a) for 5×5PAM constellations.
(c) Compare the two constellations and explain why 5×5PAM is not considered for wireless applications.

Problem 9:

The multi-carrier modem of the IEEE 802.11g uses 48 carriers for data transmission. If each carrier uses 64-QAM and the transmission rate of each carrier is 250 KSps, what is the overall transmission rate of the modem in symbol/second and bits/second?

Problem 10:

Consider the signal constellations for binary 16-PSK and 16-QAM modulations.

(a) Determine the average energy in each constellation E_s as a function of the minimum distance between the points in the constellation, d.
(b) Starting with Equation (3.6)

$$P_s \approx \frac{1}{2}\text{erfc}\left(\frac{d}{\sqrt{N_0}}\right) \geq \frac{1}{2}e^{-\frac{d}{\sqrt{N_0}}},$$

gives an approximate equation for calculation of error rate of each of these modems which relate the probability of the symbol error P_s to the average energy in the constellation E_s and the variance of the background noise, N_0.

(a) For the same expected error rate, determine the difference in power requirement for the two modems in dB.
(b) What are the advantages and disadvantages of 16-PSK versus 16-QAM modems?

Problem 11:

Suppose the maximum fade duration over a radio channel is 0.001 ms. Assume that all the bits are in error when a signal encounters a fade. What is the maximum number of consecutive bits in error for transmission through this channel if the data rate is 10 Kbps and if the data rate is 11 Mbps?

Problem 12:

For a 64-QAM modem

(a) Give the SNR at which the error rate over a telephone line is 10^{-5}.
(b) Give the average SNR at which the average error rate over a flat Rayleigh fading radio channel is10^{-5}.

Note: If you decided to use the erfc function to represent the error rate, you can use MATLAB erfinv function to calculate the inverse of this function.

Problem 13:

The IEEE 802.11a/g uses multiple modulation techniques in the same transmission bandwidth to provide different data rates. When the mobile terminal is close to the access point, a 64-QAM modulation is used, and, as the modem goes to the boundary of coverage of the access point, a BPSK modulation is used that requires substantially lower received signal strength to operate.

(a) If the data rate for the BPSK system is 12 Mbps, what is the data rate of the 64-QAM modem?
(b) What is the difference between the received signal strength requirement of the 64-QAM and BPSK modulation techniques?
(c) If the coverage with *64*-QAM is D meters, what is the coverage with BPSK modem when we operate in a large indoor open area with a distance-power gradient of $\alpha \approx 2$?
(d) Repeat (b) for an indoor office area with a distance-power gradient of $\alpha = 3$.

Problem 14:

(a) Use Equation (4.15) to calculate the bandwidth efficiency C/W of the 1×1, 2×2, and 4×4 MIMO systems when the SNR per bit is E_b/N_0 = 10 dB.

(b) Repeat (a) for E_b/N_0 = 6 dB.

Problem 15:

(a) Use Equation (4.16) to calculate the signal-to-noise requirement in dB for the bandwidth efficiency of C/W = 10 and a 1×1, 2×2, and 4×4 MIMO systems.

(b) Repeat (a) for C/W = 20.

Project 1: Error Rate and Phase Jitter in Phase Modulation

(a) Sketch a typical quadrature amplitude modulation (QPSK) signal constellation and assign the 2-bit binary codes to each point in the constellation. Define the decision line for the received signal constellation so that the receiver can detect received noisy symbols from each other.

(b) In Section 4.2.3, we discussed that the probability of symbol error for a multi-amplitude, multi-phase modem with coherent detection can be approximated by $P_s = 0.5\,\mathrm{erfc}\left(d/2\sqrt{N_0}\right)$, where d is the minimum distance between the points in the constellation and N_0 is the variance of the additive Gaussian noise. Use this equation to calculate the probability of error of the QPSK modems. Observe that if we consider the signal constellation and the decision lines of part (a), this equation can be modified to $P_s = 0.5\,\mathrm{erfc}\left(\delta/8\sqrt{N_0}\right)$, where δ is the minimum distance of a point in the constellation from a decision line.

(c) Use MATLAB or an alternative computation tool to plot the probability of symbol error versus SNR in dB. What are the SNRs (in dB) for the probability of symbol error of 10^{-2} and 10^{-3}? Let us refer to these two SNRs as SNR-2 and SNR-3.

(d) Simulate transmission of the QPSK signal corrupted by additive Gaussian noise for 10,000 transmitted bits. Generate random binary bits and use every 2 bits to select a symbol in the constellation of part (a), add complex additive white Gaussian noise to the symbol so that the signal to noise in dB is SNR-2, and use the decision lines to detect the symbols. Find the number of erroneous symbol decisions and divide it by the total number of symbols to calculate the symbol error rate. Compare the error rate with the expected error rate of 10^{-2}.

(e) Repeat (d) for SNR-3 and error rate of 10^{-3}.

(f) Assume that a channel produces a fixed phase error θ. Give an equation for the calculation of the probability of error for a QPSK modem operating over this channel. Use the minimum distance from a decision line, δ, and Equation $0.5\,\text{erfc}\left(\delta/8\sqrt{N_0}\right)$ for the calculation.

(g) Assume that the received signal-to-noise is 10 dB, and sketch the probability of error versus the phase error $0 < \theta < \pi/4$.

(h) Repeat (c), (d), and (e) for a channel with a phase shift of $\theta = \pi/8$.

Project 2: Error Rate and Phase Jitter in 16-QAM Modulation

(a) For a 16-QAM modem, use MATLAB to plot the probability of symbol error, P_S, versus E_S/N_0, where N_0 is the variance of the noise and E_S is the average energy per transmitted symbol.

(b) Repeat (a) for 10° phase errors at the receiver and compare the results with those of part (a). What are the SNRs (in dB) for the probability of symbol error of 10^{-2} and 10^{-3}?

(c) Repeat (a) and (b) for 64-QAM.

5

Medium Access Control

5.1 Introduction

This chapter presents an overview of the medium access methods commonly used to access the core networks. Access methods form a part of Layer 2 of the transmission control protocol/Internet protocol (TCP/IP) stack and Layer 3 of the IEEE 802 standard for local area networks (LANs) that are responsible for interacting with the medium to coordinate successful operation of multiple terminals over a shared access channel. Most multiple access methods were originally developed for wired networks and later adapted to the wireless medium. However, requirements on the wired and wireless media are different, thereby demanding modifications in the original protocols to make them suitable for the wireless medium. Today, the main differences between wireless and wired channels are the availability of bandwidth and reliability of transmissions. The wired medium includes optical media with enormous bandwidth and very reliable transmission (with error rates very close to zero all the time). Bandwidth in wireless systems is always limited because the medium (air) cannot be duplicated and also the medium is shared between all wireless systems that include multi-channel broadcast television, and a number of other bandwidths demanding applications and services. In the case of wired operation, we can always lay additional cables to increase the capacity, as needed, even if it is an expensive proposition. In a wireless environment, we can reduce the size of cells to increase capacity. With the reduction of the size of the cells, the number of cells increases, and with this, the need for improvements in the wired infrastructure to connect these cells increases. Also, the complexity of the network for handling additional handoffs and mobility management increases, posing a practical limitation upon the maximum capacity of the network. As far as transmission reliability is concerned, as we saw in Chapter 3, the wireless medium suffers from multipath and fading that causes a serious threat to reliable data transmission over the communication link. Since the wireless channel is so unreliable, as discussed in Chapters 3 and 4, people have developed several signal processing and coding techniques to improve transmission reliability over the wireless channel. In spite of these techniques, the reliability of the wireless medium is below that of the wired medium used as the backbone of the wireless networks.

Although in practice, we prefer to have the same access method and the same frame structure for wired backbone and the wireless access, wireless networks often use different packet sizes and a modified access method to optimize the performance to the specifics of the unreliable wireless medium. To avoid substantial overlap with existing literature, we use as examples the access methods used in wireless networks with a justification of why and how they are employed in different wireless networks.

The access methods adopted for access to the public switched telephone network (PSTN) and the Internet was traditionally quite different. Voice-oriented PSTN networks were designed for relatively long telephone conversations as the major application exchanging several megabytes of information in a two-way real-time conversation mode. A signaling channel that exchanges short messages between two calling components set up the call by obtaining communications resources (such as the link, switches, etc.) in the telephone network at the beginning of the conversation and terminates these arrangements by releasing the resources at the end of the call. The wireless access methods evolved for interaction with these networks assign a slot of time, a portion of frequency, or a specific code to a user preference for the entire length of the conversation. We refer to these techniques as centralized assigned-access methods. Data networks were originally designed for bursts of data and access to the Internet for which the supporting network does not have a separate signaling channel. In packet data communications, each packet carries some “signaling information” related to the address of the destination and the source. We refer to the access methods used in these networks as random-access methods accommodating randomly arriving packets of data to connect to the Internet. Certain local area data networks also *take turns* in accessing the medium as in the case of token passing and polling schemes. In some other cases, the random-access mechanisms are used to temporarily *reserve* the medium for transmitting the packet. The use of voice-over-IP (VoIP) for telephone conversations has blurred the distinction between voice-oriented and data-oriented networks. However, differentiation between the types of access schemes, namely assigned-access and random-access is still useful for teaching these methods. In the next two sections of this chapter, we provide a short description of assigned-access and random-access methods and methods to evaluate their performances, respectively.

5.2 Assigned Access to PSTN

Wireless access to the PSTN as well as the Internet with cellular networks uses assigned-access or channel partitioning techniques. In assigned-access to the medium, a fixed amount of allocated channel resources (frequency, time, or a spread spectrum code) is assigned on a predetermined basis to a single user for the duration of a communication session. The choice of the multiple-access method has a great impact on the capacity and quality of the service provided by a cellular telephone network. The three basic assigned multiple-access methods in 1G-3G cellular networks were frequency division multiple access (FDMA), time division multiple access (TDMA), and code division multiple access (CDMA), respectively. As cellular technology evolves into the 4G and 5G cellular networks the orthogonal frequency division multiple access (OFDMA) and multiple-input multiple-output (MIMO) technologies are optimized for computer communication rather than voice emerged in cellular network.

Another important design parameter related to the access method is the differentiation between the carrier frequencies of the forward (downlink – communication between the base station and mobile terminals) and reverse (uplink – communication between the mobile terminal and the base station) channels. If both forward and reverse channels use the same frequency band for communications, but the forward and reverse channels employ alternating time slots, the system is referred to as employing time division duplexing (TDD). If the forward and reverse channels use different carrier frequencies that are sufficiently separated, the duplexing scheme is referred to as frequency division duplexing (FDD). With TDD, since only one frequency carrier is needed for duplex operation, we can share more of the radio frequency (RF) circuitry between the forward and the reverse channels. The reciprocity of the channel in TDD allows for exact open-loop power control and simultaneous synchronization of the forward and reverse channels. TDD techniques are used in systems intended for low-power local area communications where interference must be carefully controlled and where low complexity and low power consumption are very important. Thus, TDD systems are often used in local area pico- or micro-cellular systems. FDD is mostly used in macro-cellular systems designed for coverage of several

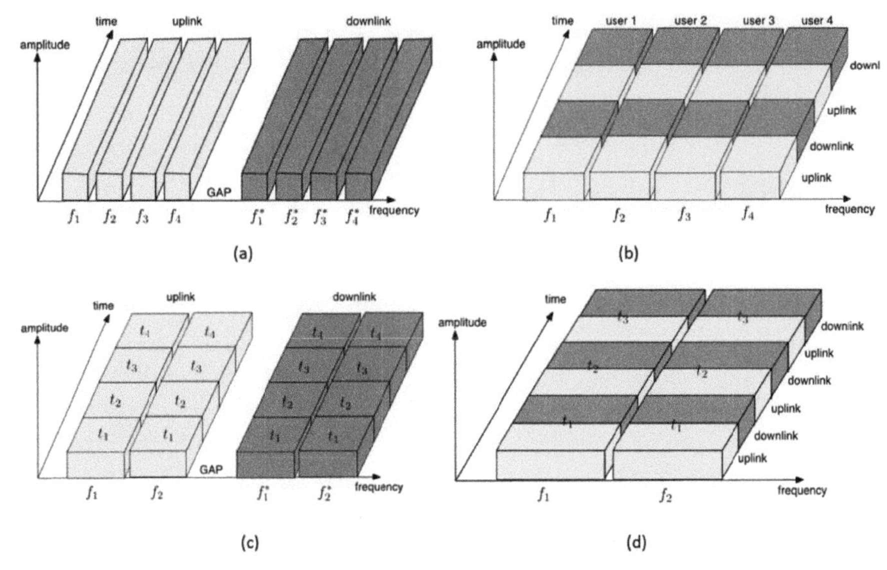

Figure 5.1 Various combinations of FDMA and TDMA with FDD and TDD options. (a) FDMA/FDD. (b) FDMA/TDD. (c) TDMA/FDD. (d) TDMA/TDD.

tens of kilometers where implementation of TDD is more challenging (see Figure 5.1).

5.2.1 1G Frequency Division Multiple Access

In an FDMA environment, all users can transmit signals simultaneously and they are separated from one another by their frequency of operation. The FDMA technique is built upon *frequency division multiplexing* (FDM). FDM is the oldest and still a commonly used multiplexing technique in the trunks connecting switches in the PSTN. It is also the choice of radio and TV broadcast as well as cable TV distribution. FDM is more suitable for analog technology since it is easier to implement. When FDM is used for channel access, it is referred to as FDMA.

Example 5.1: FDMA in AMPS with FDD:

Figure 5.2(a) shows the FDMA/FDD system commonly used in first generation analog cellular systems and several early cordless telephones. In FDMA/FDD systems, forward and reverse channels use different carrier frequencies and a fixed sub-channel pair is assigned to a user terminal during the communication session. At the receiving end, the mobile terminal filters

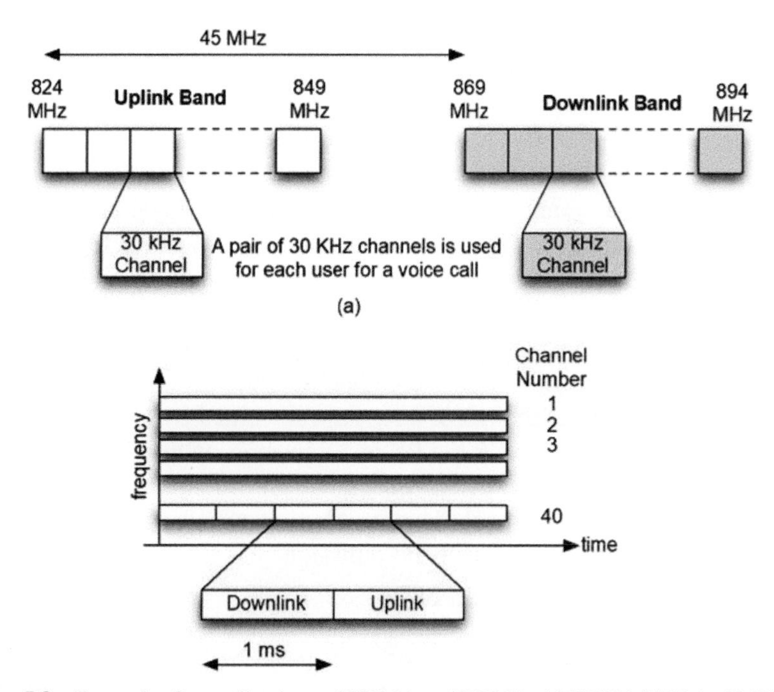

Figure 5.2 Examples for applications of TDMA and FDMA. (a) FDMA/FDD in AMPS. (b) FDMA/TDD in CT-2.

the designated channel out of the composite signal. The early analog cellular telephone system called advanced mobile phone system (AMPS) allocated 30 kHz of bandwidth for each of the forward and reverse channels. The result is a total of 421 channels in 25 MHz of spectrum assigned to each direction – 395 of these channels were used for voice traffic and the rest for signaling.

Example 5.2: FDMA in CT-2 with TDD:

Figure 5.2(b) shows an FDMA/TDD system used in the CT-2 digital cordless telephony standard. Each user employs a single carrier frequency for all communications. The forward and reverse transmissions take turns via alternating time slots. This system was designed for distances of up to 100 m and a voice conversation is based on 32 kbps voice coding. The total allocated bandwidth for CT-2 is 4 MHz supporting 40 carriers, each using 100 kHz of bandwidth.

The designer of an FDMA system must pay special attention to adjacent channel interference, in the reverse channel. In both forward and reverse channels, the signal transmitted must be kept confined within its assigned band, at least to the extent that the out-of-band energy causes negligible interference to the users employing adjacent channels. The operation of the forward channel in wireless FDMA networks is very similar to wired FDM networks. In forward wireless channels, in a manner like that of wired FDM systems, the signal received by all mobile terminals have the same received power and interference is controlled by adjusting the sharpness of the transmitter and receiver filters for the separate carrier frequencies. The problem of adjacent channel interference is much more challenging on the reverse channel. On the reverse channel, mobile terminals will be operating at different distances from the base station (BS). The received signal strength (RSS) at the BS of a signal transmitted by a mobile terminal close to the BS, and the RSS at the BS of a transmission by a mobile terminal at the edges of the cell are often substantially different, causing problems in detecting the weaker signal. This problem is usually referred to as the near–far problem. If the out-of-band emissions are large, they may swamp the actual information-carrying signal.

Example 5.3: Near–Far Problem:

a) *What is the difference between the RSS of two terminals located 10 m and 1 km from a base station in an open area?*
b) *Explain the effects of shadow fading on the difference in the RSSs.*
c) *What would be the impact if the two terminals were operating in two adjacent channels? Assume out-of-band radiation that is 40 dB below the main lobe.*

Solution:

a) *As we saw in Chapter 3, the RSS falls by around 40 dB per decade of distance in open areas. Therefore, the received powers from a mobile terminal that is 10 m from a BS and another, that is at a distance of 1 km, are 80 dB apart.*
b) *In addition to the fall of the RSS with distance, we also discussed the issues of multipath and shadow fading in radio channels that cause power fluctuations in the order of several tens of dBs. Therefore, the difference in the received powers due to the near–far problem may exceed even 100 dB.*
c) *If the out-of-band emission is only 40 dB below that of the transmitted power, it may exceed the strength of the information-bearing signal by almost 60 dB.*

To handle the near–far problem, FDMA cellular systems adopt two different measures. First, when frequencies are assigned to a cell, they are grouped such that the frequencies in each cell are as far apart as possible. The second measure employed is power control to keep terminals close to the base station at low transmission and only mobile devices at the edge of a cell transmit at the highest power. This approach reduces the overall interference from all mobile terminals and saves the battery life of the mobile. In addition, whenever FDMA is employed, *guard bands*[1] are also used in between frequency channels to further reduce adjacent channel interference. This, however, has the effect of reducing the overall spectrum efficiency.

5.2.2 2G Time Division Multiple Access

In TDMA systems, several users share the same frequency band by taking assigned turns in using the channel. The TDMA technique is built upon the *time division multiplexing* (TDM) scheme commonly used in the trunks for telephones systems. The major advantage of TDMA over FDMA is its format flexibility. Because the format is completely digital and provides the flexibility of buffering and multiplexing functions, time-slot assignments among multiple users are readily adjustable to provide different access rates for different users. This feature is particularly adopted in the PSTN and the TDM scheme forms the backbone of all digital connections in the heart of the PSTN. The hierarchy of digital transmission trunks used in North America is the so-called T-carrier system that has an equivalent European system (the E-carriers) approved by the international telecommunication union (ITU). In the hierarchy of digital transmission rates standardized throughout North America, the basic building block is the 1.544 Mbps link known as the T-1 carrier. A T-1 transmission frame is formed by time division multiplexing 24 pulse code modulation (PCM) encoded voice channels each carrying 64 kbps of user data. Service providers often lease T-carriers to interconnect their own switches and routers and for forming their own networks.

With TDMA, a transmit controller assigns time slots to users, and an assigned time slot is held by a user until the user releases it. At the receiving end, a receiver station synchronizes to the TDMA signal frame and extracts the time slot designated for that user. The heart of this operation is synchronization that was not needed for FDMA systems. The TDMA concept was developed in the 1960s for use in digital satellite communication

[1] When we say each AMPS channel is 30 kHz wide in Example 5.5, this also includes guard bands.

systems and first became operational commercially in telephone networks in the mid-1970s.

In cellular telephone networks, the migration to TDMA from FDMA took place in the 2G systems. The first cellular standard adopting TDMA was GSM. The GSM standard was initiated to support international roaming among Scandinavian countries in particular and the rest of Europe in general. The digital voice adoption in TDMA format facilitated the network implementation, resulted in improvements in the quality of the voice, and provided a flexible format to integrate data services in the cellular network. The FDMA systems in the US very quickly observed a capacity crunch in major cities and among the options for increasing capacity, a standard for American TDMA (IS-54/IS136) evolved and deployed, but it was ultimately replaced either by the IS-95 CDMA or the GSM. TDMA was also adopted in cordless telephones such as CT-2 and digital enhanced cordless telephony (DECT) to provide format flexibility and to allow more compact and low-power terminals.

Example 5.4: TDMA in GSM:
Figure 5.3 shows an FDMA/TDMA/FDD channel used in GSM. The particular example shows the eight-slot TDMA scheme used in the GSM system. Forward and reverse channels use separate carrier frequencies (FDD). Each carrier can support up to eight simultaneous users via TDMA, each using a 13 kbps encoded digital speech, within a 200 MHz carrier bandwidth. A total of 124 frequency carriers (FDMA) are available in the 25 MHz allocated band in each direction. 100 kHz of the band is allocated as a guard band at each edge of the overall allocated band.

Due to the near–far problem, the received signal on the reverse channel from a user occupying a time slot can be much larger than the received power from the terminal using the adjacent time slot. In such a case, the receiver will have difficulties distinguishing the weaker signal from the background noise. In a manner like FDMA systems, TDMA systems also use power control to handle this near–far problem.

Capacity of 1G/2G TDMA/FDMA Cellular Systems:
The capacity of a cellular telephone network is the maximum number of cellular telephones available per one cell of the service provider network. In the case of 1G TDMA or 2G FDMA systems, capacity depends on the reuse factor (how many cells apart can the same frequencies be reused), N_f. The total number of carriers owned by a service provider is given by the

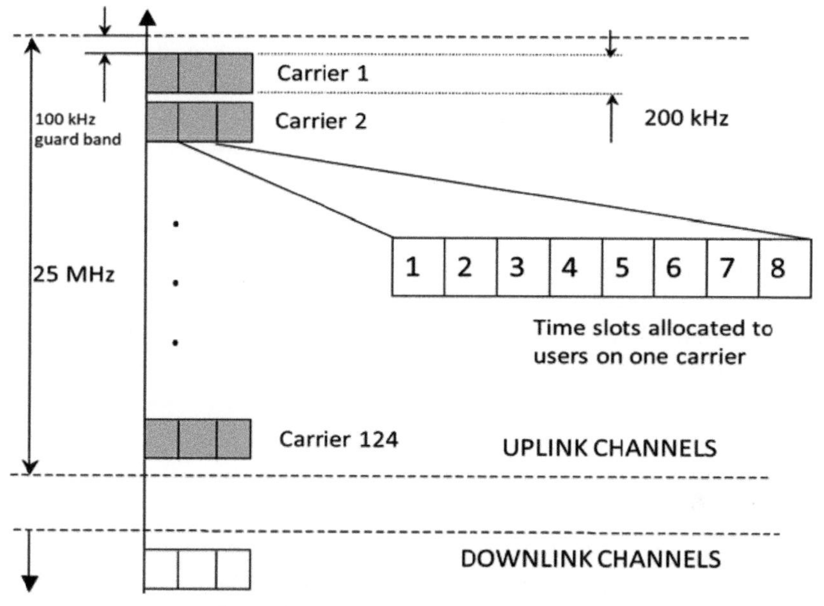

Figure 5.3 FDMA/TDMA/FDD in GSM.

total available bandwidth, W, divided by the bandwidth of one carrier, B. The number of carriers per cell is the number of carriers divided by the frequency reuse factor. In FDMA, each carrier carries one user, and in TDMA networks, each carrier carries m users. So, the number of simultaneous users supported will be $M = m \times W/B$. If the frequency reuse factor is, N_f , then the number of simultaneous users per cell will be

$$M = \left(\frac{W}{B}\right)\frac{m}{N_f}. \tag{5.1}$$

Example 5.5: Comparison of the Capacity of Different AMPS and GSM:

Each service provider in the US was originally assigned with 12.5 MHz of bandwidth using AMPS analog FDMA standard. In AMPS, a bandwidth of the user carrier is 30 kHz and the frequency reuse factor is 7. Later on, some of the service providers resorted to the GSM TDMA standard with a bandwidth per carrier of 200 kHz accommodating eight users per carrier with a frequency reuse factor of 3. What was the capacity gain by moving from AMPS to GSM?

Solution:

For the GSM standard with a carrier bandwidth of B = 200 kHz, and the number of users per carrier of m = 8, and frequency reuse factor of N_f = 3 for W = 12.5 MHz of bandwidth provides for

$$M = \frac{W}{B}\frac{m}{N_f} = \frac{12.5\,\text{MHz}}{200\,\text{kHz}}\frac{8}{3} = 165$$

users per cell. For the 1G analog system with carrier bandwidth of B = 30 kHz, number of users per carrier of m = 1 and frequency reuse factor of N_f = 7 (commonly used in these systems) each W = 12.5 MHz of bandwidth provides for

$$M = \frac{W}{B}\frac{1}{N_f} = \frac{12.5\,\text{MHz}}{30\,\text{kHz}}\frac{1}{7} = 59$$

users per channel. Therefore, GSM increases the network capacity close to three times, 165/59 = 2.8, in addition to providing an integrated voice and data network with SMS services.

5.2.3 3G Code Division Multiple Access (CDMA)

CDMA was first used by QUALCOMM as an alternative to the 2G American digital cellular IS-54 around the 1990s. Then, it was adopted by Europeans as the 3G standard evolving from 2G GSM. With the growing interest in higher data rates and integration of voice, data, and video traffic in the late 1990s, CDMA appeared increasingly attractive as the wireless access method of choice for 3G cellular. Fundamentally, integration of various types of traffic is readily accomplished in a CDMA environment since coexistence in such an environment does not require any specific coordination among user terminals. In principle, CDMA can accommodate various wireless users with different bandwidth requirements, switching methods, and technical characteristics without any need for coordination. Of course, since each user signal contributes to the interference seen by other users, power control techniques are essential in the efficient operation of a CDMA system. The QUALCOMM CDMA implemented on direct sequence spread spectrum (DSSS) technology transmitted with a carrier bandwidth of 1.25 MHz and W-CDMA adopted by EU had a minimum bandwidth per carrier of 3 MHz. As compared with GSM with 200 kHz bandwidth and American digital cellular IS-136 with a bandwidth of 3 kHz, CDMA's wider bandwidth and format flexibility allows the implementation of a variety of data rates per user channel. The wide bandwidth and format flexibility for the integration of multimedia made CDMA the choice of 3G to achieve data rates on the order of 2 Mbps.

To illustrate CDMA and how it is related to FDMA and TDMA, it is useful to think of the available band and time as resources we use to share among multiple users. In FDMA, the frequency band is divided into slots, and each user occupies that frequency throughout the communication session. In TDMA, a larger frequency band is shared among the terminals, and each user uses a slot of time during the communication session. As shown in Figure 5.4, in a CDMA environment, multiple users use the same band at the same time and the users are differentiated by a code that acts as the key to identify those users. These codes are selected so that when they are used at the same time in the same band, a receiver knowing the code of a particular user can detect that user among all the received signals. In CDMA/FDD (Figure 5.5(a)), the forward and reverse channels use different carrier frequencies. If both transmitter and receiver use the same carrier frequency (Figure 5.5(b)), the system is CDMA/TDD.

In CDMA, each user is a source of noise to the receiver of other users and if we increase the number of users beyond a certain value, the entire system collapses because the signal received in each specific receiver will be buried under the noise caused by many other users. An important question is: how many users can simultaneously use a CDMA system before the system collapses. We investigate this answer below.

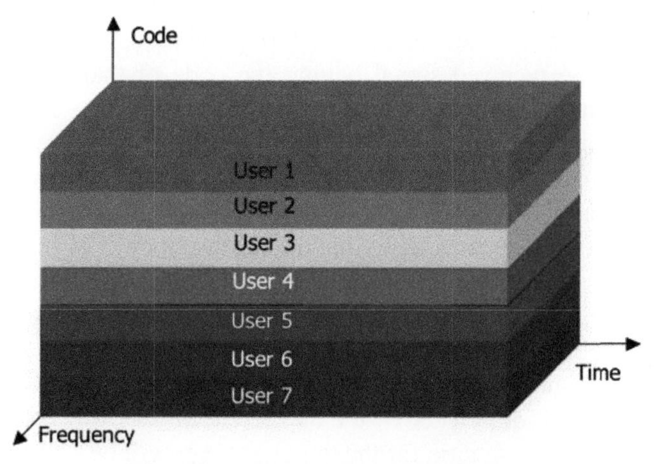

Figure 5.4 Simple illustration of CDMA.

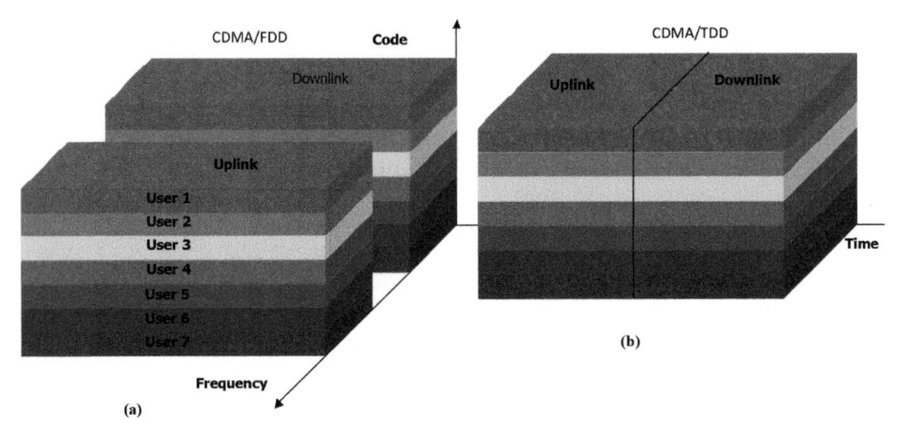

Figure 5.5 (a) Frequency and (b) time division duplexing with CDMA.

Capacity of 3G CDMA:

CDMA systems are implemented based on direct sequence spread spectrum transmission at the physical layer that was presented in Chapter 4. In its most simplified form, a spread spectrum transmitter spreads the signal power over a spectrum N times wider than the spectrum of the message signal. In other words, an information bandwidth of R_b occupies a transmission bandwidth of W, where

$$W = N_p R_b. \tag{5.2}$$

The spread spectrum receiver processes the received signal with a *processing gain* of N_p. This means that during the processing at the receiver, the power of the received signal having the code of that receiver will be increased N_p times beyond the value before processing.

Let us consider the situation of a *single cell* in a cellular system employing CDMA. Assume that we have M simultaneous users on the reverse channel of a CDMA network. Further, let us assume that we have an ideal power control enforced on the channel so that the received power of signals from all terminals has the same value P. Then, the received power from the target user after processing at the receiver is $N_p\,P$ and the received interference from $M - 1$ other terminals is $(M - 1)P$.

If we also assume that a cellular system is an interference limited and the background noise is dominated by the interference noise from other users, the received signal-to-noise ratio for the target receiver will be

$$S_r = \frac{N_p P}{(M-1)P} = \frac{N_p}{M-1}. \tag{5.3a}$$

All users always have a requirement for the acceptable error rate of the received data stream. For a given modulation and coding specification of the system, that error rate requirement will be supported by a minimum S_r requirement that can be used in Equation (5.2) to solve for the number of simultaneous users. Then, solving Equations (5.1) and (5.2) for *M*, we will have

$$M = \frac{W}{R_b}\frac{1}{S_r} + 1 \cong \frac{W}{R_b}\frac{1}{S_r} \quad (5.3b)$$

Example 5.6: Capacity of One CDMA Carrier:

Using quadrature phase keying (QPSK) modulation and convolutional coding, the IS-95 digital cellular systems require a signal-to-noise ratio in the range of 2 (3 dB) $< S_r <$ 8 (9 dB). The bandwidth of the channel is 1.25 MHz and transmission rate is R = 9600 bps. Find the capacity of a single IS-95 cell.

Solution:

Using Equation (5.3b), we can support

$$M = \frac{1.25 \text{ MHz}}{9600 \text{ bps}}\frac{1}{8} \approx 16 \qquad M = \frac{1.25 \text{ MHz}}{9600 \text{ bps}}\frac{1}{2} \approx 65$$

users for minimum and maximum signal-to-noise ratios.

Practical Considerations:

In the practical design of digital cellular systems, three other parameters affect the number of users that can be supported by the system as well as the bandwidth efficiency of the system. These are the number of sectors in each base station antenna, the voice activity factor, and the interference increase factor. These parameters are quantified as factors used in the calculation of the number of simultaneous users that the CDMA system can support. The use of sectored antennas is an important factor in maximizing bandwidth efficiency. Cell sectorization using directional antennas reduces the overall interference, increasing the allowable number of simultaneous users by a *sectorization gain factor,* which we denote by G_A. With ideal sectorization, the users in one sector of a base station antenna do not interfere with the users operating in other sectors, and $G_A = N_{sec}$ where N_{sec} is the number of sectors in the cell. In practice, antenna patterns cannot be designed to have ideal characteristics, and due to multipath reflections, users, in general, communicate with more than one sector. Three-sector base station antennas are commonly used in cellular systems, and a typical value of the sectorization gain factor is G_A = 2.5 (4 dB). The voice activity interference reduction factor G_v is the ratio of the total connection time to the active talkspurt time. On average, in a two-way conversation, each user talks roughly 50% of the time. The short pauses in the flow of natural speech reduce the activity factor further to about 40% of the connection time in each direction. As a result, the typical number used for G_v is 2.5 (4 dB). The interference increase factor H_0 accounts for users in other cells in the CDMA system. Since all neighboring cells in a CDMA cellular network operate at the same frequency, they will cause additional interference. This interference is relatively small due to the processing gain of the system and the distances involved; a value of H_0 = 1.6 (2 dB) is commonly used in the industry.

Incorporating these three factors as a correction to Equation (5.3), the number of simultaneous users that can be supported in a CDMA cell can be approximated by

$$M = \frac{W}{R_b}\frac{1}{S_r} + 1 \cong \frac{W}{R_b}\frac{1}{S_r}\frac{G_A G_v}{H_0}. \tag{5.4}$$

If we define the *performance improvement factor* in a digital cellular system as

$$K_p = \frac{G_A G_v}{H_0} \tag{5.5}$$

assuming the typical parameter values given earlier, the performance improvement factor is K_p = 4 (6 dB).

Example 5.7: Capacity of One CDMA Carrier with Correction Factors:

Determine the single carrier IS-95 CDMA capacity with correction for sectorization and voice activity. Use the numbers from Example 5.5.

Solution:

If we continue the previous example with the new correction factor included, the range for the number of simultaneous users becomes $64 < M < 260$.

5.2.4 Comparison of CDMA, TDMA, and FDMA

CDMA was, by far, the most successful multiple access schemes in 2G cellular wireless systems in the US. With wideband CDMA adopted as the multiple access scheme of choice in 3G cellular networks, one wonders why CDMA became the favorite choice for wireless access in voice-oriented networks. Spread spectrum technology became the favorite technology for military applications because of its capability to provide a low probability of interception and strong resistance to interference from jamming. In the cellular industry, CDMA was introduced as an alternative to TDMA to improve the capacity of second-generation cellular systems in the USA. As a result, much of the early debates in this area were focused on the calculation of the capacity of CDMA as it is compared with TDMA. However, capacity is not the only reason for the success of CDMA technology. As a matter of fact, calculation of the capacity of CDMA using the simple approach provided above is *not* very conclusive and is subject to several assumptions such as perfect power control that cannot be practically met. The first CDMA service providers in the US were using slogans such as "you cannot believe your ears!" to address the superior quality of voice for the CDMA. However, the superiority of voice is partially dependent on the speech coder and it is not a CDMA versus TDMA issue. To provide a good explanation for the success of complex and multi-disciplinary technology, such as a cellular network, addressing consumer market issues has always been very important.

In the rest of this section, we bring out several issues that may enlighten the reader toward a deeper understanding of the technical aspects of CDMA systems as they are compared to TDMA and FDMA networks.

Format Flexibility:

As we discussed before, telephone voice was the dominant source of income for the telecommunication industry up to the end of the past century. In the new millennium, the strong emergence of the Internet created a case for other popular multimedia applications. To support a variety of data rates with different requirements, a network needs format flexibility. As we discussed earlier, one of the reasons for migrating from analog FDMA to digital TDMA was that TDMA provides a more flexible environment for the integration of voice and data. The time slots of a TDMA network designed for voice transmission can be used individually or in a group format to transmit data from users and to support different data rates. However, all these users should be time synchronized and the quality of the transmission channel is the same for all of them. The chief advantage of CDMA relative to TDMA was its flexibility in timing and the quality of transmission. In CDMA, users are separated by their codes and unaffected by the transmission time relative to other users. The power of the user can also be adjusted with respect to others to support a certain quality of transmission. In CDMA, each user is far more liberated from the other users allowing a fertile setting to accommodate different service requirements to support a variety of transmission rates with different quality of transmission to support multimedia or any other emerging application.

Performance in Multipath Fading:
As we saw in Chapter 3, multipath in wireless channels causes frequency selective fading. In frequency selective fading, when the transmission band of a narrowband system coincides with the location of the fade, no useful signal is received. As we increase the transmission bandwidth, fading will occupy only a portion of the transmission band, providing an opportunity for a wideband receiver to take advantage of the portion of the transmission band not under fade and provide a more reliable communication link. In Chapter 4, we introduced spread spectrum, OFDM, and MIMO as technologies that can be employed in wideband systems to handle frequency selective fading. The wider the bandwidth, the better is the opportunity for averaging out the fading frequency.

These technologies were not used in the 1G analog cellular FDMA systems because they were analog systems and these techniques are digital. The Pan European GSM digital cellular system uses 200 kHz of the band and the standard recommends using channel equalization. The North American digital cellular system, IS-136, uses digital transmission over the same analog band of 30 kHz of the North American AMPS system and does not recommend equalization because the bandwidth is not very large. The bandwidth of IS-95 CDMA system is 1.25 MHz, and W-CDMA systems for 3G networks use bandwidths that are as high 10 MHz. Rake receivers are used to increase the benefits of wideband transmission by taking advantage of the so-called in-band or time diversity of the wideband signal. This is one of the reasons for having a better quality of voice in CDMA systems. As we mentioned earlier, the quality of voice is also affected by the robustness of the speech-coding algorithm, coverage of service, methods to handle interference, handoffs, and power control as well.

System Capacity:
Figure 5.6 provides a summary of capacity formulas for the FDMA, TDMA, and CDMA assigned-access methods derived in Sections 5.2.2 and 5.2.3. Comparison of the capacity depends on a number of issues including the frequency reuse factor, speech coding rate, and the type of antenna. Therefore, a fair comparison would be difficult unless we go to practical systems. The

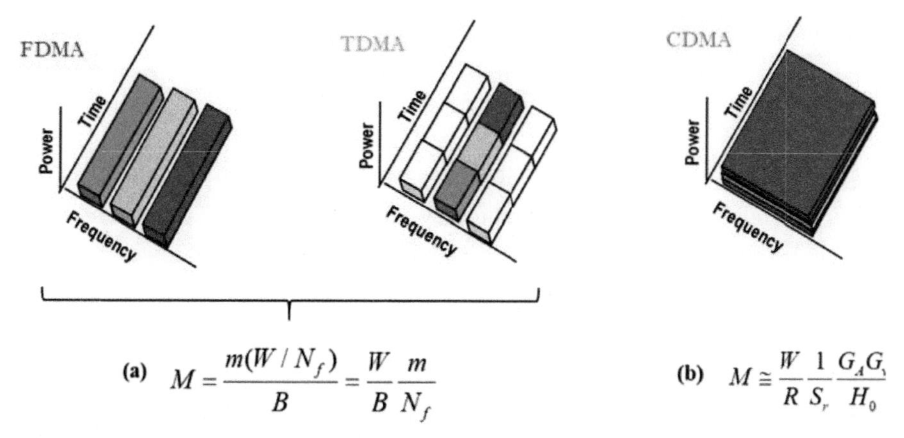

Figure 5.6 Summary of capacity for assigned-access methods: (a) for FDMA and TDMA; (b) for CDMA.

following simple example compares the capacity of FDMA, TDMA, and CDMA used in debates to evaluate alternatives for the 2G TDMA (IS-136) and CDMA (IS-95) North American cellular systems to replace the 1G analog (AMPS).

Example 5.8: Comparison of the Capacity of Different 2G Systems

Compare the capacity of 2G CDMA with 1G FDMA and 2G TDMA systems. For the CDMA system, assume an acceptable signal to interference ratio of 6 dB, the data rate of 9600 bps, a voice duty cycle of 50%, effective antenna separation factor of 2.75 (close to ideal three-sector antenna), and neighboring cell interference factor of 1.67.

Solution:

For the 2G CDMA system, using Equation (4) for each carrier with $W = 1.25$ MHz, $R_b = 9600$ bps, $S_r = 4$ (6 dB), $G_v = 2$ (50% voice activity), $G_A = 2.75$, and $H_0 = 1.67$, we have

$$M = \frac{W}{R_b}\frac{1}{S_r}\frac{G_A G_v}{H_0} = 108 \text{ users per cell.}$$

For the 2G TDMA system with a carrier bandwidth of $B = 30$ kHz, number of users per carrier of $m = 3$, and frequency reuse factor of $N_f = 4$ (commonly used in these systems) each W = 1.25 MHz of bandwidth provides for

$$M = \frac{W}{B}\frac{m}{N_f} = 31.25 \text{ users per cell.}$$

For the 1G analog system with carrier bandwidth of B = 30 kHz, number of users per carrier of m = 1 and frequency reuse factor of N_f = 7 (commonly used in these systems) each W = 1.25 MHz of bandwidth provides for

$$M = \frac{W}{B}\frac{1}{N_f} = 6 \text{ users per channel.}$$

Handoff:

Handoff occurs when a received signal in a mobile station (MS) becomes weak and another BS can provide a stronger signal to the MS. The first generation FDMA cellular systems often used the so called hard-decision-handoff in which the base station controller monitors the received signal from the BS and, at the appropriate time, switches the connection from one BS to another. TDMA systems use the so-called mobile-assisted handoff in which the mobile station monitors the received signal from available BSs and reports it to the base station controller which then makes a decision on the handoff. Since adjacent cells in both FDMA and TDMA use different frequencies, the MS has to disconnect from and reconnect to the network that will appear as a click to the user. Handoffs occur at the edge of the cells when the received signals from both BSs are weak. The signals also fluctuate anyway because they are arriving over radio channels. As a result, decision-making for the handoff time is often complex and the user experiences a period of poor signal quality and possibly several clicks during the completion of the handoff process. Since adjacent cells in a CDMA network use the same frequency, a mobile moving from one cell to another can be made "seamless" handoff by the use of signal combining. When the mobile station approaches the boundary between cells, it communicates with both cells. A controller combines the signals from both links to form a better communication link. When a reliable link has been established with the new base station, the mobile stops communicating with the previous base station, and communication is fully established with the new base station. This technique is referred to as soft handoff. Soft handoff provides a dual diversity for the received signal from two links that improve the quality of reception and eliminates clicking as well as the ping-pong problem.

Power Control:

As we discussed earlier in this chapter, power control is necessary for FDMA and TDMA systems to control adjacent channel interference and mitigate the unexpected interference caused by the near–far problem. In FDMA and TDMA systems, some sort of power control is needed to improve the quality of the voice delivered to the user. In CDMA, however, the capacity of the system depends *directly* on the power control, and an accurate power control mechanism is needed for the proper operation of the network. With CDMA, power control is the key ingredient in maximizing the number of users that can operate simultaneously in the system. As a result, CDMA systems adjust the transmitted power more often and with smaller adjustment steps to support a more refined control of power. Better power control also saves on the transmission power of the MS that increases the life of the battery. The more refined power control in CDMA systems also helps in power management of the MS that is an extremely important practical issue for users of the mobile terminals.

Implementation Complexity:

Spread spectrum is a two-layer modulation technique requiring greater circuit complexity than conventional modulation schemes. This, in turn, will lead to higher electronic power consumption and larger weight and cost for mobile terminals. Gradual improvements in battery and integrated circuit technologies, however, have made this issue transparent to the user.

Fixed assignment access methods were used with circuit-switched cellular telephone networks. In these networks, in a manner similar to the wired multi-channel environments, the performance of the network is measured by the blockage rate of an initiated call. A call does not go through for two reasons: (1) when the calling number is not available; (2) when the telephone company is out of resources to provide a line for the communication session. In plain old telephone service (POTS), for both cases, the user hears a busy tone signal and cannot distinguish between the two types of blockage. In most cellular systems, however, type (1) blockage results in a response that is a busy tone and type (2) with a message like "all the circuits are busy at this time please try your call later." In the rest of this book, we refer to blockage rate only as type (2) blockage rate. The statistical properties of the traffic offered to the network are also a function of time. The telephone service providers often design their networks so that the blockage rate at the peak traffic is always below a certain percentage. Cellular operators often try to keep this average blockage rate below 2%.

The blockage rate is a function of the number of subscribers, a number of initiated calls, and the length of the conversations. In telephone networks, the Erlang equations are used to relate the probability of blockage to the average rate of the arriving calls and the average length of a call. In wired networks, the number of lines or subscribers that can connect to a multi-channel switch is a fixed number. The telephone company monitors the statistics of the calls over a long period of time and upgrades the switches with the growth of subscribers so that the blockage rate during peak traffic times remains below the objective value. In circuit-switched cellular telephone networks, the number of subscribers operating in a cell is also a function of time. In the downtown areas, everyone uses their cellular telephones during the day, and in the evenings, they use them in their residential area that is covered by a different cell. Therefore, traffic fluctuations in cellular telephone networks are much more than the traffic fluctuations in POTS. In addition, telephone companies can easily increase the capacity of their networks by increasing their investment in the number of transmission lines and quality of switches supporting network connections. In wireless networks, the overall number of available channels for communications is ultimately limited by the availability of the frequency bands assigned for network operation. To respond to the fluctuations of the traffic and cope with the bandwidth limitations, cellular operators use complex frequency assignment strategies to share the available resources in an optimal manner.

5.2.5 Traffic Engineering Using the Erlang Equations

The Erlang equations are the core of traffic engineering for telephony applications. The two basic equations used for traffic engineering are Erlang B and Erlang C equations. The Erlang B equation relates the probability of blockage $B(N,\rho)$ to the number of channels N_u and the normalized call density in units of channels ρ. The Erlang B formula is

$$B(N_u,\rho) = \frac{\rho^{N_u}/N_u!}{\sum_{i=0}^{N_u}(\rho^i/i!)}, \tag{5.6}$$

where $\rho = \lambda/\mu$, λ is the call arrival rate and μ is the service rate of the calls.[2]

[2] The equation assumes that the arrivals are Poisson and the service rate is exponential. For details, see [BER87].

Example 5.9: Call Blocking Using Erlang-B Formula:

We want to provide a wireless public phone service with five lines to a ferry crossing between Helsinki and Stockholm carrying 100 passengers, where, on average, each passenger makes a 3-minute telephone call every 2 hours. What are the probability of a passenger approaching the telephones and none of the four lines being available?

Solution:

In practice, often, the probability of call blockage is given, and we need to calculate the number of subscribers. Here, we need an inverse function for the Erlang equation that is not available. As a result, several tables and graphs are available for this inverse mapping. Figure 5.7 shows a graph relating the probability of blockage $B(N_u, \rho)$ to the number of channels N_u and the normalized traffic per available channels ρ. From this graph, we can estimate the blocking probability. The traffic load is 100 users $\times$ 1 call/user $\times$ 3 minutes/call per 120 minutes = 2.5 Erlangs. Since there are five lines available and the traffic is 2.5 Erlangs, the blocking probability is roughly 0.07.

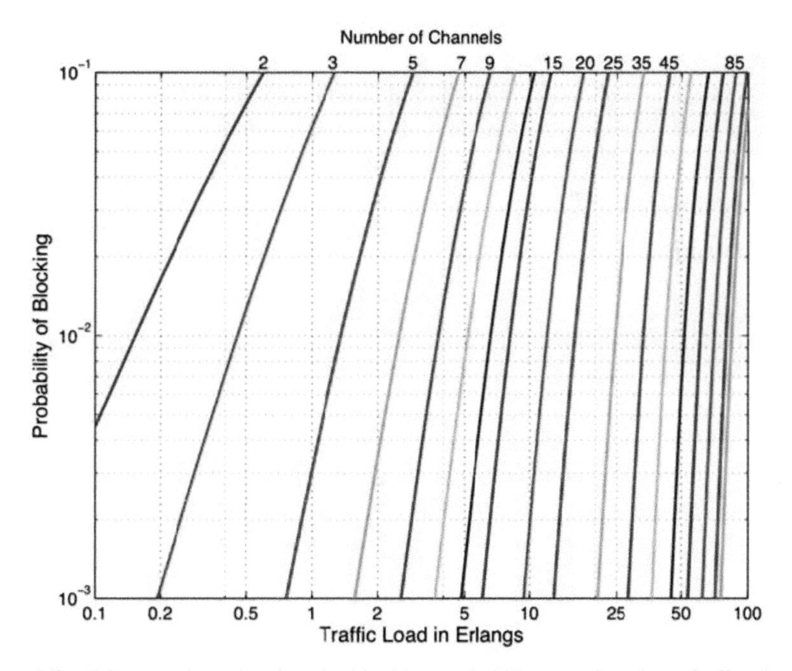

Figure 5.7 Erlang B chart showing the blocking probability as a function of offered traffic and number of channels.

Example 5.9: Capacity Using Erlang B Formula:

An IS-136 cellular phone provider owns 50 cell sites and 19 traffic carriers per cell each with a bandwidth of 30 kHz. Assuming each user makes 3 calls per hour and the average holding time per call of 5 minutes determine the total number of subscribers that the service provider can support with a blocking rate of less than 2%.

Solution:

The total number of channels is $N_u = 19 \times 3 = 57$ per cell. For $B(N_u, \rho) = 0.02$ and $N_u = 57$. Figure 5.8 shows that the $\rho = 45$ Erlangs. With an average of five calls per minute, the service rate is $\mu = 1/5$ minutes, and the acceptable arrival rate of the calls is $\lambda = \rho \times \mu = 1/5\ (\text{min}^{-1}) \times 45$ (Erlang) = 9 (Erlang/minute). With an average of three calls per hour, the system can accept 9 (Erlang/minute)/3 (Erlang)/60 (minute) = 180 subscribers per cell. Therefore, the total number of subscribers are 180 (subscribers/cell) $\times$ 50 (cells) = 8000 subscribers.

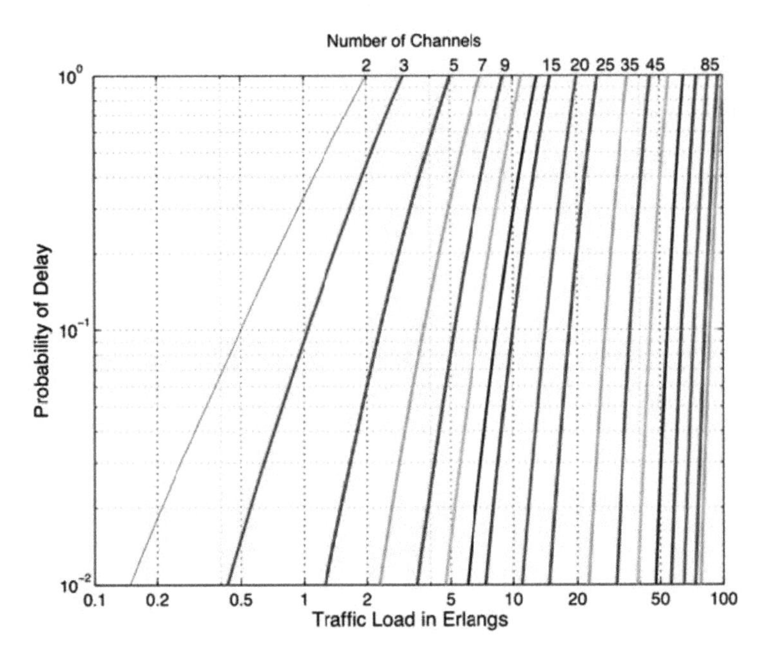

Figure 5.8 Erlang C chart relating offered traffic to number of channels and probability of delay.

The Erlang C formula relates the waiting time in a queue if a call does not go through, but it is buffered till a channel is available. These equations start with the probability that a call does not get processed immediately and gets delayed. The probability a call is delayed is given by

$$P(\text{delay} > 0) = \frac{\rho^{N_u}}{\rho^{N_u} + N_u!(1 - \frac{\rho}{N_u}) \sum_{k=0}^{N_u-1} \frac{\rho^k}{k!}}. \quad (5.7)$$

Because of the complexity of the calculation, tables or graphs are again used to provide values for this probability based on normalized values of ρ. Figure 5.8 illustrates the relationship between the probability of delay, number of channels N_u, and the normalized traffic per available channel ρ. The probability of having a delay that is more than a time *t* is given by

$$P[\text{delay} > t] = P[\text{delay} > 0]e^{-(N_u - \rho)\mu t}. \quad (5.8)$$

This indicates the exponential distribution of the delay time. The average delay is then given by the average of the exponential distribution:

$$D = P[\text{delay} > 0]\frac{1}{\mu(N_u - \rho)}. \quad (5.9)$$

Example 5.10: Call Delay Using Erlang C Formula:

For the ferry described in Example 4.17, answer the following questions:

(a) What is the average delay for a passenger to get access to the telephone?

(b) What is the probability of having a passenger waiting for more than a minute for access to the telephone?

Solution:

(a) Using Equation (7) for $N_u = 5$ and $\rho = 2.5$, we have $P[\text{Delay} > 0] = 0.13$. Using Equation (9), the average delay is $0.13/(5-2.5)/3 = 0.17$ minutes.

(b) Using Equation (8), $P[\text{delay} > 1 \text{ minute}] = 0.13 \exp[-(5-2.5)1/3] = 0.13 \exp(-0.83) = 0.0565$.

5.3 Random Access to the Internet

Random-access techniques are widely used in wired LANs and the literature in computer networking provides an adequate description of these techniques. When applied to wireless applications, these techniques often are modified from their original wired version. The objective of the rest of this section is to provide an overview of the random-access techniques that are used in wired and wireless networks. We first discuss the random-access methods used in wide-area cellular wireless data networks and then we provide some details of the access methods used in wired and wireless LAN (WLAN) applications. The random-access methods for data applications evolved in the past few decades into two groups. The first group consists of ALOHA-based access methods for which the terminals transmit their packets without any coordination between them (they contend for the medium). The second class is the carrier-sense based random-access techniques for which the terminal senses the availability of the channel before it transmits its packets. ALOHA-based methods suit better for wide-area networking when propagation delay is higher, and they were adopted by cellular telephone and satellite networks. 4G and 5G cellular networks integrated ALOHA with the traditional assigned-access methods to create hybrid methods with random access for reserving resourcing and assigned access to deliver the information. Carrier-sense multiple access (CSMA) performs better in local areas with lower propagation delay and they were adopted by IEEE 802.3, the Ethernet, and the IEEE 802.11, the Wi-Fi, and IEEE 802.15, ZigBee and other sensor networks. IEEE 802.11 also has a polling mechanism called the request to send, clear to send (RTS/CTS) and a centrally coordinated mechanism called point coordination function (PCF), which have elements of reservation in them. We begin by explaining reservation methods in 4G and 5G cellular methods and then we describe CSMA-based MACs for the local and personal area networks.

5.3.1 Reservation ALOHA and OFDMA in 4G and 5G

The original *ALOHA-protocol* is sometimes called *pure-ALOHA* to distinguish it from subsequent enhancements of the original protocol. This protocol derives its name from the ALOHA system, a communications network developed by Norman Abramson and his colleagues at the University of Hawaii and first put into operation in 1971 [ABR70]. The initial system used ground-based UHF radios to connect computers on several of the island campuses with the university's main computer center on Oahu, by use of a random-access protocol which has since then been known as the ALOHA protocol. The word ALOHA means hello in Hawaiian.

Figure 5.9(a) shows the simple basic concept of the ALOHA MAC protocol. A terminal transmits an information packet upon when the packet arrives from the upper layers of the protocol stack. Simply put, terminals say "hello" to the medium interface as the packet arrives. Each packet is encoded with an error-detection code. The BS/AP checks the parity of the received packet. If the parity checks properly, the BS sends a short acknowledgment packet to the MS. Of course, since the MS packets are transmitted at arbitrary times, there will be collisions between packets whenever packet transmissions overlap by any amount of time, as indicated in Figure 5.9(a). Thus, after sending a packet, the user waits a length of time more than the round-trip delay for an acknowledgment (ACK) from the receiver. If no ACK is received, the packet is assumed lost in a collision and it is transmitted again with a randomly selected delay to avoid repeated collisions. The advantage of the ALOHA protocol is that it is very simple, and it does not impose any synchronization between mobile terminals. The terminals transmit their packets as they become ready for transmission and if there is a collision, they simply retransmit. The disadvantage of the ALOHA protocol is its low throughput under heavy load conditions. In wireless channels where bandwidth limitations often impose serious concerns for data communications applications, this technique is often changed to its synchronized version, referred to as slotted ALOHA. The maximum throughput of a slotted-ALOHA system is double the throughput of pure-ALOHA (see Section 5.4).

In the slotted-ALOHA protocol, shown in Figure 5.9(b), the transmission time is divided into time slots. The BS/AP transmits a beacon signal for timing and all MSs synchronize their time slots to this beacon signal. When a user terminal generates a packet of data, the packet is buffered and transmitted at the start of the next time slot. With this scheme, we eliminate partial packet collision.

Assuming equal length packets, either we have a complete collision or we have no collisions. This doubles the throughput of the network. The report on collision and retransmission mechanisms remains the same as in pure ALOHA. Because of its simplicity, the slotted-ALOHA protocol is commonly used in the early stages of the registration of a mobile station to initiate a communication link with the base station. For example, in the GSM system, the initial contact between the mobile station and the base station tower to establish a traffic channel for TDMA voice communications is performed through a random-access channel using the slotted-ALOHA protocol. All other cellular systems adopt similar approaches as the first step in the registration process of a mobile station.

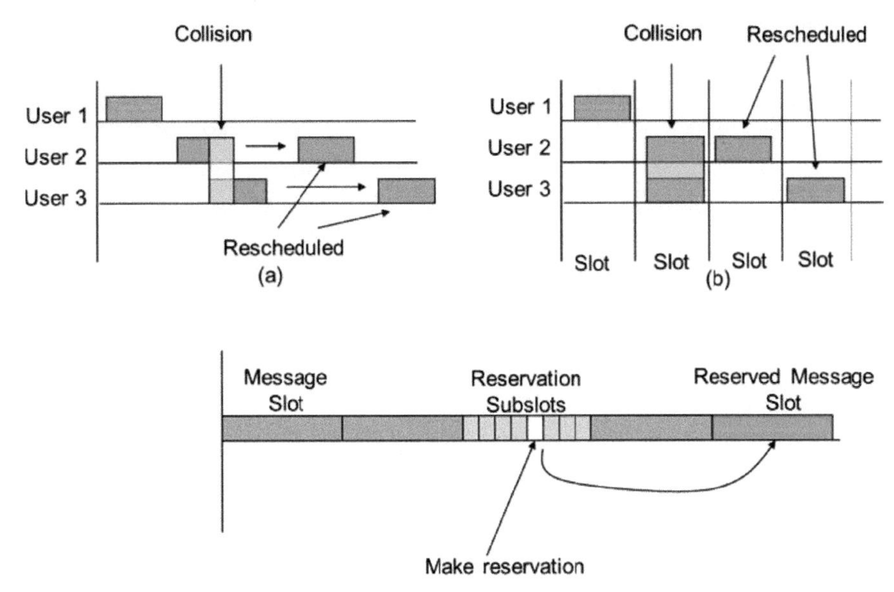

Figure 5.9 (a) Pure ALOHA protocol. (b) Slotted-ALOHA protocol. (c) Reservation ALOHA.

The throughput of slotted-ALOHA protocol is still very low for wireless data applications. This technique is sometimes combined with TDMA systems to form the so-called reservation-ALOHA protocol, shown in Figure 5.9(c). In reservation ALOHA, time slots are divided into contention periods and contention-free periods. During the contention interval, MSs use very short packets to contend for the upcoming contention-free intervals that will be used for the transmission of long information packets.

The 4G long-term evolution (LTE) cellular networks began using reservation in OFDMA systems, which behaves like a reservation in TDMA-based systems described above. In reservation OFDMA, instead of reserving time slots for individual mobile stations, *resource blocks* are reserved on the downlink. A resource block comprises a set of sub carriers and time slots. In LTE, a resource block is 1 ms long (two time slots) and 180 kHz of bandwidth comprising 12 sub-carriers, each 15 kHz wide. Resource blocks on the downlink can be allocated in a continuous manner or in chunks that are not together. Evolving next generations of cellular follow the same pattern of reservation resource allocation by defining resource blocks in terms of multiple MIMO streams as well as multiple OFDM carriers. These centrally controlled resource allocation for MAC in cellular networks allows a fair distribution of the resources among a large number of mobile terminals with multimedia contents, where we have a mixture of application data with different levels of throughput and delay requirements.

5.3.2 CSMA in Ethernet and Wi-Fi

The main drawback of ALOHA-based contention protocols is the lack of efficiency caused by the collision and retransmission process. In ALOHA, users do not consider what other users are doing when they attempt to transmit data packets and there are no mechanisms to avoid collisions. A simple method to avoid collisions is to sense the channel before the transmission of a packet. If there is another user transmitting on the channel, it is obvious that a terminal should delay the transmission of the packet. Protocols employing this concept are referred to as CSMA or listen-before-talk (LBT) protocols. Figure 5.10 shows the basic concept of the CSMA protocol. Terminal "1" will sense the channel first and then send a packet. This is followed by a sensing and packet transmission by terminal "1" again. During the second transmission time of terminal "1," terminal "2" senses the channel and discovers that another terminal is using the medium. It then delays its transmission for a later time using a backoff algorithm. The CSMA protocol reduces the packet collision probability significantly compared to the ALOHA protocol. However, it cannot eliminate the collisions entirely. Sometimes, as shown in Figure 5.10, two terminals sense the channel busy and reschedule their packets for a later time, but their transmission time overlaps with each other, causing a collision. Such situations do not cause a significant operational problem because the collisions can be handled in the same way as they were handled in ALOHA. However, if the propagation time between the terminals is very long, such situations happen more frequently, thereby reducing the effectiveness of carrier sensing in preventing collisions.

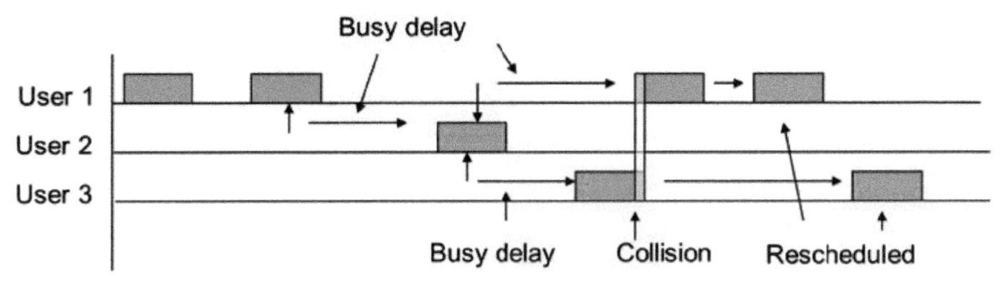

Figure 5.10 Basic operation of CSMA protocol.

As a result, several variations of CSMA have been employed in the local area applications while ALOHA protocols are preferred in wide-area applications.

Several strategies are used for the sensing procedure and retransmission mechanisms that have resulted in several variations of the CSMA protocol for a variety of wired and wireless data networks. Figure 5.11 depicts the key elements of distinction among these protocols. If, after sensing the channel, the terminal attempts another sense only after a random waiting period, the carrier-sensing mechanism is called "non-persistent." After sensing a busy channel, if the terminal continues sensing the channel until the channel becomes free, the protocol is referred to as "persistent." In persistent operation, after the channel becomes free, if the terminal transmits its packet right away, it is referred to as "1-Persistent" CSMA and if it runs a random number generator and, based on the outcome, transmits its packet with a probability "*p*," the protocol is called *p*-persistent CSMA.

In a wireless network, due to multipath and shadow fading as well as the mobility of terminals, sensing the availability of the channel is not as simple as in the case of wired channels. Typically, in a wireless network, two terminals can each be within the range of some intended third terminal, but out of range of each other, because they are separated by excessive distance or by some physical obstacle that makes direct communication between the two terminals impossible. This situation, where the two terminals cannot sense the transmission of each other, but a third terminal can sense both of them, is referred to as the *hidden terminal problem*. This is a more likely situation in

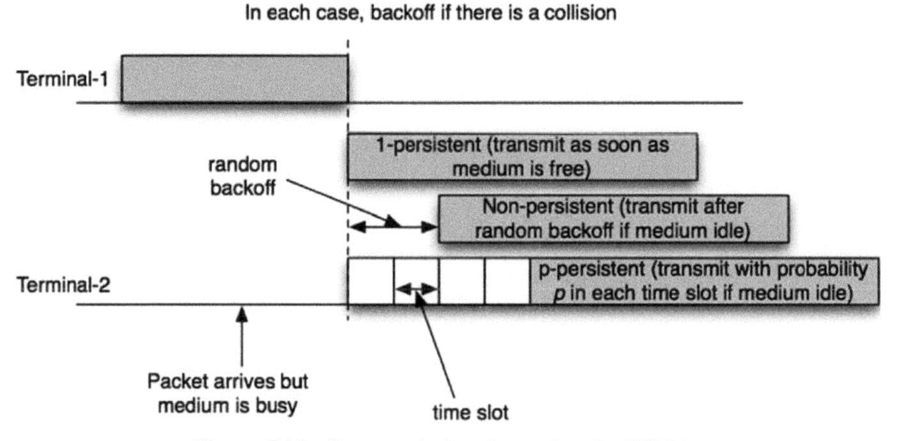

Figure 5.11 Retransmission alternatives for CSMA.

cases of radio networks covering wider geographic areas in which hilly terrain blocks some groups of user terminals from sensing other groups. In this situation, the CSMA protocol will successfully prevent collisions among the users of one group but will fail to prevent collisions between users in groups hidden from one another. CSMA is a distributed access TDD MAC ideal for local and personal area networking where propagation delay is small and the number of users in the area of coverage of the network is limited. The two most popular CSMA techniques that were adopted in the networking industry were CSMA/CD adopted for the legacy Ethernet and CSMA/CA adopted in the legacy Wi-Fi as well as ZigBee. The legacy Ethernet had a bus topology connecting computing devices with the same networking interface cards. A central control would have added complexity to the implementation of this MAC. In Wi-Fi and ZigBee, the CSMA allows flexibility to accommodate both infrastructure and *ad-hoc* connections among computers.

CSMA/CD for the Ethernet:

The first popular contention-based MAC recommended by a standards organization was the CSMA/CD adopted by the IEEE 802.3 for the legacy Ethernet. Compared to wide area networks, a LAN operates over shorter distances with smaller propagation delays and, consequently, a transmission medium that is well suited for variations of the CSMA protocol. When the length of the packets is long, it would be very useful to pay further attention to packet collisions. LANs often employ variations of the CSMA protocol that either stop transmission as soon as a packet collision is detected or add additional features to avoid the collision. The most popular version of the CSMA for wired LANs is CSMA with collision detection (CSMA/CD) adopted in the IEEE 802.3 (Ethernet) standard, the dominant standard for wired LANs. The basic operation of CSMA/CD is the same as CSMA implementations discussed earlier. The defining feature of CSMA/CD is that it provides for detection of a collision shortly after its onset and each transmitter involved in the collision stops transmission as soon as it senses a collision. In this way, colliding packets can be aborted promptly, minimizing the wastage of channel occupancy time by transmissions destined to be unsuccessful.

Unlike plain CSMA, which requires an ACK (or lack of an ACK) to learn the status of a packet collision, CSMA/CD of the IEEE 802.3 requires no such feedback information since the collision detection mechanism is built into the transmitter. When a collision is detected, the transmission is immediately aborted, a jamming signal is transmitted, and a retransmission backoff procedure is initiated, just as in CSMA. As is the case with any random-access scheme, proper design of the backoff algorithm is an important element in assuring the stable operation of the network.

The backoff algorithm recommended by IEEE 802.3 Ethernet is referred to as the binary exponential backoff algorithm that is combined with 1-persistence CSMA protocol with collision detection. When a terminal senses a transmission, it continues sensing (persistent) until the transmission is completed. After the channel becomes free, the terminal sends its own packet. If another terminal was also waiting, a collision occurs because of the 1-persistence, and the two terminals reattempt transmission with a probability of ½ after a time slot that spans twice the maximum propagation delay allowed between the two terminals. A time slot of 64 bytes (6.4 μs) that spans over twice the maximum propagation delay is selected to ensure that, in the worst-case scenario, the terminal will be able to detect the collision. If a second collision occurs, the terminals reattempt with the probability of that is half of the previous retransmission probability. If collision persists, the terminal continues reducing its retransmission probability by half up to 10 times and, after that, it continues with the same probability six more times. If no transmission is possible after 16 attempts, it reports to the higher layers that the network is congested, and transmission shall be stopped. This procedure exponentially increases the backoff time and gives the backoff strategy its name. The disadvantage of this procedure is that the packets arriving later have a higher chance to survive the collision that results in an unfair first-come last-serve environment. It can be shown that the average waiting time for the exponential backoff algorithm is 5.4T, where T is the time slot used for waiting [TAN10], [STA00].

The CSMA/CD scheme is also used in many infrared LANs, where both transmission and reception are inherently directional. In such an environment, a transmitting station can always compare the received signal from other terminals with its own transmitted signal to detect a collision. Radio propagation is not directional posing a serious problem in determining other transmissions during your own transmission. As a result, collision detection mechanisms are not well suited for radio LANs. However, compatibility is very important for WLANs, and, therefore, designers of these networks have had to consider CSMA/CD for compatibility with the Ethernet backbone LANs that dominate the wired-LAN industry.

CSMA/CA for Wi-Fi:

While collision detection is easily performed on a wired network, simply by sensing voltage levels against a threshold, such a simple scheme is not readily applicable to radio channels because of fading and other radio channel characteristics. The one approach that can be adopted for detecting collisions is to have the transmitting station demodulate the channel signal and compare the resulting information with its own transmitted information. Disagreements can be taken as an indicator of collisions, and the packet can be immediately aborted. However, on a wireless channel, the transmitting terminal's own signal dominates all other signals received in its vicinity, and, thus, the receiver may fail to recognize the collision and simply retrieve its own signal. To avoid this situation, the station's transmitting antenna pattern should be different from its receiving pattern. Arranging this situation is not convenient in radio terminals because it requires directional antennas and expensive front-end amplifiers for both transmitters and receivers.

The approach called *CSMA with collision avoidance* (CSMA/CA), shown in Figure 5.12, is adopted by the IEEE 802.11wireless LAN standard. The elements of CSMA/CA used in the IEEE 802.11 are interframe spacing (IFS), contention window (CW), and a backoff counter. The CW intervals are used for contention and transmission of the packet frames. The IFS is used as an interval between two CW intervals. The backoff counter is used to organize the backoff procedure for the transmission of packets. Figure 5.13 provides an example for the operation of the CSMA/CA mechanism used in the IEEE 802.11 standard. Stations A, B, C, D, and E are engaged in contention for transmission of their packet frames. Station A has a frame in the air when stations B, C, and D sense the channel and find it busy. Each of the three stations will run its random number generator to get a backoff time by random. Station C followed by D and B draws the smallest number. All three

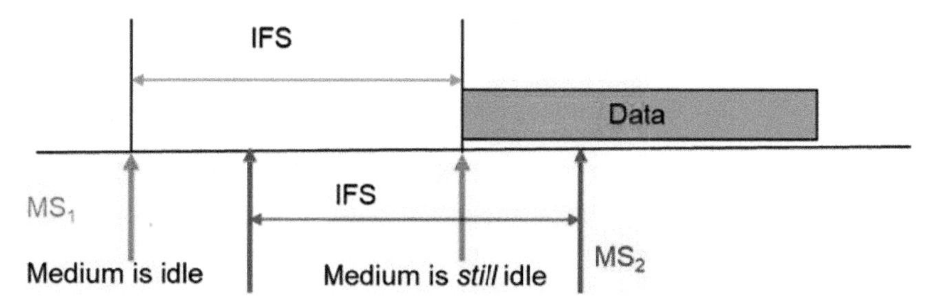

Figure 5.12 CSMA/CA adopted by the IEEE 802.11.

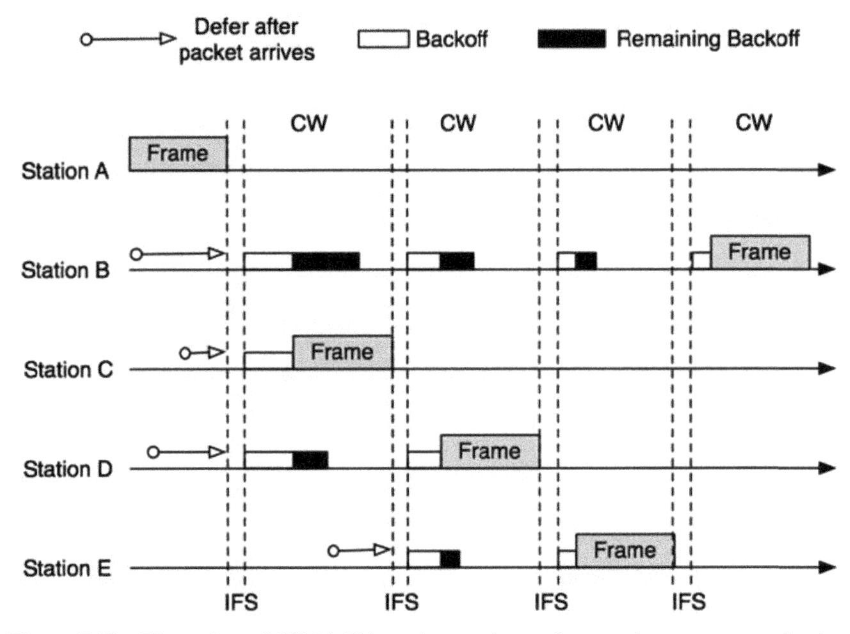

Figure 5.13 Illustration of CSMA/CA carrier sensing and contention counter mechanism.

terminals persist on sensing the channel and defer their transmission until the transmission of the frame from terminal A is completed. After completion, all three terminals wait for the IFS period and start their counters immediately after completion of this period. As soon as the first terminal, station C in this example, finish counting its waiting time, it starts transmission of its frame. The other two terminals, B and D, freeze their counter to the value that they have reached at the start of transmission for terminal C. During transmission of the frame from station C, station E senses the channel, runs its own random number generator that, in this case, ends up with a number larger than the remainder of D and smaller than the remainder of B, and defers its transmission for after the completion of station C's frame. In the same manner, as the previous instance, all terminals wait for IFS and start their counter. Station D runs out of its random waiting time earlier and transmits its own packet. Stations B and E freeze their counters and wait for the completion of the frame transmission from terminal D and the IFS period after that before they run start running down their counters. The counter for terminal E runs down to zero earlier and this terminal sends its frame while B freezes its counter. After the IFS period following completion of the frame from station

E, the counter in station B counts down to zero before it sends its own frame. The advantage of this backoff strategy over the exponential backoff used in IEEE 802.3 is that the collision detection procedure is eliminated and the waiting time is fairly distributed in a way that, on average, a first-come first-serve policy is enforced.

One of the drawbacks of the CSMA/CA protocol employed in WLANs is that the waiting times employed severely limit the throughput of the network. Recall that no user data is sent for the interframe spacing time, during the backoff slots, or during the transmission of the ACK frames. These times form an overhead for the transmission of every single MAC layer frame in IEEE 802.11. Especially as the data rates increase, the waiting times become an increasingly large part of the transmission of a frame. Consider that the actual size of a frame reduces as the data rate increases. If the waiting times remain fixed, eventually, they dominate the transmission of frames so that the throughput in a WLAN cannot exceed a certain threshold irrespective of how high the physical transmission rate is. To overcome this problem, in IEEE 802.11n,ac, additional MAC layer features have been introduced to reduce the overhead. These include aggregating several frames for transmission and block ACKs for several frames [Xia05][Sko08].

RTS/CTS for Wi-Fi:

A modification to CSMA adopted by IEEE 802.11 is the RTS/CTS the mechanism is shown in Figure 5.14. A terminal ready for transmission, after sensing the carrier to discover an opportunity for transmission, sends a short RTS packet identifying the source address, destination address, and the length of the data to be transmitted. The destination station will respond with a CTS packet. The source terminal will send its packet with no contention. After acknowledgment from the destination terminal, the channel will be available for other users. This approach is reservation access combined with CSMA, and the legacy IEEE 802.11 supports this feature to accommodate the hidden terminal problem while CSMA/CA was the main MAC protocol for networking. The RTS/CTS provides a unique access right to a terminal to transmit without any contention. As directional antennas and MIMO technology with focused beams for multiple streaming became the favorite choice for PHY in IEEE 802.11ac,ad,ax,af, the RTS/CTS became the favorite choice for medium access of Wi-Fi devices because sensing a carrier for all terminals in a directional environment was not feasible. In this latest standard, a broadcast channel using omni-directional antennas establishes the link and, the data is transferred through the RTS/CTS. This may remind the reader

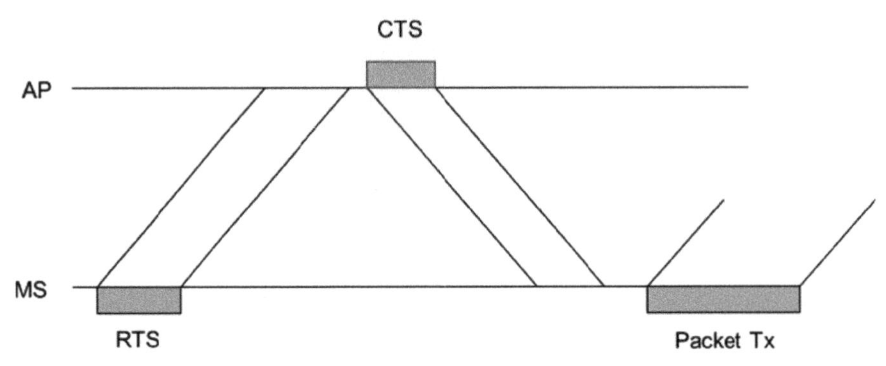

Figure 5.14 The RTS/CTS mechanism as a MAC option for the IEEE 802.11.

of centralized access methods for circuit switching, which are two separate planes for control/management and data transfer.

Quality of Service in Wi-Fi:

Quality of Service (QoS) is founded on mechanisms to differentiate traffic types. In IEEE 802.11, priority is assigned by dividing the IFS into several different-sized intervals associated with different priority levels. In combining, priority can be arranged by assigning different classes of numbers to the codes. The lower priority packets will receive earlier zero codes and higher priority packets will have a zero in their codes in the later intervals.

With the emergence of VoIP as an important application over random-access networks, it became important to provide support for QoS for voice and other real-time traffic over such networks. The IEEE 802.11e standard was an attempt to provide QoS at the MAC layer. The basic idea in IEEE 802.11e was to provide different priorities for different classes of traffic by extending the existing priority levels offered in legacy IEEE 802.11. Priority in IEEE 802.11e and the legacy IEEE 802.11 refers to the ability of frames from certain classes of traffic to access the channel earlier than other classes of traffic. This helps in reducing the delay for such classes of traffic, as well as in providing higher throughput. One way of providing different priorities to different classes of traffic in IEEE 802.11 is to enable different waiting times (interframe spaces) or different backoff intervals for different classes of traffic. This way, voice traffic, for example, could wait for a much shorter period than web traffic reducing latency and improving throughput for voice packets.

This, however, implies that access to the channel is not fair (flows from some classes of traffic may be starved of bandwidth). Fair access to the medium while assuring throughput and latency is possible by using techniques that are distributed and, yet, provide fair access to the medium. Several techniques have been proposed to address this issue (see [Pat03] for more details).

5.3.3 MAC Protocols for Wireless Sensor Networks and IoT

Wireless sensor networks and the Internet of Things (IoT) are a new class of networks that may include numerous low-cost devices deployed in a sensor field for monitoring and sensing specific phenomena. In such networks, sensors may actually transmit and receive useful data from the Internet (typically sensed quantities that have been processed locally) very infrequently. Sensors may be deployed in areas that may be inaccessible to humans. Consequently, it is important to ensure that the limited batteries in such devices last for years. Medium access schemes designed for high-speed wireless LANs are quite unsuitable for sensor networks. The coordinating sleep schedules of sensors to reduce the amount of idle listening is one method of enhancing the battery life in sensors. The reader is referred to [Dem06] for a survey of MAC protocols for sensor networks. In Chapter 9, we describe some of these methods.

5.4 Performance of Random-Access Methods

In voice-oriented circuit-switched networks, performance is measured by the probability of blockage (blockage rate) of initiating a call. If the call is not blocked, a fixed rate full-duplex channel is allocated to the user for the entire communication session. In other words, the interaction between the user and the network takes place in two steps. First, during the call establishment procedure, the user negotiates the availability of a line with the network and, if successful (not blocked), the network guarantees a connection with a certain QoS (data rate, delay, error rates, etc.) to the user. For real-time interactive applications such as telephone conversations or video conferencing, if the user does not talk, the resource allocated to the user is wasted. If these facilities, originally designed for the two-way voice application, are used for data application, (1) for bursty data file, transfers during the idle times between transmission of two packet bursts allocated resources are wasted,

Table 5.1 Throughput of various random-access protocols.

Protocol	Throughput
Pure ALOHA	$S = Ge^{-2G}$
Slotted ALOHA	$S = Ge^{-G}$
Unslotted 1-persistent CSMA	$S = \frac{G[1+G+aG(1+G+aG/2)]e^{-G(1+2a)}}{G(1+2a)-(1-e^{-aG})+(1+aG)e^{-G(1+a)}}$
Slotted 1-persistent CSMA	$S = \frac{G[1+a-e^{-aG}]e^{-G(1+a)}}{(1+a)(1-e^{-aG})+ae^{-aG(1+a)}}$
Unslotted non-persistent CSMA	$S = \frac{Ge^{-aG}}{G(1+2a)+e^{-aG}}$
Slotted non-persistent CSMA	$S = \frac{aGe^{-aG}}{1-e^{-aG}+a}$

and (2) large file transfers suffer long delay or waiting time for the transfer because resources allocated to each user is more restricted.

Users of packet-switched networks are always connected and there is neither an initiation (negotiation) procedure to be blocked nor a fixed QoS at the MAC to be allocated. Performance of these networks for data applications is often measured by the average throughput, S, and average delay, D, versus the total offered traffic, G. The *channel throughput*, S, is the average number of successful packet transmissions per time interval T. The offered traffic, G, is the number of packet transmission attempts per packet time slot T that includes new arriving packets as well as to retransmissions of old packets. The average delay, D, is the average waiting time before successful transmission, normalized to the packet duration T. The standard unit of traffic flow is Erlang that can be thought of as the number of packets per packet duration time T. The throughput is always between zero and one Erlang, while the offered traffic, G, may exceed one Erlang.

The analyses of the relationships between S, G, and D for a variety of medium access protocols have been a subject of research for a few decades. This analysis depends on the assumptions on the statistical behavior of the traffic, number of terminals, relative duration of the packets, and the details of the implementation. Assuming a large number of terminals generating fixed length packets randomly with an arrival rate of λ packets per unit time, it is common in the literature to model the arrival of k-packets in an arrival time of Δt by a Poisson distribution:

$$P(k) = \frac{(\lambda \Delta T)^k}{k!} e^{-\lambda \Delta T}. \tag{5.10}$$

Using Poisson arrival, we can relate the throughput S and of the offered traffic G by calculating the probability of having no collision during the transmission time of a packet. As shown in Figure 5.15(a), if a packet occupies the medium in the interval (0,T), that packet will be successful if no other packet arrives during ($-T$, T). When packets are arriving with Poisson distribution, the probability of no packet arriving in $\Delta t = 2T$ is

$$P(0) = e^{-\lambda \Delta t} = e^{-2\lambda T} = e^{-2GT}.$$

Because the total traffic is a multiplication of the arrival rate and length of the packet, $G = \lambda T$, the throughput of the ALOHA is then calculated by multiplying the offered traffic with the probability of success (no other packets during collision interval):

$$S = P(0)G = Ge^{-2G}. \quad (5.11)$$

We can follow the same argument for calculation of throughput in slotted-ALOHA protocol. As shown in Figure 5.15(b), for the slotted ALOHA, the collision interval is half of that of ALOHA because either packets collide totally or they do not; unlike ALOHA, here, we have no partial collision. As a result, $\Delta t = T$ and probability of successful transmission and the throughput are

$$\begin{cases} P(0) = e^{-\lambda \Delta t} = e^{-\lambda T} = e^{-GT} \\ S = P(0)G = Ge^{-G} \end{cases}. \quad (5.12)$$

This analysis can be extended to CSMA with different retransmission strategies and they were the subject of research during the 1970s and the 1980s by computer scientists. Table 5.1 summarizes the throughput expressions for ALOHA, and 1-persistent and non-persistent CSMA protocols, including the slotted and unslotted versions of each. The expressions for p-persistent protocols are very involved and are not included here. The interested reader should refer to [Kle75] [Tob75] [Tak85], where the derivations of the other CSMA expressions can also be found. The expressions in the table are also derived in [Ham86] and [Kei89]. The parameter a in this table corresponds to the normalized propagation delay defined as $a = \tau / T_p$ where τ is the maximum propagation delay for the signal to go from one end of the network to the other end.

Figure 5.16 shows plots of throughput S versus offered traffic load G for the six protocols listed in Table 5.1, with a normalized propagation delay of $a = 0.01$. All curves follow the same pattern. Initially, as the offered traffic G increases the throughput, S also increases up to a point where it reaches a maximum S_{max}. After the throughput reaches its maximum value, an increase in the offered traffic reduces the throughput. The first region depicts the stable

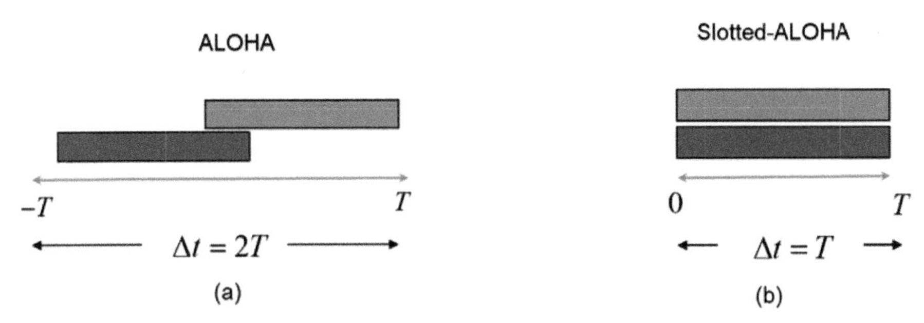

Figure 5.15 Collision interval for (a) ALOHA and (b) slotted ALOHA.

operation of the network in which an increase in aggregating traffic G that includes arriving traffic as well as packet retransmissions due to collisions, increases the total successful transmissions and thus S. The second region represents unstable operation where an increase in G reduces the throughput S because of congestion and eventually halting of the operation. In practice, as we saw in the last section, retransmission techniques adopted for the real

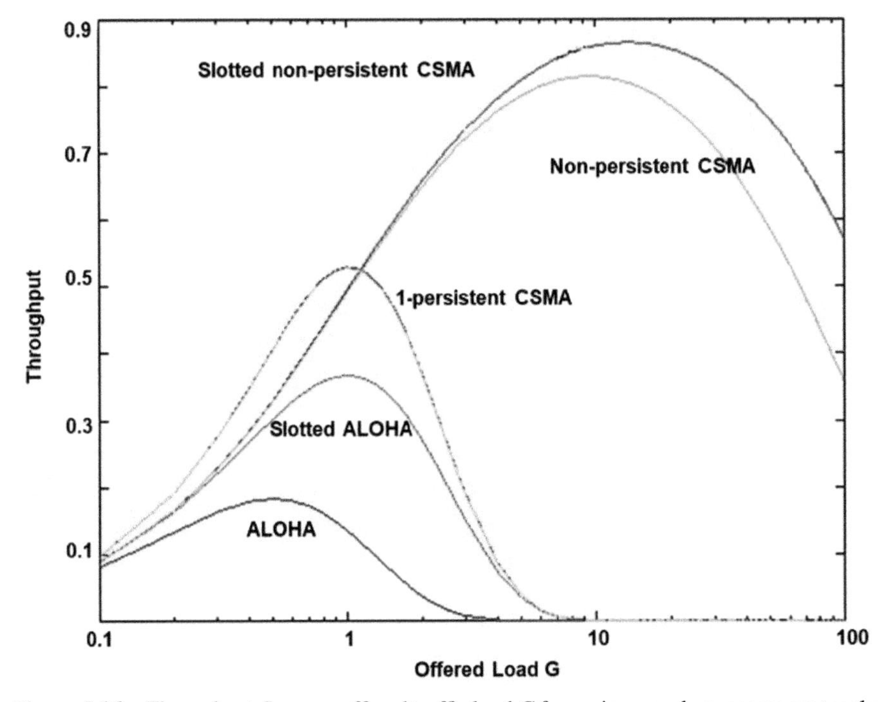

Figure 5.16 Throughput S versus offered traffic load G for various random-access protocols.

implementation include backoff mechanisms to prevent operation in unstable regions.

The throughput curves for the slotted and unslotted versions of 1-persistent CSMA is essentially indistinguishable. It can be seen from the figure that for low levels of offered traffic, the 1-persistent protocols provide the best throughput, but at higher load levels, the non-persistent protocols are by far the best. It can also be seen that the slotted non-persistent CSMA protocol has a peak throughput almost twice that of persistent CSMA schemes.

The equations in Table 5.1 can also be used to calculate capacity, which is defined as the peak value S_{max} of throughput over the entire range of offered traffic load G. An example is helpful to show how to relate the curve to a system.

Example 5.11: Relating Throughput and Offered Traffic to Data Rates:

To relate throughput and offered traffic to data rates, assume that we have a centralized network that supports a maximum data rate of 10 Mbps and serves a large set of user terminals with the pure ALOHA protocol.

(a) What is the maximum throughput of the network?
(b) What is the offered traffic in the medium and how is it composed?

Solution:

(a) Since the peak value of the throughput is $S = 18.4\%$, the terminals contending for access to the central module can altogether succeed in getting at most 1.84 Mbps of information through the network.
(b) At that peak, the total traffic from the terminals is 5 Mbps (because the peak occurs at $G = 0.5$), which is composed of 1.84 Mbps of successfully delivered packets (some mixture of new and old packets) and 3.16 Mbps of packets doomed to collide with one another.

Plots of capacity versus normalized propagation delay are plotted in Figure 5.17 for the same set of ALOHA and CSMA schemes. The curves show that for each type of protocol, the capacity has a distinctive behavior as a function of normalized propagation delay a. For the ALOHA protocols, capacity is independent of a and is the largest of all the protocols (compared when a is large). As we discussed earlier, this is the case where the area of coverage is large and propagation delays are comparable to the length of packets. The plots in Figure 5.17 also show that the capacity of 1-persistent

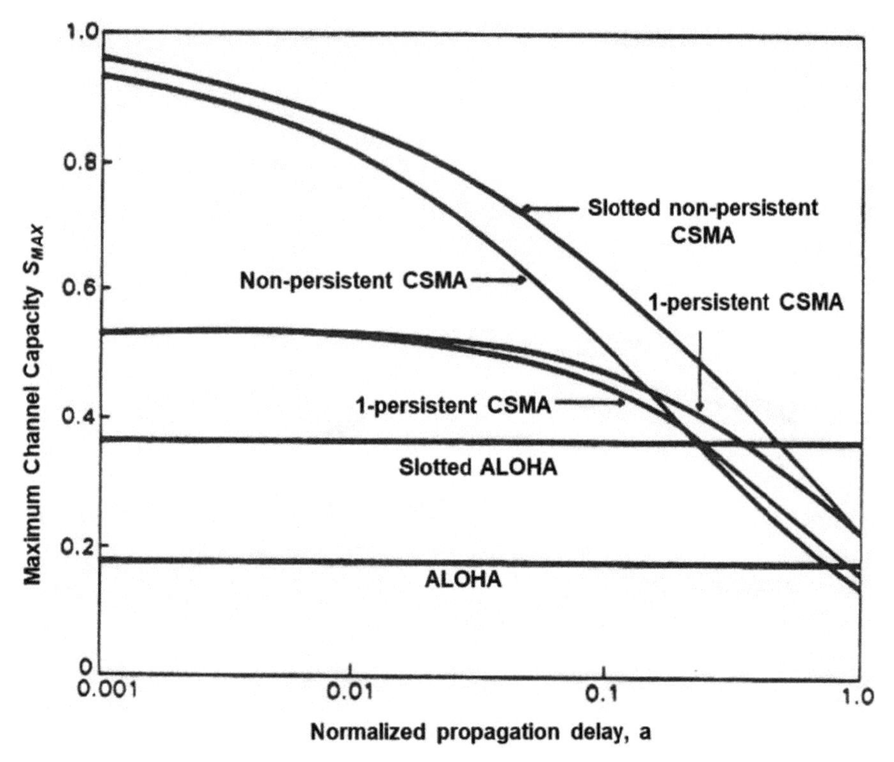

Figure 5.17 Capacity versus normalized propagation delay for various random-access protocols.

CSMA is less sensitive to the normalized propagation delay for small a, than is non-persistent CSMA. However, for small a, non-persistent CSMA yields a larger capacity than does 1-persistent CSMA, though the situation reverses as a approaches the range of 0.3–0.5.

Another important performance measure for packet data communications is the delay characteristics of the transmitted packets. For real-time applications such as voice conversations, if the delay is more than a certain value (several hundred milliseconds), the packet is not useful and it is dropped. Therefore, we need to analyze the delay characteristics of the channel to determine the capacity of the access method. In the data transfer applications, the delay characteristic is usually related to the throughput of the medium and it usually follows a hockey-stick shape. At low traffic when a small fraction of the maximum throughput is utilized, the delay often remains the same as the transmission delay. As the throughput increases, the number

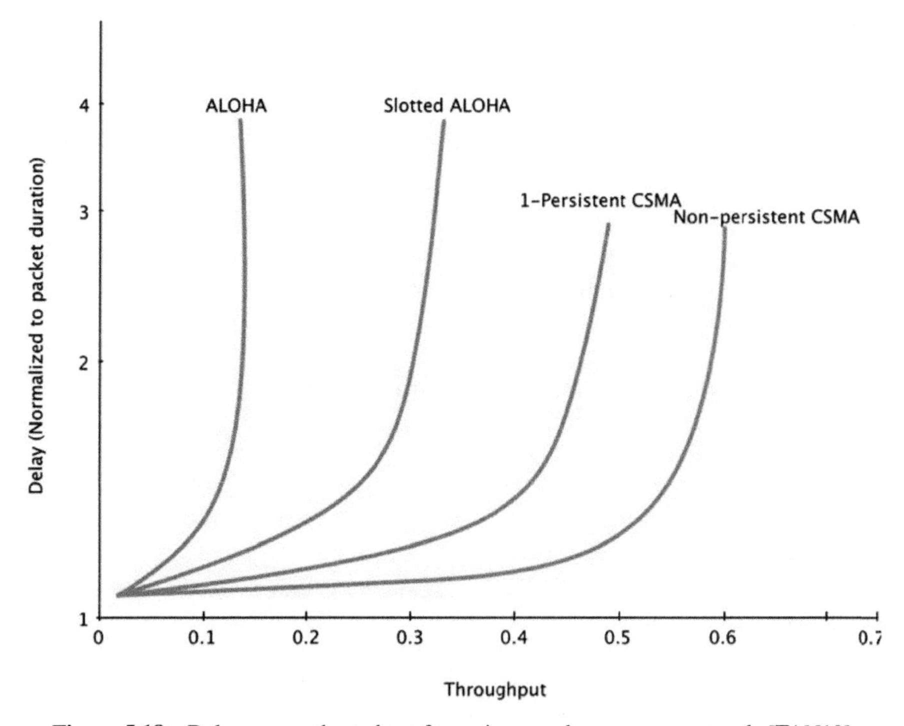

Figure 5.18 Delay versus throughput for various random-access protocols [TAN10].

of retransmitted packets increases, resulting in a higher average delay for the packets. Around the maximum throughput, the delay retransmissions grow rapidly, pushing the network toward an unstable conditions where the channel is dominated by retransmissions and the packet delays grow extremely large. Figure 5.18 shows the delay-throughput behavior of the ALOHA, S-ALOHA, and CSMA protocols.

In Table 5.1, we presented the performance of ALOHA and slotted ALOHA as we derived them in Equations (5.11) and (5.12), and the rest of the table presented the throughput of CSMA methods without any detailed derivation by referring to sources that the readers can trace. These derivations are involved in a variety of modeling, assumptions, and analytical approaches, and they are not directly applied to any popular standard. In the following sections, we derive the performance of CSMA-based MAC protocol applied to the Ethernet and Wi-Fi with an adequate amount of details to make the reader familiar with the analytical approach, modeling, and assumptions to derive the performance of a MAC protocol.

5.4.1 Performance of CSMA/CD in Ethernet

Multiple users share the medium which receives a certain data rate offered by PHY. The responsibility of the MAC layer is to control the medium access to manage the packet traffic arriving from multiple devices to share the PHY resources implemented on the medium. Depending on this sharing process, certain idle times occur between the packet transmissions, which differ among different MAC protocols. We define channel utilization or efficiency of a MAC protocol as the ratio of the average packet lengths of T_P to its addition with the average idle time T_I:

$$\text{Channel Efficiency (Utilization)} = \frac{T_P}{T_P + T_I}. \tag{5.13a}$$

This value reflects the efficiency of the MAC protocol in utilizing the PHY resources. Packets arriving at the PHY have a variety of overheads from different layers (see Figure 1.10); also, before transmission of a packet, we always leave an interframe spacing to ensure all devices at different distances sense the transmission of the packet. Figure 5.19 shows the definition of all these timing parameters involved in the transmission of a packet carrying L_D units of data bits. We define throughput as the ratio of time we spend on sending data to all the other overheads:

$$\text{Channel Throughput} = \frac{T_D}{T_P + T_I} = \frac{T_D}{T_D + T_{\text{oh}} + T_{\text{IFS}} + T_I}. \tag{5.13b}$$

If we multiply the throughput with the transmission rate of the PHY, we get the data rate that the users experience over the channel. Performance analysis

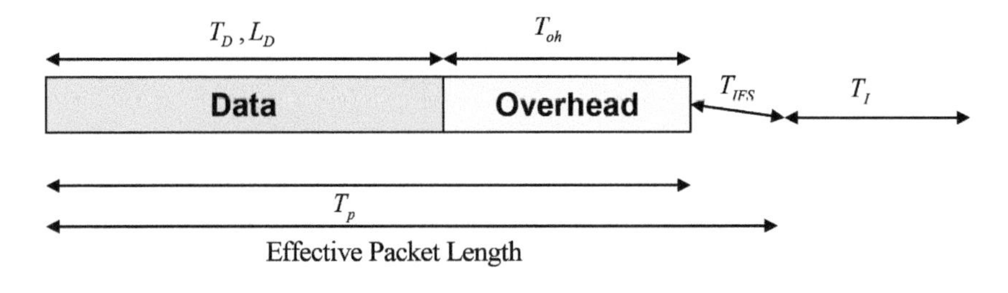

T_p : Packet length T_I : Idle time

T_D : Data time T_{oh} : Overhead time

T_{IFS} : Interframe spacing L_D : Data length in byte

Figure 5.19 A flow chart for CSMA/CD operation.

of MAC often involves queuing theory; it can become very complex, and it was a popular area of research for several decades (see Chapter 5). In this section, we address the throughput of CSMA/CD recommended by the IEEE 802.3.

Although there have been numerous rigorous calculations of the throughput of the CSMA/CD for a finite and infinite number of users, we resort to a relatively simple and practical calculation presented in [Met76]. This analysis is based on the fixed probability of contention p for N terminal contending to access a slot in the medium. As shown in Figure 5.20 for a particular terminal, the probability of successful transmission in the slot is given by $p \times (1-p)^{N-1}$which implies that the specific terminal is transmitting and the other $N-1$ terminals are not. Since we have N terminal each with $p \times (1-p)^{N-1}$probability to occupy the slot, the probability of a packet from all users occupy the slot which is actually the probability of successful transmission in the slot, A, would be given by

$$A(N,p) = \mathrm{N} \times p \times (1-p)^{N-1}. \tag{5.14}$$

As $N \to \infty$, the function $A(N, p)$ maximizes for $p = \frac{1}{N}$ for which

$$A(N,p) = (1-N)^{N-1}\big|_{N\to\infty} = 1/e.$$

As shown in Figure 5.21, given probability of successful transmission in a slot, A, and the probability of having an idle slot, $1-A$, one can determine

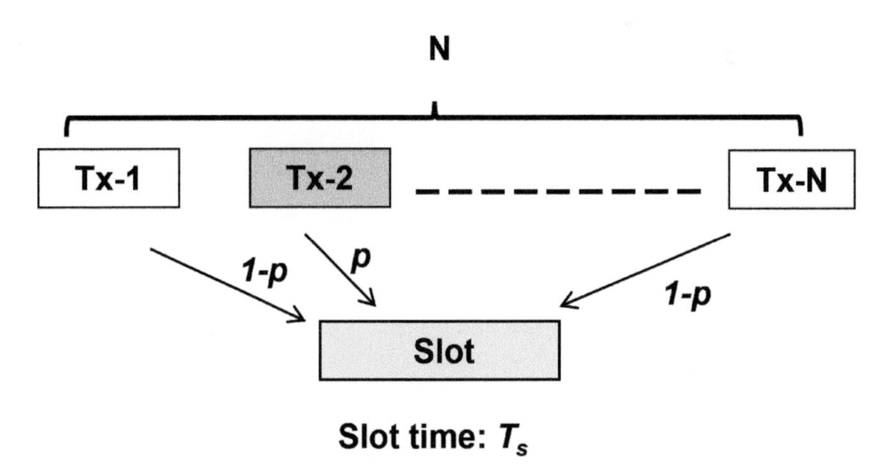

Figure 5.20 N-terminals transmitting packets with probability of p *contending* access to a transmission slot. A successful transmission takes place when one terminal sends a packet while all other terminals do not. Probability of not transmitting is $(1-p)$.

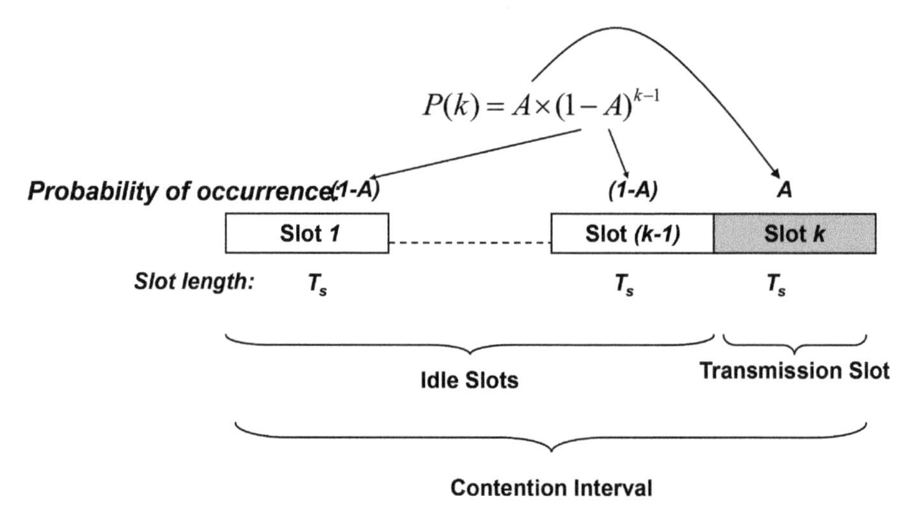

Figure 5.21 Contention interval and frame transmission.

the probability of successful transmission at the kth slots

$$P(k) = A \times (1 - A)^{k-1}. \tag{5.15}$$

This is indeed the probability of having k-slot contention from which we can calculate the average number of slots of waiting idle before successful transmission of a packet. Using the probability density function of the contention slots, we can calculate the average number of slots per contention[3]

$$\begin{aligned} E\{k\} &= \sum_{k=1}^{\infty} k \times P(k) = \sum_{k=1}^{\infty} k \times A \times (1 - A)^{k-1} \\ &= A + 2A(1 - A) + 3A(1 + A)2 + ... = \frac{1}{A}. \end{aligned} \tag{5.16}$$

If we assume the length of the slot is T_s, the idle time is then calculated from

$$T_I = \frac{T_s}{A} - T_s = T_s\frac{(1 - A)}{A}.$$

If we consider the bounds as number of users approaches infinity, $A \rightarrow 1/e$, and the average idle time is

$$T_I = T_s\frac{(1 - A)}{A} < T(e - 1) = 1.72T_s. \tag{5.17}$$

[3]Note that: $\sum_{i=0}^{\infty} ia^i = \frac{a}{(1-a)^2}$.

Substituting in Equation (5.13a) we have

$$\begin{aligned}\text{Channel Efficiency (Utilization)} &= \frac{T_D + T_{\text{oh}} + T_{\text{IFS}}}{T_D + T_{\text{oh}} + T_{\text{IFS}} + \frac{T_s(1-A)}{A}} \\ &< \frac{T_D + T_{\text{oh}} + T_{\text{IFS}}}{T_D + T_{\text{oh}} + T_{\text{IFS}} + 1.72T_s}.\end{aligned} \tag{5.18}$$

In the MAC packet of legacy Ethernet, which is longer than the minimum requirement (no pads) has 26 bytes of additional overhead plus 12 bytes (96 bits) of the interframe gap. In addition, the length of the slot is 64 bytes (512 bits):

$$\begin{cases} L_{\text{oh}} + L_{\text{IFS}} = 26(\text{overhead}) + 12(\text{IFS}) = 38\,\text{bytes} \\ L_s = 64\,\text{bytes} \end{cases}.$$

Therefore, the actual channel utilization is

$$\text{Channel Efficiency} = \frac{L_D + 38}{L_D + 38 + \frac{64(1-A)}{A}} < \frac{L_D + 38}{L_D + 148} \tag{5.19a}$$

in which L_D is the length of the data packet in bytes. Figure 5.22 shows the channel utilization of the legacy Ethernet for different lengths of the data packets. For large packet lengths close to the maximum allowed length, the efficiency is close to 90%, and for short packets, it reduces approximately three times.

The channel throughput given by Equation (5.13b) in this case becomes

$$\text{Channel Throughput} = \frac{L_D}{L_D + 38 + \frac{64(1-A)}{A}} < \frac{L_D}{L_D + 148}. \tag{5.19b}$$

Figure 5.23 shows the throughput versus packet length for different packet lengths. With large packet lengths close to maximum length, channel utilization and throughput are very close. As packet length becomes close to a minimum of 64 bytes, throughput drops substantially below the channel efficiency.

5.4.2 Performance of CSMA/CA and RTS/CLS in Wi-Fi

The CSMA/CA was the most popular medium access mechanism for Wi-Fi and RTS/CTS was originally adopted by 802.11 to support the hidden

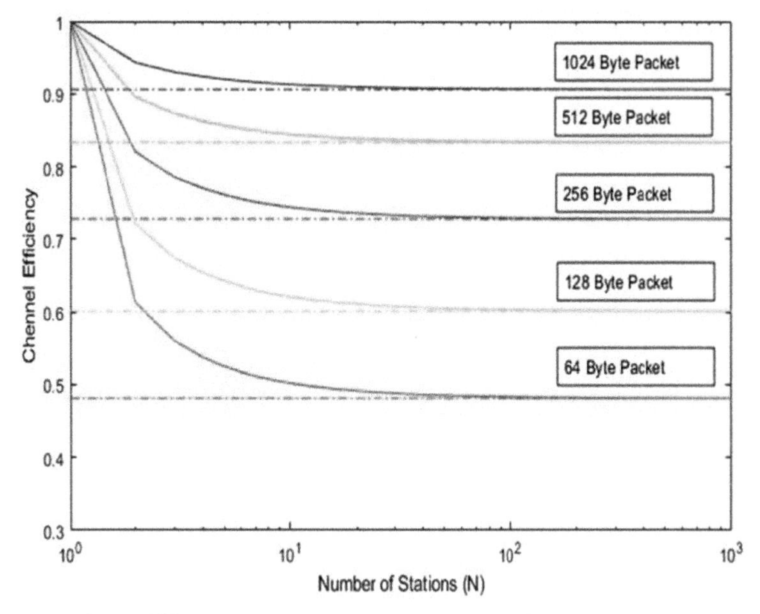

Figure 5.22 Channel utilization as a function of number of users.

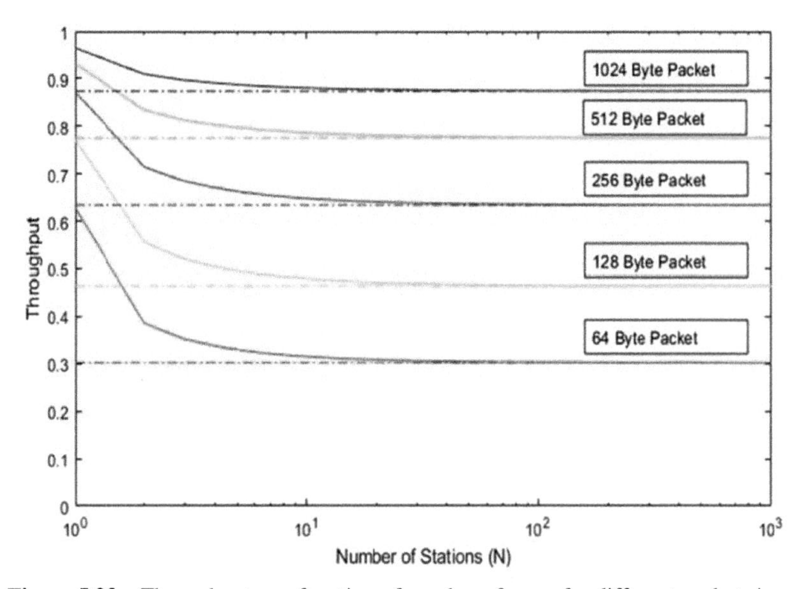

Figure 5.23 Throughput as a function of number of users for different packet sizes.

terminal problem. This problem occurs when a terminal is in a much longer distance from the access point from others, and because of the weakness of its received signal, devices closer to the access point block or reduce the far-end device throughput. With the emergence of MIMO and directional communication, this mechanism becomes popular for data communication because the broadcast medium needed for CSMA loses its essence in directional communication. As a result, mmWave IEEE standards such as IEEE 802.11ad adopted a hybrid approach with two channels (similar to the telephone networks). One channel at 2.4 or 5.2 GHz established the link with CSMA/CA and the other channel at 60 GHz transported the multimedia content with RTS/CTS.

Performance for Small Number of Users:

In residential homes or in small office and home office (SOHO), often, we have a single access point with one or a few users. In these situations, we can approximate the throughput by throughput of a single user divided by the number of users. Performance of a single user also provides a bound on the performance of Wi-Fi representing the maximum achievable data rate from an access point. Figure 5.24 shows the details of the transmission of a Wi-Fi packet with the frame format of the DSSS defined by legacy IEEE 802.11. Packet transmission begins with a CW followed with physical layer convergence protocol (PLCP) header, MAC header, MAC data (payload), and cyclic redundant codes (CRC) code as frame control sequence (FCS). Then, the packet is sent to the destination with a propagation time of δ, and the destination device waits for a short inter frame space (SIFS) interframe spacing and sends an ACK packet. After propagation delay δ, the transmitter receives the ACK and waits for a distributed coordination function interframe spacing (DIFS) before it can transmit the next packet. If we represent all overheads by H and the information packet by I, duration of the packet is given by

$$T = \mathrm{CW} + H + I + \mathrm{ACK} + \mathrm{SIFS} + \mathrm{DIFS} + 2\delta. \tag{5.20a}$$

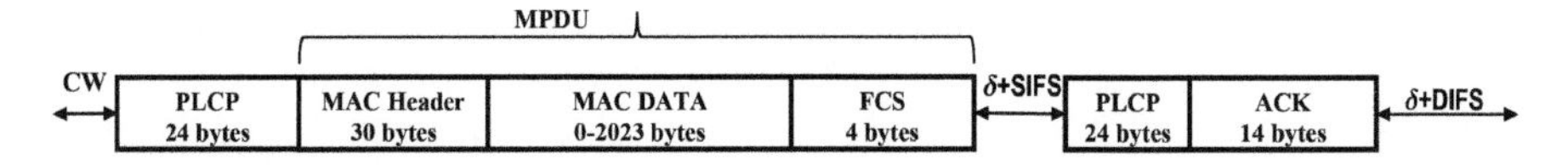

Figure 5.24 Details involved in packet transmission in Wi-Fi using legacy DSSS 802.11 and 802.11b standards.

The throughput for a single user experiencing no contention, CW = 0, is then given by

$$S = \frac{I}{T} = \frac{I}{H + I + \text{ACK} + \text{SIFS} + \text{DIFS} + 2\delta}. \tag{5.20b}$$

With multiple users, we have a contention. The contention length interval is governed by the selection of a uniformly distributed random variable between $(0, 2^m \text{CW}_{\text{min}})$, where *m* is the number of collisions and its maximum value as well as CW_{min} are specified by the standard. In low traffic, when we assume collisions are negligible, $m = 0$, we select a number between $(0, \text{CW}_{\text{min}})$ and this selection on the average comes to a value of $\text{CW} = (\text{CW}_{\text{min}} + 1)/2$. The length of the slot, σ, CW_{min}, and *m* are specified by the standard. The throughput is then given by

$$S = \frac{I}{\text{CW} + T} = \frac{I}{(\text{CW}_{\text{min}} + 1)/2 + H + I + \text{ACK} + \text{SIFS} + \text{DIFS} + 2\delta}. \tag{5.21}$$

This equation provides an upper bound to the throughput for all contending users. In a worst-case scenario, all terminals may experience maximum contention. In this case, we replace the CW_{min} in Equation (5.21) with CW_{max} to obtain a lower bound for the throughput. To calculate the throughput with any of the above equations, we need to bring all the parameters in the same unit of time. An example will help to clarify this concept.

Example 5.12: Throughput of the DSSS IEEE 802.11b:

In multi-rate IEEE 802.11 standards, the header is always transmitted at the lowest data rate and the data is transmitted at the highest possible data rate for the distance between the transmitter and the receiver. The minimum and maximum data rates for the IEEE 802.11b are 1 and 11 Mbps, respectively. The SIFS, DIFS, and slot time σ, are 10, 50, and 20μs, respectively, and CW_{min} = 31. Table 5.2 shows a detailed calculation of the parameter for the IEEE 802.11b required for calculation of the throughput for IEEE802.11 in μs. Considering that propagation delay is 3 ns per meter and Wi-Fi devices are covering a few tens of meters, the propagation delay is below 1 μs, while SIFS and DIFS are in the order of tens of ms; so we can neglect the effects of

```
%% Sample of Plotting
L_Data=0:2000;
% T_cw=0
S_0=8*L_Data./(8*L_Data+5267.9);
% T_cw=31
S_31=8*L_Data./(8*L_Data+8677.9);
% Plot
plot(L_Data,S_0,'r','LineWidth',2);
hold on
plot(L_Data,S_31,'b','LineWidth',2);
hold off
grid on
legend('CW=0','CW_{min}=31','Location'
,'NorthWest');
xlabel('Length of Data in Bytes: L');
ylabel('Throughput: S');
```

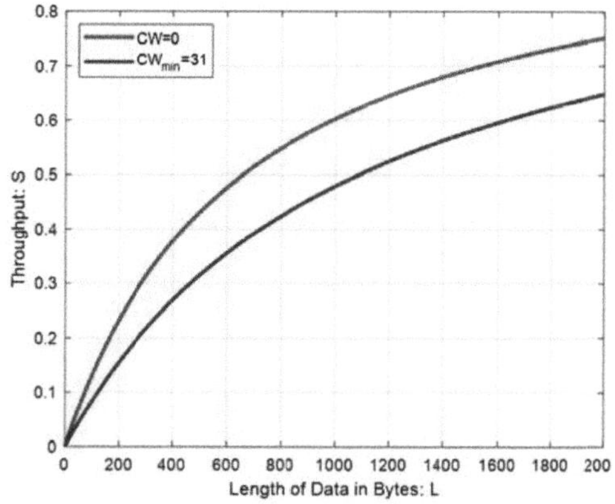

Figure 5.25 Throughput versus packet length bounds for CSMA/CD for the legacy DSSS and 802.11b standard.

propagation. Substituting these values in Equation (5.8), we have

$$S = \frac{I}{\text{CW}+H+I+\text{ACK}+\text{SIFS}+\text{DIFS}+2\delta}$$
$$= \frac{8\times L_{\text{Data}}/11(\text{Mbps})}{\text{CW}+8\times L_{\text{Data}}/11(\text{Mbps})+(50+192+24.7+10+192+10.2)(\mu s)}$$
$$= \frac{8L_{\text{Data}}}{\text{CW}+8L_{\text{Data}}+5267.9}$$
$$\text{For } T_{\text{CW}} = 0 \text{ (single user)} : S = \frac{8L_{\text{Data}}}{8L_{\text{Data}}+5267.9}$$
$$\text{For } \text{CW}_{\min} = 31, \quad \text{upper bound} \Rightarrow S = \frac{8L_{\text{Data}}}{8L_{\text{Data}}+8677.9}.$$

Figure 5.25 shows the MATLAB code and plot for the throughput versus packet length for a single user and upper bound in multiple contending users.

Performance for Large Number of Users:

The most popular method for the performance analysis of the CSMA/CA and RTS/CTS contention-based access in Wi-Fi was introduced by Bianchi [Bia00] over a decade after the beginning of the IEEE 802.11 and right after completion of this standard in 1997. This analysis begins by defining the probability of successful transmission in a slot of contention interval, P_s, and the probability of transmitting at least one packet in a slot time, P_{tr} and relating these probabilities with a length of information payload, I, duration of slot tine, σ, the average time the channel is sensed busy,T_s, and the average

Table 5.2 Throughput parameters for IEEE802.11b.

Field	Duration
DIFS	50 ms
PLCP	24 × 8 = 192192/1 Mbps = 192 ms
MPDU overhead	34 × 8 = 272272/11 Mbps = 24.7 ms
Data	*L*/11 Mbps
SIFS	10 ms
ACK PLCP	192 ms
ACK Data	14 × 8 = 112112/11 Mbps = 10.2 ms
Minimum CW	31/2 × 20 ms = 310 ms
Maximum CW	1023/2 × 20 = 10,230

time the channel is sensed busy by each station during the collision:

$$S = \frac{P_s P_{tr} I}{(1 - P_{tr})\sigma + P_s P_{tr} T_s + (1 - P_s) P_{tr} T_c}. \tag{5.22}$$

For the average information payload of I, the average amount of payload information successfully transmitted in a slot time is $P_s P_{tr} I$ since a successful transmission occurs in a slot time with probability of $P_s P_{tr}$. The average length of a slot time consists of actual several events with different probabilities. First, the event is the occurrence of an empty slot of the length, σ, which happens with the probability of $(1 - P_{tr})$ for the times that there is no transmission. When we have a successful transmission with a probability of $P_s P_{tr}$, it occupies a time interval of T_s that includes the transmission time of the information as well as other physical and MAC layer overheads. These overheads include framing data, propagation delay, and transmission of acknowledging packets. Another event is the occurrence and processing of the collision that takes T_c unit of time and its probability of occurrence is $(1 - P_s) P_{tr}$.

The value of σ is set by the IEEE 802.11 standard and it specifies the time needed by any other device in the coverage area of the transmitter to detect the transmission. Table 5.3 shows the typical values of slot time along with DIFS and SIFS interframe spacings specified by a variety of IEEE 802.11 standards. In using the above equation, the user should make sure all parameters and delays are using the same unit. It is convenient to transfer all the delays and time scales in micro-seconds (μ s) to be compatible with the parameters of Table 5.3. In this table MAC packet data unit (MPDU) overhead is calculated for the 34 byte overhead of the IEEE 802.11 packet format.

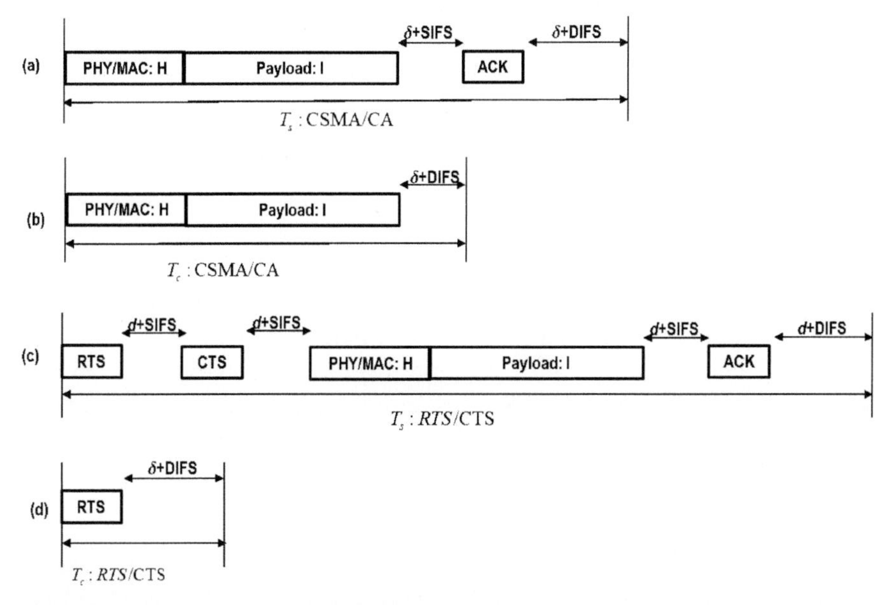

Figure 5.26 Definition of transmission times for (a) success in CSMA/CA, (b) collision in CSMA/CA, (c) success in RTS/CTS, and (d) collision in RTS/CTS.

Figure 5.26 shows the details of successful and collided packet transmission in CSMA/CA and RTS/CTS. The two timing parameters T_C and T_S differ in the two MAC access mechanisms. In successful transmission with CSMA/CA (Figure 5.26(a)), after completion of transmission of the packet that includes header H and information payload I, an ACK will arrive from the destination. Before ACK, destination should wait for a SIFS, and after the arrival of the ACK, sources need to wait for DIFS before sending a new packet. On top of this, the transmission also involves two delays of δ, one for

Table 5.3 Parameters of different IEEE standards.

Standard	Band (GHz)	DIFS (μ s)	SIFS (μ s)	Slot time (μ s)
802.11 (FHSS)	2.4	128	28	50
802.11 (DSSS)	2.4	50	10	20
802.11b	2.4	50	10	20
802.11a	5	34	9	16
802.11g	2.4	28 or 50*	10	9 or 20*
802.11n	2.4	28 or 50*	10	9 or 20*
802.11n	5	34	9	16
802.11ac	5	34	9	16

*Options for *g* and *n* are in the presence and absence of "*b*" devices.

the packet and one for the ACK and

$$T_S = H + I + \text{SIFS} + \text{ACK} + \text{DIFS} + 2\delta. \tag{5.23a}$$

If the packet collides (Figure 5.26(b)), the receiver does not send an ACK and collided packet duration becomes

$$T_C = H + I + \text{DIFS} + \delta. \tag{5.23b}$$

In successful transmission with RTS/CTS (Figure 5.26(c)), we have four packet transmissions for RTS, CTS, $H + I$, and ACK with three SIFS and one DIFS. Therefore, there is a lot more overall delay. However, collision is recognized after a short RTS transmission (Figure 5.26(d)). Therefore, transmission times for successful and collided events are given by

$$\begin{aligned} T_S &= H + I + \text{RTS} + \text{CTS} + \text{ACK} + 3 \times \text{SIFS} + \text{DIFS} + 4\delta \\ T_C &= \text{RTS} + \text{DIFS} + \delta. \end{aligned} \tag{5.24}$$

To calculate the throughput defined by Equation (5.22), we need two additional parameters, probability of successful transmission in a slot of contention interval, P_s, and the probability of transmitting at least one packet in a slot time, P_{tr}. These values depend on a number of devices contending for access, n, and the probability of transmission in a slot time, p. The probability of successful transmission is one minus probability of no device attempting to transmit:

$$P_{tr} = 1 - (1 - p)^n. \tag{5.25a}$$

The probability of a successful transmission occurring on the channel, P_s, is the probability that exactly one station transmits on the channel, conditioned on the fact that at least one station transmits:

$$p_s = \frac{np(1-p)^{n-1}}{P_{tr}} = \frac{n\tau(1-p)^{n-1}}{1-(1-p)^n}. \tag{5.25b}$$

Another important issue is the relation between the exponential backup algorithm and these parameters. The IEEE 802.11 defines the same exponential backoff algorithm for CSMA/CA and the RTS/CTS medium access methods. As a result, P_s and P_{tr} are the same for both mechanisms. In the IEEE 802.11 standard, the minimum CW is defined as CW_{min}; for m-collisions, the length of this window becomes $2^m CW_{\text{min}}$. The maximum value of m is specified by the standard and it was 10 for the legacy IEEE 802.11. The minimum window length for the frequency hopping spread spectrum (FHSS) and DSSS were 16 and 32 slots. At each packet transmission, the backoff time is uniformly chosen in the range of $(0,\ 2^m CW_{\min})$ and it depends on the number of transmissions failed for the packet. At the first transmission attempt, the CW is set to the minimum CW. After each unsuccessful transmission, it is doubled, up to a maximum value of $CW_{\max}$. Therefore, for legacy 802.11, $\mathrm{CW}_{\max} = 1024 \times \mathrm{CW}_{\min}$ because the maximum for m was 10. Therefore, the probability that a station transmits in a randomly chosen slot time, *p*, is also related to the exponential backoff algorithm details and the probability of transmission in a slot, P_{tr}. A closed form relation between these parameters is derived in [Bia00] by modeling the entire process with a two-dimensional Markov model. Details of this derivation are beyond the scope of this book and the reader is referred to [Bia00] for that. Here, we present the results of this derivation to demonstrate the relation among exponential backoff algorithm parameters, *p*, and P_{tr}. Bianchi shows that this relation is given by

$$p = \frac{2(1 - 2P_{tr})}{(1 - 2P_{tr})(CW_{\min} + 1) + CW_{\min}(1 - (2P_{tr})^m)P_{tr}}. \tag{5.25c}$$

The three sets of equations, Equations (5.16a), (5.16b), and (5.16c), have three parameters, (P_{tr}, P_s, p) and we can solve them to calculate (P_{tr}, P_s) for Equation (5.13). Using Equations (5.16a) and (5.16c), we can calculate the *p* in terms of channel parameters and then replace it in Equations (5.16a) and (5.16b).

Example 5.12: Plot of Bianchi's Method for Legacy 802.11 DSSS: The following code generates the plots of Bianchi's analysis for legacy DSSS IEEE 802.11; for 10 and 50 nodes, it can be easily extended to do it for RTS/CTS. Figure 5.27 shows the plot generated for both MAC mechanisms. In this figure, the performance of CSMA/CA is compared with that of the RTS/CTS mechanism for 10 and 50 nodes. In CSMA/CA, the performance substantially drops when the number of nodes increases from 10 to 50. RTS/CTS has very little sensitivity to the change in packet length. When packets are less than a few thousand of bits, CSMA performs better, and as packets become longer, RTS/CTS has a better performance. This comparison clearly suits that RTS/CLS is a better mechanism under heavy loads such as those in the conference or athletic events. However, in most applications in SOHO, CSMA/CD is a better choice.

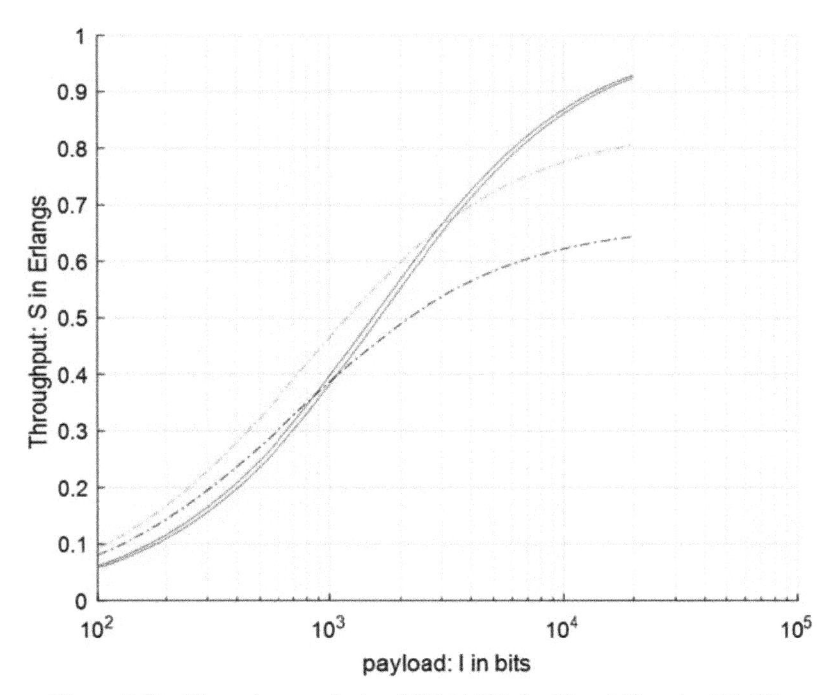

Figure 5.27 Throughput analysis of CSMA/CA for 10 and 50 nodes [Bia00].

```
%%%%%%%%%%%%%%%%%%%%%%%%%%%%%%%%%%%%%%%%%%%
% Author: Bader Alkandari,
function bianchi_main
clear all;
close all;
clc;
n_values = [10, 50]; % number of contending nodes
figure; % create and hold semi-log figure
grid
set(gca,'xscale','log')
xlabel('payload I (bits)')
ylabel('Throughput S: CSMA/CA(DSSS)')
hold on

for i = 1:length(n_values)

        W = 32; % window size CWmin or CWmax
```

```
    m = 5; % typical values are 5 or 6
    n = n_values(i); % number of stations
    calculate_CSMA_CA_DSSS % calculate p and tau given n
    sig = 20; % slot duration (usec). depends on PHY
    Ptr = 1 - (1-tau)^n;
    num_Ps = n*tau*(1-tau)^(n-1);
    den_Ps = 1 - (1-tau)^n;
    Ps = num_Ps/den_Ps;

    % assume 1 Mbit rate (all units get converted to usec)
    H = 400; % header size (bits)
    I = 100:1:20000; % information payload I (bits)
    d = 1; % propagation delay "delta"
    SIFS = 28; % usec
    DIFS = 128; % usec
    H_phy = 128; % PHY header
    RTS = 160 + H_phy; % usec
    CTS = 112 + H_phy; % usec
    ACK = 112 + H_phy; % usec

    % Ts and TC for CSMA/CA. Change this for CSMA/CA
    Ts = H + I + SIFS + d + ACK + DIFS + d;
    Tc = H + I + DIFS + d;

    % Calculate throughput S
    num_S = Ps*Ptr*I;
    den_S_1 = (1-Ptr)*sig;
    den_S_2 = Ptr*Ps*Ts;
    den_S_3 = Ptr*(1-Ps)*Tc;
    den_S = den_S_1 + den_S_2 + den_S_3;
    S = num_S./den_S;

    % plot throughput of CSMA/CA vs payload in bits
    semilogx(I,S,'-.')
end
%————
    function calculate_CSMA_CA_DSSS

    fun = @eqns2;
```

```
        x0 = [0 0];
        x = fsolve(fun,x0);

        tau = x(1);
        p = x(2);

  %————————————
          function fcns=eqns2(z)
          t = z(1);
          p = z(2);

          num = 2*(1-2*p);

          k1 = (1 - 2*p)*(W+1);
          k2 = p*W*(1-(2*p)^m);
          den = k1 + k2;

          fcns(1)= t - num/den;
          fcns(2)= p - 1 + (1 - t)^(n-1) ;
          end
        end
  end
```

Assignments

Questions

1. Name two duplexing methods and one example standard that uses each of these technologies.
2. What are the popular access schemes for data networks? Classify them.
3. Name a cellular telephony standard that employs FDMA.
4. What is binary exponential backoff algorithm, which standard uses that, and what is the purpose of using it? What is its weakness?
5. What is the purpose of the IEEE 802 standard committee and what are the steps taken to make its recommendations an international standard?
6. What is the difference between the access techniques of the IEEE 802.3 and IEEE 802.11?
7. Why most PCS standards use TDD and most cellular standards use FDD?

8. Why in the PSTN backbone hierarchy the FDM multiplexing lost its popularity in TDM multiplexing?
9. Why did the 2G cellular systems shift from analog FDMA to digital TDMA and CDMA?
10. Name three standards using TDMA/TDD as their access method.
11. What are the advantages of the CDMA access techniques?
12. What is the difference between performance evaluation of voice-oriented fixed assignment and data-oriented random-access methods?
13. Explain the difference between the effects of power control on the capacity of TDMA and CDMA systems.
14. In a radio ALOHA network, how does a terminal learn that its packet collides?
15. What is the difference between the maximum throughputs of ALOHA and slotted-ALOHA networks? What causes this difference?
16. What is the difficulty of implementing CSMA/CD in a wireless environment?
17. Explain the difference between carrier sensing mechanisms in wired and wireless channels.
18. Explain what the hidden terminal problem is and how it impacts the performance of a CSMA-based access method.
19. Explain what the capture effect is and how it impacts the performance of the random-access methods.
20. Explain the differences between the integration of data into a voice-oriented network and integration of voice into a data-oriented network.
21. Explain the relation between the receiver buffer size and packet error rate in VoIP applications.

Instructor's solution available on River Publishers' website:
https://www.riverpublishers.com/book_details.php?book_id=919

Problem 1

To provide public telephone access to commercial ferries, a telephone company installs a multi-channel wireless telephone system in a ferry. This wireless radio system connects to a base station on the shore through the air. The base station is connected to the PSTN using wires.

a) If the telephone company installs a four-channel system, what is the probability of having a person come to the telephone and none of the lines are available? Assume that the average length of a telephone call is 3 minutes, and 150 passengers of the ferry, on average, make one call per hour.

b) What is the average delay for accessing the telephones?

c) How many channels were needed to keep the blockage probability below 2%?

Problem 2

a) Neglecting the frequency spectrum used for control channels, what is the maximum number of two-way voice channels that can fit inside the frequencies allocated to the AMPS system.

b) What is the number of channels in each cell? Note that $N_f = 7$ was originally used in the AMPS.

c) Repeat (2) for IS-136/IS-54 in which $N_f = 4$ and the number of slots per TDMA channel is three.

d) Repeat (2) for IS-95 CDMA assuming the minimum required Eb/No is 6 dB. Include the effects of antenna sectorization, voice activity, and extra CDMA interference.

e) Repeat (4) for broadband CDMA where 5 MHz bands are used in each direction.

Problem 3

a) Sketch the throughput versus offered traffic G for a mobile data network using slotted non-persistent CSMA protocol. The packets are 20 ms long and the radius of coverage of each BS is 10 km. Assume the radio propagation speed is 300,000 km/s and use the worse delay for calculation of the "a" parameter.

b) Repeat (a) for slotted-ALOHA protocol.

c) Repeat (a) for 1-persistent CSMA protocol.

d) Repeat (a) for a wireless LAN with access point coverage of 100 m.

e) Repeat (a) for a satellite link with 20,000 km from the earth.

Problem 4

A cellular carrier has established 100 cell sites using AMPS with 395 channels and $K = 7$.

a) Use the provided graphs to calculate the total number of subscribers for a blocking probability of 0.02, an average of 2 calls per hour, and average telephone conversation of 5 minutes.

b) Use a computer (MathLAB, MathCAD, etc.) to calculate the same values using the Erlong B equation directly.
c) Determine (either from plot or calculation) the average delay for a call.
d) Repeat (a) for a blocking probability of 0.01.
e) Repeat (a) if IS-54 (IS-136) was used with $N_f = 7$.
f) Repeat (a) if IS-54 (IS-136) was used with $N_f = 4$.

Problem 5

We want to use a GSM system with sectored antennas ($N_f = 4$) to replace the existing AMPS system ($N_f = 7$) with the same cell sites. In the existing AMPS system, the service provider owns 395 duplex voice channels.

a) Determine the number of voice channels per cell for the AMPS system.
b) Determine the number of voice channels per cell for the GSM system.
c) Repeat (b) if we were using a W-CDMA system with a bandwidth of 12.5 MHz for each direction. Assume a signal-to-noise ratio requirement of 4 (6 dB) and include the effects of antenna sectorization (2.75), voice activity (2), and extra CDMA interference (1.67).

Problem 6

We provide a wireless public phone with four lines to a ferry crossing between Helsinki and Stockholm carrying 100 passengers where, on average, each passenger makes a 3-minute telephone call each 2 hours.

a) What are the probability of a passenger approaching the telephones and none of the four lines are available?
b) What is the average delay for a passenger to get access to the telephone?
c) What is the probability of having a passenger waiting for more than 3 minutes for access to the telephone?
d) What would be the average delay if the ferry had 200 passengers?

Problem 7

A WLAN hop accommodates 50 terminals running the same application. The transmission rate is 2 Mbps and the terminals are using slotted-ALOHA protocol. The commutative traffic produced by the terminals is assumed to form a Poisson process.

a) Give the throughput versus offered traffic equation for the system and determine the maximum throughput in Erlang.
b) What is the maximum throughput in bits per second?
c) What is the maximum throughput in bits per second for each terminal?

Problem 8

A local 3-hour tour boat with 50 passengers has one AMPS radio phone to connect to the shore. On average, each user places one call per hour and the average holding time for the calls is 3 minutes.

a) What is the probability that a person attempts to use the phone and he/she finds it occupied?
b) Repeat (a) if the AMPS phone is replaced by three IS-54 phones using the three slots of the existing IS-54 TDMA over the same band.
c) Repeat (a) if this phone is replaced by six upgrades IS-54 phones using six-slot upgraded IS-54 TDMA over the same band.

Problem 9

In a datagram packet-switched network with

P: packet size in bits

N: number of hops between two given systems

B: data rate in bps on all links

H: overhead (header in bits per packet)

T: end-to-end delay

Np: number of packets

L: message length in bits

D: propagation delay per hop

a) Give Np in terms of *L*, *P*, and *H*.
b) Give *T* in terms of *L*, *P*, *H*, *N*, *B*, and *D*.
c) What value of P, as a function of *N*, *B*, and *H*, results in minimum end-to-end delay *T*? Assume that the message length is much larger than the packet size and propagation delay is negligible ($D = 0$).

Problem 10

An *ad-hoc* 2 Mbps wireless LAN using ALOHA protocol connects two stations with a distance of 100 m from one another each, on average, generating 10 packets per second. If one of the terminals transmits a 100-bit packet, what is the probability of successful transmission of this packet? Assume that the propagation velocity is 300,000 km/s and the packets are produced according to the Poisson distribution.

6

IEEE 802.3 – The Ethernet

6.1 Introduction

Local area networks (LAN) connect computer terminals, main frames, printers, and other equipment in small geographic areas in the offices and homes. The main differences between LANs and wide area networks (WANs) are the higher data rates available to the user, smaller geographic range, and ownership of the network. A LAN is usually owned privately, while WANs are owned by a service provider leasing the service to different people or organizations. Today, the most popular wired LAN technology is the Ethernet operating over unshielded twisted pair (TP) cabling, but a variety of other technologies such as a token ring or token bus were used and competed with the Ethernet in the early days of this industry.

The cost of infrastructure in WANs is very high and the coverage is very wide. As a result, WANs are offered as a charged *service* to the user. The service provider invests a large capital for the installation of the infrastructure and generates revenue through monthly service charges. Local networks are sold as end products to the user and there is no service payment for local communications. The operation of LANs is very similar to a PBX in that the user owns them and pays monthly charges to the wide-area Internet service providers for wide-area communications.

The LAN industry emerged during the 1970s to enable sharing of expensive resources like printers and to manage the wiring problem caused by an increasing number of terminals in offices. By the early-1980s, three standards were developed: Ethernet (IEEE 802.3), token bus (IEEE 802.4), and token ring (IEEE 802.5); they specified three distinct medium access control (MAC) and physical (PHY) layers and different topologies for networking over thick cable medium but shared the same management and bridging (IEEE 802.1) and logical link control (LLC), IEEE 802.2. With the growing popularity of LANs in the mid-1980s, high installation costs of

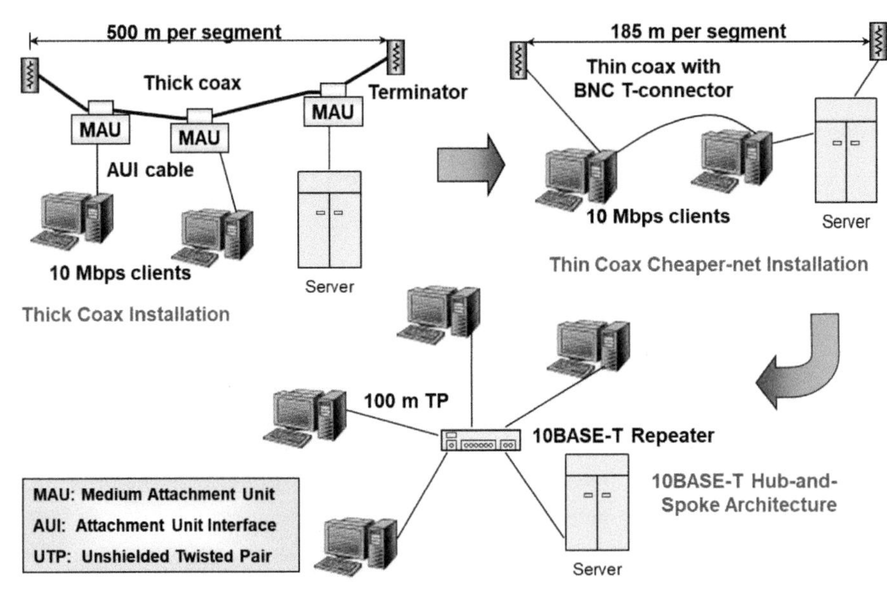

Figure 6.1 Early day evolution of Ethernet topologies from bus to star topology.

thick cable in the office buildings moved the LAN industry toward using thin cables that were also referred to as "cheapernet." Cheapernet covered shorter distances of up to 185 referred to as Hub and Spoke LANs using easy-to-wire TP wiring with coverage of 100 m was introduced.

Figure 6.1 graphically describes the early days of the evolution of the Ethernet. The interesting observation is that this industry has made a compromise on the coverage to obtain a more hierarchical structured solution that is also easier to install. TP wiring, also used by public switched telephone network (PSTN) service providers for telephone wiring distribution in homes and offices for over a hundred years, is much easier to install. The star network topology opened an avenue for structured hierarchical wiring, also similar to the telephone network topology. Today, IEEE 802.3 (Ethernet) using TP wiring is the dominant wired LAN technology and it is the focal point of this chapter.

The data rates of legacy LANs – thick, thin, and TP – were all 10 Mbps. The need for higher data rates emerged from two directions: (1) there was a need to interconnect LANs that are located in different buildings of campus to share high-speed servers, and (2) computer terminals became faster and capable of running high-speed multimedia applications. To address these needs, several standards for higher data rate operations were introduced.

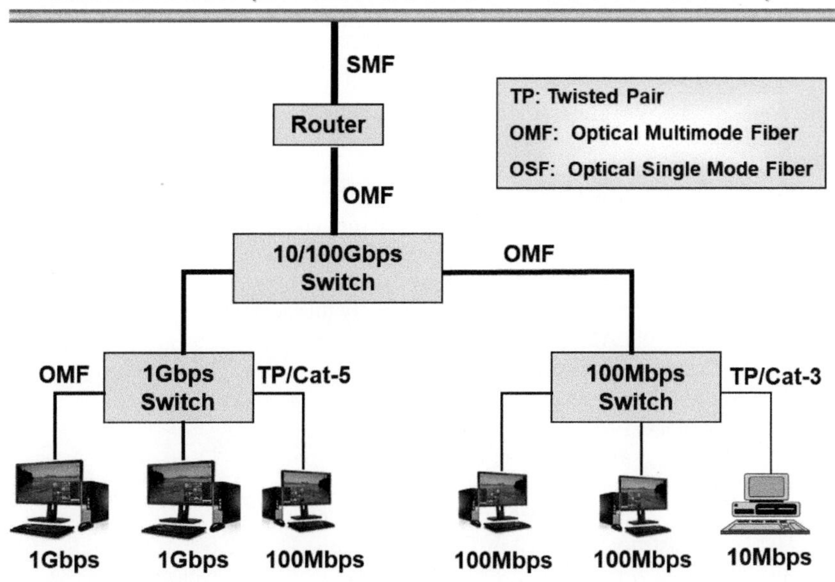

Figure 6.2 Hierarchical Ethernet LAN with star topology.

The first fast LAN operating at 100 Mbps was the fiber distributed data interface (FDDI) that emerged in the mid-1980s as a backbone medium for interconnecting LANs. The American National Standards Institute (ANSI) published this standard directly. In the mid-1990s, 100 Mbps fast Ethernet was developed under IEEE 802.3. In the late 1990s, IEEE 802.3 approved the Gigabit Ethernet followed by 10 and 100 Gbps Ethernet around 2010. These ultra-high speed Ethernets are now attracting market both in the local and wide-area networking. All these high-speed LANs use fiber optics, high-quality TP, and multiple TP wirings to support faster transmission. Figure 6.2 shows an example of hierarchical wiring of a LAN. A variety of 100 and 1000 Mbps terminals relate to two levels of switches and repeaters to a router that connects the LAN to the rest of the world. In the core networks, Ethernet is integrating with SONET/SDC for optical long-haul transmission.

In summary, the LAN industry has developed several standards, mostly under the IEEE 802 community. Table 6.1 lists the name of several IEEE 802 standard series, and Figure 1.15 related the protocol stack of the IEEE 802 standards with those of IETF Internet and OSI protocol stack reference models. The 802.1 (HILIs for bridging) and 801.2 (LLC) parts are common for all the standards, 802.3, 802.4, and 802.5 are wired LANs, and 802.9 is

Table 6.1 Overview of the early IEEE 802 standards.

IEEE 802 Standard Series
802.1: Higher level interface (HILI)
802.2: Logical link control (inactive)
802.3: CSMA/CD (Ethernet in commerce)
802.4: Token bus (disbanded)
802.5: Token ring (inactive)
802.6: MAN (disbanded)
802.7: Broadband Technical Advisory Group (disbanded)
802.8: Fiber Optics Technical Advisory Group (disbanded)
802.9: Integrated service LAN interface (disbanded)
802.10: Standard for interoperable LAN security (disbanded)
802.11: Wireless LANs (Wi-Fi in commerce)
802.12: Demand priority (inactive)
802.14: Cable TV based broadband communication networks (disbanded)
802.15: Wireless PAN (Bluetooth, ZigBee, UWB)
802.16: Wireless MAN (WiMax in commerce)
802.17 Resilient Packet Ring (RPR) Working Group
802.18 Radio Regulatory Technical Advisory Group
802.19 Coexistence Technical Advisory Group
802.20 Mobile Wireless Access Working Group
802.21 Media Independent Handover Working Group
802.22 Wireless regional area network

the so-called ISO-Ethernet that supports voice and data over the traditional Ethernet mediums. IEEE 802.6 corresponds to metropolitan area networking and IEEE 802.11, 15, and 16 are related to wireless local networks. The IEEE 802.14 is devoted to cable modem based networks providing Internet access through cable TV distribution networks operating over coaxial cable wiring and fiber originally installed for TV distribution. The IEEE 802.10 is concerned with security issues and operates at higher layers of the protocols. The three major drives of this industry have always been the ease of installation, an increase of data rate, and popularity of the technology to support mass production at lower costs.

6.2 The Legacy Ethernet

Legacy Ethernet is the original IEEE 802.3 standard complete in 1983. The standard defines the topology, frame format, PHY, and the MAC for networking over think cables. The legacy Ethernet is obsolete technology, but it is important to study because the next generations of the Ethernet and the Wi-Fi build on the same frame format and MAC technologies. Our objective in explaining legacy Ethernet is to address the fundamental issues involved in wired and wireless LAN technologies.

LANs were originally designed to solve the wiring problem in the offices where a huge number of terminal cables were drawn to a box to share expensive resources such as main frame computers, printers, and mass storage devices. The original legacy Ethernet solution was based on the idea of a shared coaxial cable acting as a broadcast transmission medium to access all terminals. Figure 6.1 (top left) shows the overall architecture of the legacy Ethernet using a bus topology connecting all terminals and a server to a thick coaxial cable with a maximum length of 500 m. The end of the network needs a terminator at each end for a proper operation which is a drawback for maintenance of the network because accidental removal of the terminator tops the network operation. The standard allows four repeaters which extend the length to 2.5 Km. When the signal is sent from one end of the cable, it becomes attenuated as it travels through the cable. For thick coaxial cables used in the legacy Ethernet, at distances of more than 500 m, the strength of the signal against the background noise is so low that the error rate of the received bits is not acceptable anymore and a repeater is needed to boost up the level of the signal. Each time that the signal level is boosted, the accompanying noise is also boosted and the accumulation of the noise for more than four repeaters does not allow proper operation over the cable. Another parameter affecting the decision on the length of the cable is the overall delay of the cable which we will discuss later when we time about the MAC of the Ethernet.

Each terminal hanged over the cable using a vampire connector shown in Figure 6.3. This approach is somehow like wireless systems with the difference that the multipath fading in wireless channels makes detection of a collision between the packets much more difficult than the cables. The common cable providing the communication channel played a role like the "ether," a substance once thought to fill all space to carry electromagnetic waves, and for that reason, the network was referred to as "Ethernet." Each device is connected to the Ethernet vampire connector with a network interface card (NIC) shown in Figure 6.4 whose role is to play as a buffer between the computer and the network to regulate the data transmission among different terminals operating with different clocks. The computer data buses usually operate as parallel lines with certain data rate, and the network interface card (NIC) card reads/writes the data from the terminal bus buffer, adds some header to create a frame with a specific format, and transmits the frame at the specified data rate (10 Mbps for legacy Ethernet) for the LAN in

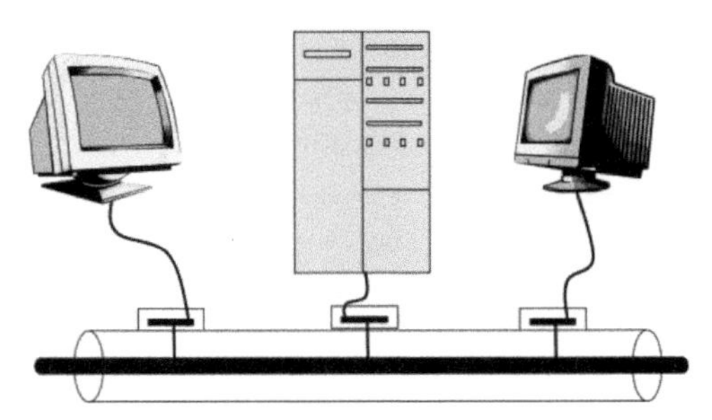

Figure 6.3 Vampire connectors in legacy Ethernet; contact with wire is made by a sharp needle.

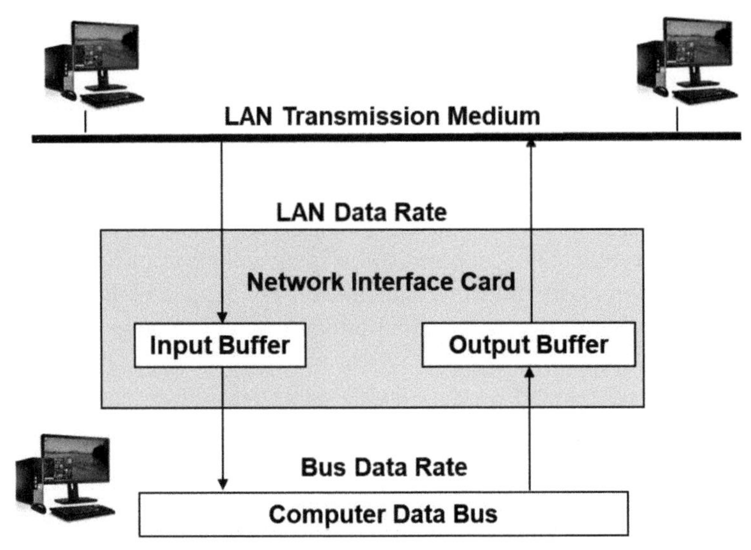

Figure 6.4 Connections to legacy Ethernet using the network interface card.

Minimum 64-bytes

Preamble (7)	Start Delimiter (1)	DA (2 or 6)	SA (2 or 6)	Length of data (2)	Data (0-1500)	Pad (0-46)	Checksum (4)

PLCP (Preamble, Start Delimiter) — MPDU (DA through Checksum)

Figure 6.5 Frame format of the IEEE 802.3 Ethernet.

the cable medium. The legacy Ethernet is also known as 10Base-5 reflecting 10 Mbps baseband line coding covering up to 500 m per segment of the cable.

6.2.1 The Packet Format and the PHY Layer

The IEEE 802.3 recommended frame format for the Ethernet is shown in Figure 6.6. Different parts of the frame are defined as follows:

Preamble: A sequence of 56 bits having alternated 1 and 0 values that are used for synchronization. They serve to give components in the network time to detect the presence of a signal and read the signal before the frame data arrives.

Start Frame Delimiter: A sequence of 8 bits having the bit configuration 10101011 indicates the start of the frame.

DA/SA: The destination/source MAC address field 802.3 standard permits either 2-bytes or 6-bytes, but, virtually, all Ethernet implementations use 6-byte addresses. A destination address may specify either an "individual address" destined for a single station or a "multicast address" destined for a group of stations. A destination address of all 1 bits refers to all stations on the LAN and is called a "broadcast address."

Length/Type: If the value of this field is less than or equal to 1500, then the Length/Type field indicates the number of bytes in the subsequent MAC Client Data field. If the value of this field is greater than or equal to 1536, then the Length/Type field indicates the nature of the MAC client protocol (protocol type).

MAC Client Data: This field contains the data transferred from the source station to the destination station or stations. The maximum size of this field is 1500 bytes. If the size of this field is less than 46 bytes, then the use of the subsequent "Pad" field is necessary to bring the frame size up to the minimum length.

Pad: If necessary, extra data bytes are appended in this field to bring the frame length up to its minimum size. A minimum Ethernet frame size is 64 bytes from the destination MAC address field through the frame check sequence.

Frame Check Sequence: This field contains a 4-byte cyclic redundancy code (CRC) remainder for dividing into a 33-bit number. When a source station assembles a MAC frame, it performs a CRC calculation on all the bits in the frame from the destination MAC address through the pad fields (that is, all fields except the preamble, start frame delimiter, and frame check sequence). The source station stores the value in this field and transmits it as

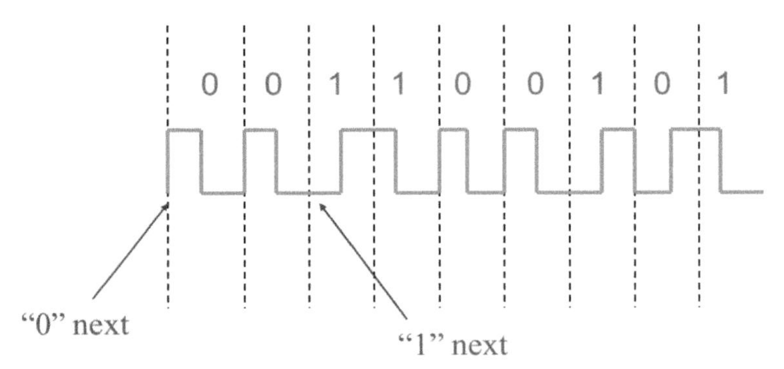

Figure 6.6 Differential Manchester coding used as the PHY of the legacy Ethernet.

part of the frame. When the frame is received by the destination station, it performs an identical check. If the calculated value does not match the value in this field, the destination station assumes that an error has occurred during transmission and discards the frame.

The minimum length of the packet is 64 bytes and the maximum is 1518 bytes allowing 1500 bytes of data plus 18 bytes of overhead. As we explain later, the minimum length is needed to ensure collision detection process in the MAC protocol. Ethernet frames allow 96 bits (9.6 μs for 10 Mbps or 96 ns for 1 Gbps) interframe time gap or space to allow the device a brief recovery time and prepare for the next frame. The PHY layer of the legacy Ethernet uses differential Manchester coding shown in Figure 6.7. As we described in Chapter 2, this line coding technique encodes the data stream in the transitions at the beginning of each bit. If the transmitted bit is a "0," we have a transition at the start of the bit, and if it is a "1," we have no transitions at the start of the bit. To have adequate transitions for synchronization at the receiver, this line coding technique enforces one transition in the middle of every bit transition interval. This is a very simple coding technique that could support the desired 10 Mbps data rate over the thick cable. The efficiency of this coding technique is 50% because each bit is transmitted with two neighboring pulses at the symbol transmission rate of 20 MSps. As we will see later in this chapter, the need for higher data rates has introduced more efficient line coding techniques for the implementation of PHY during the evolution of the Ethernet technology.

6.2.2 CSMA/CD for the MAC Layer

Legacy Ethernet used the so-called "thick-cable" as the shared medium winding around a building or campus to every attached terminal and machine.

This terminal shared the medium using carrier sense multiple access with collision detection (CSMA/CD) algorithm. This algorithm was simpler than the other two legacy LAN competing technologies, the IEEE 802.5 token ring supported by IBM for office areas and the IEEE 802.4 token bus supported by HP and others for the manufacturing environment. In a CSMA/CD MAC protocol, when a terminal wants to send a frame, it starts to sense the channel by simply reading the voltage level on the line. If there is no activity on the medium and the medium is idle, it sends its frame. If not, the terminal waits until the medium becomes ready, and after the interframe gap period of 9.6 μs in 10 Mbps legacy Ethernet, it sends the frame. If a collision occurs, the terminal goes through the collision detection process, stops the transmission of the packets, and uses a backup algorithm to retransmit the packet with a higher probability of success. Figure 6.7 shows a simple example. The User 1 senses the channel and it is idle; so it transmits its first frame and later on again senses that channel and there is no activity, so it sends the second frame. During the transmission of the second frame, User 2 senses the channel and finds it busy; so it waits until the channel becomes available. It prepares itself in 9.6 μs and sends its frame. While User 2 transmits its frame, both Users 1 and 3 sense the channel and find it busy. As soon as the medium becomes available, both users send their frames, and, shortly after, they find out that there is a collision because the voltage levels across the cable have gone above the normal values. As soon as transmitters discovered the occurrence of the collision, they stop transmission of the rest of the packet, send a 32-bit jamming signal, and reschedule the frame for the next slot. The jamming bits can have any value other than the CRC value of the colliding packet, and it ensures that the collision lasts long enough to be detected by all stations on the network. If retransmission is done immediately with the probability of one, in

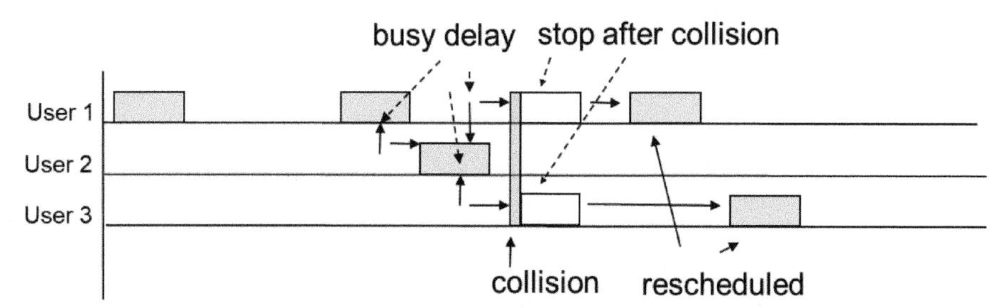

Figure 6.7 Principal of operation of the CSMA/CD specified as the MAC layer of the IEEE 802.3 Ethernet.

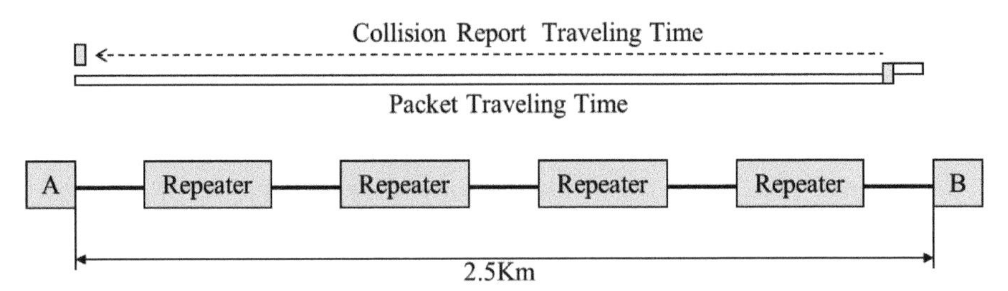

Figure 6.8 Collision detection and the length of the packet.

the next slot, packets will collide again; to avoid this collision, Ethernet uses an algorithm referred to as binary exponential backoff algorithm.

Legacy Ethernet uses the exponential backoff algorithm which is based on a random waiting time exponentially increasing with the number of collisions. The unit of waiting time is a slot and that is the minimum length of a packet so that one can detect the collision before transmission of the entire packet is completed. In the legacy Ethernet, this slot time is specified as 51.2 ms which is the time needed for transmission of 512 bits a (64 bytes) of data at the rate of 10 Mbps. Figure 6.8 shows the worst-case scenario for the collision report. Two terminals A and B are located at two ends of the network with a maximum length of 2.5 Km. Terminal A sends a frame and when that frame is about to reach terminal B, that terminal sends its own packet and a collision occurs. The collision signal should travel the 2.5 Km back to terminal A before that terminal knows that a collision has occurred. Therefore, we need 25 ms to detect a collision for the worst-case scenario. Selection of 51.2 ms which is almost double of this value allows another 25 ms for the delay associated with the four repeaters. In other words, maximum delay for collision detection and, consequently, minimum length of the packet to dctcct a collision is almost twice round trip propagation time and the extra time is needed for delay caused by repeaters.

The exponential backoff timing algorithm of the Ethernet operates based on the probability of transmission of the frame after a given number of collisions. After the first collision, the frame is sent in the next time slot with the probability of 1/2 which means that we pick from (0,1), after the second probability of transmission in the next slot reduces to 1/4 which means that we pick from (0,1,2,3), and after third, we pick from (0, 1, ..., 7) with a probability of 1/8, and we continue reducing the probability of transmission after the collision to half until the tenth collision. After ten collisions, we send a packet with the same probability of 1/1024 by repeatedly

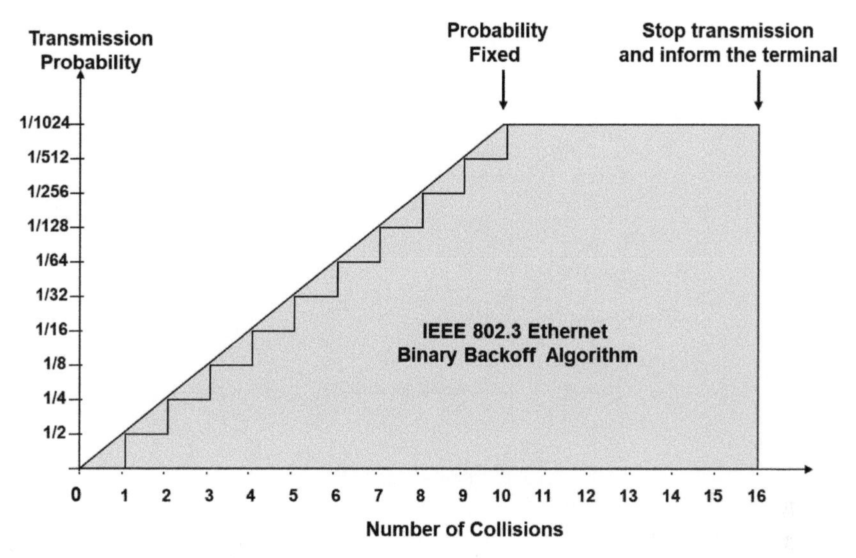

Figure 6.9 Backoff timing of the binary exponential backoff algorithm used in the Ethernet.

picking from (0, 1, ..., 1023) for six more times. After 16 collisions, MAC reports a failure to the computer and further recovery will remain up to higher layers. Figure 6.9 illustrates the relationship between transmission probability and the number of collisions for the binary exponential backoff algorithm employed in the Ethernet. This algorithm automatically prevents the instability of the network. Figure 6.10 provides a flow chart summarizing the CSMA/CD algorithm used in the Ethernet. The weakness of the exponential backoff algorithm used in the IEEE 802.3 is that it is a first-in last-out system. If a terminal arrives earlier and experiences a few collisions, it has a lower probability of access to the channel than a terminal that arrived later and has experienced its first collision. As we will see in the IEEE 802.11, this problem is resolved.

6.2.3 Early Enhancements to Legacy Ethernet

The legacy 10Base5 Ethernet soon evolved into thin-cable Ethernet also referred to as “cheapernet” with a coverage of 185 m which is also referred to as 10Base2 in which 2 reflects the approximately 200 m coverage. The IEEE 802.3a specified the thin cable medium for the 10Base2 in 1985. This adjustment in the medium would facilitate installation and reduce the cost of the cable resulting in a cheaper solution facilitating the rapid growth of the LAN installations in educational institutions and mid-size industrial

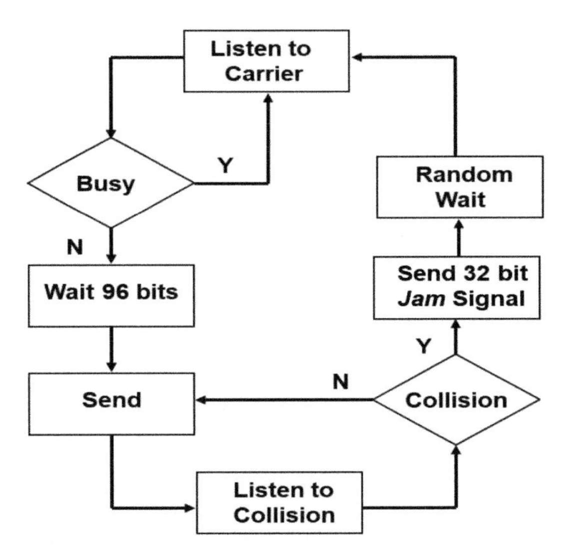

Figure 6.10 A flow chart for CSMA/CD operation.

institutions nationwide. In the mid-1980s, most mid-size universities started installing 10Base2 to connect computer terminals inside a department and a backbone FDDI at 100 Mbps to connect the department LANs. The cost of wiring for these networks would exceed several million dollars and the industry quickly exceeded several billion dollars stimulating further growth of the industry and nurturing new technical innovations to support this growth. Another advantage of the cheapernet was that it used Bayonet Neill-Concelman (BNC) connectors which were more rugged than vampire connectors, and, this way, using a power splitter, the BNC cable would hook to the Ethernet. BNC connector, originally designed to carry radio frequency (RF) signals, had a longer life than vampire connectors, reducing the maintenance cost of the network.

After moving from a few giant companies such as IBM, Intel, DEC, and Xerox and a few major research universities to mid-size industry and academic building, the LAN industry aimed at smaller buildings for small businesses for which installation of new wiring would pose a cost and speed factor. TP wiring used for telephone networks attracted attention in these environments. Installation of the TP wiring was simple and inexpensive and telephone companies had the expertise and the crew to install them efficiently at a low cost. The first implementation of the Ethernet over TP telephone wiring was StarLAN. Developed in the mid-1980s as 1Base5, later on formed the basis for the 10Base-T which was defined by IEEE 802.3i in 1990. The letter *T* used in the acronym refers to TP and the maximum length of the line was 100 m. The topology of this new revolutionary technology of its time was star used in the telephone network for many years and, for that reason, it needed a hub to connect the computer terminals with one another. In the telephone wiring, we simply connect the wires together, but in the Ethernet, the medium is also used for collision detection. Therefore, the MAC protocol needed an adjustment. This architecture also was referred to as hub-and-spoke because a terminal goes to the hub before it communicates with other terminals. This architecture is in contrast with the traditional listen-before-talk architecture of the legacy Ethernet. In the hub-and-spoke architecture, the hub detects the collision among packets arriving from different ports and transmits the jam signal into all ports. Collision is detected when the hub senses signal at two different ports at the same time and the detection mechanism does not involve analog threshold settings used in NIC cards of the legacy Ethernet. Connectors used for 10Base-T were telephone like RJ-45 connectors, which are still popular in connecting Ethernet to desktops. USB connectors are gradually replacing Rj-45.

Since the early 1990s, attention of the Ethernet community shifted from changing the medium and connectors toward higher speeds using more complex transmission technologies and accommodation of more diversified transmission medium with some modifications in the use of frame format and CSMA/CD operation. In the mid-1990s, the data rates moved to 100 Mbps by introducing the so-called "fast Ethernet" which replaced FDDI technology. Later, Gigabit Ethernet followed by 10, and, more recently, 100 Gbps Ethernet started penetrating WAN wirings traditionally dominated by T-carrier and SONET's optical carrier (OC) line is discussed in Chapter 2. In the frame format and medium access, virtual LAN (VLAN) was introduced to partition the network, the full-duplex operation became possible eliminating the use of CSMA/CD operation, and link aggregation allowed more flexibility for transmission to find more cost-efficient solutions. Table 6.2 highlights the important IEEE 802.3 standards born out of the evolution of the Ethernet. In the next two sections, we discuss technical aspects of the evolution of the physical layer followed by the frame and the access in the Ethernet.

6.3 Evolution of the PHY

In the mid-1990s, 10Mbps Ethernet technology and the hub-and-spoke architecture had dominated the market, while FDDI was used for 100 Mbps

Table 6.2 Overview of some important IEEE 802.3 standard series.

Year: Name and descriptions
1973: The first experiment at 2.94 Mbps.
1980: DEC, Intel, and Xerox released 10Base5.
1983: IEEE released the first IEEE 802.3 standard for Ethernet technology.
1985: IEEE 802.3a defined "thin" Ethernet, "cheapernet," or 10Base2.
1985: IEEE 802.3b defined 10Broad36 for "broadband" cable system.
1987: IEEE 802.3d defined 2-fiber optic inter-repeater link (FOIRL) for up to 1 Km.
1990: IEEE 802.3i defined 10Base-T.
1993: IEEE 802.3j defined 10Base-F (FP, FB, and FL) for up to 2 Km.
1995: IEEE 802.3u defined 100Base-T (fast Ethernet).
1997: IEEE 802.3x defined "full-duplex" Ethernet operation at 20 Mbps/200 Mbps.
1997: IEEE 802.3y defined 100Base-T2 for two pairs of Category 3.
1998: IEEE 802.3z defined 1000Base-X (gigabit Ethernet) on fiber/shielded twisted pair (STP)
1998: IEEE 802.3ac defined virtual LAN (VLAN) tagging on Ethernet networks.
1999: IEEE 802.3ab 1000Base-T standard defined 1 Gbps operation over four pairs of cat-5.
2000: IEEE 802.3ad link aggression allowing several NIC to connect to one port of a switch.
2003: IEEE 802.3ae 10Gbps Ethernet over fiber for connecting routers/switches up to 80 Km.
2003: IEEE 802.3af, power over Ethernet for IP phones, webcams, WLAN Aps, etc.
2004: IEEE 802.3ak, 10GBASE-CX4, four copper media up to 15 m.
2006: IEEE 802.3an, 10GBASE-T, for unshielded twisted pair (UTP).
2008: IEEE 802.3ba, 100 Gigabit Ethernet.
2020: IEEE 802.3cm, 400 Gbps over 4 and 8 multi-mode fiber lines for 100 m.

operation to connect these LANs as well as its closest competitor at that time, the IEEE 802.5 token ring. The first Ethernet advancement toward higher data rates was the so-called "fast Ethernet" which operated at 100 Mbps which was a series of standards defined by IEEE 802.3u in 1995 for a variety of fiber and TP media. This was followed by IEEE 802.3z defining the Gigabit Ethernet for fiber and shielded TP mediums in 1998 followed by IEEE 802.3ab which extended the Gbps Ethernet to the Cat-5 TP medium. In 2003, IEEE 802.3ae introduced 10 Gbps Ethernet for long-distance fiber operations which were extended to wire media by IEEE 802.3ak/an in 2004 and 2006, respectively. The advancements are summarized in Table 6.2. Currently, IEEE 802.3 standards are working on further penetration in WAN by defining a variety of standards for long-haul communications using optical line and wave division multiplexing carrying multiple streams over one fiber channel. The evolution of these standards reflects the penetration of Ethernet technology into WAN applications.

In principle, increasing the data rate is based on improving the medium explained in Chapter 3 or by using more sophisticated transmission technologies introduced in Chapter 4. To carry higher data rates over a medium, one can improve the bandwidth of the wire or the fiber, increasing the number of wires and shortening the length of the wire. To increase the data rate using transmission technique, one can see more efficient line coding (rather than Manchester coding), use multi-level and multi-dimensional transmission techniques, use signal processing techniques (such as equalization and echo-cancellation), and use more effective channel coding techniques (scramblers, convolutional coding, trellis coded modulation, turbo coding, etc.). As we will see in the following sections, these techniques have been adopted by a variety of Ethernet standards in one way or another.

Therefore, a review of the evolution of the physical layer and medium options of Ethernet provides an excellent overview of applied digital communications and information theory. These reviews also provide an example of how standards evolve with a sort of natural selection. In natural selection for living species, continuity of life is preserved by food, and in the natural selection for technologies and standards, life is preserved by the size of the market.

6.3.1 Fast Ethernet at 100 Mbps

Figure 6.11 provides an overview of the 100Base-T series of IEEE 802.3T standards for fast Ethernet. There are 100Base-X standards that are based on the fiber distributed data interface (FDDI) technology using a token ring

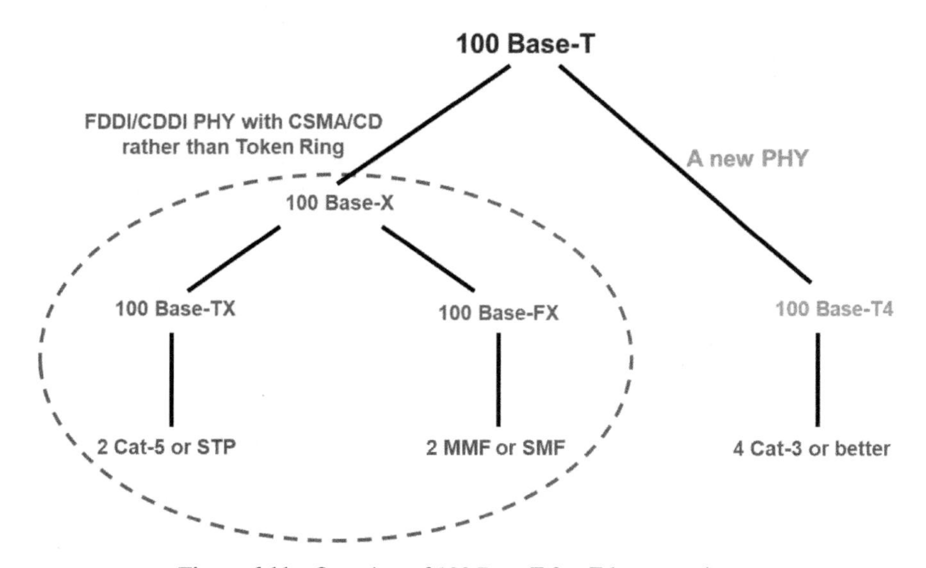

Figure 6.11 Overview of 100 Base-T fast Ethernet options.

Table 6.3 4B/5B encoding technique originally for FDDI and then used in 100Base-X.

Information	Coded block	Information	Coded block
0000	11110	1011	10111
0001	01001	1100	11010
0010	10100	1101	11011
0011	10101	1110	11100
0100	01010	1111	11101
0101	01011	Idle	11111
0110	01110	Start delimiter	11000
0111	01111	Start delimiter	10001
1000	10010	End delimiter	01101
1001	10011	End delimiter	00111
1010	10110	Transmit error	00100

which is adapted to the Ethernet environment using CSMA/CD medium access protocol. FDDI was originally designed for the fiber and later was extended to cable operation which was also sometimes referred to as copper distributed data interface (CDDI) replacing fiber “F” with copper “C” and it was very popular for connecting legacy Ethernet LANs. The 100Base-X standard would allow integration of the existing FDDI wirings into a simpler and growing Ethernet environment. The other branch is a newly defined PHY and medium using four pairs of Cat-3 or better wires, 100Base-T4, which turned out to become the most popular Ethernet used everywhere.

100Base-X: Both fiber and copper medium specifications for the 100Base-X series use the so-called 4B/5B line coding technique. The incoming data stream is formed into 4-bit blocks of information with 16 different possibilities. Each 4-bit block is mapped to one of the 32 possible 5-bit blocks so that the coded block has at least two “1” in the 5 bits. If the line coding technique is used with the 4B/5B coding codes “1” into transition in the level of the transmitted signal, these ones can support synchronization at the receiver. Table 6.3 shows all 16 possible codes and their related five bit code. The 5-bit output of the coding has 32 combinations as only 16 of them are used for the incoming data. Two of these extra codes are used for starting delimiter 2 for ending delimiter and one for reporting error, and the remaining 11 codes are not used [Sta00]. The coding rate of this line coding is 4/5 as compared with 1/2 for the Manchester line coding. Figures 6.12 and 6.13 show the details of the implementation of 100Base-FX and 100Base-X, respectively.

The 100Base-X fast Ethernets also support full-duplex operation. In traditional half-duplex Ethernet, the NIC only sends in one direction, and

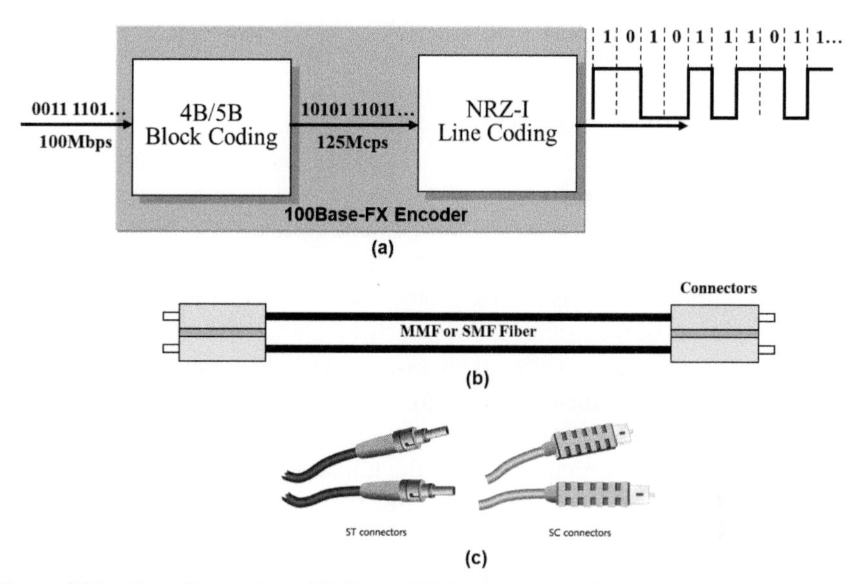

Figure 6.12 Overall operation of 100Base-XF. (a) 4B/5B and NRZ-I transmission. (b) Cable and connectors overview. (c) Two types of connectors.

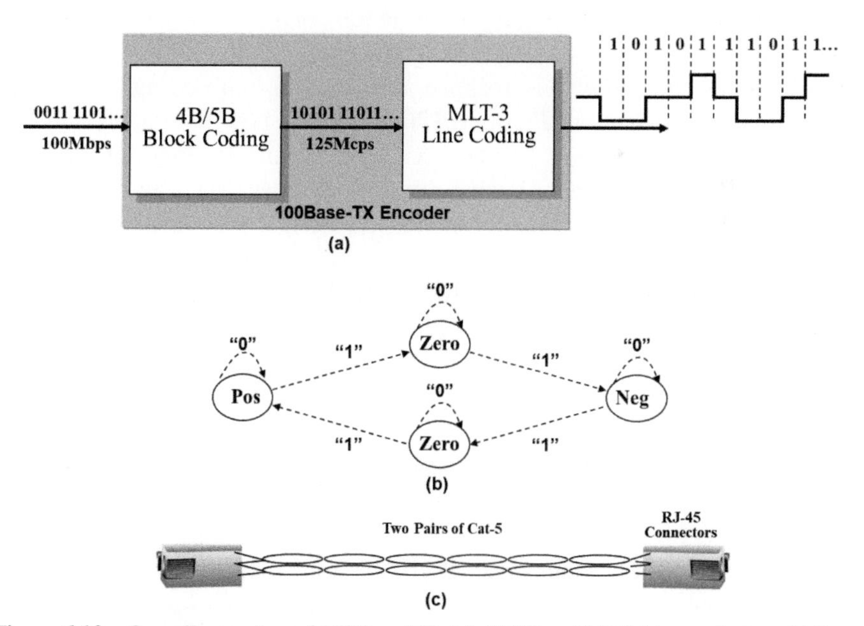

Figure 6.13 Overall operation of 100Base-TX. (a) 4B/5B and MLT-3 transmission. (b) State diagram of the MLT-3 line coding. (c) Physical connection.

in full-duplex, a NIC can transmit and receive at the same time doubling the aggregated rate of information of the LAN. Full-duplex operation curbs CSMA/CD protocol because transmission does not need to sense the availability of the medium. The problem with this approach is that the receiver buffer may overflow and we need a method to stop transmission if the buffer of the receiver is full. We discuss the details of how it operates later in this chapter when we address modifications to a legacy Ethernet packet format. Another feature introduced with fast Ethernet standard is auto-negotiation. Auto-negotiation protocol allows two connected devices to choose common transmission modes such as speed and duplex operation. Using auto-negotiation protocol, devices first share their capabilities and then choose the fastest transmission mode they both support.

100Base-FX uses the resulting 5-bit coded block after 4B/5B coding for NRZ-I line coding technique to create the transmitted waveform for the fiber line. As we explained in Section 4.2.1, for the NRZ-I line coding, each "1" is coded to a transition at the beginning of the bit. This way we have at least two transitions for every transmitted five coded bits, which is enforced by definition of 4B/5B coding, and it reflects the fact that in 4/5 (80%) of the transmitted bits carry information bits, while in Manchester coding only 50% of the transmitted bits carry information. Figure 6.12 gives an overall perspective of how this coding is implemented. Figure 6.12(a) shows the 4B/5B coding and how it is mapped into NRZ-I line coding to produce the transmitted waveform. Figure 6.12(b) and (c) show the cables and two types of connectors used in 100Base-XF. The maximum length of the cable recommended by the standard using multi-mode fiber (MMF) with 62.5/125 diameters at 1300 ns are specified at 412 m. Longer lengths are achievable using single-mode fiber (SMF).

100Base-TX is designed for 100 m high-quality Cat-5 or better TP wiring which was also used as the medium for copper or cable version of FDDI. Figure 6.13 gives the overall structure details of the 100Base-TX. Figure 6.13(a) shows the coding and transmission; the 100 Mbps data is first coded into a 125 Mbps coded stream which is then mapped into a three-level line signal referred to as MLT-3. The MLT-3 line coder shown by the state diagram in Figure 6.13(b) basically maps each "1" into transmission in the transmitted waveform. Each time that a "1" arrives, the signal level climbs the three-level signal ladder once. It climbs until there is no more place to go up and then it starts to climb down until there is no more step to switch the direction. The advantage of three-level MLT-3 line coding over the two-level NRZ-I is that it has three points rather than two in the constellation which results in a longer average distance between the points for the same average transmitted power. As we discussed in Chapter 2, this addition of the distance for the transmitted symbols results in a reduction in the error rate of the transmitted bits. Although MLT-3 has better performance, it cannot be implemented on fiber because signaling lights in the fiber transmission only take two levels representing the light in an "on" or "off" position. Figure 6.13(c) shows the two TPs and the RJ-45 connectors used for the physical operation of the 100Base-TX.

Example 6.1: PHY Constellation in 100Base-Fx and -Tx:
Figure 6.14 shows the signal constellation for transmission of the two-level signal used in 100Base-FX and the three-level signal used in the 100Base-TX. In the 100Base-TX, the points in the corners of the constellation represent the positive and negative voltage levels, and the middle points associate with the zero-level transmission. Since we have two zero-level states, the probability of occurrence of a zero-level signal (1/2) is twice that of the corner points (1/4). The average energy in the constellation shown in Figure 6.14(a) is

$$\overline{E} = \frac{E_0 + E_1}{2} = \frac{(d/2)^2 + (d/2)^2}{2} = \frac{d^2}{4}.$$

The average energy of the constellation shown in Figure 6.14(b) is

$$\overline{E} = \frac{E_1}{4} + \frac{E_0}{2} + \frac{E_{-1}}{4} = \frac{d^2}{4} + \frac{0}{2} + \frac{d^2}{4} = \frac{d^2}{2}.$$

P =1/2 P =1/2

d

(a)

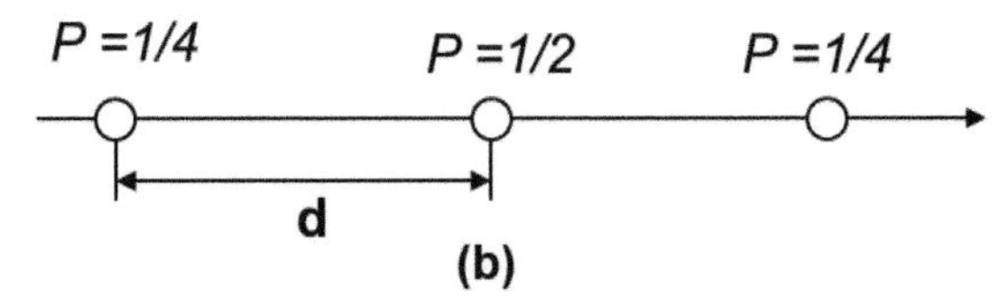

Figure 6.14 (a) Signal constellation for transmitted symbols in 100Base-FX. (b) Signal constellation for transmitted symbols in 100Base-TX.

Therefore, the three-level transmission has two times or 3 dB advantage over the two-level transmission technique, which makes it a more desirable technique.

100Base-T4 supports a 100 Mbps data rate for 100 m over four pairs of Cat-3 or better TP cabling and it is the most popular fast Ethernet commonly used to connect computer terminals in offices and homes. The main motivation for the creation of this standard was to allow 100 Mbps Ethernet to be carried over inexpensive telephone grade Cat-3 cabling as opposed to the Cat-5 cabling required by 100Base-TX. However, in practice, today, all new installations use Cat-5 because the cost of labor for installation is much higher than the cost of wire justifying the difference between the price of Cat-3 and Cat-5. Figure 6.15 shows the details for the overall operation of the 100Base-T4.

As shown in Figure 6.15(a), the input data stream is first encoded using "8B6T" encoding scheme in which 8 bits of binary data are converted into

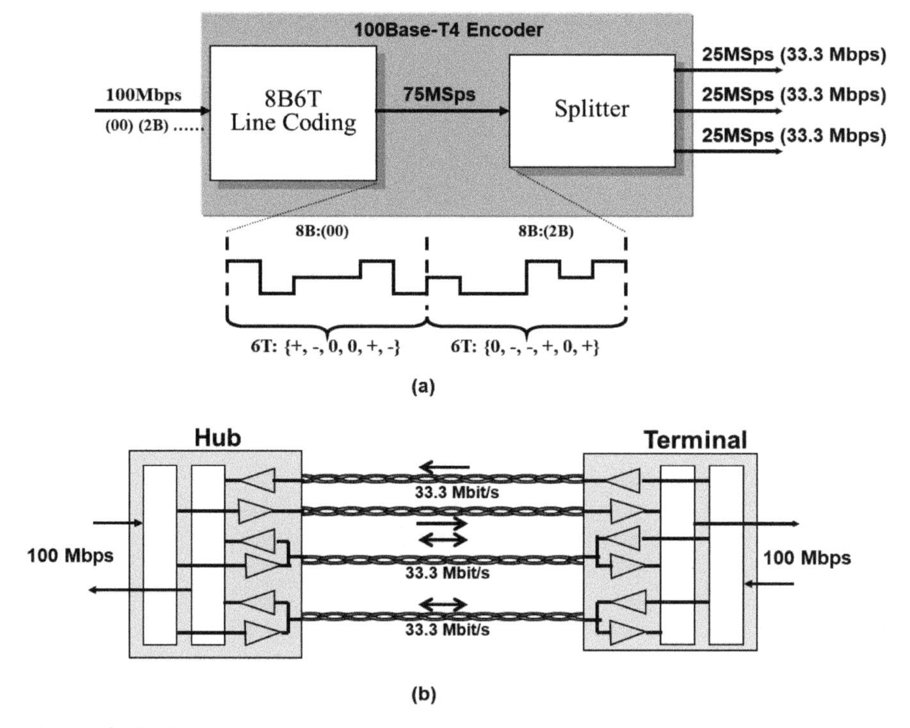

Figure 6.15 Overall operation of 100Base-T4. (a) 8B6T transmission. (b) Four-pair connections of wires.

Table 6.4 Samples of 8B6T code words used in 100Base-T4.

Data block-code word	
(00)	(+,−,0,0,+,−)
(0A)	(−,+,0,+,−,0)
(2B)	(0,−,−,+,0,+)
(2E)	(−,0,−,0,+,+)
(32)	(+,−,0,−,+,0)
(3C)	(+,0,−,0,−,+)
(3F)	(+,0,−,+,0,−)
(B8)	(−,+,0,0,+,0)
(AB)	(+,−,+,−,−,+)
(EB)	(+,−,+,−,0,+)

six “ternary” signals. For eight input bits, we need $2^8 = 256$ symbols and with six ternary digits, we can make $3^6 = 729$ symbols. Therefore, only a portion of six ternary symbols is used to map the 8-bit data block to six ternary symbols. The ternary symbols are selected so that the probability of occurrence of all three symbols is the same. Table 6.4 shows samples of the 8B6T codes; the complete set is available at [Pat96]. The 8-bit data block is represented in the octet form on the left and the ternary coded signal shown on the right side has three values: −, 0, and + resulting in three levels in a PAM-3 signal constellation.

Example 6.2: The PHY Layer of 100Base-T4:

Figure 6.16 shows the signal constellation for transmission of three-level signal used in the 100Base-T4 if all three levels have the same probability of transmission[1]. The average energy in the constellation shown in Figure 6.16 is

$$\overline{E} = \frac{E_1}{3} + \frac{E_0}{3} + \frac{E_{-1}}{3} = \frac{d^2}{3} + \frac{0}{3} + \frac{d^2}{3} = \frac{2d^2}{3}.$$

Compared with the binary constellation used in 100Base-TX shown in Figure 14(b), the constellation in 100Base-T4 has 1.33 times or 1.25 dB advantage.

Since each 8 bit is transmitted with six ternary symbols, the symbol transmission rate of the ternary signal is

$$100\,\text{Mbps} \times \frac{6\,(\text{ternary symbols})}{8\,(\text{bits})} = 75\,\text{MSps}.$$

[1]This is a valid assumption because the number of three levels used in the code word table are approximately the same.

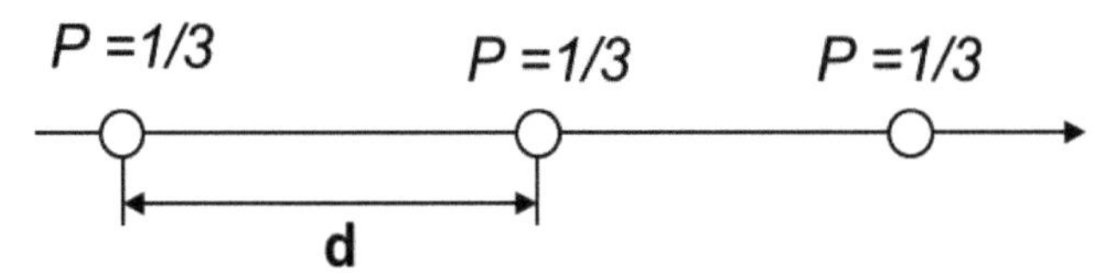

Figure 6.16 Signal constellation for transmitted symbols in 100Base-T4.

The 75 MSps stream of ternary symbols are then splinted into three streams, shown in Figure 6.15(a), each carrying a 25 MSps of ternary data. The actual binary data stream carried with each of these three lines is

$$25\,\mathrm{MSps} \times \frac{8\,(\mathrm{bits})}{6\,(\mathrm{ternary\,symbols})} = 33.3\,\mathrm{Mbps}.$$

This scheme effectively splits the 100 Mbps data rate over three streams of 33.3 Mbps so that it can carry each of these streams over a Cat-3 TPs wiring which has a bandwidth capable of carrying 25 MSps. Cat-3 wiring carries 20 MSps for the 10Base-T and its maximum capacity is 32 MSps. If binary signaling was used instead of ternary, each line should have carried 33.3 MSps which was beyond the capability of the Cat-3 wiring. The designers of the 100Base-T4 32 map these three streams of data to the four pairs of TP wires inside a bundle using the format shown in Figure 6.15(b). Two pairs are used for full-duplex communication mode carrying data only in one direction and the other two for half-duplex communication mode carrying data in both directions. This scheme provides an efficient method for utilizing the transmission capabilities of the four pairs of wires inside a bundle to support both full-duplex and half-duplex data transmission. But, in practice, 100Base-T4 is always used in half-duplex mode and it cannot support full-duplex operation. Like 100Base-TX, the 100Base-T4 also used telephone jack like click-on RJ-45 connectors.

6.3.2 Gigabit Ethernet

By the late 1990s, Ethernet had evolved into the most widely implemented physical and link layer protocol. Fast Ethernet increased speed from 10 to 100 Mbps replacing the existing backbone FDDI installations and 100Base-T4, in particular, replaced 10 Mbps legacy Ethernet as the main technology to access the end user. The next natural step was Gigabit Ethernet to increase the speed to 1000 Mbps and extend the Ethernet technology to wide-area networking. The initial standard for Gigabit Ethernet was the IEEE 802.3z, ratified in

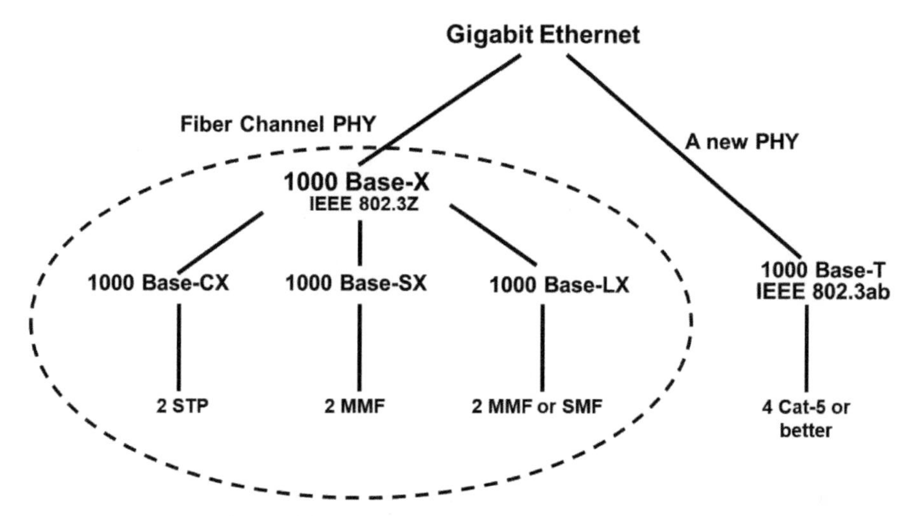

Figure 6.17 Overview of Gigabit Ethernet options.

1998, for two pairs of fiber and STP. The physical layer of this standard follows the ANSI's Fiber Channel standard for gigabit storage networking. This is in a manner similar to the adoption of ANSI's FDDI physical layer by 100Base-X standard and reflects the growth of Ethernet technology to assimilate other standards. In 1999, IEEE 802.3ab defined gigabit Ethernet for four pairs of unshielded Cat-5 or better media. This standard brought gigabit Ethernet to the desktop terminals because this standard is defined for the Cat-5 wiring already in place in many offices. As a result, some computer manufacturers started to use Gigabit Ethernet connections in their products, and the Gigabit Ethernet originally designed for the backbone became available to the end users. Figures 6.17 and 6.18 provide a summary of Gigabit Ethernet standards. Different media supports a variety of lengths designed to address different applications. The 25 m distance for 1000Base-CX supports short distances such as connecting mass storage to the main computer. The 100 m distance of the 1000Base-T supports desktop access and longer distances a support variety of backbone applications as Ethernet starts to penetrate metropolitan area applications. This pattern of supporting higher data rates for longer distances continues into higher rate Ethernets allowing this technology to penetrate wide area networking.

The *1000Base-X* operates similarly to 100Base-FX as shown in Figure 6.18. In 1000Base-X series, instead of 4B/5B, we use 8B/10B coding, and instead of NRZ-I line coding, NRZ line coding is used which was

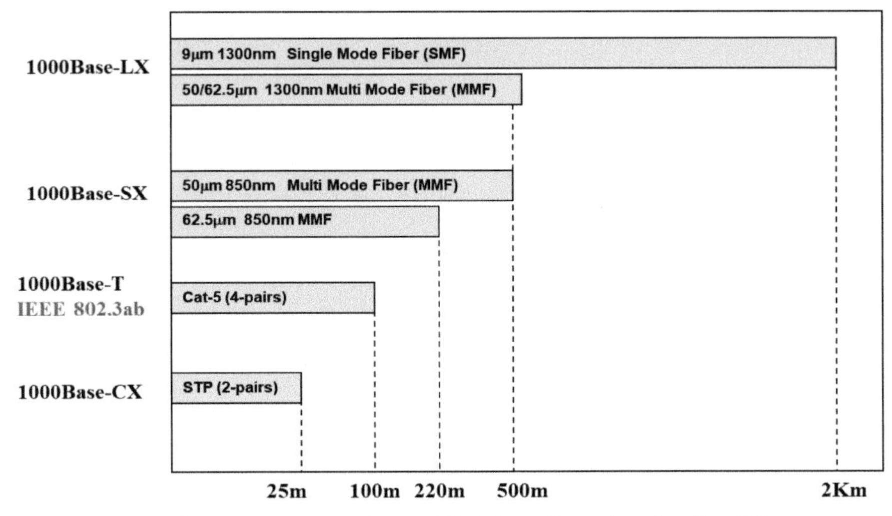

Figure 6.18 IEEE 802.3z and IEEE 802.3ab media options for Gbps Ethernet.

described in Chapter 2. In 8B/10B coding, every 8 bits of user data is mapped into a 10-bit symbol prior to transmission over the media using NRZ line coding. The efficiency of the code is 80% demanding a 1.25 Gbps line for a data rate of 1 Gbps. Since the 8-bit data takes $2^8 = 256$ possibilities while there are $2^{10} = 1024$ choices for the coded sequence, several coded words are not used for mapping the data. Similar to 4B/5B and 8B/6T coding techniques, the coded words, originally selected by the ANSI group working on fiber channel, have been selected so that the coded words are "DC balanced" so that valid coded symbols have five "1"s and five "0"s. This allows the receiver to align the symbols easily and ensure that incoming bits have frequent transitions to secure synchronization. In addition, some of the extra symbols are used to define transfer control signals such as starting and ending delimiters and idle signals. The details of the coding table are beyond the scope of this book. Like 100Base-X series, the 1000Base-X supports full-duplex and auto-negotiation operation.

1000BASE-T defined by IEEE 802.3ab is shown in Figure 6.19. Like the 100Base-T4, this standard also uses four pairs of wires to cover 100 m distance, but the minimum grade of the wires for the 1000Base-T is Cat-5, while the minimum grade of wiring for 100Base-T4 was Cat-3. The format of wiring for the 1000Base-T shown in Figure 6.18 is different from the 100Base-T4 shown in Figure 6.15. The 1000Base-T distributes the load over all four pairs evenly using a 4D five PAM encoding technique with additional

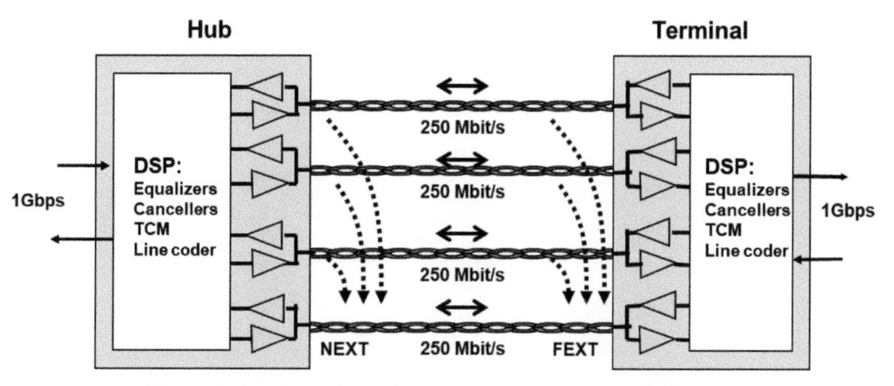

Figure 6.19 Overview of transmission system of 1000Base-T.

traditional digital communication coding techniques. Each Cat-5 or better line carries 125 MSps using five-level (PAM-5) in contrast with 100Base-TX which carried a binary NRZ-I signal at 125 MSps. Since 1000BASE-T uses all four pairs of wires in one bundle for simultaneous transmission in both directions, as shown in Figure 6.18, we will have near-end and far-end cross-talks, NEXT and FEXT, which need to be controlled using echo cancelation techniques. To implement a five-level pulse amplitude modulation (PAM-5), we need to improve the channel using an equalizer.

6.3.3 10 Gigabit Ethernet and Beyond

In the early 2000s, different Gbps Ethernets became a popular product in the backbone of the metropolitan area networks and several computer manufacturers included it as one of the connections in their desktop computer products. This success encouraged standardization of 10 Gbps Ethernet also referred to as 10GbE for desktop and further penetration into WANs. Several standardization groups within the IEEE 802.3 community defined a number of 10 Gbps Ethernet standards for different media options to support a variety of applications. These standards began commercial acceptance and integration in the core backbone for connecting routers and in LANs to support bandwidth hungry multimedia applications. As shown in Figure 6.20, there are four major standards, IEEE 802.3ae and IEEE 802.aq which are defining 10 Gbps for fiber media and, IEEE 802.3ak and IEEE 802.11an which are addressing 10 Gbps over copper cables. The 10 Gbps Ethernet is defined only for full-duplex operation which does not use CSMA/CD. Therefore, the MAC which was the first differentiating icon of the IEEE 802.3

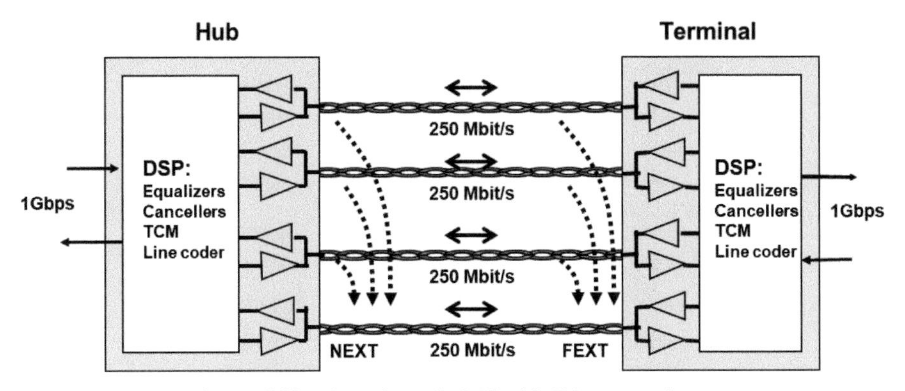

Figure 6.20 Overview of 10 Gigabit Ethernet options.

Ethernet standard is not used anymore and the specification of the physical layer to specify the transmission technique and the media options became the differentiation of different subgroups working on 10 Gbps Ethernet.

IEEE 802.3ae ratified in 2003 defines four physical layer specifications to support the transmission of Ethernet frames over MMF up to 300 m and SMF up to 40 Km. In addition, to supporting for 0 Gbps LANs, the IEEE 802.3ae also supports OC-192/ STM-64 (SONET/SDH) at 9.958 Gbps for wide-area networking and defines interoperation between them. The first of the IEEE 802.3ae specifications is the "short-range" ***10GBase-SR*** defined for 2-MMF line operating at 850 nm wavelength. As shown in Figure 6.21 for the existing 62.5 mm fiber, it can cover up to 82 m, and for new 50 μm fibers, it can cover up to 300 m. The second of the IEEE 802.3ae specifications is the "long-range" ***10GBase-LR*** supporting up to 25 Km on 1310 nm SMF. The third of the IEEE 802.3ae standards is the "extended range" ***10GBase-ER*** defined for 2-SMF lines operating at 1550 nm wavelength supporting distances up to 40Km. The fourth of this standard series is the long-range ***10GBase-LX4*** using four lasers with wavelength division multiplexing (WDM) to support up to 300 m over deployed MMF. Each line operates at 3.125 Gbps on a unique wavelength. This standard also supports 10 Km over 1310 nm SMF. The **10Base-LX4** uses the popular 8B/10B coding used in Gigabit Ethernet, fiber channel, and many other applications which is a table lookup coding like 4B/5B or 8B/6T we described before. With 8B/10B coding technique, the effective data transmission rate is $4 \times 3.125\,\text{Gbps} = 12.5\,\text{Gbps}$which supports an effective data transmission rate of $12.5\,\text{Gbps} \times (8\text{B}/10\text{B}) = 10\,\text{Gbps}$. The rest of the IEEE 802.3ae standard specifications uses another

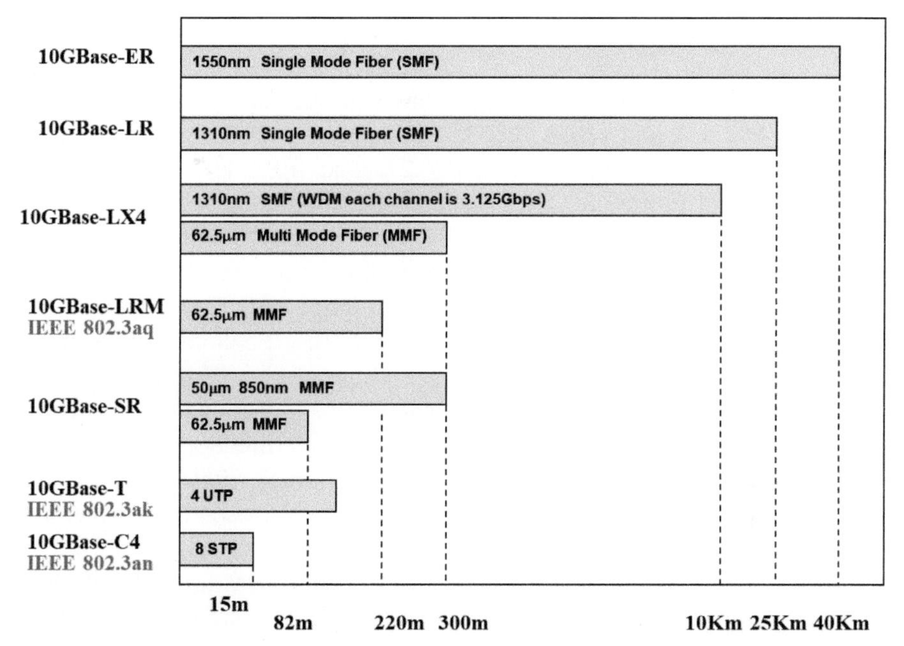

Figure 6.21 Overview of the IEEE 802.3ae and other 10 Gigabit Ethernet options.

line coding technique referred to as 64B/66B coding which was also used in the fiber channel for storage area networking.

WAN technologies defined by the **IEEE 802.3ae** specifies physical transport for 10G Ethernet across telecom OC-192/STM-64 SONET/SDH systems without having to directly map the Ethernet frames into SDH/SONET at 9.953 Gbps. The main three standards for different wavelength ranges are mapped to their corresponding SDH/SONET. In this series, **10GBASE-SW** corresponds to **10GBASE-SR**, **10GBASE-LW** corresponds to **10GBASE-LR**, and **10GBASE-EW** corresponds to **10GBASE-ER**, each supporting the same type of fiber and its associated distance. There is no wide-area physical layer standard corresponding to 10GBASE-LX4 because the original SONET/SDH standard requires a serial implementation.

Another related standard for 10 Gbps Ethernet over the fiber lines is **10GBASE-LRM** ratified by the **IEEE 802.3aq** in 2006. This standard supports up to 220 m over installed FDDI-grade 62.5 μm MMF operating at 1310 nm. The difference between this standard and the 10GBase-LX4 is that the latter uses wave division multiplexing, while 10GBase-LRM has a single wavelength.

10GbE over four pairs of copper has two standards options specified by ***IEEE 802.3ak*** and ***IEEE 802.3an*** compared with the other five fiber solutions shown in Figures 6.20 and 6.21. The IEEE 802.3ak's standard **10GBASE-CX4** ratified in 2004 is the cheapest solution with no complex signal processing for distances up to 15 m. There are four pairs in each direction to support the full-duplex operation. Like 10GBase-LX4, this standard uses four lanes, each carrying 3.125 Gbps 8B/10B coded data, and each carrying an effective data rate of $3.125 \times 8/10 = 2.5$ Gbps is NRZ coded for transmission. The 10GBase-LX4 uses 62.5 mm fiber SMF and MMF to cover long distances up to 10 Km for wide-area applications and 10GBase-C4 uses four pairs of popular Cat-5 wirings in each direction to cover short distances up to 15 m for stacking and adjacent intra-rack connection and other short distance applications. In the application of this technology, shielding interference between two sets of wires needs expensive cable assembly.

10GBase-T specified by **IEEE 802.3an** in 2006 **uses** four TPs of high-quality wirings Cat-6 and Cat-7 with a minimum symbol transmission rate of 800 MSps, 55 dB echo, 40 dB NEXT, and 25 dB FEXT in a wiring structure shown in Figure 6.22 which is similar to 1000GBase-T wiring shown in Figure 6.19. The transmission technique for this standard is PAM-16 encoded in a two-dimensional double square to create a 4D constellation with 128 points. With unscreened Cat-6, this standard covers up to 56 m, and with screened Cat-6 or Cat-7, it covers up to 100 m. It uses the popular RJ-45 connectors at the two ends of the wirings. Like 1000Base-T, implementation of 10GBase-T requires several complex signal processing to implement adaptive equalization, echo cancelation, and constellation mapping at a very high sampling rate and 10 bit per sample precession.

In 2009, the IEEE 802.3ba defined the **100 Gigabit Ethernet** or **100GbE** as an Ethernet standard with other data rates such as 40 Gbps as options. This standard will include 100GbE optical fiber Ethernet standards to support from 100 m up to 40 Km with full-duplex operation using Ethernet frame format. In the past decade, several 100 Gbps standards evolved following the same pattern of evolution as 0.1, 1, and 10 Gbps standards. The latest of these series of standards for 100 Gbps Ethernet transmission is the IEEE 802.3cm standards that were ratified in January 2020 and others are emerging; a good survey of these standards is available in Wikipedia under IEEE 802.3 [IEEE2020].

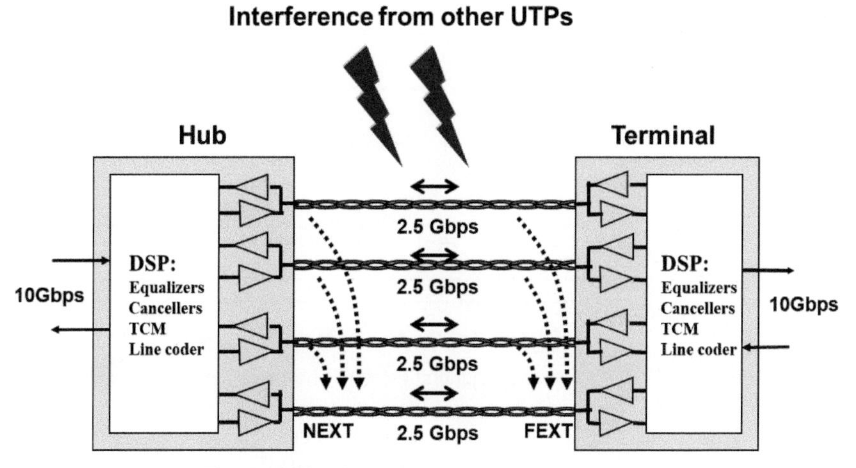

Figure 6.22 Overall operation of 1000Base-T.

It is worth mentioning that the fundamental science behind complex technologies such as adaptive equalization, echo cancelation, trellis code modulation (TCM), multi-dimensional modulations have been originally designed for voice-band data communications at 2400 Sps and it took over half a century to develop [Pah88]. With the advancement in computing speeds and micro- and nano-technologies, these fundamental sciences are becoming available to high-speed data communications over wired and wireless channels. Therefore, the physical layer design of the networks which has been the center of recent advancements in this field owes itself to two roots the applied information theory for the design of voice-band modems and the advancement in computation power and memory size of the recent years. Our emphasis in this book is on the fundamental changes in modern network technologies which are affected by applied transmission techniques and understanding and design of new media for data transmission.

6.4 Ethernet II Packets and Applications

With the emergence of Ethernet as the dominant LAN technology and its penetration into metropolitan and wide area networking, the IEEE 802.3 committee added more features to the Ethernet to support a wider variety of applications. The original Ethernet frame format (Figure 6.5) defines only one frame type and it does not support any priorities or services for packet transmission. The minimum 64 bytes of the length of the packet which was established to enable analog collision detection over legacy Ethernet

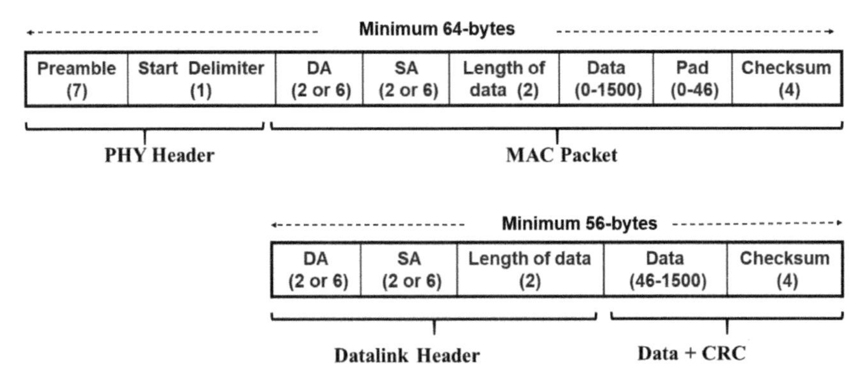

Figure 6.23 MAC packet data unit (MPDU) frame of the Ethernet II.

cable, now, is the universal packet format for computer communications, and collision detection mechanism on cables is not playing any role in that. In the past couple of decades, a few changes were made to the interpretation of Ethernet packet format to accommodate a broader range of applications. This new interpretation of the original Ethernet packet is referred to as the Ethernet II packet format. Figure 6.23 illustrates how Ethernet II packet interpretation relates to the legacy Ethernet packet. The first 8 bytes of the legacy Ethernet packets are PHY headers to establish the digital link. The rest of the packets are also divided into link layers, representing the source and destination address and the type of packet, and the data plus error detection CRC codes.

The fundamental idea behind the Ethernet II interpretation is to rename the 2 byte "length of data" field to "packet type." The length of data is actually 46–1500 byte and 1500 in hexadecimal is "05DC"; the maximum 2byte hex number is "FFFF" that is 65,535. This provides us an opportunity to use 65,535 -1500 -1 = 64,034 numbers for this field that is more than 1500 and use them for different types of packets to implement new applications on the Ethernet packets. If the length of the data field that follows the source address is less than O5DC hex, it is a normal data packet. Otherwise, it is something else, which can be a control or signaling packet. In the remainder of this section, we provide a few applications benefitting from this new interpretation of the legacy Ethernet packets.

6.4.1 Implementation of VLAN

The basic concept behind a VLAN is shown in Figure 6.24. A VLAN partition the broadcast domain allowing different groups to operate as virtual

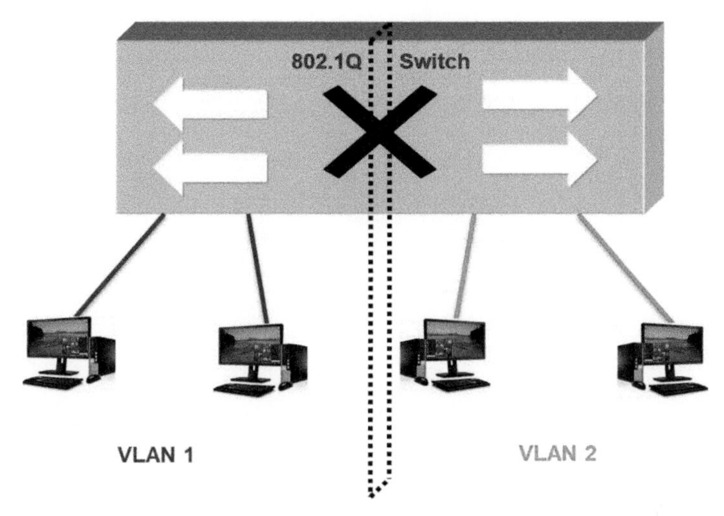

Figure 6.24 Basic concept behind a VLAN partitioning.

Ethernet LANs. In other words, VLAN breaks a single broadcast domain into multiple domains to allow having multiple logical Ethernet switches on a single physical switch. The major benefits of VLAN are easing network administration, allowing the formation of work groups, enhancing network security, and providing a means of limiting the broadcast domain of the Ethernet. More details on VLAN are discussed in Chapter 6 when we address the details of bridges. Here, we focus on the implementation of VLAN tags on the legacy Ethernet frames. In order to implement the VLAN, we need to change the frame format so that the bridges can recognize VLAN tags from common Ethernet frames. The change in the packet format should be done in the IEEE 802.3 community and the IEEE 802.1 working on issues related to bridging needs to approve the changes so that the bridge manufacturers following IEEE 802.1 standards can implement the changes in the bridges. For the implementation of the VLAN, the IEEE 802.3ac defines the protocol and changes in the frame format to support VLAN tagging on Ethernet networks and gets its frame format changes approved by the IEEE 802.1Q. The VLAN protocol permits the insertion of an identifier, or "tag," into the Ethernet frame format to identify the VLAN to which the frame belongs. It allows frames from stations to be assigned to logical groups.

As we discussed in Section 2.1, the legacy Ethernet standard defined the minimum frame size is 64 bytes and the maximum frame length as 1518 bytes, 18 bytes of which are for the overhead. If we consider the common

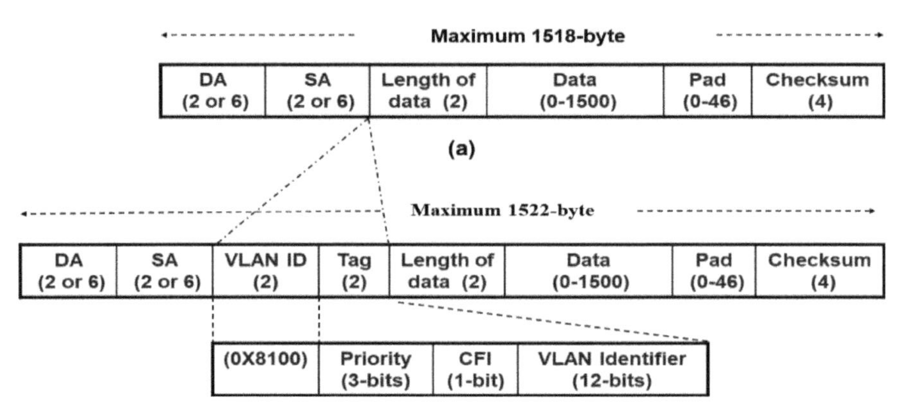

Figure 6.25 Frame format of (a) the IEEE 802.3 Ethernet and (b) the IEEE 802.1Q with VLAN tag.

practice of 6 bytes, addressing these 18 bytes is used for MAC addressing, length of data, and the frame check sequence. The minimum addressing length of 2 bytes includes the preamble and start of frame delimiter fields. In 1998, the IEEE 802.3ac standard released the VLAN tag specification and extended the maximum allowable frame size to 1522 bytes.

Figure 6.25 shows the frame format of the details of the IEEE 802.3 and IEEE 802.1Q frame formats. The 4-byte VLAN tag is inserted into the 802.3 frames between the source MAC address field and the length of the data field. The first 2 bytes of the VLAN tag is the fixed number 0X8100 which is greater than 1500 bytes, the maximum length of the data. The 0x8100 value is actually reserved by the IEEE 802.1Q to indicate the presence of the VLAN tag, and it indicates that the traditional length of the data field can be found at an offset of 4 bytes further into the frame. The last 2 bytes of the VLAN tag contain 3 bits as the user priority field to assign priority to Ethernet packets, 1 bit as canonical format indicator (CFI) to indicate the presence of routing information field, and 12 bits are the VLAN identifier uniquely identifying the VLAN to which the Ethernet frame belongs to. The additional 4 bytes increase the maximum length of the packet from 1518 to 1522 bytes. A LAN switch or bridge complying with IEEE 802.1Q standard recognizes these tags from the normal frames when it reads the 0X8100 as the length of the packet in which case it looks for the appropriate VLAN address to direct the packet.

6.4.2 Full-Duplex Operation and Pause Frame

In 1997 IEEE 802.3x released the standard for full-duplex Ethernet packet transmission which bypasses the CSMA/CD protocol. Figure 6.26 illustrates

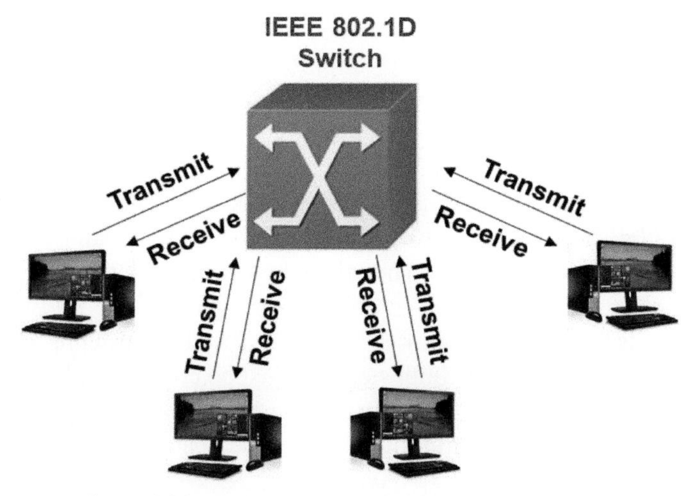

Figure 6.26 Basic concept behind full-duplex operation.

the basic concept behind the full-duplex operation. Each terminal is connected to the hub by two separate lines used for transmission and reception. Therefore, each terminal has a separate collision domain and there is no need for CSMA/CD. One pair of TP wire or one fiber strand is used exclusively for transmission and another pair or strand for the reception. This wiring with a full-duplex connection resembles telephone network wiring and switches, except that, here, we have packet switching rather than circuit switching. The CSMA/CD was a half-duplex protocol in which each terminal is allowed to either transmit or receive data but never at the same time. The half-duplex nature is inherited in the CSMA/CD protocol for which simultaneous transmission and reception of data is an indicator of collision between the two packets. Full-duplex mode allows two stations to simultaneously exchange data over a point-to-point link with two independent physicals transmitting and receiving wiring paths. In CSMA/CD, the embedded half-duplex operation for collision detection purposes would not allow full utilization of the transmission medium even when we have two separate pairs of wires in a point-to-point connection. The half-duplex in the presence of two transmission media connecting the terminals reduces the utilization of the transmission medium capacity to 50%. In full-duplex transmission over two pairs of transmission medium resources, each station can simultaneously transmit and receive data, and aggregated throughput of the link is effectively double of the half-duplex operation. This means that a full-duplex 100 Mbps station provides for 200 Mbps of bandwidth if the physical medium can support full-duplex with simultaneous transmission and reception without interference from other terminals. A star topology and two separate links are needed for full-duplex operation. Among the IEEE 802.3 media specifications, we discuss in this chapter, the ones that follow these specifications are: 10-Base-T, 100Base-X, 100Base-T2, 1000Base-X, and all 10GbE options. The bus topology of the 10Base5 and 10Base2 and uneven wiring of 100Base-T4 do not allow full-duplex operation. Full-duplex operation is restricted to point-to-point links connecting only two stations when there is no contention for the shared medium and, consequently, no need for collision detection. Both stations must be configured for full-duplex operation and frames may be transmitted at will only be limited by the required separation of the minimum interframe spacing of 96 bits required for all Ethernet packets.

In addition to doubling the maximum capacity, the efficiency of the link is improved by eliminating the potential for collisions. Elimination of the collision detection requirement lifts the segment length restriction as well. Segment lengths are no longer limited by the timing requirements of half-duplex Ethernet to ensure collisions are propagated to all stations within the required 512-bit times. This change allows support of longer transmission lengths. For example, 100Base-FX is limited to 412 m segment length in half-duplex CSMA/CD operation mode, but with the full-duplex operation mode, it can support segment lengths of 2 Km. The adaptor card and the hub must have full-duplex connectors. Therefore, full-duplex is used only with switches and traditional repeaters cannot support full-duplex operation.

The full-duplex mode defined by the IEEE802.3x includes an optional flow control operation implemented by a new frame called the "PAUSE" frame. PAUSE frame in a full-duplex link between two end stations allows one of the stations to temporarily stop transmission of traffic from the other end station. Figure 6.27 shows the application of the PAUSE frame in the full-duplex operation. Packets arriving from station A are at a rate that causes congestion in station B so that there is no input buffer space to receive additional frames. Station B transmits a PAUSE frame to station A to stop transmission for a specific period defined in the packet. When the PAUSE frame arrives in station A, this station suspends frame transmission for the specified time unless another PAUSE frame arrives and changes the suspension period for station A. After completion of the suspension period, station A resumes its normal packet transmission. This operation

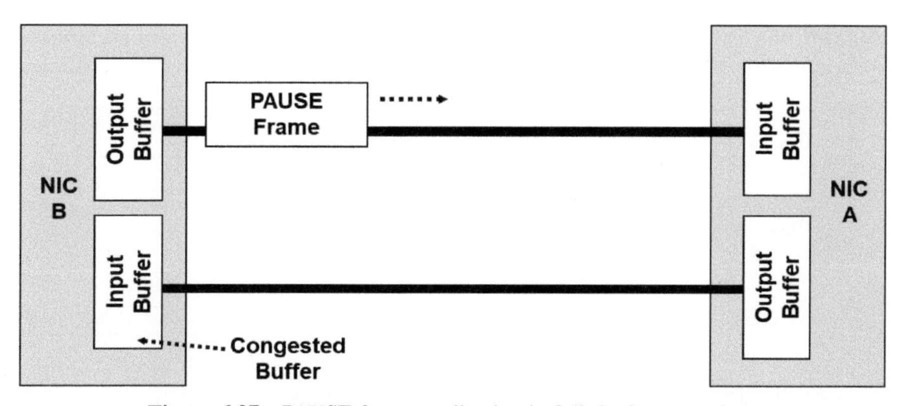

Figure 6.27 PAUSE frame application in full-duplex operation.

allows station B to recover from congestion during the specified PAUSE time requested from station A. If the congestion problem is completed sooner than the specified period, station B has a choice to send another PAUSE frame with zero waiting time to activate terminal A. During the PAUSE state, the station can send only PAUSE frames allowing two stations to PAUSE each other at the same time. The PAUSE frame can be sent by a device not supporting PAUSE operation. Stations use the auto-negotiation protocol which was defined as a part of the fast Ethernet standard to learn the PAUSE capability of the station at the other end of the link.

PAUSE frame is a control signaling frame and to implement it, one needs to find a way to differentiate among different Ethernet frames which were originally defined as a universal single formatted frame. Figure 6.28 shows the details for the implementation of the PAUSE frame. In a manner like VLAN frames, the IEEE 802.3x uses the length of the address field to differentiate between normal and control frames. This time, the standardization committee assigns 88-08 (34,814 > 1500) to indicate that the packet is a MAC control packet. The next two bytes (00-01) defines the opcode for the PAUSE frame. The MAC control opcode field of 00-01 indicates that the type of MAC control frame being used is a PAUSE frame. The PAUSE frame is the only type of MAC control frame currently defined; other combinations of opcodes are available for future control signals. The PAUSE opcode is followed by another 2 bytes (00-00 to FF-FF) to specify the duration of the PAUSE. These 2-byte MAC control parameters fields specify the duration of the PAUSE event in units of 512-bit times. If an

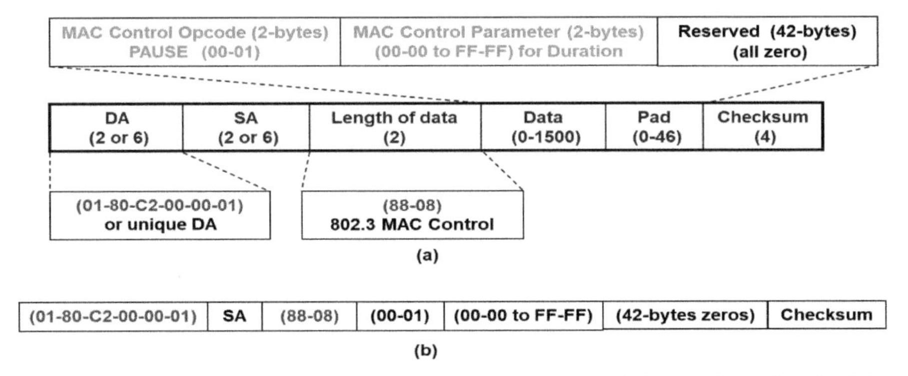

Figure 6.28 Details of the PAUSE frame. (a) Modifications in fields. (b) Overall look of the packet.

additional PAUSE frame arrives before the current PAUSE time has expired, its parameter replaces the current PAUSE time; so a PAUSE frame with parameter zero allows traffic to resume immediately.

The next 42 bytes are padded with zeros to keep the minimum length of 46 bytes for the data for Ethernet frames.

The destination address of the PAUSE frame may be set to either the unique destination address of the station to be paused or to the globally assigned multicast address (01-80-C2-00-00-01). This multicast address has been reserved by the IEEE 802.3 standard for use in MAC control PAUSE frames. It is also reserved in the IEEE 802.1D bridging standard as an address that will not be forward by bridges. This ensures the frame will not propagate beyond the local link segment.

6.4.3 Implementation of Link Aggregation

In 2000, IEEE 802.3ad started working on the backbone link aggregation which allows several NICs in a computer to connect to different ports of a switch. Figure 6.29 shows the basic concept behind link aggregation which applies only to full-duplex operation. Link aggregation between two Ethernet stations increases the link bandwidth by combining multiple physical links into a single logical link. For example, it allows several less expensive 1 Gbps Ethernet solutions for long distances to support a higher data rates of up to 8 Gbps to avoid using more expensive 10 Gbps Ethernets. At the first glance, it seems that having multiple links between two stations should be possible anyway. The difficulty for having multiple links was that, as we

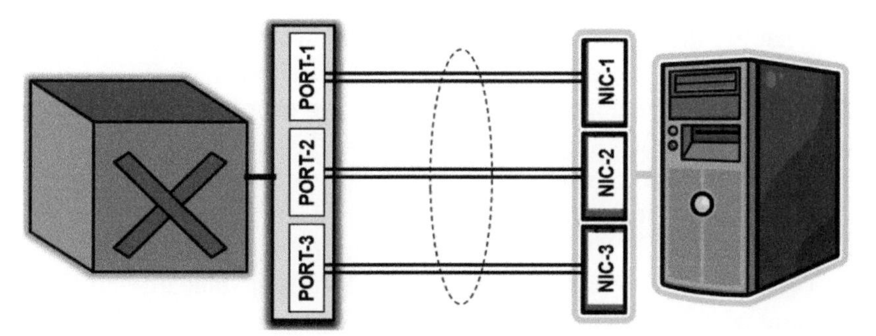

Figure 6.29 Basic concept behind link aggregation.

discussed in Section 2.5.1, LAN switches use spanning-tree algorithm to avoid loops that eliminates parallel paths between two stations. Therefore, without link aggregation standardization, the only way to have multiple links between two stations in a single network was to use each link in a different VLAN. Link aggregation standard allows multiple links between servers, switches, and end-user stations to resolve this short coming of the Ethernet networks. Each of the Ethernet ports involved in the link aggregation at the switch and the terminal has its own unique MAC address. Link aggregation is implemented by a new layer of function between MAC and the higher layer protocols. As frames pass, these additional layer MAC addresses are changed so that the aggregated ports appear as a single link with a unique MAC address.

This method of operation masks link aggregation function to all higher layer protocols and functions, such as spanning tree algorithm or VLAN. Link aggregation protocol ensures that the arrival of the frame sequence numbers distributes among the multiple links.

Assignments

Questions

1. What are the main objectives of the IEEE 802.1 standardization activity and how does it relate to standards for the LANs?
2. What is the length of the contention slot (window) in the 802.3? Why is this length selected?
3. What is binary exponential backoff algorithm, which standard uses that, and what is the purpose of using it?

4. What are the options for the length of the MAC address field in IEEE 802 standard? Which length is commonly used in practice?
5. What are the advantages of 4B/5B encoding over Manchester coding?
6. What is the difference between 100BASE-TX and 100BASE-FX?
7. What is PAM 5X5, which standard uses it, and how does it compare with differential Manchester coding used in the legacy Ethernet?
8. What category of TP wiring is used in the 100BASE-T2 and 100BASE-T4? How many pairs of wires are used in each standard and how are these wires connected? Give a diagram for wiring connections.
9. What are NEXT and FEXT and how do they relate to the selection of the maximum length of wire in LANs?
10. What is the pause frame in Ethernet and how does it get implemented without changing the existing Ethernet frames?
11. What is the purpose of the PAUSE frame and when is it used in a full-duplex operation?
12. What is a VLAN and how does it differ from traditional Ethernet in terms of frame format and services?
13. How many levels of priority and how many different VLANs are supported by 802.1Q tag? How many bytes does this tag add to the 802.3 traditional MAC frames?
14. What is link aggregation and how is it useful to local network designers?
15. How does 1000Base-T provide 1 Gbps over four pairs of Cat-5 TPs and what is the relation between PAM 5 modulation and 1000Base-T transmission technique?
16. What is the maximum coverage of a Gigabit Ethernet and which medium is supporting this maximum coverage?
17. Compare the complexity of the digital signal processing for designing the 1000BASE-T and that of 1000BASE-F transceivers.
18. Draw the wiring format of the 100BASE-T4 and 1000BASE-T and explain their differences.
19. Explain the differences between 10GBASE-XR and 10GBASE-XW standard specifications.
20. What is the difference between the 10GBASE-LR and 10GBASE-LRM standards for Ethernet?
21. What are the similarities and differences between 10GBASE-C4 and 10GBASE-LX4 technologies?

Instructor's solution available on River Publishers' website:
https://www.riverpublishers.com/book_details.php?book_id=919

Problems

Problem 1

a) If we use differential Manchester coding on a 10 Mbps legacy Ethernet for the following 8-bit data stream: | 0 | 0 | 1 | 0 | 1 | 0 | 1 | 1 |:

   a-1) Give the transmitted waveform.
   a-2) Give the coded symbol transmission rate.

b) If, instead of Manchester coding, a 4B/5B coding with the same coded symbol transmission rate was used to code the data stream:

   b-1) Give the transmitted waveform.
   b-2) Give the user's information transmission rate.

c) Repeat (b) if we use 8B6T rather than 4B/5B coding.
d) What is the Shannon–Hartley bound on data rate if the acceptable signal-to-noise ratio (SNR) was 10 dB? Explain this limit compared with the data rates achieved by Manchester coding, 4B/5B coding, and 8B6T coding.
e) Repeat (d) for SNR of 6dB.

Problem 2

We want to install a LAN in a five-storey office building with identical floor plans. Each floor of the building is an 80 m $\times$ 80 m square with a height of 4 m. There are 60 terminals on each floor of the building and the external wiring comes to the first floor.

a) What is the total cost of wiring, equipment, and installation of the entire network if an IEEE 802.3 star network with one 240 port switch on the first floor connects all terminals to each other and the external connection? Assume a charge of $150 per run of wiring between two locations, a $3,000 cost for the switch, and a $20 cost for the NIC per terminal.
b) To avoid wiring costs, assume that we use four access points per floor to support the terminals using a wireless solution. What is the total cost of the wireless LAN solution if each access point costs $400 and each NIC is sold for $50? Note that we still need a switch and wiring from the switch to each access point.
c) Compare the advantages and disadvantages of the two solutions.

Problem 3

Assume that we have an environment with five users connecting to a 100BASE-T LAN, all running the same application that produces 1500-byte length paths at a rate of five packets per minute.

a) Determine data transmission time T and normalized propagation parameter "a" assuming that the propagation speed is 200,000 Km/s and the terminals are located at the maximum length allowed by the standard for the network span.
b) Determine the maximum utilization (throughput) S of the network assuming that the retransmission probability is $p = 1/N$.

Problem 4

If in the 100BASE-T2 standard, rather than PAM5X5, a 16-QAM constellation was used:

a) What would be the difference in the number of signal levels and the symbol transmission rate in each of the two pairs of TP wires?
b) What would be the difference in received SNR requirement in dB?
c) If we assume a signal loss of 10 dB per 100 m for the TP wiring, what would be the maximum length of the cable if we were using 16-QAM instead of PAM5X5?

Problem 5

In a network N, user terminals are contending to send their packets in an available time slot. Each terminal transmits its packet with a probability of p.

a) Determine, A, the probability of successful transmission of a packet in the slot.
b) If a terminal that does not send in the first slot continue contending to send its packets with the same probability p, in the next slots, determine $P(k)$, the probability of successful transmission of that packet in the kth slot in terms of k, the number of contention slots.
c) Calculate the average contention slot, $E\{k\}$ as a function of p and N.
d) If we have and interframe spacing of $T_{IFS} = 12$ bytes and the length of the slots is $T_s = 64$ bytes, give an equation for calculation of the channel utilization in terms of time slot, interframe spacing, and length of the packet in Byte, T_p.

Problem 6

a) Draw the 5-PAM and 5X5-PAM constellations. Show the probability of transmission and number of bits per symbol for each symbol in each of the two constellations.
b) For each constellation, calculate the average energy per constellation assuming the minimum distance between the points is $d = 1$.
c) What is the difference in average energy in the two constellations in dB (10log of the ration of the two energies)?
d) What is the average number of bits per symbol for each constellation?

Problem 7

The PAM5X5 is commonly used in Ethernet LAN standards. Another alternative for this constellation, usually used in wireless LANs, is 16-QAM.

a) Draw the signal constellations for both. Show the probability of transmission and the number of bits associated to each symbol in each of the two constellations.
b) What is the average number of bits per symbol for each constellation?
c) If we use these constellations over two pairs of Cat-7 twisted pair (TP) wires with a bandwidth that accommodates transmission of 700Msymbol/sec, what would be the effective bit rate for each constellation?

7

IEEE 802.11 – The Wi-Fi

7.1 Introduction

In the previous chapter, we considered IEEE 802.3 or Ethernet, the primary wired technology that is used as the access network to the Internet. Ethernet also started as a local area networking technology that connected hosts that belonged to the same organization. The IEEE 802.11, commercially known as Wi-Fi, is the primary wireless local area networking (WLAN) technology that is built on Ethernet and we may call it wireless Ethernet. The goal of this chapter is to provide an overview of these technologies and how it evolved to become the most popular wireless technology carrying approximately 70% of the IP traffic [Pah20B]. Standardization of the legacy Ethernet, the IEEE 802.3, took only a few months and the length of the document is around 30 pages. Standardization of the legacy IEEE 802.11 took ten years and it created approximately 500 pages of documents. This was because of the restriction and complexity of wireless communications caused by the government regulation of the available frequencies, the unreliability of the radio-frequency (RF) propagation medium due to multipath radio propagation, and authentication and security issues demanded networking in a broadcast medium available for eavesdropping. These difficulties make the design of wireless access much more complicated than wired medium. Although Wi-Fi is a wireless Ethernet, its impact in our daily lives surpasses those of Ethernet because in addition to a method for access to the Internet, the Wi-Fi supports portable mobile access to the Internet, and that mobility of access enabled numerous applications essential for our daily routines. Another important aspect of Wi-Fi is that all the important physical layer technologies evolved for wireless communications were first implemented in Wi-Fi.

In this chapter, we begin with a short description of the importance of Wi-Fi in our daily lives is followed by an overview of the evolution of Wi-Fi technology and standards. Then, we describe how legacy Wi-Fi emerged from the legacy Ethernet, and, at last, we explain how Wi-Fi technology evolved from a 2 Mbps to a Gbps technology. One of the major reasons for studying Wi-Fi is that all the popular transmission techniques for wireless communication such as optical communications, spread spectrum technology, orthogonal frequency division multiplexing (OFDM), multiple-input multiple-output (MIMO) antenna system, and mmWave pulse transmission systems, were first used in Wi-Fi and then spread to cellular and wireless sensor network. From a technical point of view, Wi-Fi is the wireless Ethernet and its packet format and MAC follows that design, while it has a more complex PHY to accommodate unreliable wireless communications over multipath fading channels. From a practical point of view, Ethernet and Wi-Fi protocols are dominant in connecting the IP traffic and growing to become the wired and wireless access methods to the core IP network. Ethernet is growing to penetrate the core networking standards for long-haul transmission and RF propagation from Wi-Fi devices are hosting new cyberspace intelligence applications to relate the device's location to the environment surrounding them and possibly bringing another revolution in human–computer interfacing and location intelligence.

7.1.1 Importance of Wi-Fi in Our Daily Life

The importance of Wi-Fi is well described in this quote, "Wi-Fi is the fastest and most cost-effective way of wireless Internet connectivity. Nowadays, almost all mobile phones and an increasing number of home entertainment systems are Wi-Fi-enabled. Being the key enabler of the "Internet of Everything", Wi-Fi brings including people, processes, data, and devices, together and turns data into valuable information that makes life better and business thrive. With all mobile devices, wearable gadgets, home entertainment systems, and home automation systems connected and linked to the Internet, devices can now interact with one another and data be shared among the devices" [Fon16]. Figure 7.1, adapted from [Pah20], shows an artistic illustration of the relation of Wi-Fi to human basic needs using Maslow's hierarchy of human needs with an additional lowest layer called Wi-Fi [Mcl07]. Usually, Maslow's hierarchy is shown as a pyramid, but to illustrate the crucial importance of Wi-Fi, the hierarchy is shown using the inverted version of the common symbol for Wi-Fi signal strength with Wi-Fi as the most basic of human needs. The page is entitled "is fast Wi-Fi the most basic human need?" [Ber17]. People enter this argument by asking questions

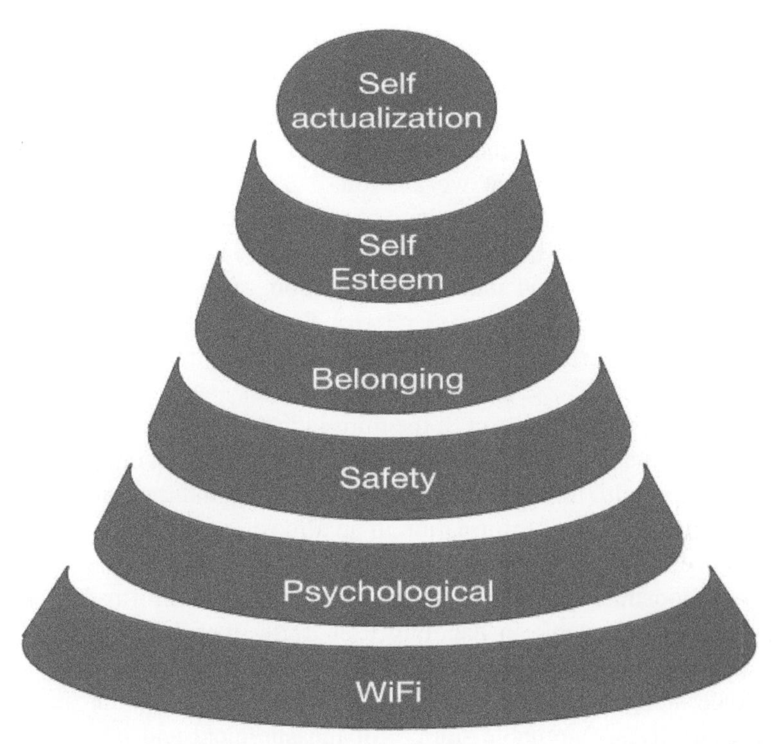

Figure 7.1 Maslow's hierarchy of human needs with an additional layer referring to Wi-Fi as enabler of these needs, adapted from [Pah20].

like "how long can you live without air?," "how about water?," and "how about food?," expecting a sequence of answers that progresses as "minutes," "days," and "weeks," respectively. Then, they continue with "how long can you live without Wi-Fi?," and they related it to the answer to the previous questions. The answer reveals itself better in examples like, when you arrive at a hotel, what do you ask? Most probably, "where is my room?," "where is the restaurant?," and "how can I connect to Wi-Fi?" These layman examples illustrate the importance of Wi-Fi in our daily lives.

In more scientific terminology, to understand the importance of Wi-Fi technology, we begin by describing important innovations in technology and how we measure their importance. In the same way that the importance of literature is measured by its citations, the importance of technical innovation is measured by the number of people that are using that innovation in their daily lives. As such, fire and wheel are considered the most important technical innovation in the history of humanity. We use **fire** to generate heat and light, and heat as the energy source has enabled humans to process food to generate a variety of nutrients and reduce disease by killing organisms inside the ingredients of food. Light has enabled us to work in the dark and that substantially extends our period of productive activity. Without fire, we would not be what we are today. The wheel is also considered one of the most important inventions of all time because of its fundamental impact on transportation, agriculture, and all other industries. All innovations after the first and second industrial revolution, such as the steam engine, the internal combustion engine, electricity, the telegraph, and the telephone, radio, television, airplane, and rockets, had profound impacts on the way we live and have affected many other industries (such as entertainment). However, the Internet, the fruit of the third industrial revolution, enabling the emergence of the "information age," has had a wider impact on our daily lives. In comparison with those of fire and the wheel, the Internet provides access to unlimited amounts of information in an almost instant manner, anywhere, and that is enabled by Wi-Fi technology.

To demonstrate that Wi-Fi is the enabler of the information age, we need to measure its role in handling user traffic worldwide. IP traffic is a good measure of information exchange on the Internet. IP traffic includes text, voice, images, and videos that will comprise the communication needs in our daily lives. A reliable source for measurement and prediction of IP traffic is the Cisco Visual Networking Index: Global Mobile Data Traffic Forecast [For19]. Figure 7.2, adopted from this source, shows the breakdown of this data from mobile, Wi-Fi, and fixed access in different years. We use their prediction of traffic in 2022 as a measure to demonstrate the role of Wi-Fi in handling traffic. The traffic is divided into wireless (Wi-Fi and cellular mobile) and wired (Ethernet) with wireless carrying 70.6% and wired carrying 29.4%. Because of its flexibility for connection, wireless traffic carries more than twice the wired traffic. Fixed devices generate 58% of the traffic and mobile devices generate 42%. Wi-Fi carries 22.9% of the traffic from mobile devices with a cellular connection and 28.1% of traffic from Wi-Fi only devices for a total 51% of the entire traffic. This means that by the year 2022, Wi-Fi will carry the majority of IP global traffic soaring to reach the unbelievable high value of zettabytes (10^{21} bytes). The reason for the success of Wi-Fi over Ethernet, carrying 29.2% of the traffic, is Wi-Fi's connection flexibility, and the reason for success over cellular, carrying 19.6% of traffic, is Wi-Fi's higher speed, and cheaper connection cost. We use these numbers as a proxy metric to now explain why we need Wi-Fi. This discussion clarifies

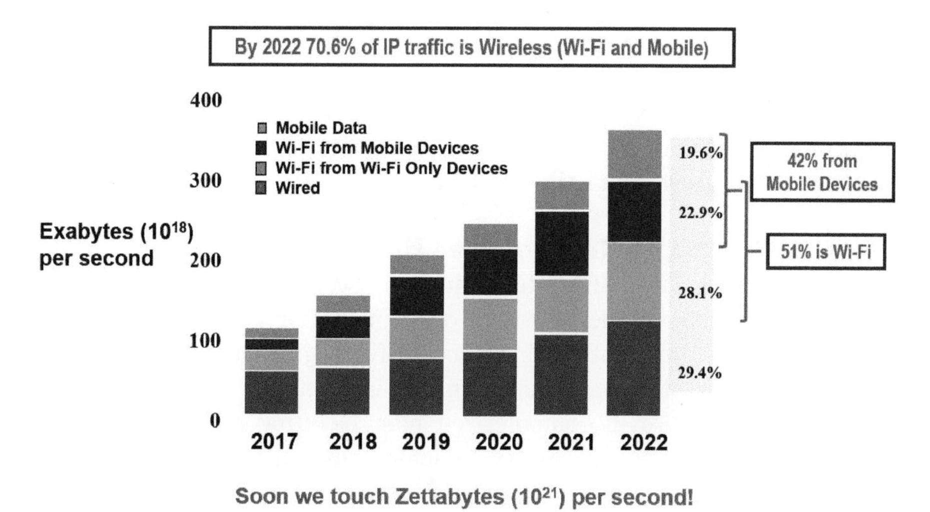

Figure 7.2 Global monthly IP traffic for different Internet access methods. Exabyte is 10 to the power of 18 (adopted from [For19]).

our artistic expression at the beginning of this section, about the impact of Wi-Fi on our daily needs, in a broader scientific context.

7.2 Evolution of Wi-Fi

In this section, we provide a holistic overview of the evolution of Wi-Fi technology and its applications as the author experienced it in the past few decades [Pah20]. It covers the history of Wi-Fi from three angles. First, how the physical (PHY) and medium access control (MAC) of Wi-Fi technology evolved and what were the novel technologies that were introduced in this endeavor. Second, how Wi-Fi positioning emerged as the most popular positioning technology in indoor and urban areas and how it has impacted our daily lives. Third, how other cyberspace applications, such as motion and gesture detection as well as authentication and security, are emerging to revolutionize human–computer interfacing using the RF signal radiated from Wi-Fi devices. Finally, we present an overview of protocols for Wi-Fi as an introduction for the following sections for technical details of this technology.

7.2.1 Evolution of Wi-Fi Technology and Standards

The PHY and MAC technologies of Wi-Fi emerged over the past few decades from the basic idea to become the most important communication technology affecting human life. We present this evolution by dividing them into three eras, before 1985, when the technology was originated, between 1985 and 1997, when the legacy IEEE 802.11 standard finalized, and after 1997 until present, when OFDM and MIMO technologies enabled Wi-Fi to enhance the supported data rates from 2Mbps to Gbps.

The Origin of WLAN Technologies Before 1985:
Around the year 1980, IBM Rueschlikon Laboratory, Zurich, Switzerland began to research and development on using infrared (IR) technology to design WLANs for manufacturing floors [Gfe79]. At that time, wired LANs were popular in office areas and large manufacturers such as GM were considering their use in computerized manufacturing floors. To wire the inside of offices, it was necessary to snake wires in walls and low-height suspended ceilings. In manufacturing floors, there are limited numbers of partitioning walls, and, moreover, ceilings are high. Consequently, WLAN technology offered itself as a practical alternative. Around the same time frame, HP Palo Alto Laboratory, California, reported a prototype WLAN using direct sequence spread spectrum (DSSS) with surface acoustic wave (SAW) devices (for implementation of a matched filter at the receiver) [Fre80]. HP Laboratories at that time had open space offices without partitioning with walls, which again created challenges for wiring through the walls. Dropping wires from the ceiling to desktops was not aesthetically pleasing. It was at this time that optical wireless and spread spectrum, combined with CDMA emerged, that could support huge amounts of capacity for indoor wireless LANs (WLANs) to connect desktops and printers to the computers [Pah84][Pah85].

Prior to all this, Norm Abramson at the university of Hawaii had designed the first experimental wireless data network, the ALOHA system [Abr70]. The difference between the two approaches was that ALOHA was academic experimentation of wireless packet data networks with an antenna deployed outdoors with relatively low data rate modems at a speed of around 9600 bps. However, the concept had inspired low-speed wireless data networking technologies such as Motorola's ARDIS, and Ericsson's Mobitex (which we refer to as mobile data services now generally subsumed by cellular data services in 3G and 4G). For WLAN technologies, the antenna is installed indoors, and the data rate needed to be at least 1 Mbps (at that time) to be considered by IEEE 802 standards organization community like a LAN [Pah94]. The MAC of both wireless data services was contention based, originally experimented in ALOHA, and later evolving into carrier-sensing-based contention access adopted by IEEE 802.3 standard, for the wired Ethernet.

The main obstacle for commercial implementation of the early WLANs was interference and availability of a low-cost wideband spectrum in which the WLAN could operate. Indoor optical WLANs did not need to consider regulation by the Federal Communications Commission (FCC) and it could potentially provide extremely wide bandwidths. However, optical communications were restricted to single rooms. Spread spectrum was an anti-interference technology which, at that time, could potentially manage the interference problem allowing multiple users to share a wideband [Pah84][Pah85][Pah88A]. In the summer of 1985, Mike Marcus, the chief engineer at FCC at that time, released the unlicensed industrial, scientific, and medical (ISM) bands with restrictions of using spread spectrum technology for interference management [Mar85]. For WLANs to become a commercial product, there was a need for large bandwidths (at that time) and modem technologies that could overcome the challenges of indoor RF multipath propagation to achieve data rates beyond 1 Mbps required to be considered by the IEEE 802 committee as a LAN. The ISM bands and spread spectrum technology could address both issues.

WLAN Technologies and Standards Between 1985 and 1997: The summer of 1985 was a turning point for the entrepreneurship for implementation of the WLAN industry, suitability of spread spectrum and IR for implementing wireless office information networks had captured the cover page of the IEEE Communication Magazine in June [Pah85] and FCC had released the ISM bands for commercial implementation of low power spread spectrum technology in May [Mar85]. Suddenly, several startup companies and a few groups in the large companies, almost exclusively in North America, emerged to begin developing WLAN using spread spectrum and IR technologies. IR devices do not need FCC regulations because they operate above the 300 GHz, governed by this agency. Among the exceptions for the location of these companies was a small group in NCR, Netherlands, which designed the first DSSS technology to achieve 2 Mbps [WPI'91]. Other companies, such as Proxim, Mountain View, CA, USA, resorted to frequency hopping spread spectrum (FHSS), and a third group led by Photonics, resorted to IR solution for WLAN, both groups achieving 2 Mbps. These three groups originally started in the late 1980s, laid the foundation of the first legacy IEEE 802.11 standard, finalized in 1997. The final standard for DSSS and FHSS operated at 2.4 GHz ISM bands. Other exceptions in technologies were an effort by Motorola, IL, USA, and WINDATA, Marlborough, MA, USA. Motorola introduced a revolutionary WLAN technology operating at 18 GHz licensed bands achieving 10 Mbps using sectored antennas [WPI'91], and WINDATA, achieved 6 Mbps in a dual band mode with 2.4 GHz for uplink and 5.2 GHz at the downlink.

The first academic research in the physical layer of WLAN began at the Worcester Polytechnic Institute, Worcester, MA, USA, in the fall of 1985 [Pah00A]. The early academic research literature in this area began by empirical modeling of the multipath indoor radio propagation [Pah89][How90A,B][Gan91][How92], examining decision feedback equalize (DFE) [Sex89], and M-ary orthogonal coding [Pah90] to achieve data rates beyond 2 Mbps studied by the IEEE 802.11 to achieve rates on the orders of 20 Mbps, and integration of voice and data for WLAN [Zha90]. A breakthrough, first patented in this era, was an application of OFDM to WLAN, filed by CSIRO research laboratory, Sydney, Australia [Osu96]. A form of M-ary orthogonal coding is adopted by IEEE802.11b, DFE was adopted by HIPERLAN-I, and a form of OFDM was implemented in HIPERLAN-2 and IEEE 802.11a. The origin of these transmission technologies was first discovered for commercial voice band communications [Pah88A].

The IEEE group for WLAN standardization was first formed as IEEE 802.4L. The IEEE 802.4 was devoted to Token Bus LANs for the manufacturing environment and was on the verge of disbanding. The rational is that new IEEE standards usually begin in a relevant standard and after going through the establishment procedure form their own. IEEE 802.11 was the same group formed later in July 1990 [Hay91]. In the early days of these standards, the important issue was to find the correct direction for the future. In 1991, the standard group participated in the first IEEE sponsored conference to decide on the future of these technologies among several acceptable technologies [WPI'91]. The early IEEE standards for wired LANs were differentiating from each other by their MAC method. IEEE 802.11 was the first with three MAC mechanisms and three PHY methods. The legacy IEEE 802.11 standard completed in 1997 with three PHY recommendations, DSSS and FHSS operating at 2.4 GHz ISM bands and diffused infrared (DFIR) wireless optical options. All three options employed three options for MAC, carrier sense multiple access (CSMA), *Request-to-Send* (RTS) Clear-to-Send (CLS), and point coordination function (PCF) and operated at a PHY with data rate options of 1 and 2 Mbps.

The high-performance LAN (HIPERLAN) was another standardization activity, sponsored by the european telecommunication standards institute

(ETSI), which began its work in 1992. The HYPERLAN-1 was the first attempt to achieve data rates above 10 Mbps using decision DFE technology and at 5 GHz ISM bands [Wil95][Sex89][How92]. This standard was completed in 1997, but it failed in developing a market. Another more popular but abundant standardization activity for wireless indoor networking in this era was the Wireless ATM [Aya96][Liu98]. A comparison of this technology with Wi-Fi is available in [Pah97].

In summary, it is fair to say that during the 1985–1977 era, the WLAN industry was discovering technologies for wideband indoor wireless communications and it examined spread spectrum, IR, licensed bands at 18 GHz, and DFE technologies and the importance of analysis of the effects of multipath to achieve higher data rates. The spread spectrum and IR technologies of the legacy IEEE 802.11 standard were the only technologies that survived in the market and a modified form of these technologies have remained in other popular standards such as Bluetooth, using FHSS, and ZigBee, using the DSSS. This era also opened channels for dissemination of research and scholarship by publishing pioneering textbooks [Pah95][Rap96] and establishing the first scientific journals and magazines of the field [IJW94][Wir94].

WLAN Technologies and Standards After 1997:

The IEEE 802 standards define MAC and PHY specifications of networks as standards. Figure 7.3 shows the evolution of PHY and MAC of the

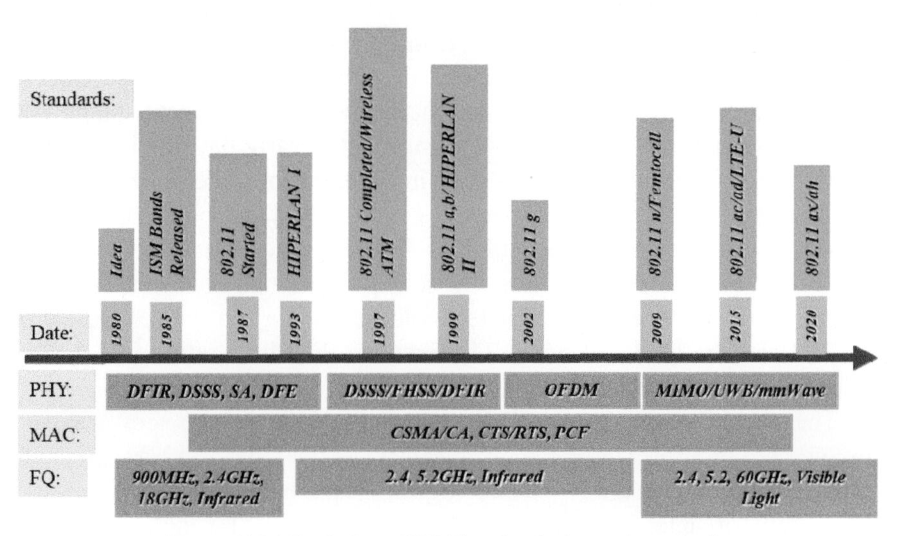

Figure 7.3 Evolution of Wi-Fi technologies and standards.

IEEE 802.11 standards from the beginning to the present. The first step of evolution of the standard after completion of the legacy 802.11 in 1997 was the IEEE802.11b using a complex M-ary orthogonal coding known as complementary code keying (CCK). The IEEE802.11b standards were completed in 1999 and it operated at speeds up to 11 Mbps with a fall back to 1–2 Mbps to legacy 802.11. Both 11b and legacy 11 operated at 2.4 GHz; in 1999, the organization also completed specifications for the IEEE 802.11a operating at 5.2 GHz using the OFDM transmission technology to achieve up to 54 Mbps. The IEEE 802.11a at PHY layer was coordinating with HIPERLAN-2 standard in Europe [Khu99]. In comparison with Wi-Fi, the centralized MAC of the HIPERLAN-2 [Dou02] would allow better management of quality of service, vital to the cellular industry. Perhaps, that was the motivation of Ericsson to pursue the leadership of this effort. However, like wireless ATM, this standard was abundant later. This could be because, in wide-area networking, we have a large number of users with less bandwidth resources, and rationalizing this scarce resource requires centralized supervision by enforcing quality of service laws. In local areas with abundant availability of bandwidth and only a few users, a distributed MAC would be more practical at that time frame. Although a new standard for integration of local and metropolitan area networks, such as HIPERLAN-2 or wireless ATM, did not become a reality, the need for this integration cellular with Wi-Fi became the reality. The concept of integration of Wi-Fi with cellular using vertical handoff and mobile IP technology emerged in the early days of commercial popularity of Wi-Fi [Pah00B] and it has prevailed all the way along up to the time of writing. Continuation of an ideal of new standards for local operation continued into Femtocell [Has09] and LTE-U [Zha16], but it has not created any serious challenge to Wi-Fi technology yet. In the same way that the WLAN industry in the early days of its survival resorted to point-to-multipoint outdoor installation for wider area coverage, it can be thought that WiMax [And07] emerged as a successful application of local centralized control technologies. These technologies filled in several pages of conference proceedings and scientific journals, but they failed to keep investors in developing these technologies as happy as those invested in Wi-Fi. Later, in 2003, IEEE 802.11g defined OFDM operation in 2.4 GHz with the same data rate as 11a.

The breakthrough in wireless communications in this era was the discovery of multiple streaming using MIMO technology. The basic contents of these technologies are two technologies, adaptive antenna arrays to focus the pattern of antennas and space time coding (STC) which enable the creation of multiple streams. The benefits of multiple transmitting and receiving antennas have existed in the antenna and propagation society since the 1930s [Jen16]. This is the STC [Ala89][Tar89][Ala01], which enables multiple streaming and that is why it is considered as one of the most important innovations around the year 2000. Multiple streaming using MIMO technology opened a new horizon in scaling the physical layer transmission in multipath fading channels [Fos89]. The next giant step in the evolution of technology for IEEE 802.11 community was the introduction of the IEEE 802.11n in 2009, using MIMO technology to enable multiple streaming to achieve data rates on up to 600 Mbps both at 2.4 and 5.2 GHz. Other standards such as IEEE 802.11ac,ax, followed the same OFDM/MIMO technology.

Another major hype in physical layer technologies for wireless communications were the mmWave pulse transmission technology. The IEEE 802.11ad adopted mmWave transmission technology at 60 GHz with an ultra-wideband transmission with bandwidth exceeding 2 GHz and speeds on the orders of Gbps. Although mmWave technology created a huge hype in the evolution of the 5G cellular networking industry [Rap13], the IEEE 802.11ad and ay, as the first completed standards using these technologies, were not successful in attracting a huge share of the WLAN market. The reader should notice that mmWave propagation cannot penetrate walls and that restricts its practical indoor applications for Wi-Fi, however it is well suited for beam forming and outdoor coverage for the cellular 5G or other applications.

Regarding the MAC of the IEEE802.11, the main two techniques which became dominant were CSMA/CA and the RTS/CLS. The CSMA/CA is a practical extension of CSMA/CD, which was adopted for IEEE 802.3, the Ethernet, to the wireless medium. The IEEE 802.11 grows as the wireless Ethernet and CSMA/CA would enable Ethernet to become wireless. The RTS/CTS mechanism was originally designed to address the hidden terminal problem, but it became more popular for application with directional antennas for IEEE 802.11ac, ad. Analytical comparison of these two MAC techniques is a challenging problem that received a very thorough and popular analysis in the year 2000 [Bia00].

A good survey of all these standards is presented in Wikipedia [IEEE802.11]. Here, we try to establish that all major PHY technologies, optical wireless, SS, M-ary orthogonal coding, OFDM, MIMO, and mmWave technologies, were first introduced in the IEEE 802.11 standardization community. The DSSS in 3G, OFDM/MIMO in 4G, and mmWave in 5G/6G cellular telephone technologies came after the adoption of these technologies in IEEE802.11 standards. The MAC of cellular telephone industry was centralized and different from the WLAN to accommodate a high density of users in larger coverage of cellular networks. The IEEE 802.15 wireless personal area networks followed a similar pattern by the adoption of FHSS for Bluetooth and DSSS for ZigBee, after they were first introduced by the IEEE802.11 standard. The MAC of Bluetooth and, in particular, ZigBee carries similarities with those of the MAC of IEEE 802.11. Therefore, it is fair to say that the WLAN industry pioneered the design of the dominant PHY technologies of today's wireless networking industry, and this is a huge technological impact.

7.2.2 Evolution of Wi-Fi Applications and Market

Applications fuel the market and they are connected to the network through devices running these applications. Local area networks were networking computers to share common access devices such as printers or storage memories. In the late 1980s and the early 1990s, when the WLAN industry began to test the market, PC and Workstations were competing to capture the market of mini-computers. Laptops emerged in this market a little later. From the networking point of view in that era, engineers were searching for wiring solutions for the growing market of these devices. The early WLAN startups were thinking of wireless as a replacement to wired LANs to connect PCs in open areas such as manufacturing floors or open offices without partitions. These companies were assuming that these small computers will grow on office desks in manufacturing areas in clusters. The idea was that if we connect this cluster of desktop computers to a hub and then we connect the hub to a central node connected to the Ethernet backbone, we will avoid snakes hanging from the ceilings of manufacturing floors and offices.

In a typical startup proposal for venture capital, these companies were arguing that close to half of the cost of LAN industry associates with installation and maintenance of these networks, which can be vanished when we go wireless. As a result, the first WLAN products were shoe box size hubs and central units, and following the argument above, these companies were estimating that a few billion-dollar markets will emerge for these devices in the early 1990s. Based on this idea, a typical startup company or a small group in a large company could raise up to $20M at that time, adequate to support a design and marketing team to get going. Therefore, the early products from Proxim, Aironet, WINDATA, Motorola, NCR, Photonics, and others appeared in the market (see the left part of Figure 7.4). The reader can find a variety of photos of these historical WLANs in the proceedings of the *1st IEEE Workshop on WLAN*, Worcester, MA, May 1991 [WPI'91]. This workshop was held in Worcester, MA, in parallel to the IEEE 802.11 official meeting to decide on the future of this industry. Around the year 1993, these products were in the market, but the expected few billion-dollar markets developed only to a few hundred million. These sales were mostly for selected vertical applications and by research laboratories discovering the technology, not for the horizontal market for connecting desktop computers everywhere. This resulted in a retreat in the original few tens of companies, searching for a new market domain.

During the market crash of 1993 for WLAN industry designed for clustering desktop computers, two new marketing solutions emerged. The first solution was point-to-point or point-to-multipoint WLAN bridges. The idea is to take WLAN outdoor and add a strong roof top antenna to take advantage of free space propagation and antenna gain to extend the expected 100 m indoor coverage to a few miles in outdoor coverages. As examples of these markets, two hospitals in Worcester, which were a few miles away, could connect their networks with low-cost private WLAN, instead of using expensive leased lines from the telephone companies. Or, Worcester Polytechnic Institute could connect the dormitories to the main campus local network. The other idea was to design smaller wireless personal computer (PC) cards for the emerging laptop market. The lower left corner of Figure 7.4 shows the picture of Roamabout access point (AP) box and the laptop PCMCIA cards of the first successful product of that type, design at DEC, Maynard, MA, USA. These devices were the showcase of the *2nd IEEE Workshop on WLAN*, October 1996 [WPI'96].

Examples of the practical markets for laptop operations were like large financing corporates, for example, Fidelity in Boston, who would purchase a laptop for their marketing and sales and other staff and they wanted them to be connected to the corporate network, when in office. The next wave of market demand for WLAN was for small office/home office (SOHO), which began around 2000. The authors believe that this story began in the mid-1990 with the penetration of the internet to homes with America Online (AoL) for small indoor area distributions. Penetration of the internet in home fueled the cable and digital subscriber loop (DSL) modems for high data rate home services, and with that came the growth of a number of home devices demanding home networking. In that era, several ideas such as using home wiring or electricity wiring for implementation of Home-LAN were studied, but Wi-Fi emerged as the natural solution. In that time period, the price of a Wi-Fi AP, such as Linksys, has fallen below $100 and a PC card could

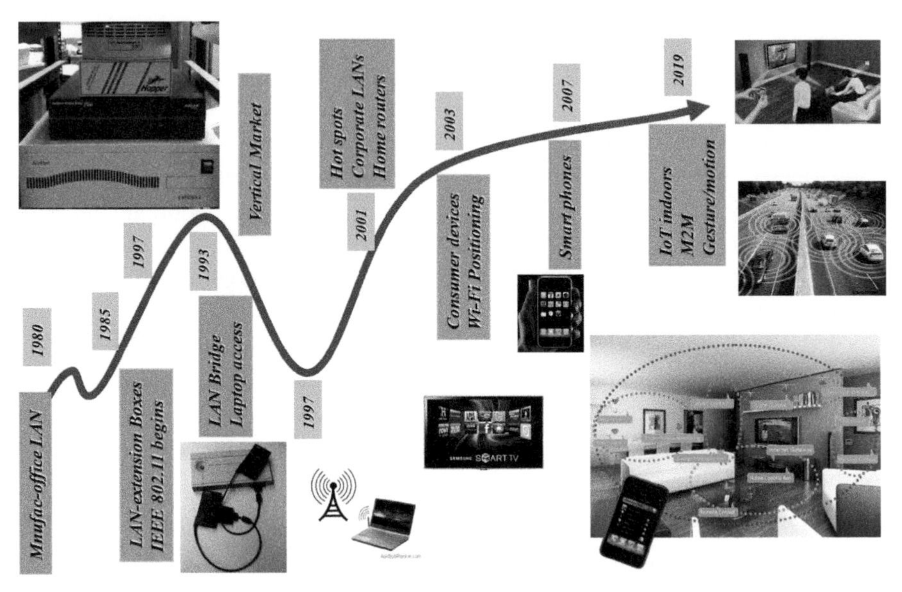

Figure 7.4 Evolution of Wi-Fi applications and market.

be purchased at a reasonable price of a few tens. The original shoe boxes were selling for a few hundred for the hub and up to a few thousand for the AP! With these prices, coffee shops and others could afford to provide free Wi-Fi and homeowners could bring Wi-Fi home. This was perhaps the first large market bringing Wi-Fi from office to home. During the 2000s, in spite of the crash of .com industry, the Wi-Fi market began to grow exponentially.

The exponential growth of Wi-Fi for SOHO encouraged consumer product manufacturing to consider Wi-Fi for integration in their products. But to the authors' opinion, the integration of Wi-Fi in the iPhone was the next major marketing breakthrough for Wi-Fi popularity and market growth. In 2007, Steve Jobs announced the iPhone and with that introduced integration of Wi-Fi and top of that Wi-Fi Positioning [Pah10] in the iPhone. When he opens his talk on the positioning engine of the iPhone, he describes how Skyhook, a small startup in Boston, had this "neat idea" presented to the market. This had two major impacts. First, now that Wi-Fi was integrated in smartphones, a potential for selling billions of Wi-Fi chipsets emerges for the market. Second, Wi-Fi positioning emerges as the first commercial application of widespread deployment of Wi-Fi infrastructure worldwide.

7.2.3 Overview of IEEE 802.11 Standard Protocols

Today, the Ethernet packet is the most popular packet format used for communication networking at the lower layers with the TCP/IP for higher layer communications. The Ethernet protocol and medium access mechanisms run over a variety of wired physical media and the star topology has overtaken the bus topology for local access. The IEEE 802.11 or Wi-Fi was originally designed as a wireless Ethernet and its packet format and MAC builds on those in IEEE 802.3, the wired Ethernet, while the physical layer has gone through a complete redesign to fit the substantially more difficult wireless medium.

The medium access in Ethernet is based on CSMA with collision detection (CSMA/CD) which was described in Section 6.2.2. If two hosts on the same LAN transmit packets such that they may be on the medium at the same time, a collision occurs. In the legacy Ethernet for wired medium, the network interface cards detected collisions by an increase in voltage of the baseband signal above the level of transmission expected from a single terminal transmission. Unlike the wired medium, where collisions can be simply detected by observing voltage increases above a threshold, due to extensive power variations caused by fading and antenna feedback, it is extremely challenging to detect collisions on the wireless medium. As a result, in the IEEE 802.11 Standard, rather than CSMA/CD, a modified version of this protocol, carrier sensing with collision avoidance or CSMA/CA is implemented. Overall, it is preferable from a performance and cost stand point to use collision avoidance, rather than detecting collisions in the air. This is one of the primary differences between the MAC of Ethernet and that of Wi-Fi. In the IEEE 802.11, several approaches for avoiding collisions are used as we will see later in this chapter.

The packet format of the Wi-Fi also builds on the Ethernet packet format. The Ethernet packet comprises a header consisting of a preamble and a starting delimiter. The preamble is used to synchronize the receiver with the timing of the transmitted packet and the starting delimiter is a special sequence denoting the start of the actual information contents of the MAC packet. The information contained in the MAC packet starts with the destination address, followed by the source address, followed by a field for the actual payload and a cyclic redundant code (CRC) error detecting code to examine the sanity of the received packets and to determine if retransmission is necessary. In IEEE 802.11, the header is separated from the MAC information and it is called the physical layer convergence protocol

(PLCP). In addition to the preamble and the starting delimiter, the PLCP carries more information such as the data rates which can vary depending on the channel and protocol type that we will explain later. The MAC packet of IEEE 802.11 has four address fields because each terminal connects to an AP. So, we need two MAC address points to address a wireless terminal. Additional fields to support control packets as well as facilitating other features needed for the wireless environment will be discussed later.

Figure 7.5 shows the protocol stack associated with IEEE 802.11. The physical layer in Wi-Fi is far more complex than the simple baseband signals employed in Ethernet. We have discussed some of these physical layer transmission and coding schemes in Chapter 4. We consider the details of these physical layer schemes in this chapter. The 802.11 standards, like most LAN standards, is concerned only with the lower two layers of the open systems interconnection (OSI) protocol stack model, namely the PHY and MAC layers. The MAC and PHY layers operate under the IEEE 802.2 logical link control (LLC) layer that supports many other LAN protocols. In the case of IEEE 802.3, the Ethernet, there are several physical layers that correspond to the same MAC specifications and a variety of mediums. This standard was originally designed for thick coaxial cable but was subsequently revised to

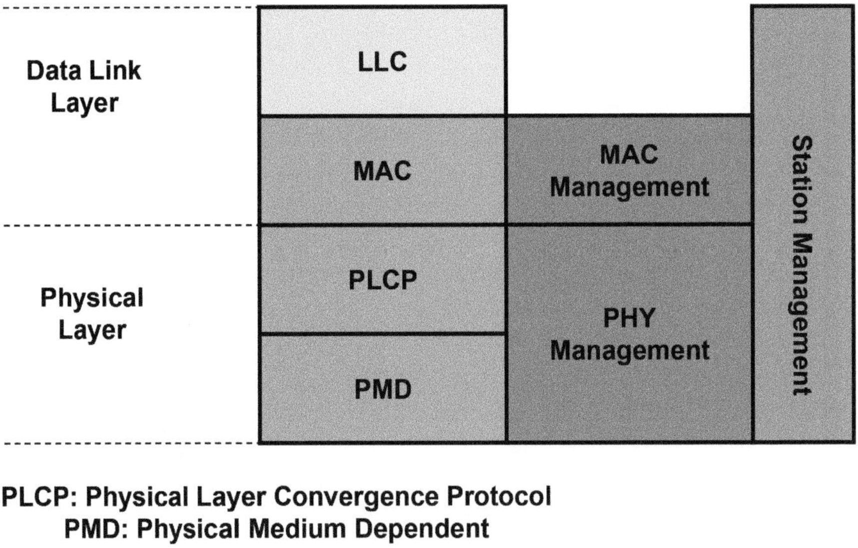

Data Link Layer | LLC | MAC | MAC Management
Physical Layer | PLCP | PMD | PHY Management
Station Management

PLCP: Physical Layer Convergence Protocol
PMD: Physical Medium Dependent

Figure 7.5 Protocol stack of IEEE 802.11.

include thin coaxial cable, a variety of twisted pair cables, and fiber optic links. In the same way, the IEEE 802.11 standard specifies a common MAC protocol that is used over many different PHY standards. The PHY standards are the "base" IEEE 802.11 standard, the 802.11b and g standards, and the 802.11a standard. The 802.11n, ac,ad,ax,af physical layers were released by the 802.11 working group later on. The MAC protocol is based on CSMA with collision avoidance (CSMA/CA). An optional polling mechanism called PCF is also specified. In addition to the MAC and PHY layers, the IEEE 802.11 standard also specify a management plane that transmits management messages over the medium and can be used by an administrator to tune the MAC and PHY layers. The MAC layer management entity (MLME) deals with management issues such as roaming and power conservation. The PHY layer management entity (PLME) assists in channel selection and interacts with the MLME. A station management entity (SME) handles the interaction between these management layers.

The base IEEE 802.11 standard specifies three different PHY layers – two using RF and one using IR communications. The RF PHY layers are based on spread spectrum – either direct sequence (DS) or frequency hopping (FH) while the IR PHY layer is based on pulse position modulation (PPM). Two different data rates are specified – 1 and 2 Mbps for each of the three PHY layers. The RF physical layers are specified in the 2.4 GHz ISM unlicensed frequency bands.

The IEEE 802.11b standard specifies the physical layer at 2.4 GHz for higher data rates – 5.5 and 11 Mbps. The PHY layer makes use of a modulation scheme called complementary code keying. The transmission rate depends on the quality of the signal and it is backward compatible with the DSSS based base-802.11 standard. Depending on the signal quality, the transmission rates could fall back to lower values. The 802.11g standard further increases the data rates to up to 54 Mbps in the 2.4 GHz ISM bands using OFDM. The IEEE 802.11a standard [Kap02] deals with the PHY layer in the 5 GHz unlicensed national information infrastructure (U-NII) bands. Once again, data rates up to 54 Mbps are specified in these bands with OFDM as the modulation technique. Depending on the PHY layer alternative, the frequency band is divided into several channels. Each channel supports the maximum data rate allowed by that PHY layer alternative. The proposal for a very high rate PHY layer (>100 Mbps) called 802.11n,ac,ax employs multiple inputs and output antennas at the transceivers. The technology is popularly called MIMO and also uses OFDM as the modulation scheme. The IEEE 802.11ad uses mmWave ultra-wideband (UWB) pulse transmission for PHY. These latest standards enabled this technology to achieve Gbps transmission.

In the next few sections, the IEEE 802.11 standard is discussed in a top-down manner. First, the different topologies possible in IEEE 802.11 are considered with the focus on understanding some of the management functions. Then detailed discussions of the MAC layer of 802.11 and different PHY layer alternatives are presented. Once the basic operation of the 802.11 WLAN has been considered, security issues in IEEE 802.11 will be discussed as also the recent ongoing activities to extend the standard further.

In the remainder of this chapter, we describe the operational aspects, the MAC layer, the physical layer, security aspects, and deployments of the Wi-Fi networks.

7.3 Topology and Architecture of IEEE 802.11

The topology of an IEEE 802.11 WLAN can be one of the two types – infrastructure or *ad hoc* (see Figure 7.6). In the infrastructure topology, an AP covers an area called the basic service area (BSA) and mobile stations (MSs) communicate with each other or with the Internet through the AP [Cro97]. The AP is connected to a LAN segment and forms the *point of access* to the network. All communications go through the AP. So, an MS that wants to communicate with another MS first sends the message to the AP. The AP looks at the destination address and sends it to the second MS. The AP along with all the MSs associated with it is called a basic service set (BSS). In the *ad hoc* topology (also called independent BSS or IBSS), MSs that are in the range of each other can communicate directly with one another without a wired infrastructure. However, it is not possible for an MS to forward packets meant for another MS not in the range of the source MS. Figure 7.6 shows the schematics of both topologies. MSs and APs are identified by a 48-bit MAC address that is like other MAC addresses at the link layer. In an infrastructure topology, the MAC address of the AP also forms the BSSID, a unique identifier of the BSS.

If we assume that the range of communication of any WLAN device, be it an MS or an AP, is a region of radius R, we can look at the comparative advantages of the two topologies. An MS can communicate with another MS that is up to $2R$ away using an AP, provided both MSs are within a distance R of the AP. The cost here is the additional transmission from the AP to the destination. In the *ad hoc* topology, a destination MS cannot be more than a

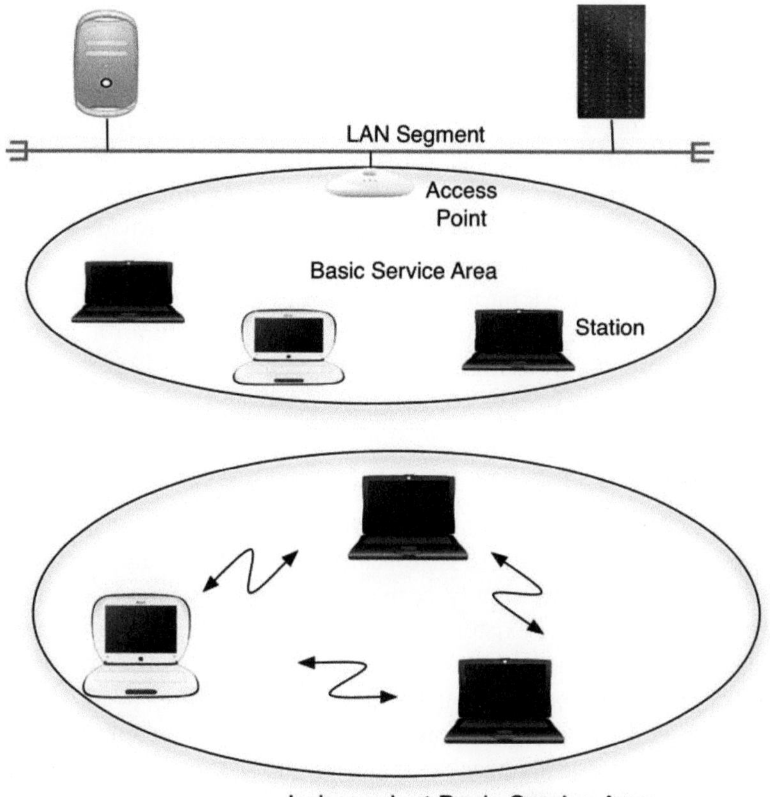

Figure 7.6 Topologies in IEEE 802.11.

distance *R* from the source MS. The advantage is that the information can be received in one hop.

7.3.1 Extending the Coverage in Infrastructure Topology

Depending on the environment in which it is deployed and the transmit powers that are used, an AP can cover a region with a radius anywhere between 30 and 250 feet. The coverage depends upon radio propagation characteristics in the environment and the antenna features (see Chapter 2). The presence of obstacles such as walls, floors, equipment, and so on can reduce the coverage. Many new 802.11 equipped devices have integrated antennas that additionally reduce coverage. To cover a building or a campus, it often becomes necessary to deploy multiple APs that are connected to the same LAN. A group of such APs and the member MSs is called an *extended service set* (ESS). The coverage area is called the extended service area (ESA). The wired backbone that connects the different APs along with services that enable the ESS is called the *distribution system.* The distribution system, for example, supports roaming between APs so that MSs can access the network over a wider coverage area than before. This is like cellular telephone systems where multiple base stations provide coverage to a region, each base station covering only a cell. Note, however, that cellular telephone systems have a far more complex infrastructure to handle roaming and handoff. In 802.11 WLANs, it is easy to roam within a single LAN and requires support from higher layers (such as mobile IP) to roam across different LANs.

7.3.2 Network Operations in an Infrastructure Topology

When an MS is powered up and configured to operate in an infrastructure topology, it can perform a passive scan or an active scan. In the case of a passive scan, the MS simply scans the different channels to detect the existence of a BSS. The existence of a BSS can be detected through *beacon* frames that are broadcast by APs pseudo-periodically. The reason why it is called pseudo-periodic is that the beacon is supposed to be transmitted regularly at certain intervals. However, the AP cannot preempt an ongoing transmission to transmit a beacon. When we discuss the MAC layer, we will see that any device must wait for the medium to be free before transmitting a frame. If the medium is busy, the AP will transmit the beacon after the medium becomes free in which case, the beacon may not be precisely periodic. The beacon is a management frame that announces the existence of a network. It contains information about the network – the BSSID and the capabilities of the nctwork (the PHY alternatives it supports, if security is mandatory, whether the MAC layer supports polling, the interval at which beacons are transmitted, timing parameters, and so on). The beacon is like certain control channels in cellular telephone systems (for instance, the broadcast control channel – BCCH in GSM). The MS also performs signal strength measurements on the beacon frame. In the case of an active scan, the MS already knows the ID of the network that it wants to connect to. In this case, the MS sends a *probe request* frame on each channel. APs that hear the probe request respond with a *probe response* frame that is similar to the beacon. In either case, the MS can create a scan report that provides it with information about the available BSSs, their capabilities, their channels, timing parameters, and other information. The MS makes use of this information to determine a compatible network that it can associate itself with.

To associate itself with an AP, the MS must authenticate itself if this is part of the capability of the network and we will look at this in a later section. Otherwise, if the MS satisfies the announced capabilities of the network, it can send an *association request* frame to the AP. The association request informs the AP of the intention of the MS to join the network and it also provides additional information about the MS such as its MAC address, how often it will listen to the beacon (called the *listen interval*), the supported data rates, and so on. If the AP is satisfied with the capabilities of the MS, it will reply with an *association response* frame. In this message, the MS is given an association ID and this frame confirms that the MS is now able to access the network. During this association phase, the MS can be authenticated by the network and vice versa. Unlike the *ad hoc* mode of operation, administrators can control access to the network in the infrastructure mode of operation.

If an MS moves across BSSs or if it moves out of coverage and returns to the BSA of an AP, it will have to reassociate itself with the AP. For this purpose, it will use a *reassociation request* frame similar in form to the association request frame, except that the MAC address of the old AP will be included in the frame. The AP will respond with a *reassociation response* frame. There are three mobility types in IEEE 802.11. The "No Transition" type implies that the MS is static or moving within a BSA. A "BSS Transition" indicates that the MS moves from one BSS to another within the same ESS. The most general form of mobility is "ESS Transition" when the MS moves from one BSS to another BSS that is part of a new ESS. In this case, upper layer connections may break (it will need Mobile IP for continuous connection). An MS moving from one BSS to another will have to detect the drop in signal strength from the old AP and detect the beacon of the new AP before the reassociation request. It could also use a probe request message instead of detecting the beacon from the new AP. This simple handoff between two APs is MS initiated. In cellular telephone systems, the MS is instructed by entities in the network (such as base station controllers) to perform the handoff from one base station to another. They may use information supplied by the MS such as the signal strength from different base stations. The handoff procedures in a WLAN are as shown in Figure 7.7.

One of the important issues in wireless networks is roaming between different points of access to the wired network. This is possible only if

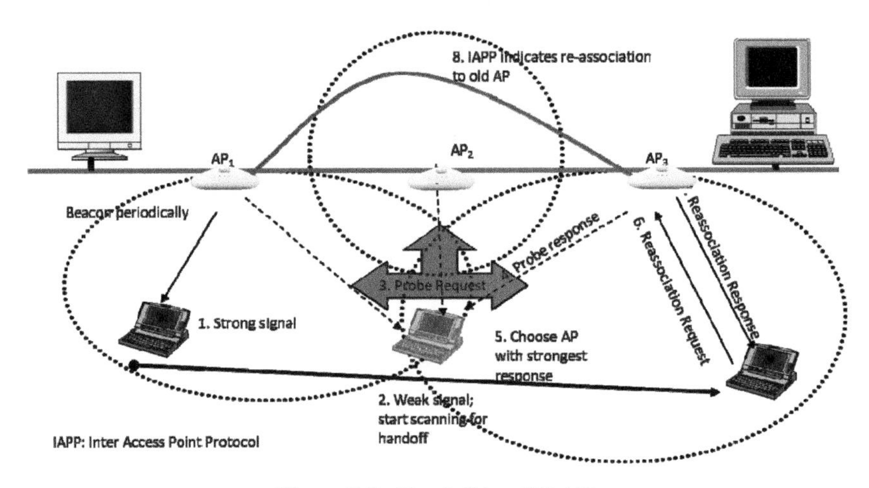

Figure 7.7 Handoff in a WLAN.

equipment from different vendors supports the same set of protocols and is interoperable. Previously, there existed a Task group F that had proposed an inter-access point protocol (IAPP) to achieve multi-vendor interoperability. For example, when an MS moves from one AP to another and sends a reassociation request, the new AP must be able to converse with the old AP over the distribution system to inform it of the handoff and to free the resources in the old AP. This is achieved using the IAPP now standardized as 802.11f in July 2003. The 802.11f standard specifies the information and format of the information to be exchanged between APs and includes the recommended practice for multi-vendor AP interoperability via the IAPP across distribution systems.

Power management is an important component of network operations in an IEEE 802.11 WLAN. The idle receive state dominates the LAN adaptor power consumption. The challenge is how we can power off during the idle periods and maintain the session. The IEEE 802.11 solution is to put the MS in sleeping mode, buffer the data at AP, and send the data when the MS is awakened. Compared to the continuous power control in cellular telephones, this is a solution tailored for bursty data applications. When MSs have no frames to send, they can enter a sleep mode to conserve power. If an MS is sleeping when frames arrive at an AP for it, the AP will buffer such frames. A sleeping MS wakes up periodically and listens to the beacon frames. How often it wakes up is specified by the listen interval mentioned

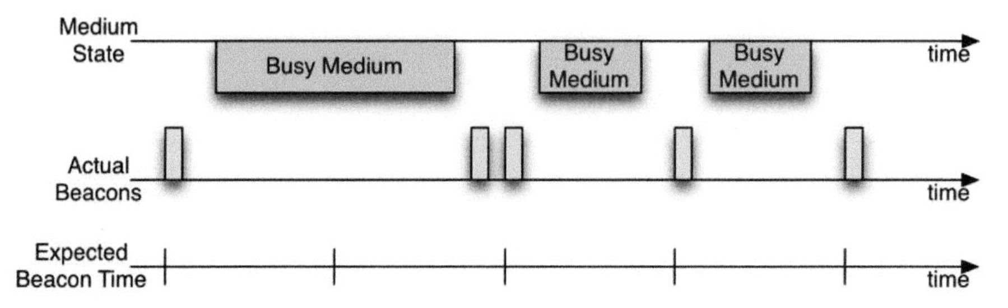

Figure 7.8 Power management in IEEE 802.11.

earlier (Figure 7.8). The beacon frame also contains a field called the traffic indication map (TIM). This field contains information about whether or not packets are buffered in the AP for a given MS. If an MS detects that it has some frames waiting for it, it can wake up from the sleep mode and receive those frames before going back to sleep. The MS uses a *power-save poll* frame to indicate to the AP that it is ready to receive buffered frames. The AP sends the buffered data when the station is in active mode.

If the MS chooses to leave the network or shut down, it will send a *dissociation* frame to the AP. This frame will terminate the association between the MS and the network enabling the network to free resources that were previously reserved for the MS (such as the association ID, buffer space, etc.).

7.3.3 Network Operations in an Ad Hoc Topology

In an *ad hoc* topology, there is no fixed AP to coordinate transmissions and define the BSS. An MS that operates in *ad hoc* mode will power up and scan the channels to detect beacons from other MSs that may be in the vicinity and that may have set up an IBSS. If it does not detect any beacons, it may declare its own network. If it does detect a beacon, then the MS can join the IBSS in a manner like a process in the infrastructure topology. MSs in an IBSS may choose to rotate the responsibility of transmitting a beacon. Power management works similarly, except that the source MS itself must send an announcement traffic indication map (ATIM) frame to the recipient MS.

7.3.4 Network Operations in Mesh Topology

Recently, wireless mesh networking has received a lot of attention as it enables the deployment of wireless networks over a large area without the need for an extensive fixed (wired) infrastructure. A wireless mesh network consists of entities that connect to each other over an air interface and relay packets in the network thus created. This eliminates the need for a wired backbone to relay packets. Some of these entities may act like APs creating infrastructure WLANs and becoming points of access to the mesh network. Other entities will connect to the Internet and enable any device in the mesh network to access the Internet. Wireless mesh networks can use a variety of technologies such as IEEE 802.16 or WiMAX based devices (see Section 7.3) or IEEE 802.11 based devices. In 2004, a task group "S" of 802.11 was set up to investigate mesh networking with 802.11 and propose a standard for using a *wireless distribution system* unlike the wired distribution system in the ESS. While the standard is not finalized as of this writing, some elements of operations in a mesh network have been proposed [Lee06].

In a mesh network, APs (or MSs) are required to be capable of relaying packets to one another using the air interface so that packets may be delivered from a source MS to a destination MS through multiple wireless hops. Entities with relay capabilities are called mesh points. A mechanism for determining a path from one mesh point to another is also necessary and it is expected to be implemented at the MAC layer. Mesh portals enable connectivity to other mesh networks, LANs, or the Internet. Multicast and broadcast capability at the MAC layer is also another aspect that the IEEE 802.11s standard is expected to address. This task group is also supposed to develop enhancements to the current IEEE 802.11 standard to provide a method to configure the distribution system using the four MAC addresses, thereby enabling some form of mesh networking between APs. This could be a wired or wireless mesh network that allows for automatic topology learning and dynamic path configuration over self-configuring multi-hop topologies. There are already proprietary protocols performing this task to extend coverage within homes, but the standard will allow for different scenarios with different requirements (e.g., quick setup and tear down, maximizing throughput, etc.).

7.4 The IEEE 802.11 MAC Layer

MSs in an IEEE 802.11 network must share the transmission medium, which is air. If two MSs transmit at the same time and the transmissions are both in the range of the destination, they may collide, resulting in the frames being lost. The MAC layer is responsible for controlling access to the medium and ensuring that MSs can access the medium in a fair manner with minimal collisions. The medium access mechanism is based on CSMA, but there is no collision detection unlike the wired equivalent LAN standard (IEEE 802.3). In IEEE 802.3, sensing the channel is very simple. The receiver reads the peak voltage on the wire of the cable and compares that against a threshold. Collisions are extremely hard to detect in RF because of the dynamic nature of the channel. Detecting collisions also incur difficulties in hardware implementation because an MS must be transmitting and receiving at the same time. Instead, the strategy adopted is to avoid collisions to the extent possible. In IEEE 802.11, there are two types of carrier sensing – physical sensing of energy in the medium and virtual sensing. Physical sensing is through clear channel assessment (CCA) signal produced by PLCP in the physical layer of the IEEE 802.11. The CCA is generated based on "real" sensing of the air interface either by sensing the detected bits in the air or by checking the RSS of the carrier against a threshold. Decisions based on the detected bits are made slightly slower, but they are more reliable. Decisions based on the RSS may create false alarms caused by high interference levels. Best designs take advantage of both carrier sensing and detected data sensing. In addition to physical sensing, the IEEE 802.11 also provides for virtual carrier sensing. Virtual sensing is implemented by decoding a duration field in the 802.11 frame that allows an MS to know the time for which a frame will last. A "length" field in the MAC layer is used to specify the amount of time that must elapse before the medium can be freed. This time is stored in a *network allocation vector* (NAV) that counts down to zero to indicate when the medium is free again. To illustrate the IEEE 802.11 MAC layer, we will use the *ad hoc* topology as an example. However, the procedures are identical in an infrastructure topology as well.

7.4.1 Distributed Coordination Function and CSMA/CA

We will first describe the basic medium access process in IEEE 802.11 called the distributed coordination function or DCF. Consider Figure 7.9 that shows the basic method for accessing the medium in IEEE 802.11. An MS will initially sense the channel before transmission. If the medium is free, the MS will continuously monitor the medium for a period called the DCF *inter frame space* (IFS) or DIFS. If the medium is still idle after DIFS, then the MS can transmit its frame without waiting. Otherwise, the MS will enter a backoff process. The rationale is that if another MS senses the medium after the first MS, it will also wait for DIFS. However, before a time DIFS expires, the first MS would have started its transmission. Upon hearing the transmission, the second MS will have to back off. The wireless medium is harsh and unreliable, and, hence, all transmissions are acknowledged. The destination of the frame will send an acknowledgment (ACK) back to the source of the frame is successfully received as follows. It will wait for a time called the *short inter frame space* (SIFS) and transmits the ACK. The SIFS value is smaller than the DIFS value.

All IFS values depend on the physical layer alternative. Thus, any other MS that senses the channel as idle after the original frame was transmitted will still be waiting and ACK frames have priority over their transmissions. To maintain fairness and avoid collisions, the MS that senses the medium as free for a time DIFS and transmits a frame will have to enter the backoff process if it wants to transmit another frame immediately. The exception is when it is transmitting one frame in many fragments. In such a case, the MS can indicate the number of fragments in the first frame to be transmitted and occupy the channel till the frame is completely transmitted.

The backoff process works as follows. Once an MS enters the backoff process, it picks a value called the *backoff interval* (BI) which is a random value uniformly distributed between zero and a number called the *contention window* (CW). The MS will then monitor the medium. When the medium is free for at least a time DIFS, the MS will start counting down from the BI value as long as the medium is free. The counter is decremented every so often (called a slot). If the medium is sensed as occupied before, the counter goes down to zero, the MS will freeze the counter and continue to monitor the medium. As soon as the counter becomes zero, the MS can transmit its frame. This process is shown in Figure 7.10.

The IEEE 802.11 MAC supports *binary exponential backoff* like IEEE 802.3. Initially, the CW is maintained at a value called CW_{min} which is typically $2^5 - 1 = 31$ slots. So the BI will be uniformly distributed between 0 and 31 slots. The slot time varies depending on the physical layer alternative. For example, it is 20 μs in the IEEE 802.11b standard and 9 μs is the 802.11a standard. If a packet is not successfully transmitted (this could be due to collisions or a channel error), the value of CW is essentially doubled. The MS will now pick a BI value that is uniformly distributed between 0 and $2^6 - 1 = 63$ slots. This process can be continued till CW reaches a value that is CW_{max} (usually 1023 slots). The rationale behind this approach is as follows. If there are many MSs contending for the medium, it is likely that one or more MSs may pick the same BI value. Their transmissions will then collide. When increasing the value of CW, it is likely that this probability will go down, thereby reducing collisions.

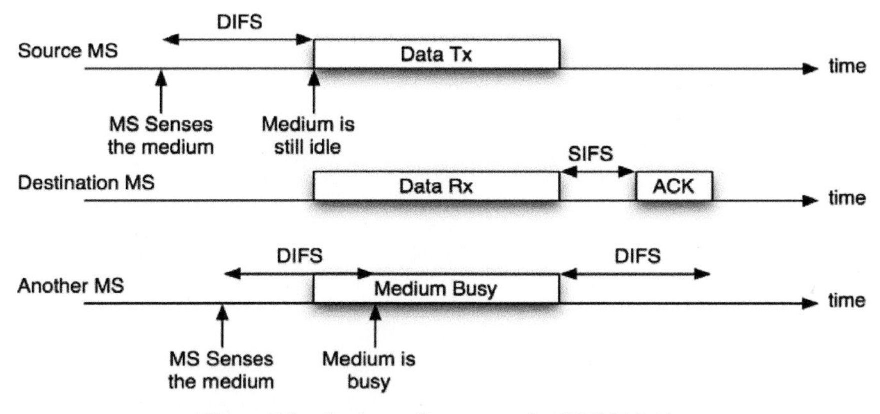

Figure 7.9 Basic medium access in IEEE 802.11.

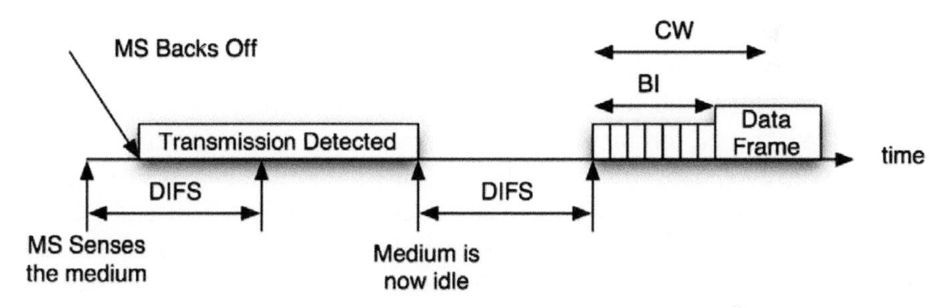

Figure 7.10 Backoff process in IEEE 802.11.

Frames may be lost due to channel errors or collisions. A positive ACK from the destination is necessary to ensure that the frame has been successfully received. In IEEE 802.11, each MS maintains retry counters that are incremented if no ACKs are received. After a retry threshold is reached, the frame is discarded as being undeliverable.

7.4.2 RTS/CTS for Hidden Terminals

In wireless networks that use carrier sense, there is a unique problem called the hidden terminal problem. Suppose all MSs are identical and have a transmission and reception range of R as shown in Figure 7.11. The transmission from MS-A can be heard by MS-C but not MS-B. So when MS-A is transmitting a frame to MS-C, MS-B will not sense the

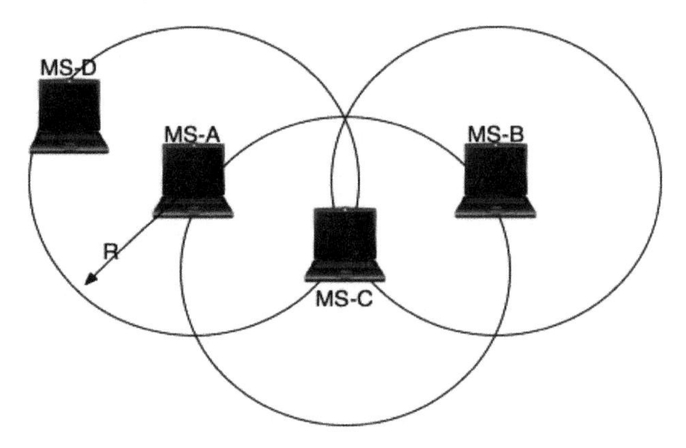

Figure 7.11 Illustrating the hidden and exposed terminal problems.

channel as busy and MS-A is *hidden* from MS-B. If both MS-A and MS-B transmit frames to MS-C at the same time, the frames will collide. This problem is called the hidden terminal problem. There is a dual problem called the exposed terminal problem. In this case, MS-A is transmitting a frame to MS-D. This transmission is heard by MS-C, which then backs off. However, MS-C could have transmitted a frame to MS-B and the two transmissions would not interfere or collide. In this case, MS-A is called an *exposed terminal.* Both hidden and exposed terminals cause a loss of throughput.

To reduce the possibility of collisions due to the hidden terminal problem, the IEEE 802.11 MAC has an optional mechanism at the MAC layer as shown in Figure 7.12. Suppose MS-A wants to transmit a frame to MS-C. It will first transmit a short frame called the RTS frame. The RTS frame is heard in the transmission range of MS-A and includes MS-C and MS-D but not MS-B. Both MS-C and MS-D are alerted by the fact that MS-A intends to transmit a frame and they will not attempt to simultaneously use the medium. This is achieved by the virtual carrier sensing process that sets the NAV to a value equal to the time it will take to complete the exchange of frames. In response to the RTS frame, MS-C will send a *clear-to-send* (CTS) frame that will be heard by all MSs in its transmission range. This includes MS-B and MS-A but not MS-D. The CTS frame lets MS-A know that MS-C is ready to receive the data frame. It also alerts MS-B to the fact that there will be a transmission from some MS to MS-C. Consequently, MS-B will defer any frames that it wishes to transmit in anticipation of the completion of the communication to

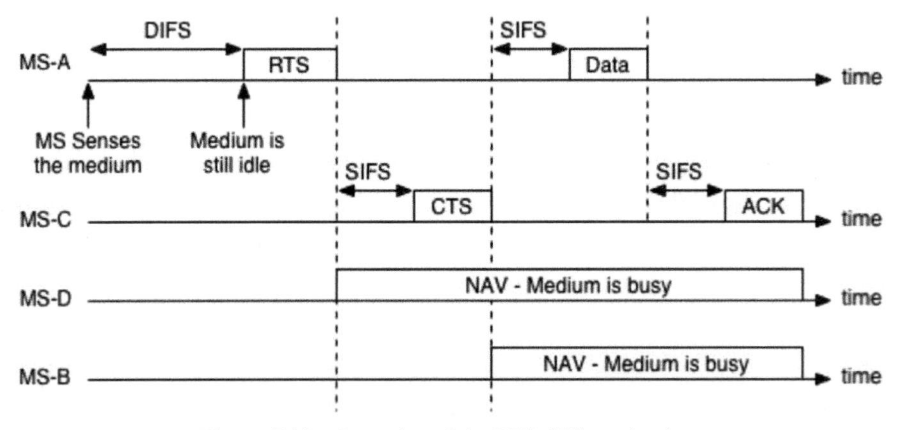

Figure 7.12 Operation of the RTS-CTS mechanism.

MS-C. This way, even though MS-B is outside the transmission range of MS-A, the CTS message can be used to *extend* the carrier sensing range, thereby reducing the hidden terminal problem. Of course, it is quite possible that the RTS frame itself collided with a transmission from MS-B. In such a case, both MS-A and MS-B will have to enter the backoff process and retransmit their frames.

The RTS-CTS mechanism can be controlled in IEEE 802.11 by using an RTS threshold. All unicast and management frames larger than this threshold will always be transmitted using RTS-CTS. By setting this value to 0 bytes, all frames will use RTS-CTS. The default value is 2347 bytes which disable RTS-CTS for all packets. When RTS-CTS signals are used, the CTS frame is transmitted by the destination MS after waiting simply for a time equal to SIFS. This way, the CTS frame has priority compared to all other transmissions that must wait for at least a time DIFS and perhaps an additional waiting time in backoff. Using the RTS-CTS signals reduces the throughput of a WLAN, but it may be essential to use this in dense environments.

7.4.3 The Point Coordination Function

One consequence of using CSMA/CA as described above with DCF is that it is impossible to have any bounds on the delay or jitter suffered by frames. Depending on the traffic load and the BI values that are picked, a frame may be transmitted instantaneously or it may have to be buffered till the medium becomes free. For real-time applications such as voice or multimedia, this can result in performance degradation especially when strict delay bounds are necessary. To provide some bounds on the delay, an optional MAC mechanism called the *point coordination function* or PCF is part of the IEEE 802.11 standard [Cro97]. PCF provides contention-free access to frames using a polling mechanism described below.

The process starts when the AP captures the medium by sending a beacon frame after it is idle for a time called the *PCF inter frame space* (PIFS). The PIFS is smaller than the DIFS and larger than SIFS. In the beacon frame, the AP, also called the point coordinator, announces a *contention-free period* (CFP) where the usual DCF operation will be preempted. All MSs that use only DCF will set a NAV to indicate that the medium will be busy for the duration of the CFP. The AP maintains a list of MSs that need to be polled during the CFP. MSs get onto the polling list when they first associate with the AP using the association request. The AP then polls each MS on the list for data. The polls are sent after a time SIFS and the ACKS to the poll and any associated data will be transmitted by the corresponding MS also after a time SIFS. If there is no response from an MS to a poll, the AP waits for a time PIFS before it sends the next poll frame or data. The AP can also send management frames whenever it chooses within the CFP. An example of PCF operations is shown in Figure 7.13. The AP indicates the culmination of the CFP via a message called CFP-End. This is a broadcast frame to all MSs and frees the NAV in MSs that are only DCF based. Following the CFP, a contention period starts. In this period, it must be possible for an MS to transmit at least one maximum length frame using DCF and receive

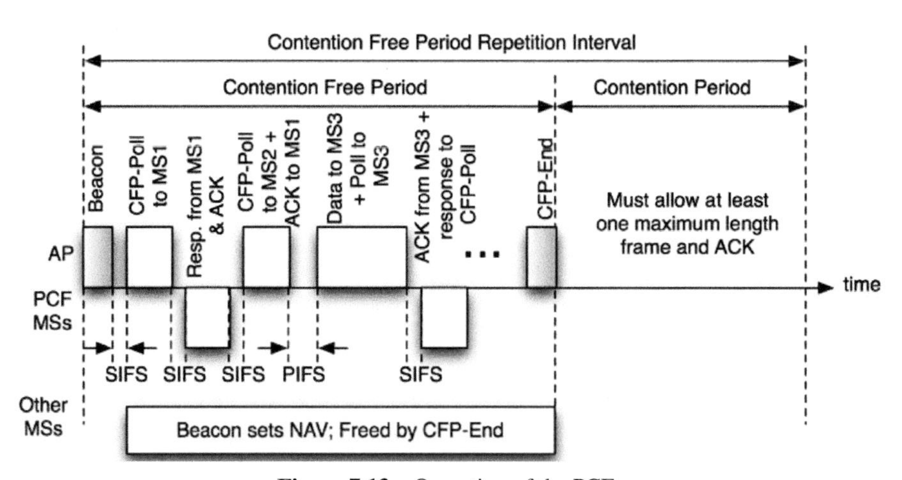

Figure 7.13 Operation of the PCF.

an ACK. The CFP can be resumed after the completion of the contention period. The PCF mechanism is optional in IEEE 802.11. Most commercial systems deployed today do not support PCF and real-time services do not have very good support in WLANs today. Note that polling has a lot of overhead especially if MSs do not have frames to send when they are polled.

7.4.4 MAC Frame Formats

While this article will not define all of the different frame formats of an IEEE 802.11 MAC frame and discuss the fields in great detail, it will consider some examples to illustrate the MAC frame formats. Figure 7.14 shows the general format of a MAC frame. The most significant bit is last (right most) and the bits are transmitted from left to right. The *frame control* field has 2 bytes and comprises many fields. It carries information such as the protocol version, the type of frame (management – probe request, association, authentication, and so on; control – RTS, CTS, and so on; or data – pure data, CFP poll and data, null, and so on), the number of retries, and whether the frame is encrypted (discussed later). The duration field is important to set the NAV during virtual carrier sensing.

There can be up to four address fields in the frame [Gas02]. The addresses can be different depending on the type of frame. Common addresses used are the source and destination addresses, the receiver address if the destination is different from the receiver (e.g., the receiver is the AP, but the destination is a wired node on the LAN segment), the transmitter address (once again if the transmitter is different from the source – it is the AP), and the BSSID.

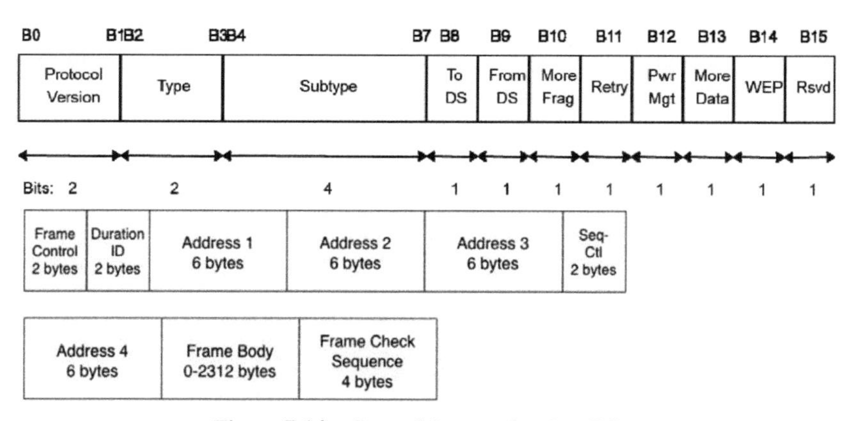

Figure 7.14 General format of an MAC frame.

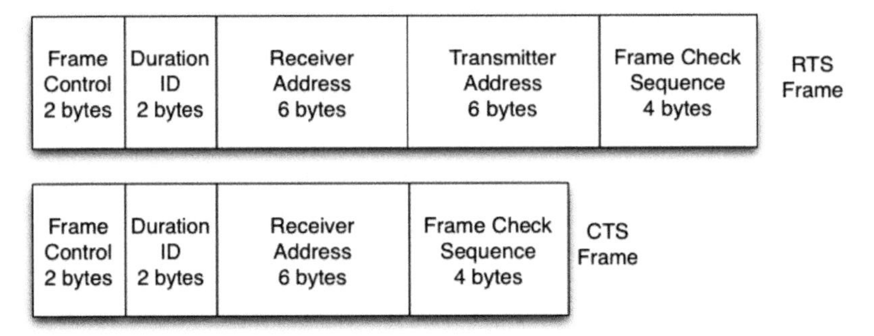

Figure 7.15 RTS and CTS frame formats.

The sequence control field is used in case there is a fragmentation of frames. The frame body carries the payload from upper layers and the frame check sequence is a 32-bit cyclic redundancy check used to verify the integrity of the frame at the receiver. The frame format in Figure 7.14 is used in an infrastructure topology. In an IBSS, only three address fields are used.

The RTS and CTS frames are very short frames – 20 and 14 bytes, respectively, and are shown in Figure 7.15. The ACK frames are very similar to the CTS frame. Compared to Ethernet, IEEE 802.11 is a wireless network that needs to have control and management signaling to handle the registration process, mobility management, power management, and security. To implement these features, the frame format of the 802.11 should accommodate several instructing packets, similar to those we described in wide-area networks. The capability of implementing these instructions is embedded in the control field of the MAC frames. Figure 7.16 shows the overall format of the control field in the 802.11 MAC frame with a description of all fields except type and subtype. These two fields are very important because they specify various instructions for using the packet. The 2-bit ***Type*** field specifies four options for the frame type:

Management Frame (00)

Control Frame (01)

Data Frame (10)

Unspecified (11)

The 4-bit ***Subtype*** provides an opportunity to define up to 16 instructions for each type of frame. Table 7.1 shows all used 6 bits for the Type and

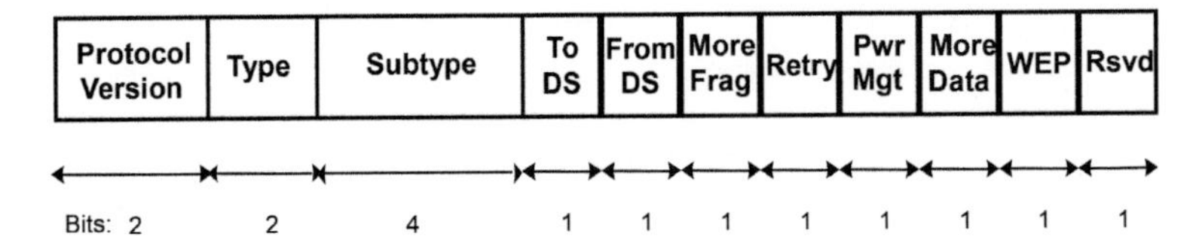

Protocol Version: currently 00, other options reserved for future

To DS/from DS: "1" for communication between two APs

More Fragmentation: "1" if another section of a fragment follows

Retry: "1" if ack is retransmitted

Power Management: "1" if station is in sleep mode

Wave Data: "1" more packets to the terminal in power-save mode

Wired Equivalent Privacy: "1" data bits are encrypted

Figure 7.16 Details of the frame control field in the MAC header of IEEE 802.11.

Table 7.1 Type and subtype fields and their associated instructions.

- **Management type (00)**
  - Association request/response (0000/0001)
  - Reassociation request/response (0010/0011)
  - Probe-request/response (0100/0101)
  - Beacon (1000)
  - ATIM: announcement traffic indication map (1001)
  - Dissociation (1010)
  - Authentication/deauthentication (1011/1100)
- **Control type (01)**
  - Power save poll (1010)
  - RTS/CTS (1011/1100)
  - ACK (1101)
  - CF End/CF End with ACK (1110/1111)
- **Data type (10)**
  - Data/data with CF ACK/No Data (0000/0001)
  - Data poll with CF/data poll with CF and ACK (0010/0011)
  - No data/CF ACK (0100/0101)
  - CF poll/CF poll ACK (0101/0110)

Time Stamp 8 bytes	Beacon Interval 2 bytes	Capab. Info. 2 bytes	SSID Variable	FH Parameter Set 7 bytes	DS Param Set 2 bytes	CF Parameter Set 8 bytes	IBSS Parameter Set 4 bytes	TIM Var.

Figure 7.17 Frame body of the beacon frame.

Subtypes in the frame control field. Combinations that are not used provide an opportunity to incorporate new features in the future.

Figure 7.17 illustrates the frame body of the beacon frame. The time stamp allows MSs to synchronize to a BSS. The beacon interval says how often the beacon can be expected to be heard. It is typically 100 ms but could be changed by an administrator. The capability information (2 bytes) provides information about the topology (whether infrastructure or *ad hoc*), whether encryption is mandatory, and whether additional features are supported. One such feature is *channel agility* where the AP hops to different channels after a predetermined amount of time.

We have not discussed PHY layer alternatives yet. The parameter sets in the beacon provide information about the PHY layer parameters that are necessary to join the network. For instance, if FH is used, the FH parameter set will specify the hopping pattern. The TIM field is used to support MSs that may be sleeping as described earlier.

7.5 The IEEE 802.11 PHY Layer

The IEEE 802.11 standards body has standardized several different PHY layer alternatives. When it was first standardized in 1997, there were three PHY layer options. We will call these options as the "base" IEEE 802.11 PHY layer alternatives. The IEEE 802.11b supports up to 11 Mbps in the 2.4 GHz ISM bands, the IEEE 802.11g standard supports up to 54 Mbps in the 2.4 GHz ISM bands, and the IEEE 802.11a standard supports up to 54 Mbps in the 5 GHz U-NII bands. Before we discuss these alternatives, let us look at the PHY layer in IEEE 802.11.

The PHY layer in IEEE 802.11 is broken up into two sub-layers – the PLCP and the *physical medium dependent* (PMD) layers. The PLCP includes a function that adapts the underlying medium-dependent capabilities to the MAC level requirements. The PLCP would, for instance, add some additional fields to the frame to enable synchronization at the physical layer. The PMD determines how information bits are transmitted over the medium. When the MAC protocol data units (MPDU) arrive at the PLCP layer, a header is attached that is designed specifically, for the PMD of the choice for transmission. The PLCP packet is then transmitted by PMD according to the specification of the signaling techniques.

7.5.1 PHY in the Legacy IEEE 802.11

The legacy IEEE 802.11 standard specifies three different PHY layer alternatives. Two of these use RF transmissions in the 2.4 GHz ISM bands and one uses DFIR. The RF PHY implements on FHSS and DSSS technologies.

The FHSS Option:

The first option for transmission in the 2.4 GHz ISM bands makes use of FHSS. The entire band is divided into 1 MHz wide channels and the specification makes it important to confine 99% of the energy to one such channel during transmission to reduce interference to the other channels. These restrictions are also due to the rules imposed by the FCC in the US. The standard specifies 95 such 1 MHz wide channels and they have numbered accordingly. In the US, only 79 of these channels are allowed. Devices that use the FH option hop between these channels when transmitting frames. The dwell time in each channel is approximately 0.4 s or the minimum hop rate of the IEEE 802.11 FHSS system is 2.5 hops per second that are rather slow. The hop sequences (the channel hopping pattern) depends on mathematical functions. An example hopping pattern is $\{3, 26, 65, 11, 46, 19, 74, \ldots\}$. In the US, each set of hopping patterns can have at most 26 different channels. This means that it is possible to create three orthogonal hopping sets (since there are 79 channels in the US). If three APs use these three orthogonal hopping sets, there will be no interference between these networks. The modulation scheme used with FHSS is called Gaussian frequency shift keying (GFSK). This modulation scheme makes use of the frequency information to encode data. It is possible to use either two frequencies within the channel or four frequencies within the channel. In the former case, the data rate will be 1 Mbps, and in the latter case, the data rate will be 2 Mbps. The advantage of the FHSS system is that the receivers are less complex to implement. The PLCP for the FHSS PMD introduces an 80-bit field for synchronization, a frame delimiter, and some fields to indicate the data rate. Depending on this field, the data rate can be modified in steps of 500 Kbps from 1 to 4.5 Mbps. However, the standard only supports 1 and 2 Mbps. The values of SIFS and the slot for backoff in this option are 28 and 50 μs, respectively.

Figure 7.18 shows the details of the PLCP header, which is added to the whitened MAC PDU to prepare it for transmission using the FHSS physical

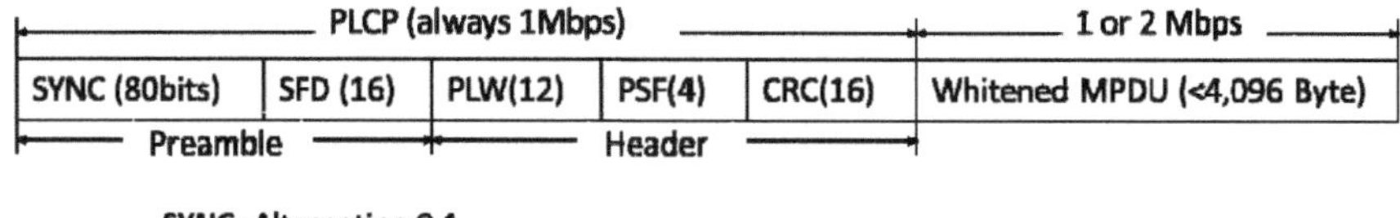

SYNC: Alternating 0,1
SFD: 0000110010111101
PLW: Packet Length Width
PSF: Data rate in 500Kbps steps
CRC: PLCP header coding

Figure 7.18 PLCP frame for the FH option in IEEE 802.11.

layer specifications of the IEEE 802.11. The PLCP additional bits consist of a preamble and a header. The ***Preamble*** is a sequence of alternating 0 and 1 symbol for the 80 bits that are used to extract the received clock for carrier and bit synchronization. The start of the frame delimiter (***SFD***) is a specific pattern of 16 bits, shown in the figure, indicating the start of the frame. The next part of the PLCP is the header that has three fields. The 12-bit packet length width (***PLW***) field identifies the length of the packet that could be up to 4 Kbytes. The 4 bits of the packet-signaling field (***PSF***) identifies the data rate in 0.5 Mbps steps starting with 1 Mbps.

Example 7.1: Specification of Data Rate on the Physical Layer:

The existing 1 Mbps is represented by 0000 as the first step. The 2 Mbps by 0010 that is 2×0.5 Mbps + 1 Mbps = 2 Mbps. The maximum 3-bit number represented by this system is 0111 that is associated with $7 \times 0.5 + 1 = 4.5$ Mbps. If all 4 bits are used, we have $15 \times 0.5 + 1 = 8.5$ Mbps. These limitations imply that data rates cannot even reach 10 Mbps.

The rest of the rates are reserved for the future. The 16-bit CRC code is added to protect the PLCP bits. It can recover from errors of up to 2 bits and otherwise identify whether the PLCP bits are corrupted or not. The total overhead of the PCLP is 16 bytes (128 bits) that is less than 0.4% of the maximum MPDU load justifying the low impact of running the PCLP at lower data rates. The received MPDU is passed through a scrambler to be randomized. Randomization of the transmitted bits, which is also called whitening because the spectrum of a random signal is flat, eliminated the DC bias of the received signal. A scrambler is a simple shift register finite state machine with special feedback that is used both for scrambling and descrambling of the transmitted bits.

The DSSS Option:

The DSSS modulation technique has been the most popular commercial implementation of IEEE 802.11. DSSS has some inherent advantages in multipath channels and can increase the coverage of an AP for this reason [Tuc91]. We will briefly discuss the features of this PMD layer.

In a DSSS system, the data stream is "chipped" into several narrower pulses (chips), thereby increasing the occupied spectrum of the transmitted signal. One common way of doing this is to multiply the data stream (typically a series of positive and negative rectangular pulses) by a spreading signal (typically another series of positive and negative rectangular pulses but with much narrower pulses than the data stream). While the data stream is random and depends on what needs to be transmitted, the spreading signal is deterministic. Figure 7.19(a) shows an example where the data stream $d(t)$ is multiplied by a spreading signal $a(t)$ to produce a signal $s(t)$ that is then modulated over an RF carrier. In this figure, 11 narrow pulses are contained within one broad data pulse. The pulses could have a positive (+) or negative (−) amplitude. This results in the bandwidth expanding by a factor of 11 and this is also called the *processing gain.* A specific pattern of pulses in the spreading signal is used. The pattern used in the IEEE 802.11 standard is a Barker sequence. The interesting property of the Barker sequence is that its autocorrelation has a very sharp peak and very narrow sidelobes as shown in Figure 7.19(b). Because of this property, it is possible for a receiver to reject interference from multipath signals and recover information robustly in a harsh wireless environment. The Barker sequence with differential binary phase shift keying (DBPSK) is used for data rates of 1 Mbps and the Barker sequence with differential quadrature phase shift keying (DQPSK) is used for data rates of 2 Mbps. In either case, the chip rate is 11 Mcps (mega chips per second).

Unlike FHSS, a signal carrying 2 Mbps now occupies a bandwidth that is as large as 25 MHz. In the IEEE 802.11 standard, 14 channels are specified for the DSSS PMD. Channel 1 is at 2.412 GHz, Channel 2 at 2.417 GHz, and so on (see Figure 7.20). Only the first 11 channels are available for use in the US. Figure 7.20 shows the channelization in the US. Since each channel occupies roughly 25 MHz bandwidth and the channel separation is only 5 MHz, there is significant overlap between channels. If two WLANs in the same vicinity were to use adjacent channels, there would be severe interference and throughput degradation. There are three *orthogonal* channels – Channels 1, 6, and 11 in the US that can be deployed without interference. The FH option is easier in terms of implementation because the sampling

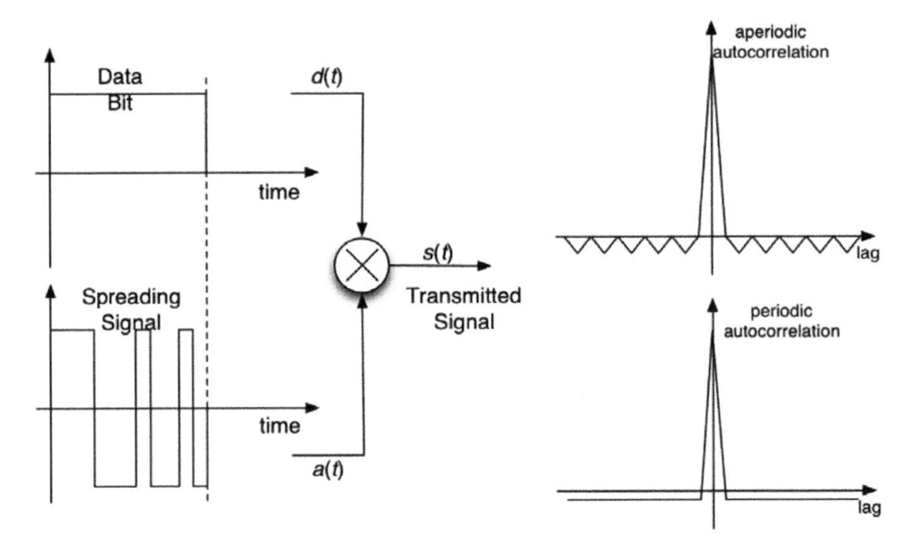

Figure 7.19 (a) Direct sequence spread spectrum and (b) autocorrelation of the Barker pulse.

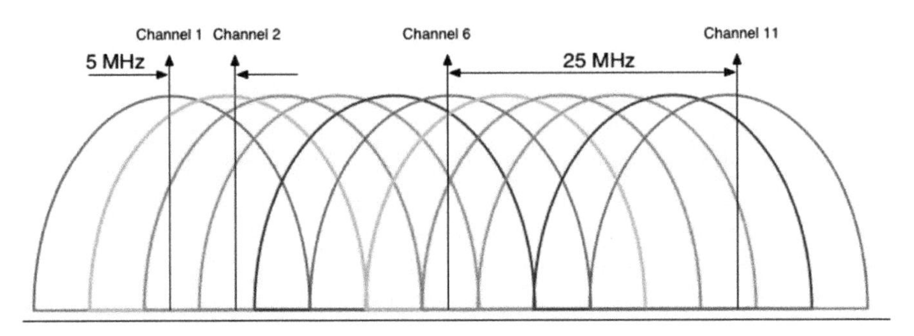

Figure 7.20 Channelization for the IEEE 802.11 DS.

rate is on the order of the symbol rate of 1 Msps. The DS implementation requires sampling rates on the order of 11 Mcps. However, because of the wider bandwidth, DSSS provides better coverage and a more stable signal.

Figure 7.21 shows the details of the PLCP frame for the DSSS version of the IEEE 802.11. The overall format is similar to the FHSS, but the length of the fields is different because transmission techniques are different and different manufacturers designed the model product for the development of the FHSS and DSSS standards. The PLCP sub-layer once again introduces some fields for synchronization (128 bits), frame delimiting, and error checking. The PLCP header and preamble are always transmitted at 1 Mbps

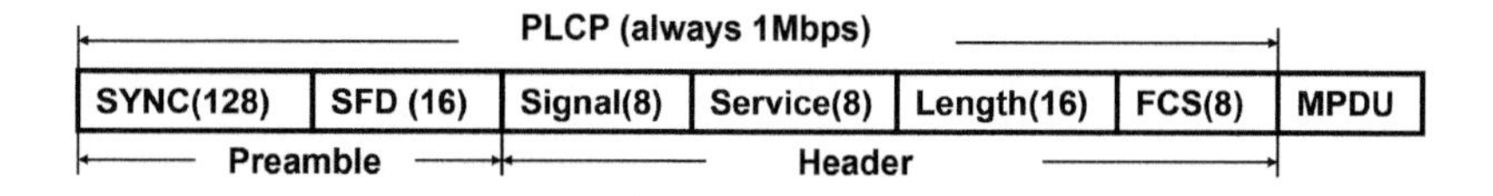

SYNC: Alternating 0,1
SFD: 1111001110100000
Signal: Data rate in 100KHz steps
Service: reserved for future use
Length: Length of MPDU in microsecond
FCS: PLCP header coding

Figure 7.21 PLCP frame for the DSSS of the IEEE 802.11.

using DBPSK. The rest of the packet is transmitted using either DBPSK or DQPSK depending on the data rate. The values of SIFS and the slot for backoff in this option are 10 and 20 μs respectively. The MPDU from the MAC layer is transmitted either at 1 or 2 Mbps; however, analogous to the FHSS version of the standard, the PLCP of the DSSS version also uses the simpler BPSK modulation at 1 Mbps all the time. The MPDU for the DSSS does not need to be scrambled for whitening because each bit is transmitted as a set of random chips that is a whitened transmitted signal. The length of the synchronization (SYNC) in the DSSS is 128 bits that are longer than FHSS because DSSS needs a longer time to synchronize. The format of the SFD of the DSSS is identical to that of the FHSS, but the value of the code, shown in Figure 7.21, is different. The PSF field of the FHSS is called the signal field and it uses 8 bits to identify data rates in steps of 100 Kbps (five times more precision than FHSS).

Example 7.2: Frame Formats for Various Data Rates in IEEE 802.11:

Using the above encoding, we represent 1 Mbps for DSSS by 00001010 (10 $\times$ 100 Kbps) and the 2 Mbps by 00010100 (20 $\times$ 100 Kbps), and 11 Mbps (used in IEEE 802.11b) by 001101110 (55 $\times$ 100 Kbps) and 01101110 (110 $\times$ 100 Kbps). The maximum number in this system is 11111111 that represents 255 $\times$ 100 Kbps = 25.5 Mbps.

The service field in the DSSS is reserved for future use and it does not exist in the FHSS version. The length field of the DSSS is analogous to the PLW in the FHSS; however, the length field specifies the length of the MPDU in microseconds. The frame correction sequence (FCS) field of the DSSS is identical to the CRC field of the FHSS.

The DFIR Option:

The third option in IEEE 802.11 is to use IR for transmission [Val98]. The spectrum occupied by the IR transmission is at wavelengths between 850 and 950 nm. The technique used for transmission is DFIR – that is, communications are omnidirectional. The range specified is around 20 m, but the transmissions cannot penetrate physical obstacles. The modulation scheme used is PPM. A data rate of 1 Mbps is supported using 16 PPM and a data rate of 2 Mbps is supported using 4 PPM. This is in comparison with the IR Data Association (IrDA) standard, which primarily allows communications at a few hundred Kbps to a few Mbps between two devices (like a laptop and a personal digital assistant) that are within a few feet of one another.

The PMD of DFIR operates based on the transmission of 250 ns pulses that are generated by switching the transmitter LEDs on and off for the duration of the pulse. Figure 7.22 illustrates the 16-PPM and 4-PPM modulation techniques recommended by the IEEE 802.11 for 1 and 2 Mbps, respectively. In the 16-PPM, blocks of 4 bits of the information are coded to occupy one of the 16 slots of a 16-bit length sequence according to their value. In this format, each 16 × 250 ns = 4000 ns carries 4 bits of information that supports 4 bits/4000 ns = 1 Mbps transmission rate. For the 2 Mbps version, every 2 bits are PPM modulated into 4 slots of duration 4 × 250 ns = 1000 ns that generates data at 2 bits/1000 ns = 2 Mbps. The peak transmitted optical power is specified at 2 W with an average of 125 or 250 mW. The PLCP packet format for the DFIR is shown in Figure 7.23. The PLCP signals

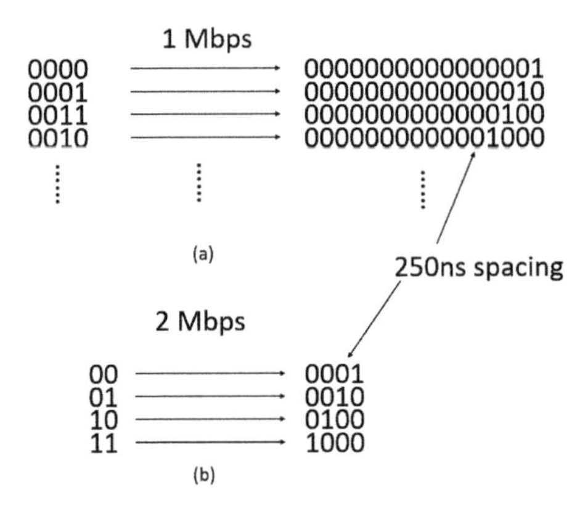

Figure 7.22 PPM using 250 ns pulses in the DFIR version of the IEEE 802.11. (a) 16 PPM for 1 Mbps. (b) 4 PAM for 2 Mbps.

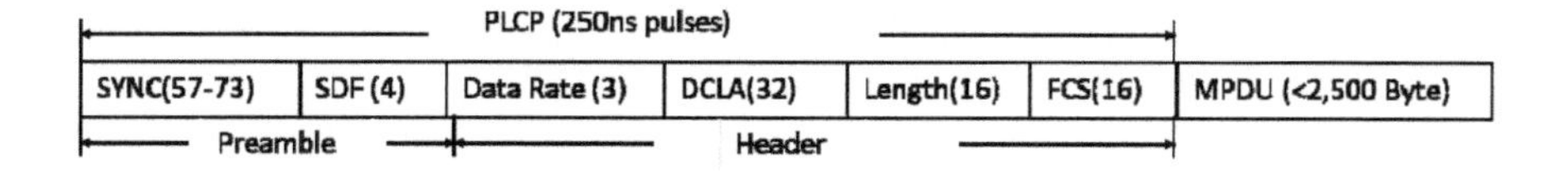

SYNC: Alternating 0,1 pulses
SFD: 1001
Data Rate: 000 and 001 for
DCLA: DC Level Adjustment sequences
Length: Length of MPDU in microsecond
FCS: PLCP header coding

Figure 7.23 PLCP frame for the DFIR of the IEEE 802.11.

are shown in the unit of slots of 250 ns for one basic pulse. The SYNC and SFD fields are shorter because non-coherent detection using photosensitive diode detectors does not need carrier recovery or elaborate random code synchronizations. The 3-slot data rate indication system starts by 000 for 1 Mbps and 001 for 2 Mbps. The length and FCS are identical to the DSSS. The only new field is the DC level adjustment (***DCLA***) that sends a sequence of 32 slots allowing the receiver to set its level of the received signal to set a threshold for deciding between received "0"s and "1"s. The MPDU length is restricted to 2500 bytes.

7.5.2 IEEE 802.11b and CCK Coding

The DSSS option for IEEE 802.11, although successful, consumed a lot of bandwidth for the given data rate. The chip rate is 11 Mcps, but the maximum data rate is 2 Mbps. That is, one Barker sequence of 11 chips, transmitted every microsecond, can at most carry 2 bits of information. To increase the data rate, the IEEE 802.11b standard adopted a slightly different method. Instead of transmitting one 11-chip sequence every microsecond, with IEEE 802.11b, the device transmits one 8-chip codeword every 0.727 μs. Each 8-chip codeword can carry up to 8 bits of information for a maximum data rate of $8/(0.727 \times 10^{-6}) = 11$ Mbps. If the codeword carries only 4 bits of information, the data rate will be 5.5 Mbps. The codewords are derived from a technique called CCK [Hal99].

CCK works as follows for the case when 8 bits are mapped into an 8-chip codeword. The incoming data stream is broken up into units of 8 bits. Suppose the least significant bit is labeled d0 and the most significant bit is labeled d7. Then, four phases are defined to correspond to the four possible values of a

Table 7.2 Mapping for CCK.

DIBIT	PHASE PARAMETER	DIBIT (d_{i+1}, d_i)	PHASE
(d1, d0)	φ_1	(0,0)	0
(d3, d2)	φ_2	(0,1)	π
(d5, d4)	φ_3	(1,0)	$\pi/2$
(d7, d6)	φ_4	(1,1)	$-\pi/2$

pair of bits as shown in the first two columns of Table 7.2. Depending on what the bits are, the phases then take on a value as shown in the third and fourth columns in Table 7.2. For example, if d5 = 0 and d4 = 1, then the phase $\varphi_3 = \pi$. Once the phases are determined, the 8-chip codeword is given by the vector:

$$c = \{e^{j(\varphi_1+\varphi_2+\varphi_3+\varphi_4)}, e^{j(\varphi_1+\varphi_3+\varphi_4)}, e^{j(\varphi_1+\varphi_2+\varphi_4)}, - e^{j(\varphi_1+\varphi_4)}, e^{j(\varphi_1+\varphi_2+\varphi_3)}, e^{j(\varphi_1+\varphi_3)}, -e^{j(\varphi_1+\varphi_2)}, 1\}.$$

This vector has elements that belong to the set $\{+1, -1, +j, -j\}$ where j is the square root of –1. These four elements can be mapped in RF to the phase of the carrier and the receiver can decode this phase information to recover the data bits. CCK can be thought of either as a modulation scheme or as a coding scheme. All the terms of the above equation share the first phase; if we factor that out, we have

$$c = \{e^{j(\varphi_2+\varphi_3+\varphi_4)}, e^{j(\varphi_3+\varphi_4)}, e^{j(\varphi_2+\varphi_4)}, - e^{j(\varphi_4)}, e^{j(\varphi_2+\varphi_3)}, e^{j(\varphi_3)}, -e^{j(\varphi_2)}, 1\}e^{j(\varphi_1)}.$$

This coding suggests that our 256 transformation matrix can be decomposed into two transformations, one a unity transformation that maps 2 bits (one complex phase) directly and the other one that maps 8 bits (4 phases) into an 8-element complex vector with 64 possibilities determined by the inner function of the above equation. The above decomposition leads to a simplified implementation of the CCK system that is shown in Figure 7.24. At the transmitter, the serial data is multiplied into 8-bit addresses. Six of the eight bits are used to select one of the 64 orthogonal codes produced as one of the 8-complex code and 2 bits are directly modulated over all elements

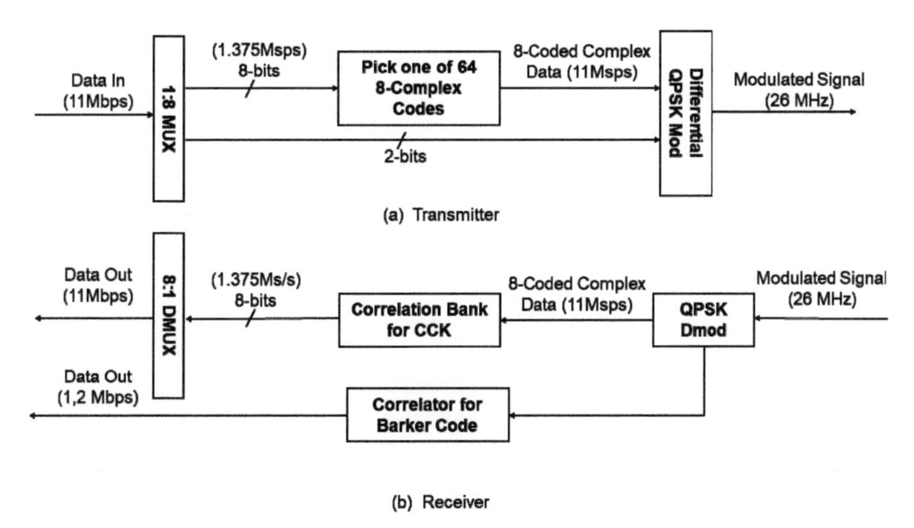

Figure 7.24 Simplified implementation of the CCK for IEEE 802.11b.

of the code that are transmitted sequentially. The receiver comprises two parts: one the standard IEEE 802.11 DSSS decoder using Barker codes and one a decoder with 64 correlators for the orthogonal codes and an ordinary demodulator for IEEE 802.11b. By checking the PLCP data rate, the field receiver knows which decoder should be employed for the received packets. This scheme provides an environment for the implementation of a WLAN that accommodates both 802.11 and 802.11b devices. The advantage of CCK is that it maintains the channelization of IEEE 802.11 while increasing the data rate by a factor of 5. CCK is also robust to the degradations caused by multipath in the wireless environment. The values of SIFS and the slot for backoff in this option are 10 and 20 μs, respectively. IEEE 802.11b also has an optional modulation method called packet binary convolutional coding (PBCC) that is not widely implemented. The advantage of PBCC over CCK is the use of powerful convolutional coding for forwarding error correction.

7.5.3 The IEEE 802.11a, g and OFDM

The IEEE 802.11g standard [Vas05] maintains backward compatibility with IEEE 802.11b and IEEE 802.11 DSSS options by adopting minimal PHY layer frame changes and by including some mandatory and optional physical layer components. In the PLCP layer, 802.11g allows for the use of short preambles to reduce packet overhead. The four physical layers specified in

Table 7.3 Data rates and associated parameters in IEEE 802.11a.

Data Rate (Mbps)	*Modulation*	*Code rate*	*Data Bits/Symbol*	*Coded Bits/ Sub-Channel*	*PLCP Rate Field*
6	BPSK	1/2	24	1	1101
9	BPSK	3/4	36	1	1111
12	QPSK	1/2	48	2	0101
18	QPSK	3/4	72	2	0111
24	16-QAM	1/2	96	4	1001
36	16-QAM	3/4	144	4	1011
48	64-QAM	2/3	192	6	0001
54	64-QAM	3/4	216	6	0011

this standard are prefixed by the term Ethernet ring protocol (ERP) that stands for *extended rate physical*. The standard specifies OFDM and CCK as the mandatory modulation schemes with a data rate of 24 Mbps as the maximum mandatory data rate. With OFDM, IEEE 802.11g also provides for optional higher data rates of 36, 48, and 54 Mbps (see Table 7.3). OFDM is the same modulation scheme that is used in IEEE 802.11a and we discuss it below. PBCC is an optional modulation scheme in 802.11g that allows for raw data rates of 22 and 33 Mbps. We discuss OFDM in the context of 802.11a below, but the discussion could also apply to 802.11g with some modifications.

One of the primary problems for huge data rates in wireless channels is what is called the coherence bandwidth of the wireless channel caused by multipath dispersion. The coherence bandwidth limits the maximum data rate of the channel to that which can be supported within this bandwidth (for example, if the coherence bandwidth is B Hz and the channel bandwidth is $W >> B$ Hz, a transmission bandwidth of W Hz will result in irrecoverable errors unless equalization or spread spectrum is used). To overcome this limitation, we can send data in several sub-channels each on the order of the coherence bandwidth or less so that many of them will get through correctly. Using several sub-channels and reducing the data rate on each channel increases the symbol duration in each channel. If the symbol duration in each channel is larger than the multipath dispersion, errors will be smaller, and it will be possible to support larger data rates. This principle can be exploited while maintaining bandwidth efficiency using an old technique called OFDM. OFDM has been used in digital subscriber lines (DSL) as well to overcome the variations in attenuation with frequency over copper lines. OFDM enables spacing carriers (sub-channels) as closely as possible and implementing the system completely in digital eliminating analog components to the extent possible. OFDM [Kap02] is used as the physical layer in IEEE 802.11a, HIPERLAN/2, and IEEE 802.11g.

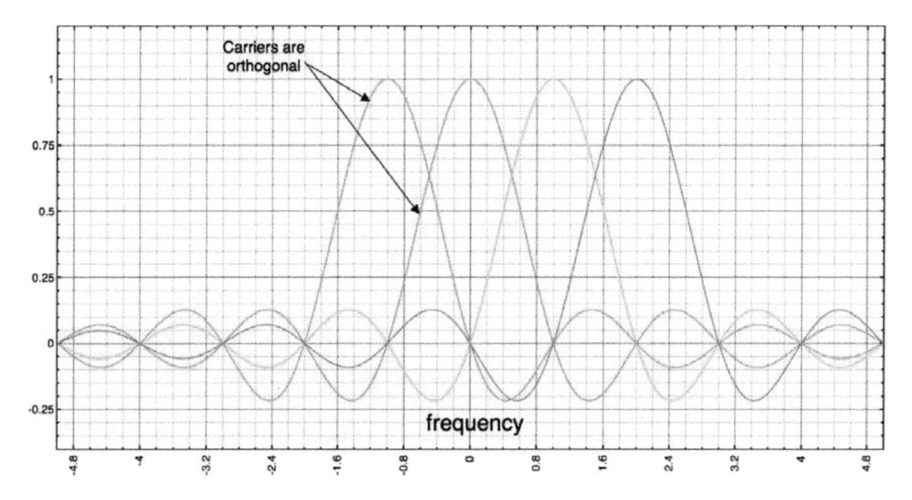

Figure 7.25 Orthogonal carriers in OFDM.

IEEE 802.11a specifies eight 20 MHz channels [vNee99]. As shown in Figure 7.25, several sub-channels are created in OFDM using orthogonal carriers in each channel. Fifty-two sub-channels are specified for each channel with a bandwidth of approximately 300 kHz each. Forty-eight sub-channels are used for data transmission and four are used as pilot channels for synchronization. One OFDM symbol (consisting of the sum of the symbols on all carriers) lasts for 4 μs and carries anywhere between 48 and 288 coded bits. For example, at 54 Mbps, the OFDM symbol has 216 data symbols. With a code rate of 3/4, the number of coded bits/symbols will be $4 \times 216/3 = 288$. This is possible by using different modulation schemes – ranging from BPSK where we have 1 bit per sub-channel to more complex modulation schemes like quadrature amplitude modulation (QAM). Error control coding also plays an important role in determining the data rate. Table 7.4 summarizes some features of the different supported data rates.

The PLCP in the case of 802.11a is a bit different in that there is no synchronization field. A rate field with 4 bits indicates the data rate that is being transmitted. This field is shown in Table 7.4 for different data rates. The preamble and header are always modulated using BPSK (lower data rates). The values of SIFS and the slot for backoff in this option are 16 and 9 μs, respectively.

Table 7.4 Summary of PHY alternatives in IEEE 802.11.[1]

Standard	*Frequency*	*Data Rates*	*PHY Technology*	*Year Completed*
Legacy 802.11	2.4 GHz	1, 2 Mbps	FHSS/DSSS	1997
	316/353 THz	1, 2 Mbps	DFIR	
802.11a	5 GHz	6-54 Mbps	OFDM	1999
802.11b	2.4 GHz	1-11 Mbps	CCK	1999
802.11g	2.4GHz	1-54 Mbps	OFDM, CCK	2003
802.11n	2.4/5 GHz	< 600 Mbps	MIMO/OFDM	2009
802.11ac, ax	2.4/5 GHz	> 1 Gbps	MIMO/OFDMA	2013/2019
802.11ah, af	< 1 GHz	< 570Kbps	MIMO/OFDMA	2014/2016
802.11ad, aj, ay	45/60 GHz	0.5-8Gbps	mmWave	2012/2018/2020
802.11bb	60/790 THz	?	Li-Fi Visible Light	2021?

[1] An update of these standards are available in the IEEE 802.11 official website (http://www.ieee802.org/11/) and an summary update is maintained at the Wikipedia (https://en.wikipedia.org/wiki/IEEE_802.11)

7.5.4 The IEEE 802.11n, ac and MIMO

The IEEE 802.11a and 802.11g MAC and PHY layers constrain the raw data rate to 54 Mbps and the throughput to a fraction of that depending on traffic load, channel conditions, and so on. Thus, a new task group started working on an IEEE 802.11n standard that would look at both MAC and PHY enhancements to improve the throughput to more than 100 Mbps (to up to 600 Mbps). Note that this throughput was not simply the raw data rate on the air, but the actual throughput of the network. For some time, there were two competing proposals in Task Group N for the PHY and MAC layers – World-Wide Spectrum Efficiency (WWiSE) and TGnSync with many vendors in each group. Both proposals used MIMO at the physical layer. Products conforming to parts of these proposals were also available in the market. Neither of the proposals was successful in obtaining 75% of the vote. In January 2006, these two proposals merged with a third proposal that was finally approved.

Some of the ideas that were floated to improve throughput and made it to the standard were to use directional antennas or beamforming, *channel bonding* of two 20 MHz channels for a wider 40 MHz channel, MIMO with OFDM, and throughput enhancements at the MAC layer. As we have mentioned in Chapter 3, MIMO enables the spectral efficiency of links to go well above the 1 bps/Hz that is usually the order in traditional systems to tens of bps/Hz. This increase in spectral efficiency is possible using *space–time techniques* such as space–time coding, beamforming, and spatial multiplexing. These techniques either increase the reliability of the link through diversity, increase capacity by canceling interference, or from the simultaneous transmission of multiple data streams from multiple antennas. The primary MAC enhancement to improve throughput in 802.11n is the use of frame aggregation to reduce overhead [Xia05]. At very high data rates, the overhead of waiting times, back off, and frame headers can reduce throughput significantly. One method of reducing this overhead is to aggregate frames – either at the MAC or PHY layer. Similarly, acknowledgments can also be delayed and aggregated. Frame aggregation can be used for single destinations, multiple destinations, and use multiple rates for multiple destinations.

The IEEE 802.11n standard provided throughput improvements. However, technology must evolve with the applications that emerge over time. Over the last few years, streaming of high-quality video has become a very important application. More people now watch streaming video on demand, not only on their televisions but also on other devices like laptops and tablet computers. Simultaneous video streams and other real-time bandwidth-intensive applications like gaming necessitated the move toward higher throughputs with WLANs.

The result of this need has been the emergence of the IEEE 802.11ac standard that is being rolled out at the time of this writing. This standard incorporates channel bonding that exceeds that of IEEE 802.11n. Instead of bonding two 20 MHz channels, IEEE 802.11ac allows the bonding of up to four channels for a total bandwidth of 80 MHz! Further, two 80 MHz channels can be used simultaneously by a single device to increase the throughput further. Of course, this is possible only in the 5 GHz bands where the bandwidth exists and not in the 2.4 GHz bands. Two other changes to the physical layer are expected to help the throughput increase. The first is the use of 256 QAM in addition to the 64-QAM modulation scheme that has been used with IEEE 802.11a/g/n. With the highest code rate of 5/6 and two spatial streams with MIMO, a link in IEEE 802.11ac can have a raw data rate of over 800 Mbps in an 80 MHz bonded channel. The second is the use of *multi-user* MIMO where the AP can transmit data over multiple spatial streams to multiple MSs *simultaneously*. This increases the overall throughput of the network since MSs need not wait for channel access and the AP can transmit at up to four times the 800+ Mbps rate.

7.5.5 IEEE 802.11ad and mmWave Technology

The frequency bands ranging from 30 to 300 GHz have wavelengths of 10 to 1 mm and are referred to mmWaves. Satellite communications systems and radars were using these frequencies since WW-2. The new hype in the application of this technology for wireless communications began by a sharp decline of the cost of circuit design at these frequencies and FCC release of 7 GHz of the unlicensed band at 57–64 GHz on August 9, 2013. The IEEE 802.15c and IEEE 802.11ad began looking into this technology to achieve Gbps. The UWB and mmWave technologies rely on direct pulse transmission that is implemented by switching an antenna in on- and off-mode at Gbps. The time of arrival of the pulse is measured at the receiver as soon as the receiver signal strength passes a threshold above the background noise indicating the arrival of a pulse. A pulse position modulated transmission scheme can benefit from this simple scheme to carry information bits at Gbps by encoding the bits in their drift bet = fore or after a certain periodical time. These are the same baseband digital communication techniques applied to optical wireless communications. However, switching at Gbps needs bandwidths on the orders of several GHz, which is only available at UWB and mmWave bands. The transmission bandwidth of traditional IEEE 802.11 technologies operating at 2.4 and 5.2 GHz center frequencies is around 20 MHz and those systems need complex OFDM/MIMO technologies operating in the upper half of the Shannon–Hartley plot (Figure 4.11). UWB and mmWave transmission systems are like SS, operating in the lower part of the Shannon–Hartley bounds supporting low energy communications.

Both UWB and mmWave FCC spectrum have around 7 GHz of bandwidth, but the center frequencies are different. mmWaves are centered at around 60 GHz and UWB bands are around 6 GHz. Since the size of antenna elements are usually one-quarter of the wavelength of the signal, antenna elements in mmWaves are around 1.25 mm allowing the formation of an 8×8 antenna array with 64 elements in a 1×1 cm^2 area. An antenna array for UWB with the same number of elements needs approximately 100 times wider surface. The IEEE 802.11ad was the first completed wireless communications standard using mmWave technology. Later, 5G wide-area cellular networks adopted mmWave for outdoor applications. Massive antenna arrays enable the implementation of more focused beams with the antenna array and, consequently, support a higher number of streams between the transmitter and the receiver that increases the data rate.

The IEEE 802.11ad defines four channels each with a bandwidth of 2160 MHz that can achieve data rates of up to 6.7 Gbps. It is indeed a heterogeneous technology that uses 2.4 and 5.2 GHz for broadcasting. However, radio propagation at 60 GHz in indoor areas is contained in a room and does not penetrate well through the walls. This restriction brings this technology in competition with Li-Fi and other IEEE 802.15 wireless personal area network (WPAN) technologies, while its power consumption is more than those technologies. For a wider area coverage, the IEEE 802.11ac with channel bonding and multiple streaming provides a better solution in popular Wi-Fi applications for residential and corporate Wi-Fi infrastructure. As a result, although mmWave has remained as important transmission technology for 5G and beyond for the cellular, the mmWave IEEE802.11ad lags significantly below the IEEE 802.11ac. With the popularity of multi-band Wi-Fi routers and APs for residential and corporate areas, this technology may develop a market.

7.5.6 IEEE 802.11 for WSN and IoT

As we discussed in Chapter 6, the success of the Ethernet sustained the evolution of this standard and it continues its evolution in penetrating the backbone hierarchy for the core network. Similarly, the success of the IEEE 802.11 has kept the continual evolution of the IEEE 802.11 to extend to other spectrums allowed by regional government agencies and adoption of other technologies in parallel with an evolution of cellular wide area networks and short-range low power wireless communication networking industry. The basic PHY technologies that we discussed in this section were FHSS, DSSS, and DFIR in legacy Wi-Fi followed by CCK M-ary orthogonal coding in IEEE 802.11b, OFDM in IEEE 802.11a, g, and orthogonal frequency division multiple access (OFDMA)/MIMO in IEEE 802.11n, ac. The IEEE 802.11ax (2020) allowed narrowband low power extension of IEEE 802.11ac to implement OFDMA that was used earlier in 4G cellular networks and enables integration of Internet of Things (IoT) devices into Wi-Fi networks. Other emerging technologies are mmWave technology first adopted by IEEE 802.11ad (2012), then extended to 5G and 6G cellular networks that we will discuss in Chapter 9 as well as IEEE 802.11aj (2018) and "ay" (2020). Other notable activities are an extension of OFDMA/MIMO technologies to frequencies under 1 GHz for wider coverage and low-energy IoT devices in IEEE 802.11af (2014) for 54–790 MHz, and IEEE 802.11 ah (2016) for 700–900 MHz, as well as the revitalization of optical wireless for visible lights also known as Li-Fi in IEEE 802.11bb. Table 7.4 summarizes different PHY layer alternatives in IEEE 802.11 that we referred to in this section.

7.6 Security Issues and Implementation in IEEE 802.11

Security in wireless networks is an important problem especially because it is extremely difficult to contain radio signals within a protected perimeter [Edn04]. Anyone can listen to radio signals and anyone can also potentially inject signals into the network. Typically, in any network wireless or wired, it is common to deploy security features or services like confidentiality, entity authentication, data authentication, and integrity, and so on to protect against security threats [Sti02]. The IEEE 802.11 standard has some mechanisms to provide confidentiality, integrity, and authentication at the link level (see Chapter 7 for more details). All data that leaves the 802.11 link will not be protected. For instance, an MS communicating with an AP can have all its IEEE 802.11 frames that are air protected. Once the AP receives the frame, all protection is removed before it is transmitted to the distribution system. So additional security at the higher layers (such as IPSec or the secure sockets layer – SSL) may be required for some applications if the payload needs to be secure.

The original mechanism for providing confidentiality and authentication in IEEE 802.11 is called *wired equivalent privacy* (WEP) [Gas02][Edn04]. Over the last few years, several techniques for compromising WEP have been published in the literature. Tools such as AirCrack, Kismet, and WEPcrack are freely available that can be used to extract the secret key used in WEP encryption. WEP makes use of the RC4 stream cipher with 40-bit keys (although there are options to use 128-bit keys in most commercial products today). Both the implementation of WEP and the RC4 algorithm itself have vulnerabilities that have rendered WEP not secure for today's applications. WEP was initially proposed in the standard as a self-synchronizing, exportable, and efficient option. While it does satisfy these three properties, its security has left much to be desired.

7.6.1 Entity Authentication with WEP

The mandatory entity authentication mechanism in IEEE 802.11 is called *Open-System Authentication.* In this case, there is no real authentication. If one IEEE 802.11 device sends a frame to another, it is implicitly accepted. For example, an MS may simply send a frame to the AP choosing "open system" as the authentication algorithm (authentication algorithm = 0). The AP will simply accept it if open system access is allowed and send a response. From this transaction, the AP will obtain the MAC address of the MS for communication purposes.

A better authentication procedure is called *shared-key authentication* where WEP is implemented [Gas02]. If the network is using WEP, shared-key authentication is mandatory. The assumption is that all devices in the network share a secret key. An MS will send a frame for authentication with sequence number 0, the authentication algorithm set to 1 (to indicate shared-key authentication). The AP will then send a challenge message (128 bits) in clear text to the MS along with its response. The MS will respond with an encrypted version of the challenge text. If the AP can verify the integrity of the reply, the MS is authenticated, and it has the shared WEP key configured in it. Sometimes, an MS will authenticate itself with several APs before associating itself with one of them. This process is called pre-authentication. This authentication scheme is still not very secure, however, and creates weaknesses in the protocol due to the way in which it is employed with a stream cipher (see Chapter 7 for details).

Several commercial products also implement address filtering where only certain MAC addresses are allowed access to the network. This is not part of the standard and it is also possible for malicious users to spoof MAC addresses easily. However, address filtering is an additional security measure that is available for IEEE 802.11 networks.

7.6.2 Confidentiality and Integrity with WEP

Confidentiality is simply provided in IEEE 802.11 by encrypting all packets using the RC4 stream cipher. Stream ciphers operate as follows. Using a secret key, a pseudo-random sequence of bits (called the key stream) is generated. If this sequence has a very long period and the algorithm is strong, it will be computationally impossible for someone to generate the sequence without knowing the secret key. A pseudo-random generator is used along with the 40-bit secret key to create a key sequence that is simply XOR-ed with the plaintext message. The pseudo-random sequence, thus, generated will be XOR-ed with the MAC frame to make the contents of the frame secure from interception. RC-4 is one algorithm to generate the pseudo-random key stream. This algorithm makes use of a secret key and, in the case of WEP, an initialization vector (IV) that is 24 bits long. Since the key is constant for all transactions, the same pseudo-random key stream is generated if the IV is not changed. An attacker could capture two streams of encrypted frames, XOR them together, and eliminate the key stream. He would then have an XOR of two data frames. If by some chance, he knows the contents of one data frame, he can get the other as well. Since the IV is only 24 bits long, it is possible for an attacker to break the encryption scheme. One well-publicized attack is the Fluhrer, Mantin, Shamir (FMS) attack on RC-4. In addition, there are several weak keys that could make the encryption scheme easier to break. It is also possible for an attacker to replay packets depending on the sequence numbers that are being used. To ensure that an attacker has not modified a message, the WEP protocol uses the in-built CRC to verify the integrity of the message.

Checking the integrity of the message using the CRC has vulnerabilities that have been publicized in recent years.

7.6.3 Key Distribution in WEP

The IEEE 802.11 standard does not specify how the shared keys must be distributed to devices (AP and MSs). It is usually a manual installation of keys where a user will type the key in the device driver software. This process is unfortunately not scalable and has several human vulnerabilities. Users may write down the key on a piece of paper when they buy a new device and lose this paper. Some vendors have automated methods of key distribution. Cisco's light extensible authentication protocol (EAP) makes use of the challenge-response mechanism to generate a key at the AP and an identical matching key locally in the MS that could then be used in successful encrypted communication.

7.6.4 Security Features in 802.11

The Task Group I of the IEEE 802.11 working group has prepared an enhanced security framework for IEEE 802.11 called 802.11i that was approved as a standard in June 2004. Several vendors have already implemented elements of this standard. This framework includes what is called a *robust security network* (RSN) that is like WEP but has several new capabilities in devices [Edn04]. It is possible for both WEP and RSN devices to coexist in a *transitional security network* (TSN).

A consortium of major WLAN manufacturers called the Wi-Fi alliance considered options to improve security in legacy devices while 802.11i was being standardized. The proposal from this alliance is called Wi-Fi protected access (WPA) that introduces enhancements to WEP called *temporal key integrity protocol* (TKIP). In this protocol, RC-4 is still used as the encryption algorithm. However, this protocol adds some features to overcome the weakness of WEP. A message integrity code is used instead of the CRC check. It changes the way in which IVs are generated. It changes the encryption key for every frame, increases the size of the IV, and adds a mechanism to manage keys.

In 802.11i, RC-4 is replaced by the advanced encryption standard (AES). In particular, the key stream and message integrity check will be generated by a counter-mode cipher-clock-chaining MAC protocol (CCMP). AES is a block cipher – it operates on fixed blocks of data, unlike a stream cipher that generates a key stream. However, any block cipher can operate in different *modes* and cipher-block-chaining (CBC) is one such mode of operation.

The counter mode is another mode of operation. It is expected that the counter mode will be used to generate the key stream and the CBC will be used to generate the message integrity check. Both these modes have been used in other systems with good security.

Both TKIP and AES-CCMP provide confidentiality and message integrity. In order to perform entity authentication, the IEEE 802.11 system has to still rely on challenge response protocols. Over the years, there have been several protocols developed for dial-up entity authentication and for port security in wired LANs. These include 802.1X, the EAP, and remote authentication dial-in user service (RADIUS). Note that all these protocols are not equivalent – for instance, both 802.1X and RADIUS could use EAP for entity authentication and key distribution. EAP itself would use some challenge-response protocol like challenge handshake authentication protocol (CHAP) or SSL to authenticate the devices. Both WPA and RSN mandate 802.1X and EAP as part of the access control mechanism for 802.11 networks. Note that access control is increasingly becoming an important problem with the emergence of hot spot networks in airports, cafes, and so on.

Assignments

Questions

1. How does the current state-of-the-art data rate of the wired (Ethernet) and wireless (Wi-Fi) compare with one another?
2. Why are unlicensed bands essential for the Wi-Fi industries?
3. Explain the difference between the wireless inter-LAN bridges and WLANs.
4. What three topologies can IEEE 802.11 WLANs operate in? What are the differences?
5. Name four major transmission techniques implemented in Wi-Fi devices and give the standard activity associated with each of them.
6. Compare OFDM and SS technology as PHY alternatives for Wi-Fi.
7. What are the MAC services of Wi-Fi (IEEE 802.11) that are not provided in the Ethernet (IEEE 802.3)?
8. Why does the MAC layer of Wi-Fi have four address fields compared to Ethernet that has two?
9. What is the PCF in 802.11, what services does it provide, and how is it implemented?
10. Explain the difference between a hidden terminal and an exposed terminal.
11. Explain why a Wi-Fi AP (IEEE 802.11) also acts as a bridge?

12. What is the purpose of PIFS, DIFS, and SIFS time intervals and how are they used in the IEEE 802.11?
13. What is the difference between a probe and a beacon signal in 802.11?
14. Explain the operation of the timing of the beacon signal in Wi-Fi.
15. How are authentication and integrity provided in IEEE 802.11?
16. How is IEEE 802.11n different from IEEE 802.11a or IEEE802.11g?
17. How is IEEE 802.11ac different from IEEE 802.11n?
18. What are the differences between the residential and corporate deployment of Wi-Fi infrastructure?
19. What are the benefits of Wi-Fi expanders and how does their location impact the user throughput in a residential environment?
20. What is the difference between discrete and continuous models for relationships between data rate and distance of a Wi-Fi AP?

Instructor's solution available on River Publishers' website:
https://www.riverpublishers.com/book_details.php?book_id=919

Problem 1

You want to transmit the information sequence 00111100 using CCK as in 802.11b. What is the CCK codeword in vector form? Show all steps. Assume that bit d0 is the left-most bit.

Problem 2

a) Use the equation for generation of CCK to generate the complex transmitted codes associated with the data sequence {0,1,0,0,1,0,1,1}.
b) Repeat (a) for the sequence {1,1,0,0,1,1,0,0}.
c) Show that the two generated codes are orthogonal.

Problem 3

The original WaveLAN, the basis for the IEEE 802.11, uses an 11-bit Barker code of [1,−1,1,1,−1,1,1,1,−1,−1,−1] for DSSS.

a) Sketch the aperiodic autocorrelation of the code (see problem 3).
b) If we use the system using random codes with the same chip length in a CDMA environment, how many simultaneous data users can we support with an omni-directional antenna and one AP?

Problem 4

a) If in the PPM-IR PHY layer used for the IEEE 802.11 instead of PPM we were using baseband Manchester coding, what would be the transmission data rate? Your reasoning must be given.

b) What is the symbol transmission rate in the IEEE 802.11b? How many complex QPSK symbols are used in one coded symbol? How many bits are mapped into one transmitted symbol? What is the redundancy of the coded symbols (the ratio of the coded symbols to a total number of choices)?

c) What is the symbol transmission rate of the coded symbols per channel in the IEEE802.11g? How does this symbol rate relate to the data rates (6, 9, 12, 18, 27, 36, and 54 Mbps) and convolutional coding rates (½, ¾, and 9/16)?

Problem 5

Redraw the timing diagram of Figure 7.8, if all MSs use RTS/CTS mechanism to send packets.

Problem 6

A voice over IP application layer software generates a 64 Kbps coded voice packet every 20 ms. This software is installed in two laptops with WLAN PCMCIA cards communicating with an AP connected to a Fast Ethernet (100 Mbps).

a) What is the length of the voice packets in ms, if the PCMCIA cards were DSSS IEEE 802.11?

b) If the two terminals start to send voice packets almost at the same time, give the timing diagram to show how the first packets are delivered through the wireless medium to the AP using CSMA/CA mechanism.

c) Repeat (b) and (c) if 802.11b at 11 Mbps was used instead of DSSS 802.11. How would this change if 5.5 Mbps was the data rate?

Problem 7

Figure P7.1 shows the layout of an office building. If the distance between the AP and the MSs 1, 2, and 3 are 50, 65, and 25 m, respectively, determine the path loss between the AP and MSs:

a) Using path loss per wall model and free space loss (assume that the wall loss is 3 dB per wall).

b) Using the 802.11 path loss model from Chapter 3.

Using the 802.11 transmitted and received power specifications, determine whether a single AP can cover the entire building.

Problem 8

You are designing a WLAN for an office building. You are not able to perform measurements or site surveys and have to rely on statistical models and certain other information. There are also certain constraints on where you can place the access point(s). You have the following information available to you:

- Maximum number of walls between an access point and a mobile terminal = 4
- Maximum number of floors between an AP and the mobile terminal = 2
- Transmit power possibilities = 250 and 100 mW
- Sensitivity of the receiver is –90 dBm
- Maximum distance from AP to building edge = 30 m
- The building has office walls, brick walls, and metallic doors
- Shadow fading margin = 8 dB

What would be a conservative estimate of the number of APs required for the WLAN setup? Why? State your assumptions, models, and provide reasons for all your assumptions and calculations. *Hint: Use path loss models from Chapter 3 that are applicable to indoor areas.*

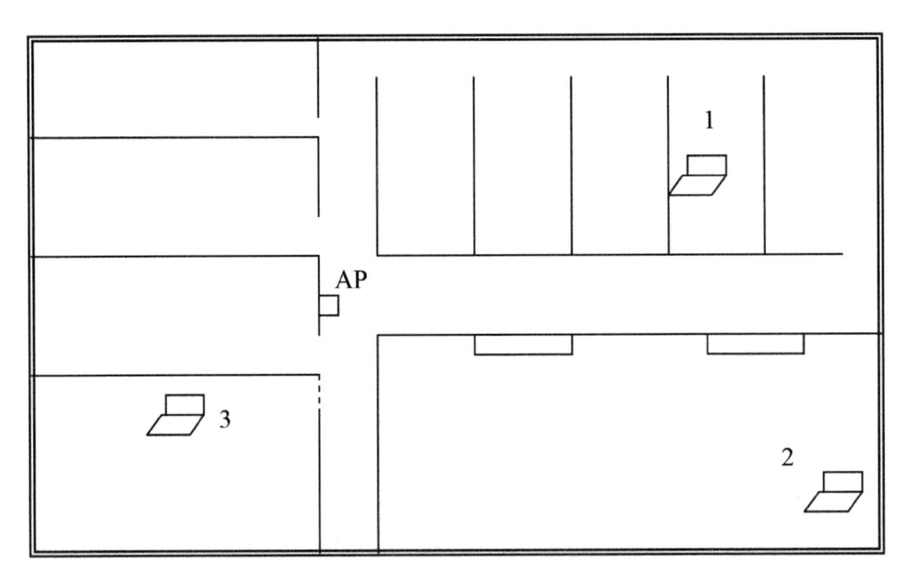

Figure P7.1 Layout of an office building.

Problem 9

Suppose the coverage areas where 11, 5.5, 2, and 1 Mbps data rates are reliably available to have radii of 20, 30, 40, and 50 m. What is the spatial capacity of the AP? How would the spatial capacity change if the 1 Mbps data rate was available up to 75 m? Plot the spatial capacity vs. the range of the 1 Mbps coverage area as it varies from a radius of 50 to 100 m assuming all other values remain the same.

Problem 10

Figure P7.2(a) shows the overhead for packet formation and applications using TCP packets. Each TCP packet can have a length of up to 65,495 byte that should be fragmented to fit the maximum MAC packet of 2312 byte. The TCP/IP header is 40 byte, the 802.2 LLC/SNAP header is 8 byte, and the 802.11 MAC and PLCP headers and synchronization preamble are 34 and 24 bytes respectively. The TCP ACK is a TCP header with no application data and the MAC ACK is shown in Figure P7.2(b). Assuming SIFS and DIFS intervals of 10 and 50 s, respectively, determine the application throughput of the 802.11b for data rates of 11, 5.5, 2, and 1 Mbps for data packets of length 100 and 1000 bytes.

Problem 12

The throughput of a WLAN is a function of the channel characteristics and it fluctuates in time. Figure P7.2 shows a typical application throughput of an 802.11b terminal in a 1 minute observation time. Due to the channel fading and other imperfections, this throughput varies in time as we measure it at a certain distance between the transmitter and the receiver. As the distance between the transmitter and the receiver increases and the RSS reduces, this average throughput also reduces. The throughput (Mbps) versus distance (m) relation of an IEEE 802.11b in an office building is empirically determined to follow the following approximated equation:

$$S_u(r) = -0.2r + 5.5 \tag{P7.1}$$

a) Determine the maximum throughput in Mbps and maximum coverage in meters?

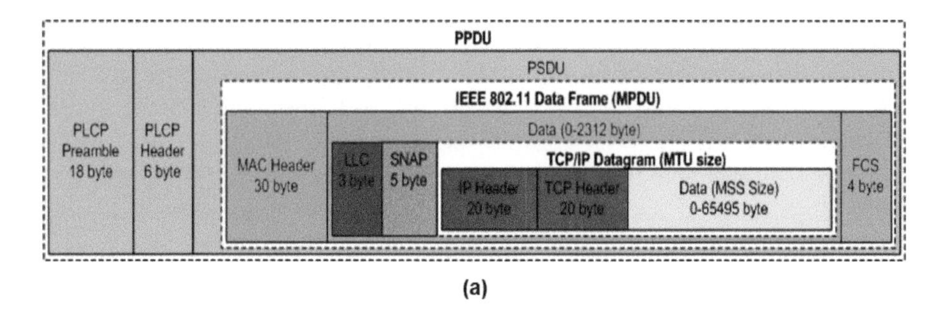

(a)

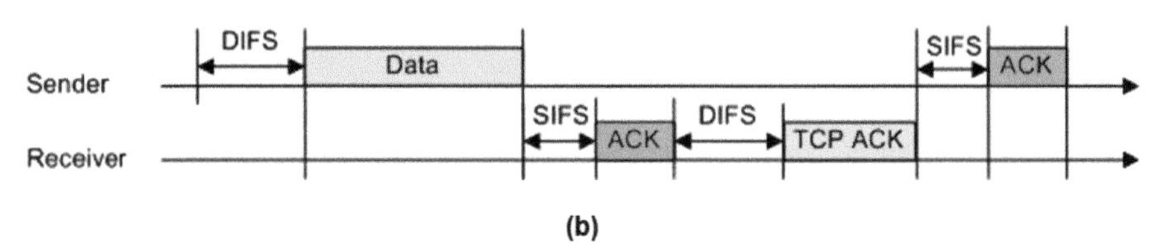

(b)

Figure P7.2 Packet transmission in the IEEE 802.11b. (a) Overheads for the formation of a packet. (b) Overheads for successful transmission of a TCP packet.

b) Show that we can find the average throughput of a user randomly walking in the coverage area of an AP by

$$\overline{S_u} = \frac{2\int_0^{R_L} rS_u(r)dr}{{R_L}^2} \text{ [Mbps]} \tag{P7.2}$$

in which R_L is the distance for which the throughput of the WLAN approaches zero and the WLAN has no coverage anymore.

c) Use Equations (P7.1) and (P7.2) to calculate the average throughput of a user randomly walking in the coverage area of the WLAN.

d) Compare your results with the minimum and maximum nominal data rates of the IEEE 802.11b. Explain the difference between your results.

Problem 13

The average throughput versus distance relationship of an IEEE 802.11g in a typical office building is measured to fit the following function:

$$S_u(r) = \begin{cases} 22\,; & 0 < r < 1 \\ -22\log 10r \; + \; 25\,; & r > 1 \end{cases}. \tag{P7.3}$$

a) Use Equations (P9.1) and (P9.2) to calculate the average throughput of a user randomly walking in the coverage area of the WLAN.

b) Compare your results with the minimum and maximum nominal data rates of the IEEE 802.11g. Explain the difference between your results.

Project 1: The RSS in IEEE 802.11:

There are several software tools (e.g., WirelessMon by PassMark), which can be used to gather information about APs in the proximity of an MS. Many tools come with the operating system on a laptop. These tools provide multiple features, but we are going to use them to log the received signal strength (RSS) from chosen APs at different locations and to compare these measurements with the IEEE 802.11 models. The following steps can be used to make an RSS measurement using these tools:

- Install a software tool for measurement of RSS (e.g., you can download wirelessmon.exe from http://www.passmark.com/products/wirelessmonitor.htm)
- On your laptop, set the software to monitor the AP of your choice; the APs can be distinguished from each other by their MAC addresses and sometimes their SSIDs.
- Modify the logging options of the software for recording the characteristics of an AP.
- Record the RSS readings from a specific AP (see a typical expected sample result in Figure P7.3).

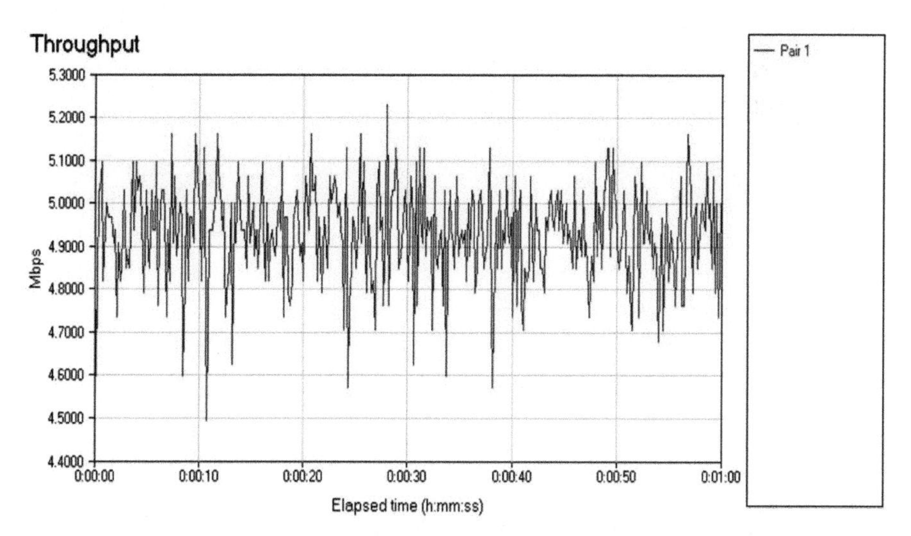

Figure P7.3 Throughput variations in one location for 802.11b.

a. Do "war driving" on a specific floor of your building, for which you have a floorplan schematic available, to find the exact location of the AP in that floor. Show the locations in the schematics of the building.
b. Select five different locations on the floor of your choice which are approximately 1, 5, 10, 20, and 30 m away from your AP of choice. Spread the points over the entire floor and mark them on your schematic floor plan. Determine the distance from the selected points to each of the AP locations on that floor.
c. Measure the RSS at each location for at least 1 minute. Calculate the *average* RSS received from each AP in each location and record them in a table, which relates the distance to the RSS from your target AP.
d. Use the table to generate a scatterplot of the average RSS (in dBm) vs. the distance (in logarithmic scale) for all APs in your target floor.
e. Find the best fit 802.11 model for your data.
f. Use www.speakeasy.net/speedtest to record the measured data rate in each of the five locations.
g. Explain the correlation among the throughput from speakeasy and the power and distance at each location.

Project 2: Coverage and Data Rate Performance of IEEE 802.11 WLANs

I. Modeling of the RSS

To develop a model for the coverage of the IEEE 802.11b/g WLANs, a group of undergraduate students at WPI measured the RSS in six locations on the third floor of the Atwater Kent Laboratory (AKL) at WPI, shown in Figure P7.4. After subtracting the RSS from the transmitted power recommended by the manufacturer, they calculated the path loss for all the points that are shown in Table P7.1.

To develop a model for the coverage of the WLANs, they used the simple distance-power gradient model:

$$L_p = L_0 + 10\alpha \log d$$

where d is the distance between the transmitter and the receiver, L_p is the path loss between the transmitter and the receiver, L_0 is the path loss at the first meter, and α is the distance-power gradient. One way to determine L_0 and α

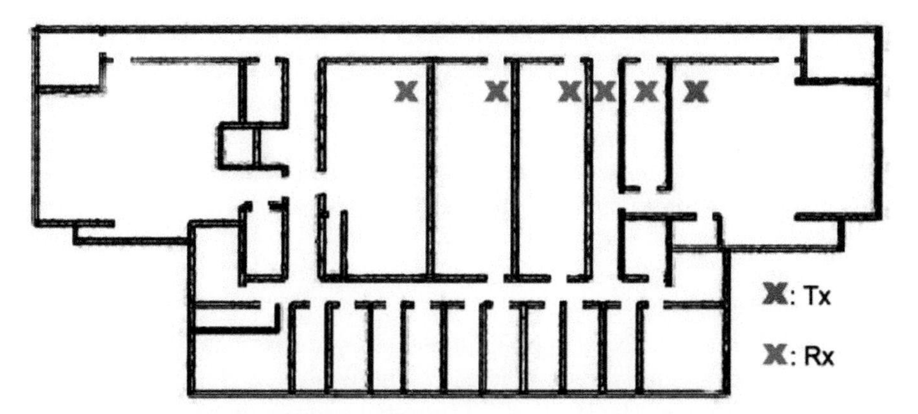

Figure P7.4 Location of the transmitter and first five locations of the receiver used for calculation of the RSS and path loss.

Table P7.1 Distance and the associated path loss for the experiment.

Distance (m)	*Number of Walls*	L_p *(dB)*
3	1	62.7
6.6	2	70
9.5	3	72.75
15	4	82.75
22.5	5	90
28.8	6	93

from the results of measurements is to plot the measured L_p vs. log d and find the best fit line to the results of measurements.

a. Use the results of measurements by the students to determine the distance-power gradient, α, and path loss at the first meter from the transmitter, L_0. In your report, provide the MATLAB code and the plot of the results, and the best-fit curve.
b. Manufacturers often provide similar measurement tables for typical indoor environments. Table P7.2 shows the RSS at different distances for open areas (an area without a wall), semi-open areas (typical office areas), and closed areas (harsher indoor environments) provided by Proxim, one of the manufacturers of WLAN products. Use the results

of measurements from the manufacturer and repeat part (a) for the three areas used by the manufacturer. Which of the measurement areas used by the manufacturer resembles the third floor of the AKL where the students took measurements? Assume that the transmitted power used for these measurements was 20 dBm. In your report, include the plots used for calculations of the distance-power gradient at different locations.

II. Coverage study

IEEE 802.11b/g WLANs support multiple data rates. As the distance between the transmitter and the receiver increases, the WLAN reduces its data rate to expand its coverage. The IEEE 802.11b/g standards recommend a set of data rates for the WLAN. The first column of Table P7.2 shows the four data rates supported by the IEEE 802.11b standard and the last column represents the required RSS to support these data rates. Table 7.3 shows the data rates and the RSS for the IEEE 802.11g provided by Cisco.

a. Plot the data rate versus coverage (staircase functions) for IEEE 802.11b WLANs for closed, open, and semi-open areas using Table P7.2 provided by Proxim. Discuss the coverage vs. data rate performance in different areas and relate them to the value of α in different areas, calculated in part I of the project.
b. Using α and L_0 found for the third floor of the Atwater Kent Labs, plot the data rate vs. coverage (staircase functions) for IEEE 802.11b and g WLANs operating in that area. Discuss the differences in data rate vs. coverage performance of 802.11b and g on the third floor of the Atwater Kent Labs.

Table P7.2 Data rate, distance in different areas, and the RSS for IEEE 802.11b (Source: Proxim).

Data Rate (Mbps)	*Closed area (m)*	*Semi-Open area (m)*	*Open area (m)*	*Signal Level (dBm)*
11	25	50	160	-82
5.5	35	70	270	-87
2	40	90	400	-91
1	50	115	550	-94

8

IEEE 802.15 for WSN and IoT

8.1 Introduction

As we described in Chapter 1, the air-interface of the wireless networks evolved around two distinctive paths, one led by the cellular telephone to connect handsets to the public switched telephone network (PSTN), evolved from circuit switched telephone network access to mobile devices, and the other around Wi-Fi for wireless access to the Internet to laptops and other devices. This evolution started in the early 1980s, when the public switched telephone was the most popular medium for communication and, by far, the largest sector of the telecommunication industry. Throughout this evolution, the habits of the world population for communicating with one another gradually shifted from telephone and PSTN toward the data applications supported by the Internet. In parallel to that, the Wi-Fi industry gained increasing importance. The air-interfaces designed for Wi-Fi data applications influenced the migration of the cellular air-interface from voice centric 2G time-division multiple access (TDMA) and 3G code-division multiple access (CDMA) to the data centric orthogonal frequency-division multiplexing (OFDM) designs with centralized scheduling for 4G long-term evolution (LTE) and beyond. Besides, the wireless personal area networking (WPAN) industry emerged to complement IEEE 802.11 Wi-Fi with the IEEE 802.15 Bluetooth and ZigBee technologies for short-range low energy wireless communications, sensor network, and the Internet of Things (IoT). The emergence of the IEEE 802.11ah,af for low energy began in parallel to these technologies. Another activity that emerged in the early 2000s in the IEEE 802.15 was initiated by ultra-wideband (UWB) technology for gigabit wireless local networking to increase the data rate of Wi-Fi beyond far existing 54 Mbps of the IEEE 802.11a,g. The invention of the multiple-input multiple-output (MIMO) technology opened another path in the IEEE 802.11 to increase the data with multiple streams in IEEE

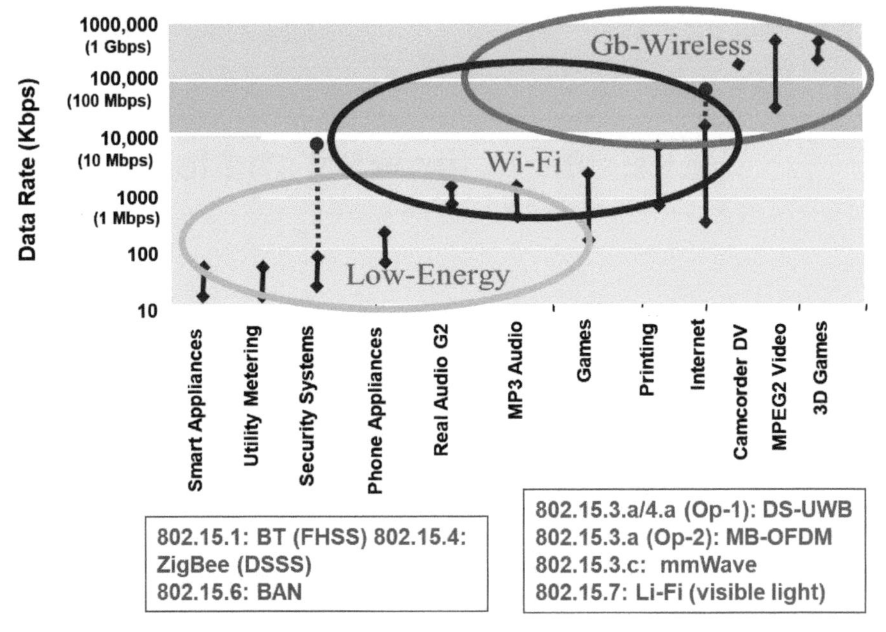

Figure 8.1 Applications, bandwidth requirements, and IEEE 802.15 for WPAN.

802.11n,g, and with mmWave in IEEE 802.11ad in parallel to these activities. Figure 8.1 illustrates the relationship between the data rate and several popular applications and how IEEE 802.15 standards evolved around them to complement the IEEE 802.11 and Wi-Fi technology.

The very first WPAN to appear in the literature was the BodyLAN that emerged from a Defense Advanced Research Projects Agency (DARPA) project in the mid-1990s [DEN96]. The BodyLAN was low power, small size, inexpensive, WPAN with modest data rates that could connect personal devices with a range of around 5 feet on and around a human being. Motivated by the BodyLAN project, a WPAN group was originally started in June 1997 as a part of the IEEE 802.11 standardization activity. In January 1998, the WPAN group published the original functionality requirements for WPANs. In May 1998, the development of Bluetooth was announced, and a Bluetooth special group was formed within the WPAN group [SIE00].

In March 1999, the IEEE 802.15 working group was approved as a separate group in the 802 community to handle WPAN standardization. At the time of this writing, the IEEE 802.15 WPAN group is a major standardization

Table 8.1 Summary of several IEEE 802.15 standards.

802.15.1 Classic Bluetooth v.1.0 (1998)
802.15.2 Co-existence (2000)
802.15.3a/4a UWB (2003-9)
802.15.4 ZigBee (2004)
802.15.4e/k Wide Area LE (2014)
802.15.6 BAN (2007-2011)
802.15.1 BLE v4.0-2 (2011-2017)
802.15.3c mmWave (2010)
802.15.7 Li-Fi (2015)

committee with several subcommittees for gigabit wireless and low energy sensor networking.

In the standardization for low power sensor networks, the IEEE 802.15 community has completed the very successful IEEE 802.15.1 Bluetooth and IEEE 802.15.4 (which specifies the PHY and MAC layers and works with ZigBee applications) standards. The IEEE 802.15.6 group was working on the emerging body area networks (BANs). The IEEE 802.15.a worked on UWB technology, the IEEE 802.15.c on mmWave, and IEEE 15.7 on Li-Fi using visible lights for wireless communications. Table 8.1 shows the chronology of popular IEEE 802.15 standards.

In this chapter, we address the technical aspects of the two most popular low power and lower data rate IEEE 802.15 WPAN technologies, Bluetooth and ZigBee, and we provide an overview of other interesting IEEE 802.15 standards for wireless sensor networks (WSN) and IoT. These technologies include WSN and IoT technologies covering large geographical areas, pulse transmission using UWB, mmWave, and visible light transmission (Li-Fi).

8.1.1 Wireless Technologies, Data Rate, and Power

In Chapter 4, we introduced optical wireless, spread spectrum, OFDM, MIMO, UWB, and mmWave technologies as major transmission technologies in wireless networks. Federal Communication Commission (FCC) regulates up to 300 GHz and the frequency of operation in optical wireless is in THz, way above these regulations. As a result, in theory, optical wireless communication can adopt bandwidths up to the infinity and bottleneck for data rate and power requirement is on implementation of devices to operate in these bands. In the rest of these technologies, UWB and mmWaves can occupy bandwidth on the orders of GHz, which allows the implementation of reasonable technologies with data rates much smaller than the bandwidth

$R_b << W$. This situation also holds in spread spectrum transmission. Shannon bounds, relating the data rate to the power requirements, help to have a broad overview of these technologies to classify them into logical categories. Figure 8.2 shows how the Shannon is instrumental in dividing the behavior of these technologies into three regions. In Figure 8.2(a), it divides single stream technologies into high data rate transmission techniques using quadrature amplitude modulation (QAM) in multi-carrier form, OFDM, which is the most popular in the IEEE 802.11 standards. In this region, the data rate is higher than the available bandwidth and they require signal to noise ratio (SNR) per bits much above one (0 dB). In classical information theory, transmission techniques operating in this region have restrictions on bandwidth allocation, but they can transmit a higher level of power. As a result, this region is also called bandwidth limited region with a high data rate and high transmission power. The lower region of Figure 8.2(a) belongs to transmission techniques with an SNR per bits of much below one and data rates that are much smaller than the available bandwidth. These are narrow pulse transmission techniques demanding UWB transmission bandwidth. Traditional UWB system was first studied in IEEE 802.15.3c in the early 2000s for UWB bands and later moved to mmWave transmission at 60 GHz. The difference between UWB and mmWave is the center frequency. Higher frequencies allow the implementation of MIMO systems with reasonable antenna array sizes to enable multiple streaming. For a single antenna system

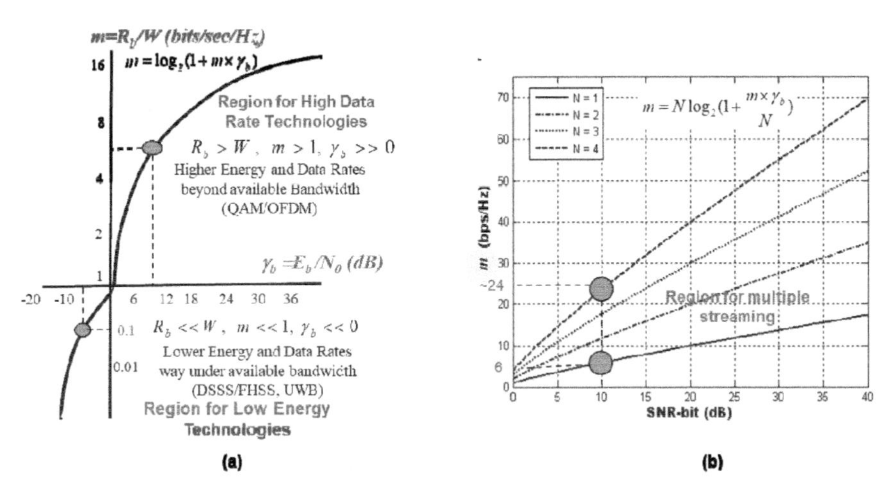

Figure 8.2 Shannon bounds for (a) high rate and low energy regions, (b) multiple streaming regions.

and from Shannon–Hartley's point of view, they can perceive the same. Wi-Fi and cellular networks use MIMO systems in a different number of antennas. In Wi-Fi devices, a 3-3 MIMO antenna system is popular and, in the cellular network at mmWave 8-8 and even higher rates is emerging in massive MIMO systems for 5G and 6G cellular. Figure 8.2(b) shows the Shannon–Hartley bounds for this region. Using these technologies, we see that IEEE 802.11a, g has achieved data rate of 54 Mbps for a single stream with 20 MHz bandwidth and 64-QAM. The IEEE 802.11n, ac using 4-8 MIMO streams and bounding multiple 20 MHz channels have achieved Gbps wireless at 2.4 and 5.2 GHz competing IEEE 802.11ad at mmWave with UWB pulse transmission. The frequency hopping spread spectrum (FHSS) and direct sequence spread spectrum (DSSS) used in IEEE 802.15 and Bluetooth and ZigBee technologies operate in the low energy region with low data rates to sustain the long battery life needed for sensor networks.

8.1.2 Spread Spectrum for WSN and IoT

Spread spectrum technology, described in Section 4.5.1, is widely used in low energy WSN and IoT IEEE 802.15 standards. Before we begin discussing the details of these standards, it is beneficial to review this technology and its benefits for low energy communications to strengthen the lifetime of batteries beyond those of the IEEE 802.11. Figure 8.3 shows a summary reminder of (a) DSSS and (b) FHSS technologies in time- and in the frequency-domain as well as their characteristics defined by Shannon–Hartley bounds. In the spread spectrum, the data rate is well below the available bandwidth and the bandwidth. The ratio of bandwidth to the data rate is the processing gain (PG) of the system:

$$R_b << W; \;\; \text{PG} = \frac{W}{R_b} >> 1; \;\; \gamma_b << 0 \;\; (\text{dB}). \tag{8.1}$$

A large processing gain reduces the SNR requirement to very small values. We can benefit from this feature to increase the battery life and coverage of an IoT device or to overlay a low speed spread spectrum feature over an existing application in a band.

Battery Life and Coverage of Spread Spectrum Devices:
For m-bits per symbol, using Shannon–Hartley bound, we have

$$\frac{R_b}{W} = \log_2 [1 + \text{SNR}] = \log_2 \left[1 + \frac{E_r(m)}{N_0} \right], \tag{8.2}$$

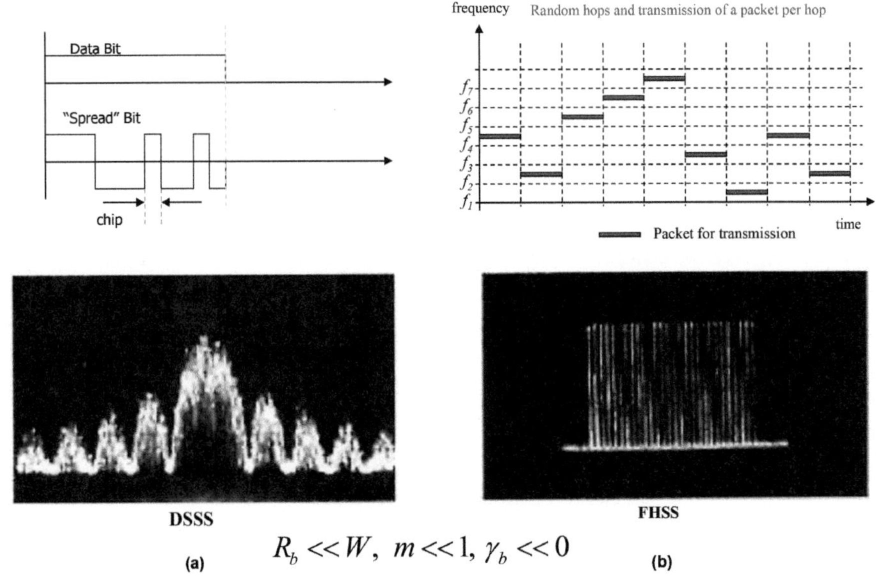

Figure 8.3 Review of Spread spectrum technologies in time- and in frequency domain, (a) DSSS, (b) FHSS

where $E_r(m)$ is the average received energy per symbol and m is the number of bits per symbol. The received energy is related to the transmitted average energy per symbol, $E_t(m)$, by

$$E_r(m) = \frac{E_0}{d^\alpha} E_t(m), \tag{8.3}$$

where E_0 is the received energy in unit distance from the transmitter. Substituting Equation (8.3) into Equation (8.2), we have

$$\begin{cases} m = \log_2(1 + \frac{E_0 E_t(m)}{d^\alpha N_0}) \\ E_t(m) = \frac{d^\alpha N_0}{E_0}(2^m - 1) \end{cases}. \tag{8.4}$$

Coverage for a multi-rate wireless device is calculated from the coverage of its lowest data rate. Finding the power of the two devices needed for the same coverage can lead us to find the difference in their battery life.

Example 8.1: Battery Life of Legacy 802.11 and 802.11g:

For the legacy DSSS IEEE 802.11, bandwidth is $W = 26$ MHzand minimum maximum data rate is $R_b = 1$ Mbps; therefore, $m = R_b/W = 1/26 = 0.04$.

The minimum rate of IEEE 802.11g is 6 Mbps and the bandwidth is 20 MHz, resulting in $m = 6/20 = 0.3$. For the same coverage for the lowest data rates, both devices need to have the same received power in the first meter. Then, using the lower part of Equation (8.5), the ratio of the transmitted powers is

$$\frac{E_t(0.3)}{E_t(0.04)} = \frac{2^{0.3} - 1}{2^{0.04} - 1} = \frac{0.23}{0.03} = 7.7.$$

Therefore, the same coverage for the 802.11g needs close to eight times more power consumption making battery life for the legacy 802.11 close to eight times more than 802.11g. This is at the expense of supporting a higher maximum data rate of 54 Mbps and a higher average spatial data rate in coverage (spatial throughput).

Example 8.2: Coverage of Legacy 802.11 and 802.11g:
Following Example 8.1, for $m = 0.04$ for legacy 802.11, we have

$$\gamma_b = \frac{1}{m}(2^m - 1) = \frac{1}{0.04}(2^{0.04} - 1) = 0.03\ (-15.2\,\text{dB})\,.$$

For minimum data rate of the 802.11g, we have $m = 0.3$

$$\gamma_b = \frac{1}{0.3}(2^{0.3} - 1) = 0.23\ (-6.4\,\text{dB})\,..$$

The difference is 8.8 dB and for that

$$10\alpha \log(d_1/d_2) = 8.8 \Rightarrow d_1/d_2 = 10^{8.8/10\alpha}.$$

For $\alpha = 2$, with the same transmitted power, the coverage ratio is close to three times (2.8) and for $\alpha = 3.5$, it is close to two times (1.8).

Examples (8.1) and (8.2) show how processing gain of the DSSS increases the battery life and coverage of spread spectrum technology at the expense of reduced data rate. The processing gain of the legacy DSSS IEEE 802.11 was 11; by increasing this value, we can extend the life of the battery or the coverage of a device. As we will see in the following sections, WSNs, and IoT device standards and technologies often use the spread spectrum technology to achieve these goals because in these devices, life of the battery or coverage of the technology is more important than the maximum achievable data rate. This discussion clarifies why, at the same time, Wi-Fi devices migrated from spread spectrum to OFDM and MIMO; these technologies found their way in Bluetooth and ZigBee technologies based on the IEEE 802.15 technology.

Spread Spectrum Overlay Feature:
In Section 5.24, we discussed the format flexibility of the spread spectrum technology to accommodate multi-rate applications to share a spectrum with CDMA technology. This feature can be generalized to overlay a spread spectrum system over another system. Since spread spectrum with high processing gain can operate under a very small value of signal well below one, while traditional wireless communication devices operate with an SNR that is much higher than one, we can overlay a spread spectrum transmission over a traditional transmission system. We keep the power of the spread spectrum low enough that the SNR of the traditional device stays at its desirable value and we keep the processing gain of the spread spectrum system high enough to operate when the traditional system is the source of the noise. An example will further clarify this technique.

Example 8.3: DSSS Over Broadcast TV:
Figure 8.4 shows the spectrum for a digital broadcast TV and a DSSS signal overlaying that band. The digital TV has an acceptable quality if the received SNR is 10 dB (10) that is satisfied when we set the DSSS power 10 dB below the power of the TV signal. Then, the SNR for the DSSS system is -10 dB (0.1). The SNR of the DSSS depends on its processing gain, and without spread spectrum coding, the SNR of the overlay signal is 0.1; if the processing gain of the DSSS is at least 100, this signal to noise after processing at the receiver becomes $0.1 \times 100 = 10$ (10 dB) that is acceptable for proper operation of the overlay channel.

If the bandwidth of the TV station is $W = 5.5$ MHz, according to Shannon–Hartley, the maximum data rate for the overlay digital transmission is

$$R_b = W \log_2 (1 + \mathrm{SNR}) \Rightarrow R_b = 5.5\,(\mathrm{MHz}) \log_2 (1 + 0.1) = 756\,\mathrm{Kbps}.$$

With a processing gain of 100 for DSSS coding, this data rate reduces to 756/100 = 7.56 Kbps.

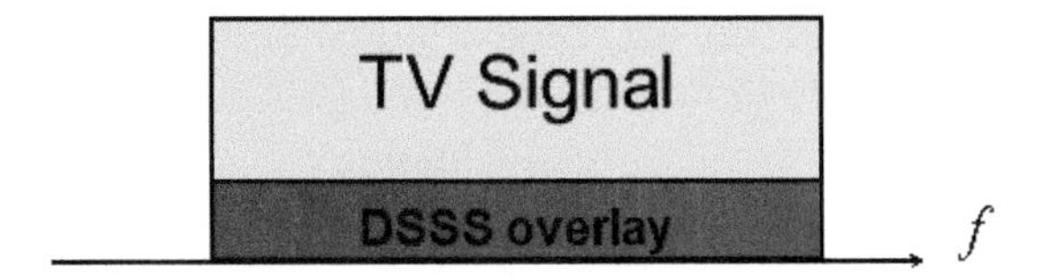

Figure 8.4 Overlay of a DSSS signal over a broadcast TV channel. DSSS signal is noise for TV and TV is noise for DSSS.

Example 8.3 shows how we can implement an overlay channel with low data rate over an existing broadcast TV channel. This channel can support additional features such as adding captions to the TV signal or any other sort of low speed overlay data. Indeed, this is an extension of reverse channel CDMA systems. In CDMA, codes separating the channels are orthogonal; more recently, the concept of non-orthogonal multiple access (NOMA) benefiting from the same concept is becoming popular for 5G/6G cellular networks [DIN16].

8.2 IEEE 802.15.1 and the Legacy Bluetooth

Bluetooth and ZigBee technologies draw upon the original spread spectrum based air-interface first implemented with the legacy IEEE 802.11 standard. Although the air-interface for IEEE 802.11 wireless local area networks (WLANs) has since evolved to use OFDM and MIMO technology, legacy spread spectrum transmission has become the technology of choice for low-power WPANs in the IEEE 802.15 standards. This is not incidental because, as we discussed in Chapter 3, spread spectrum technology is the technology of choice for low power and low data rate wireless air-interface design. At the MAC layer, Bluetooth was originally designed with the support of companies designing equipment for cellular telephony. It filled the void mostly for WPAN voice-oriented applications. ZigBee followed the IEEE 802.11 MAC as well as the PHY layer and emerged as the low-power WPAN technology for sensor data applications. In this chapter, we explain these two technologies and how they relate to the legacy IEEE 802.11 technology. This discussion provides a good overview of the centralized versus random access for voice and data applications in the personal area networks designed for covering only a few meters. The examples in the discussion also include the details of FHSS and DSSS transmission techniques in a WPAN environment.

Bluetooth is an open specification for short-range wireless voice and data communications. It was originally developed as a cable replacement in personal area networks to operate all over the world. In 1994, the initial study for the development of Bluetooth started at Ericsson in Sweden. In 1998, companies such as Ericsson, Nokia, IBM, Toshiba, and Intel formed a special interest group to expand the concept and develop a standard under the IEEE 802.15 banner. In 1999, the first specification, v1.0b, was released and then accepted as the IEEE 802.15 WPAN standard for 1 Mbps networks. Today, Bluetooth has penetrated the huge smartphone market as well as numerous consumer devices.

The story of the origin of the name Bluetooth is interesting and worth mentioning. "Bluetooth" was the nickname of Harald Blaatand, 940–981 A.D., King of Denmark and Norway. When Bluetooth was introduced to the public, a stone carving, shown in Figure 8.5, claimed to be erected from Harald Blaatand's capital city Jelling was also presented [BLU00]. This strange carving was interpreted as Bluetooth connecting a cellular phone and a wireless notepad in his hands. This picture was used to symbolize the vision of using "Bluetooth" to connect personal computing and communication devices. Bluetooth, the king, was also known as a peacemaker and a person who brought Christianity to Scandinavians to harmonize their beliefs with the rest of Europe. That fact was used to symbolize the need for harmony among manufacturers of WPANs around the world and to support the growth of the WPAN industry.

Bluetooth was the first popular technology for short-range *ad-hoc* networking that was designed for integrated voice and data applications. Unlike WLANs, Bluetooth does not strive for very high data rates. It maintains effective data rates under 1 Mbps, but it has an embedded

Figure 8.5 Image of the King Harald Blaatand, the Bluetooth, on the stone connecting a computer with a cellphone used to market the technology at its inception in the late 1990s.

architectural design that is suitable to support voice applications with the transmission of centrally controlled shorter packets. Bluetooth 3 allows the use of co-located Wi-Fi signals to increase data rates (when needed).

Since 2011, Bluetooth 4.0 has specified support for devices such as pedometers and heart-rate monitors with extremely low energy consumption, expanding the potential applications where it can be employed. The batteries in such devices are expected to last for months without replacement or recharging.

The Bluetooth group originally considered three basic application scenarios that are shown in Figure 8.6 [BLU00]. The first application scenario, shown in Figure 8.6(a), was for "wire replacement," to connect a personal computer or laptop to its keyboard, mouse, microphone, and notepad. As the name of the scenario indicates, it avoids multiple short-range wiring surrounding today's personal computing devices. The second application scenario, shown in Figure 8.6(b), was for *ad hoc* networking of several different devices in very short range of each other, such as in a conference room. As we saw in Chapter 7, WLAN standards and products also commonly consider this scenario. The third scenario, shown in Figure 8.6(c) is to use Bluetooth as an access point to wide-area voice and data services provided by the cellular networks, wired connections, or satellite links. The IEEE 802.11 community also considers this overall concept of the access point. However, the Bluetooth access point is used in an

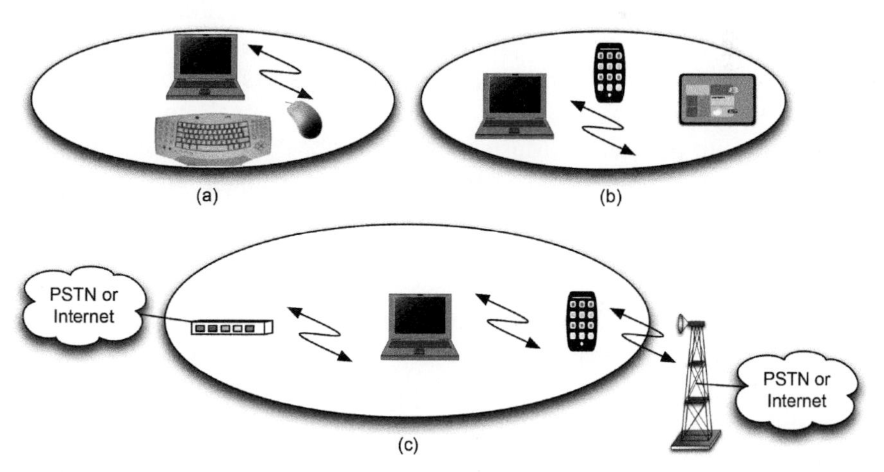

Figure 8.6 Bluetooth application scenarios. (a) Cable replacement. (b) *Ad-hoc* personal network. (c) Integrated access point.

integrated manner to connect to both voice and data backbone infrastructures. Today, Bluetooth is mostly used for the first scenario and increasingly for connecting the so-called "smart" devices such as fitness monitors to smartphones. It is also used to stream music to speakers or headphones from devices such as computers or MP3 players.

8.2.1 Overall Architecture

The topology of the Bluetooth is referred to as a *scattered ad-hoc topology* that is illustrated in Figure 8.7. In a scattered *ad-hoc* environment, several small networks, each supporting a few terminals, co-exist or possibly interoperate with one another. To implement such a network, we need a plug-and-play environment. The network should be self-configurable, providing an easy mechanism to form a small new network, as well as participation in an existing small network. To implement that environment, the system should be capable of providing different states for connecting to the network. The terminals should have options to associate with multiple networks at the same time. The access method should allow the formation of small independent *ad-hoc* connections as well as the possibility of interacting with large voice and data networks considered by Bluetooth.

To accommodate these features, the Bluetooth specification defines a small cell (similar to a basic service area in IEEE 802.11) as a *piconet* and identifies four states, Master "M," Slave "S," stand-by "SB," and Parked/Hold or Parked "P" for a Bluetooth enabled terminal. The mobile terminal that initiates a connection is a Master device in that piconet. The devices connecting to the Master are called Slaves. As shown in Figure 8.7, the Bluetooth topology, however, allows Slave terminals to participate in more than one piconet. A Master terminal in the Bluetooth can handle 7 simultaneous and up to 200 active slaves in a piconet. If access is not available, a terminal can enter the SB mode waiting to join the piconet later. A radio can also be in a "Parked" or a low power connection. In the parked mode, the terminal releases its MAC address, while in the SB state, it keeps its MAC address. Up to 10 piconets can operate in one area [BLU00]. Also note that in Figure 8.7, the coverage of the two master devices are different since Bluetooth allows for different classes of devices with different transmit powers. Bluetooth specifications have selected the unlicensed ISM bands at 2.4 GHz for operation. The advantage is the worldwide availability of the bands and the disadvantage is the existence of other users, IEEE 802.11 products in the same band. In the early

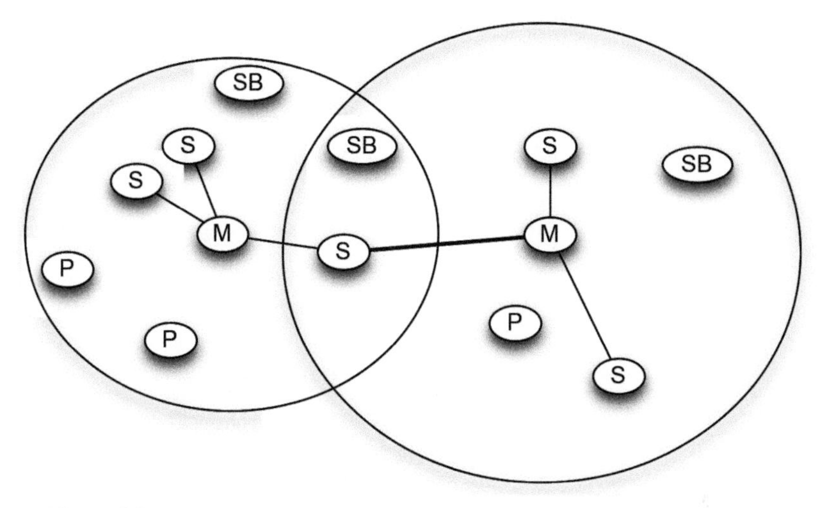

Figure 8.7 Bluetooth's scattered *ad-hoc* topology and the concept of *piconet.*

2000s, a sub-committee of IEEE 802.15 worked on the interference issues related to the Bluetooth and IEEE 802.11 and 11b that we have considered in Chapter 5.

8.2.2 Protocol Stack

One of the distinct features of Bluetooth is that it provides a complete protocol stack that allows different applications to communicate over a variety of devices. Other wireless local networks, such as the IEEE 802.11, usually specify only the lowest three lower layers for communications. The protocol stack for voice, data, and control signaling in Bluetooth is shown in Figure 8.8 [HAA00]. The *RF layer* specifies the radio modem used for transmission and reception of the information. The *Baseband layer* specifies the link control at bit and packet level. It specifies coding and encryption for packet assembly and how the frequency hopping operation should work. The *Link Management Protocol* configures the links to other devices by providing for authentication and encryption, state of units in the piconet, power modes, traffic scheduling, and packet format. The *logical link control and adaptation protocol* provide connection-oriented and connectionless data services to the upper layer protocols. These services include protocol multiplexing, segmentation and reassembly, and group abstractions for data packets up to 64 Kb in length. The audio signal is directly transferred from the application to the Baseband layer. Also, the link management protocol and

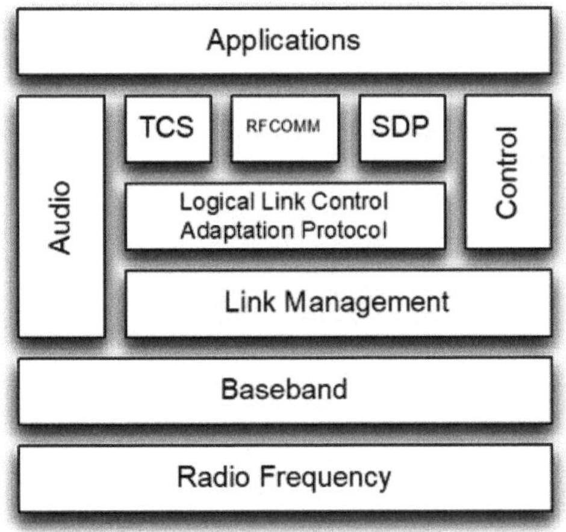

Figure 8.8 Protocol stack of the Bluetooth.

the application exchange control messages interact to prepare the physical transport for an application. There are three other protocols above the logical link control adaptation protocol. The *service discovery protocol* finds the characteristics of the services and connects two or more Bluetooth devices to support service such as streaming music, faxing, printing, teleconferencing, or e-commerce facilities. *Telephony control protocol* defines the call control signaling and mobility management for the establishment of speech for telephony applications. Using these protocols, legacy telecommunication applications can be developed.

Example 8.4: Telephony Control Protocol in Bluetooth:
Figure 8.9 shows the protocol stack for implementation of the cordless telephone application. The audio signal is directly transferred to the Baseband layer while service discovery protocol and telephony control protocol operating over LLC application protocol and link management protocol handle signaling and connection management.

The *radio frequency communication (RFCOMM)* is a "cable replacement" protocol that emulates the standard short-range wired serial interface conversion data signals over Bluetooth baseband. Using this interface

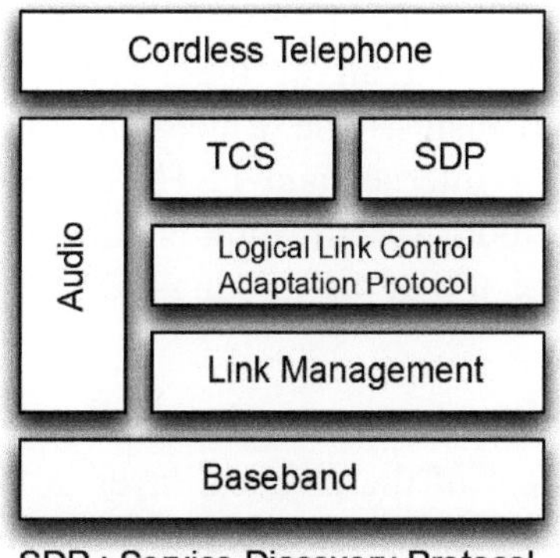

Figure 8.9 Protocol stack for implementation of cordless telephone over Bluetooth.

protocol, several non-Bluetooth specific protocols can be implemented on the Bluetooth devices to support legacy applications that use serial interfaces.

Example 8.5: Light-Weight Applications in Bluetooth:
Figure 8.10 shows the implementation of a vCard (digital business card) transfer application. This application protocol runs over object exchange protocol that is carried by the RFCOMM protocol in the Bluetooth protocol stack. As shown in Figure 8.9, the resulting packets are then passed through the logical link control adaptation protocol and then Baseband before actual radio frequency "over-the-air" transmission. The protocols defined in Figure 8.7 are either developed by the Bluetooth group exclusively or

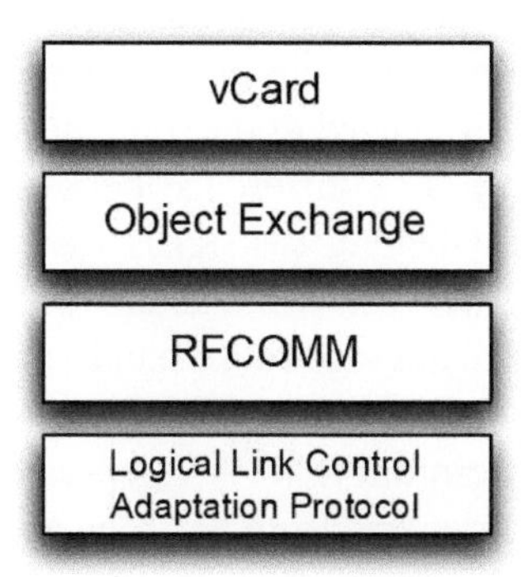

Figure 8.10 Protocol stack for implementation of vCard exchange over Bluetooth.

are modified versions of existing protocols. Different applications may use different protocol stacks, but all of them share the same physical and data link control mechanisms. The Bluetooth special interest group has adopted several popular Internet and PSTN existing protocols within its specification. The Bluetooth specification is itself open and other protocols can be accommodated on top of the existing protocol stack.

8.2.3 Physical Layer

The traditional equivalent of the physical layer in the case of Bluetooth is embedded in the RF and Baseband layers of the Bluetooth protocol stack. The physical connection of Bluetooth uses an FHSS modem with a nominal antenna power of 0 dBm (around 10 meter coverage) that has an option to operate at 20 dBm (around 100 meter coverage). The low power version of the Bluetooth provides reasonable coverage for its popular applications as cable replacement and guaranteed modest interference with 802.11 devices, which operate in the same frequency bands.

Like the 1 Mbps option of the IEEE 802.11 FHSS standard, the Bluetooth specification uses a two level Gaussian FSK modem with a transmission rate of 1 Mbps that hops over 79 channels in the unlicensed bands starting at 2.402 GHz and stopping at 2.480 GHz. As explained in Chapter 2, the rms delay spread in indoor areas for short-range coverage is under 100 ns. With that, the coherence bandwidth of the channel is around $\frac{0.1}{100 \times 10^{-9}} = 1$ MHz. Therefore, for a transmission rate at 1 MSps, the channel frequency response is relatively flat. The transmitted waveforms preserve their shape and we do not need complex signal processing techniques to equalize the channel. The two-level Gaussian FSK modem allows the implementation of simple non-coherent detection using frequency demodulators. A more complex version of this modem is one that uses Gaussian minimum shift keying (GMSK) as the modulation scheme. GMSK is used as the transmission technique in the GSM standard (see Chapter 10). The difference between frequency shift keying (FSK) and minimum shift keying (MSK) is that the two frequencies used for data transmission in MSK are twice as close as the two tones separation in FSK. Therefore, GMSK is twice more bandwidth efficient than Gaussian frequency shift keying (GFSK), but GFSK is implemented with simpler and more power-efficient circuitry. WPANs operate in the unlicensed bands, where one can access a wider spectrum and they are designed to support low-power *ad-hoc* sensor networking. In such environments, GFSK is a better solution than GMSK, used in GSM.

Although the transmission technique of the base FHSS physical layer in the original IEEE 802.11 and Bluetooth are the same, the hopping rate and pattern, and number of hops used in Bluetooth are different from those in IEEE 802.11. The Bluetooth frequency hopping rate is 1600 hops per second (625 μs dwell time) as compared to the 2.5 hops per second (400 ms dwell time) system adopted by IEEE 802.11. As we described in Chapter 2, the Doppler spread in most indoor environments is around 5 Hz. With this value of Doppler spread, the coherence time of the channel is around 200 ms. Since the slow FHSS systems used by Bluetooth and IEEE 802.11 send a packet per hop, the frame format of Bluetooth comprises much shorter packets with respect to the coherence time of the channel, which is better suited for voice-oriented networks, where the network needs to avoid retransmissions for real-time streaming of voice conversations.

The Bluetooth specification assigns a specific frequency hopping pattern for each piconet. This pattern is determined by the piconet identity and master clock phase residing in the Master terminal in the piconet. Figure 8.11 illustrates the essence of the frequency hopping strategy in Bluetooth. The overall hopping pattern is divided into 32 hop segments. The 32-hop pseudorandom hopping pattern segment is generated based on the master identity and clock phase. The 79 frequency hops at the ISM bands are arranged in odd and even classes. Each 32-hop sequence starts at some point in the spectrum and hops over the pattern that covers 64 MHz because it hops either on odd or even frequencies. After completion of each segment, the sequence is altered, and the segment is shifted 16 frequencies in the forward direction. The 32 hops are concatenated and the random selection of the odd or even index is changed for each new segment. This way, segments slide through the carrier list to maintain the average durations for which each frequency is used close to each other (i.e., like a uniform probability distribution). A change of the clock or identity of the piconet will change the frequency hopping sequence and segment mapping. This allows different piconets to operate in the same vicinity with different sets of (pseudo)random frequency hopping sequences. These frequency hopping sequences are not orthogonal to one another, but they are randomized against each other. With 79 hops, it is difficult to find many orthogonal sequences anyway [HAA00].

To protect the integrity of the transmitted data, Bluetooth uses two error-correction schemes in the baseband controllers. A forward error correcting code is always applied to the header information, and, if needed, it is extended to the payload data for the voice packets. The optional coding of the payload in data applications reduces the number of retransmissions that is desirable

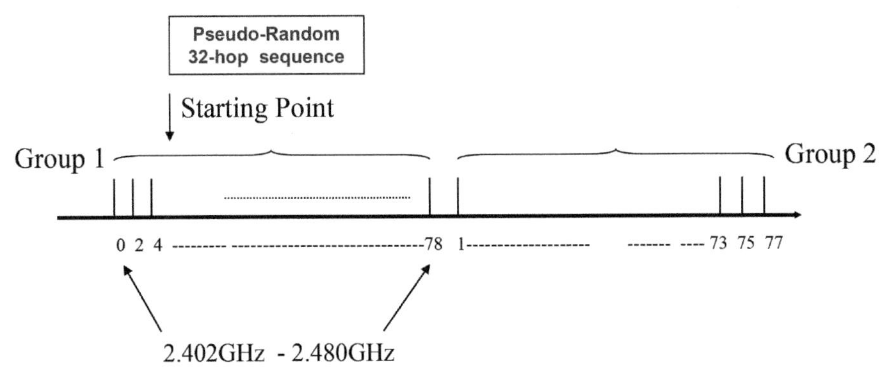

Figure 8.11 The hopping sequence mechanism in Bluetooth.

to increase the throughput. Coding of the payload for voice applications increases the integrity of the real-time streaming packets that will improve the quality of the service for the users. Coding is always applied to the header because header information is short and important. The detection of an error in the header allows a fast request for retransmission and for voice payloads, a fast decision for keeping or dropping a corrupted packet. In general, the flexibility of optionally using forward error correction for the payload provides an option to avoid overhead in favor of increased throughput when the channel is good and error-free. An unnumbered automatic repeat request scheme is also applied by the baseband layer for data packets, in which the recipient acknowledges the received data. For data transmission to be acknowledged, both the header error check and the payload check, if applied, must indicate a "no error" condition. These functionalities implemented in the baseband layer of the Bluetooth protocol stack are often implemented in the data link layer of traditional protocol reference models for networks.

8.2.4 MAC Mechanism

Although the modulation technique and frequency of operation of the Bluetooth radio system closely follow that of the FHSS 802.11, the MAC mechanism in the Bluetooth is widely different from the 802.11. The Bluetooth access mechanism is a voice-oriented innovative system that is neither identical to the data-oriented carrier sense multiple access with collision avoidance (CSMA/CA), used in the IEEE 802.11 WLANs (see Chapter 7), nor voice-oriented CDMA or TDMA access methods used in cellular networks (see Chapters 10 and 11), and yet has elements that are somehow related to these access methods.

The medium access mechanism of the Bluetooth is an FHSS/CDMA/TDD system that employs *polling* to establish the link. The relatively rapid hopping of 1600 frequency hops per second allows short time slots of 625 μs (625 bits at 1 Mbps) for the transmission of one packet that allows a better performance in the presence of interference. The medium access in the case of Bluetooth is, in some respect, a CDMA system that is implemented using FHSS. In the Bluetooth CDMA, each *piconet* has its own spreading sequence, while in the traditional DSSS/CDMA system used for digital cellular systems, each *user's* link is identified by a different spreading code. DSSS/CDMA has not been selected for Bluetooth because DSSS/CDMA needs a centralized power control (for protection against near–far effects) that is not possible in a scattered *ad-hoc* topology with many device classes as envisioned for Bluetooth applications. Without the need for centralized power control for CDMA operation, the FHSS/CDMA in Bluetooth allows tens of piconets to overlap in the same area providing an effective throughput that is much larger than 1 Mbps.

As we discussed in Chapter 7, the FHSS version of IEEE 802.11 operates over the same 79 hops as Bluetooth with only three sets of hopping patterns. The throughput of the Bluetooth FHSS/CDMA system, however, is less than the 79 Mbps that could be achieved in a coordinated FDM or OFDM system employed as in 802.11a/g. In Bluetooth, the FHSS/CDMA is selected over simple FDM or OFDM because ISM bands at 2.4 GHz only allow spread spectrum technology. The access method in each piconet of Bluetooth is based on TDMA/TDD. The TDMA format allows multiple voice and data terminals to participate in a piconet. Duplexing in time (TDD) eliminates the cross-talk between the transmitter and the receiver, allowing a single chip implementation in which a radio alternates between transmitter and receiver modes. To share the medium among a larger number of terminals, in each slot, the "Master" device decides and *polls* a "Slave" device allowing it to transmit. Polling is used rather than contention access methods because contention-based access creates excessive overhead for short packets (625 bits). Recall from Chapter 4 that contention-based access requires waiting times and back-off periods, which may be longer than the 625 μs slot used in Bluetooth.

8.2.5 Frame Formats

The Bluetooth packet format is based on one packet per hop and a basic one-slot packet that is 625 μs long. The packet can be extended to three slots

(1875 μs) and five slots (3125 μs). This frame format and the FHSS/TDMA/TDD access mechanism allow a "Master" terminal to poll multiple "Slave" terminals at different data rates for voice and data applications in the piconet.

Example 8.6: Operation of Piconets:

Figure 8.12 illustrates several examples of Bluetooth frame formats for operation in a piconet. In Figure 8.12(a), a Master terminal (M) is communicating with three Slave terminals (S1–S3). The TDMA/TDD format allows simultaneous connectivity to the three terminals assigning 625 μs (equivalent to 625 bits at 1 Mbps) slots for transmission and a time gap between the two packets in each direction. Terminals may run different applications (voice or data at different rates), but transmissions should occur on one of the one-slot detailed packet formats that are specified by the Bluetooth group standardization committee. The time gap is specified at 200 μs to allow a terminal to switch from transmitter to receiver mode for the TDD operation [HAA00]. Figure 8.12(b) shows an asymmetric communication in which the Master uses a higher speed three-slot link while the Slave operates at a lower rate with one-slot packets. Figure 8.12(c) represents a symmetric higher speed three-slot communication link and Figure 8.12(d) an asymmetric high-speed five-slot Master-to-Slave link with a lower speed one-slot link from the Slave to the Master.

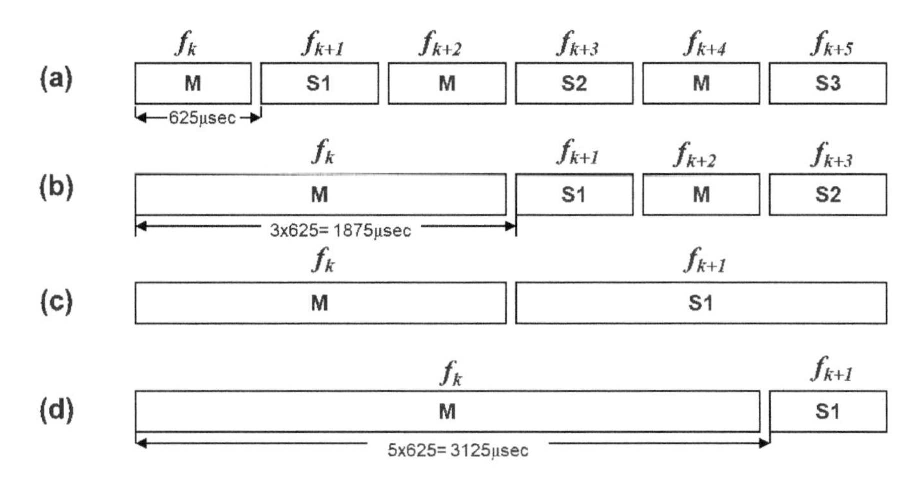

Figure 8.12 TDMA/TDD multi-slot packet formats in Bluetooth. (a) One-slot packets. (b) Asymmetric three slots. (c) Symmetric three slots (1875 μs). (d) Asymmetric five slots (3125 μs).

The overall packet structure of the Bluetooth is shown in Figure 8.13. There are 74 bits for an access code field, 54 bits for the header field, and up to 2744 bits for different payloads that can be of five slots. In IEEE 802.11 FHSS packets, the preamble and header of the physical layer were 96 and 32 bits, respectively, while the payload could be $4096 \times 8 = 32{,}768$ bits. The size of the overhead is in the same range, but the maximum payload of 802.11 is at least an order of magnitude larger. Bluetooth uses more flexible shorter packets for better performance in fading, but these gains are at the expense of a higher percentage of overhead that reduces the throughput.

The access code field consists of a 4-bit preamble and a 4-bit trailer plus a 64-bit synchronization pseudo noise (PN)-sequence with a large number of codes with good autocorrelation and cross-correlation properties (see Chapter 11 for a discussion of autocorrelation and cross-correlation). The 48-bit IEEE MAC address unique to every Bluetooth device is used as the seed to derive PN-sequence for hopping frequencies of the device. There are four different types of access codes. The first type identifies a Master terminal and its piconet address. The second type of access code specifies a Slave identity that is used to page a specific Slave. The third type is a fixed access code reserved for the inquiry process that will be explained later. The fourth type is the dedicated access code that is reserved to identify specific sets of devices such as printers or cellular phones.

As shown in Figure 8.13, the header field has 18-bits that are repeated three times to increase the reliability. The 18-bit starts with a 3-bits Slave address identifier, 4-bits packet type, 3-bits for status reports, and 8-bit error check parity for the header. The 3-bit Slave address allows addressing the seven possible active Masters in a piconet. The 4-bit packet type allows 16 choices for different grade voice services, data services at different rates, and control packets. The 3-bit status reports are used to flag the overflow of the terminal with information, acknowledgment of successful transmission of a packet, and sequencing to differentiate the sent and resent packets.

The Bluetooth special interest group specifies different payloads and associated packet type codes that allow the implementation of several voice and data services. Different master–slave pairs in a piconet can use different packet types, and the packet type may change arbitrarily during a communication session. The 4-bit packet type identifies 16 different packet formats for the payloads of the Bluetooth packets. Six of these payload formats are primarily used for packet data communications. Three of the payload formats are primarily used for voice communications. One is an

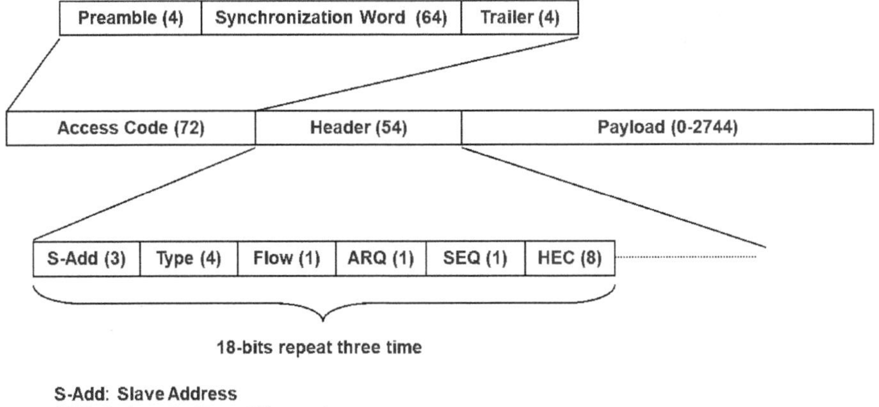

Figure 8.13 Overall frame format of the Bluetooth packets.

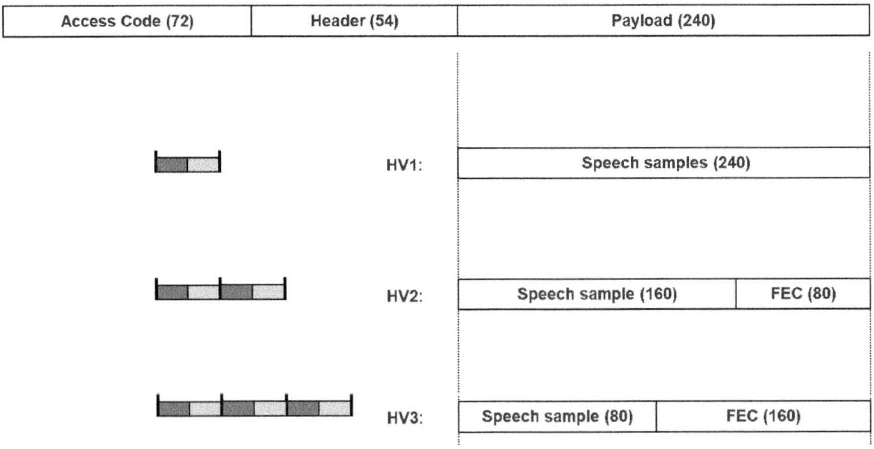

Access Code (72) | Header (54) | Payload (240)

HV1: Speech samples (240)

HV2: Speech sample (160) | FEC (80)

HV3: Speech sample (80) | FEC (160)

Figure 8.14 Three options for one-slot voice packet frame formats.

integrated voice and data packet and four are control packets common for both voice and data links.

The three voice packets, shown in Figure 8.14, are packets with different grades of protection, numbered as 1, 2, and 3, to designate the level of quality. These voice packets are all single-slot packets, the length of the payload being fixed at 240 bits. They do not use the status report bits because voice packets are sensitive to delay but not to a modest packet loss rate of less than 1%. So, they do not need to be retransmitted. However, voice is a real-time application demanding a steady data rate. As a result, voice packets are transmitted over reserved periodic duplex intervals to support 64 Kbps per voice conversation that are commonly used in the PSTN to carry digital voice. The lowest grade 1 voice packet uses all 240 bits for the user voice samples, grade 2 uses 160 bits for user voice samples and 80 bits of parity for a 1/3 forward error correction code, and grade 3 uses 80 bits of user voice samples and 160 bits of parity for a 2/3 forward error correcting code. To keep the data rate for voice samples at 64 Kbps, the grades 1, 2, and 3 packets in each direction are sent every six, four, and two slots, respectively.

Example 8.7: Data Rate of High Quality Voice Packets:
The grade 1 voice packets are 240 bits long, and they are one-slot packets sent every six slots. Slots carrying a packet are created at the rate of 1600 slots/s. Therefore, the effective data rate to support grade 1 voice is: $\frac{1600\ (\frac{\text{slots}}{\text{s}})}{6\ (\text{slots})} \times 240\ (\text{bits}) = 64$ Kbps.

Example 8.8: Integration of Voice and Data Packets:
Figure 8.15 illustrates an example of integrating grade 1 voice packets with different formats of data in a piconet. Every six slots, a two-way packet of voice is exchanged to support a symmetric grade 1 voice channel and the remaining four slots are used for the transmission of data packets in different symmetric and asymmetric formats.

Grade 1 voice occupies two slots of the six slots available for each piconet to transmit 240 bits per six slots that are equivalent to a 64 Kbps steady data flow. To support the needed 64 Kbps link for a voice connection under grades 2 and 3, with 160 and 80 bits per slot, we need four and six slots every six slots, respectively. Therefore, grade 3 voice occupies all of the resources of a

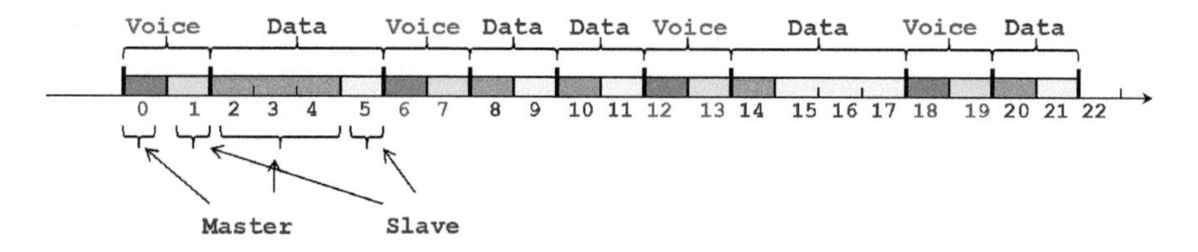

Figure 8.15 Integration of voice and data in Bluetooth medium access.

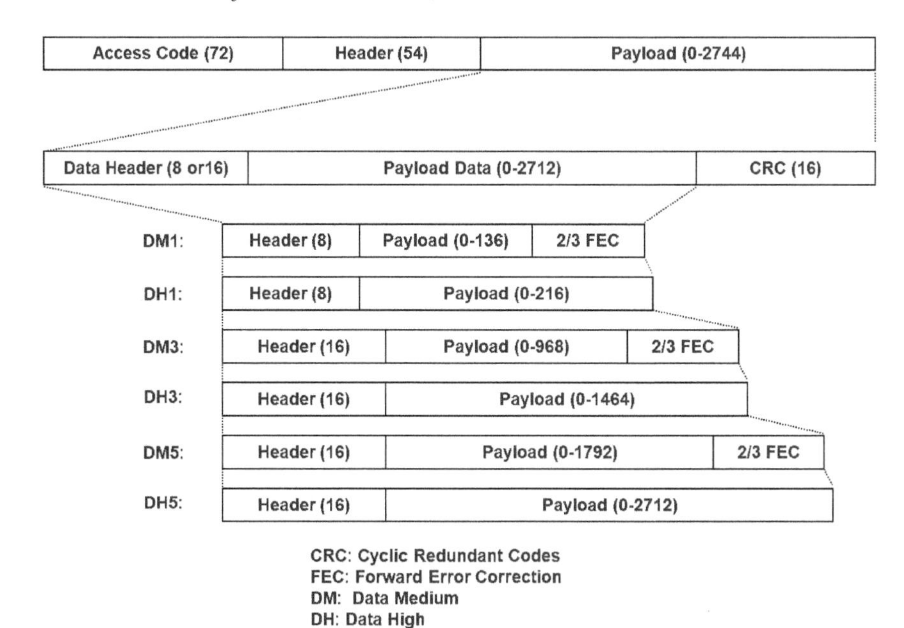

Figure 8.16 Frame format for data transmission in packets of length one, three, and five slots with two different levels of qualities.

piconet to support the best quality of voice. Most Bluetooth voice applications use this mode that is really a digital telephone wire replacement application.

The overall format of the payload for the six data packets is shown in Figure 8.16. The payload has its own 8- or 16-bit header, payload, and 16-bits cyclic redundancy check (CRC) code used as an error detection mechanism for long packets of data (see Chapter 3 for a description of block codes of which CRC codes form a subset). The header has information on the length and identity of the packet. If we want to compare the headers with those of IEEE 802.11, we may compare the overhead with the MAC overhead of 802.11. This time, the overhead of Bluetooth is significantly lower than the 34 bytes (272 bits) overhead of the 802.11 MAC frames. Most of the saving in the overhead of Bluetooth occurs because 802.11 employs four addresses – source, destination of the device, and the intermediate access points. Bluetooth uses one 48-bit IEEE MAC address to identify a device that is embedded in the access code and is not needed elsewhere.

The six data packets are divided into medium and high rate based on how much error protection they receive and there are grades for their data rate according to the number of slots: 1, 3, or 5 to carry the data. These differentiations allow for six classes of voice packets. Figure 8.16 shows the overall frame format of all six classes of data-oriented packets. The medium rate data packets use rate 2/3 forward error correcting codes that improve the reliability of the link at the cost of lower data rates. High rate data packets do not employ coding to achieve higher data rates. Using a different number of slots for a packet data payload size, exercising the coding option, and changing the nature of the transmitted packets in each direction (symmetric or not), a number of packet data links with flexible rates can be implemented according to the Bluetooth specification.

Example 8.9: High Data Rate in Bluetooth:

A symmetric data with a high rate in one-slot (DH1) link between a Master and a Slave terminal, shown in Figure 8.16, carries 216 bits per slot at a rate of 800 slots per second (every other slot) in each direction. The associated data rate is 216 (bits/slot) × 800 (slots/s) = 172.8 Kbps.

Example 8.10: Medium Data Rate in Bluetooth:

The asymmetric data with medium rate and five-slot length (DM5) link, shown in Figure 8.16, uses five-slot packets carrying 1792 bits per packet. If we assign this configuration to a Master to download data, then as shown in Figure 8.12(d), we must assign a one-slot medium rate data packet (DM1) carrying 136 bits per packet by the Slave terminal. In this situation, the number of packets per second in each direction is 1600/6 = 266.67. Therefore, the data rate of a Master is given by

$$1792\ \text{(bits/packet)} \times 1600/6\ \text{(packets/s)} = 477.8\ \text{Kbps}.$$

The data rate of the Slave terminal in this asymmetric connection is: 136 (bits/packet) × 1600/6 (packets/s) = 36.3 Kbps.

Table 8.2 shows all 12 symmetric and asymmetric data links that are supported with the frame format of the Bluetooth specification. The data rates for these links can be calculated in a manner like Examples 8.9 and 8.10. The maximum data rate of 723.2 Kbps is available in an asymmetric channel for a single user while the reverse channel supports only 57.6 Kbps. The reader should remember that data applications operate in bursts, and, therefore, even if a Master node communicates with the maximum seven Slave data terminals, most of the time, only one of the Slave terminals will communicate with the Master. When more than one Slave terminal simultaneously attempts to communicate with a Master terminal, the quality of service (QoS) provided to the Slave terminals has to be compromised either by sharing the throughput

Table 8.2 ACL packet types and associated data rates in symmetric and asymmetric modes.

Type	Symmetric (Kbps)	Asymmetric (Kbps)	
DM-1	108.8	108.8	108.8
DH-1	172.8	172.8	172.8
DM-3	256	384	54.4
DH-3	384	576	86.4
DM-5	286.7	477.8	36.3
DH-5	432.6	721	57.6

or by providing additional delays. The decision-making process to reach a compromise in the voice-oriented access methods, such as the one used in Bluetooth, needs a complex algorithm to handle the QoS as negotiated at the start of a session. Comparing this situation with CSMA/CA used in 802.11, there is no negotiation at the starting point. When more than one terminal attempts to communicate with a single access point, the medium is shared, and the compromise is made automatically through the CSMA/CA access method described in Chapters 4 and 7. For distributed data, only applications CSMA/CA is more appropriate. However, when voice applications become dominant, a TDMA/TDD type access methods can guarantee a certain QoS (e.g., steady data rate) with the fast hardware at the lower layers for voice, while CSMA/CA cannot do this easily and needs to implement it at software at the higher levels (e.g., using QoS mechanisms or polling that sits on top of CSMA/CA).

The only remaining traffic packet in Bluetooth is a data-voice packet that is a mixed voice and data packet with the same access code and overall header that must be transmitted at regular intervals. The voice part carries 80 bits of voice payload without any coding and the data part is a short packet of length 0–72 bits with a 16-bit 2/3 CRC coding and an 8-bit data payload header. This packet also uses three status report bits.

The Bluetooth specification also defines four control packets. The first packet occupies only half of a slot and it carries the access code with no data or even a packet type code. This packet is used before connection establishment to only pass an address. The second and third packets have the access code and the header, and so they have packet type codes and status report bits. The second packet is used for acknowledgment signaling and there is no acknowledgment for this packet. The third packet is used for polling and its format is similar to that of the second packet, but it has an acknowledgment. Master terminals use the polling packet to find the Slave terminals in their coverage area. The fourth packet carries all the information necessary to synchronize two devices in terms of access code and hopping timing. This synchronization packet is used in the inquiry and paging process that will be explained later.

8.2.6 Connection Management

The link management protocol layer and LLC application protocol layer of the Bluetooth, shown in Figure 8.8, perform the link setup, authentication, and link configuration. An important issue in a truly *ad-hoc* network is how to establish and maintain all the connections in a network whose elements appear and disappear in an *ad-hoc* manner and there is no central unit transmitting signals to coordinate these terminals. In both digital cellular systems and WLANs, there is a common control signal or a beacon signal that allows a new terminal to lock to the network and exchange its identity with the network's identity. The Bluetooth specification achieves initiation of the network through a unique inquiry and page algorithm.

The overall state diagram of the Bluetooth is shown in Figure 8.17. At the beginning of the formation of a piconet, all devices are in stand-by mode. Then one of the devices starts with an Inquiry and becomes the Master terminal. During the Inquiry process, the Master terminal registers all the stand-by terminals that then become Slave terminals. Note that it is not necessary for a more powerful device to be the master. A camera could be a master device and a laptop the slave device since the camera initiated the Inquiry process. After the inquiry process, identification and timing of all Slave terminals are sent to the Master terminal using the synchronization control packets. A connection starts with a page message with which the Master terminal sends its timing and identification to the Slave terminal. When a connection is established, the communication session takes place and, in the end, the terminal can be sent back to the Stand-By, Hold, Park, or Sniff states. Hold, Park, and Sniff are power saving options. The Hold mode is used when connecting to several piconets or managing a low power device. In the Hold mode, data transfer restarts as soon as the unit is out of this mode. In the Sniff mode, a slave device listens to the piconet at reduced and programmable intervals according to the application needs. In the Park mode, a device gives up its MAC address but remains synchronized to the piconet. A Parked device does not participate in the traffic but occasionally listens to the traffic of the Master terminal to resynchronize and check on broadcast messages.

The main innovative part of the inquiry and paging algorithms in Bluetooth is a searching mechanism for two terminals that are not

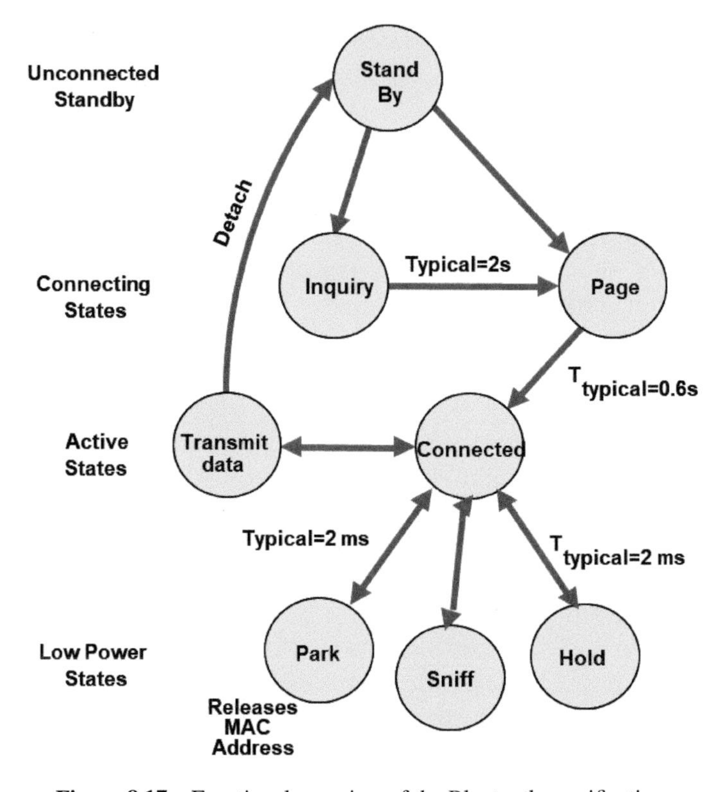

Figure 8.17 Functional overview of the Bluetooth specification.

synchronized, but they both know a common address. The following example explains this algorithm.

Example 8.11: Search Algorithm for Synchronization:

Two Bluetooth devices with a common 48-bit IEEE 802 address first use the common address to generate a common frequency hopping pattern of 32 hops and a common PN-sequence for the access code of all their packets. Then they start their operation as depicted in Figure 8.18. In the initial state, Terminal 1 sends two ID packets carrying the common access code every half slot on a different hop frequency associated with the common frequency hopping pattern and listens to the response of the slave device (Terminal 2) in the next slot. If there is no response, it continues broadcasting the ID packets on the two new frequencies in the common hop pattern and repeats this procedure eight times for a period of 10 ms (eight two-slot times). During this 10 ms, the common ID is broadcast over 16 of the total 32 different hop frequencies. If there is no response, Terminal 1 assumes that Terminal 2 is in sleep mode and repeats the same broadcast again and again until the period of transmission becomes longer than the expected sleeping time of Terminal 2. At this time, Terminal 1 assumes that Terminal 2 has scanned, but its scan frequency was not among the 16 hops, designated by A in Figure 8.18 and continues its broadcast with the second half of the 32 hop frequencies, designated by B in the figure. If Terminal 2 is in sleeping mode, it wakes up periodically for a period of 11.25 ms to scan the channel at a given frequency for its desirable access code and sleeps again. In each scan period of 11.25 ms, the sliding correlator in Terminal 2 tries to detect the desired address at 16 different frequencies. If one of these frequencies is the same as the scanning frequency, the correlator peaks and synchronization are signaled. Depending on the operation, Terminal 2 can scan the second time at the same frequency or at a new frequency for verification. In either case, the objective is to maximize the probability of hitting the same frequency as the broadcast frequency.

The basic principle explained in the above example is used during the Inquiry and Paging processes. The following two examples explain these applications for the above mechanism.

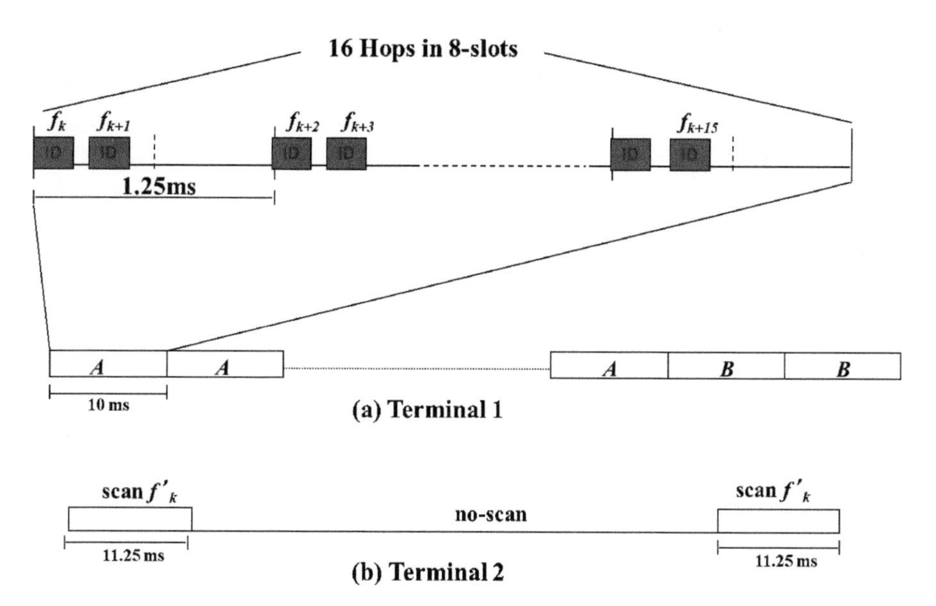

Figure 8.18 Basic search for paging algorithm in the Bluetooth.

Example 8.12: Paging:
As in the previous example, the Master terminal (Terminal 1) broadcasts repeating ID page trains carrying the access code of the paged terminal, two per slot, waits for the response in the next slot, repeats the page trains at new hopping frequencies of the paged terminal to cover 16 frequencies every 10 ms, and repeats this for the estimated length of the sleeping time. The Slave terminal scans for 11.25 ms with one of the 32 frequencies of its hopping pattern, sleep, and scans at the next hopping frequency. When frequencies are the same, a peak appears at the correlator output of the Slave terminal and the slave responds by sending its own ID packet as an acknowledgment for detection of frequency hopping timing. The Master terminal then stops broadcasting ID packets and sends a synchronization packet containing its own ID and timing information. The Slave terminal responds with another ID packet to correspond to the timing of the Master terminal and then the connection is established, and the Slave joins the piconet for information exchange. Usually, the Master terminal knows the approximate timing of the hopping pattern and the 16 most probable hops are adequate to establish the connection. In case this estimate is not correct, as in the previous example, the Master terminal resorts to the second half of the 16 hops when there is no response after the estimated sleeping time.

Example 8.13: Inquiry:
The Inquiry message is typically used for finding Bluetooth devices, including printers that have Bluetooth, fax machines, and other similar devices with an unknown address. The general format of the Inquiry process is very similar to the Paging mechanism. A unique access code and frequency hopping pattern are reserved for Inquiry. In other words, the Inquiry process is universally identified with all attributes of any device. Like Paging, Inquiry starts with an "Inquirer" broadcasting an ID packet every half slot at a different hop frequency, covering 16 frequencies every 10 ms, and repeats the same process until it receives responses. The "Inquire" scans with the sliding correlator for 11.25 ms. When frequencies are the same, the sliding correlator peaks in all devices that are scanning. To avoid the collision, a device detecting the Inquiry ID runs a random number generator and waits for the length of the outcome before it scans the channel again. When the peak appears the second time after random waiting time, the Inquirer terminal sends a synchronization packet, allowing the Inquirer to learn its ID and timing information. After this process is completed, the Inquirer's radio has the Device IDs and Clocks of all radios in its range of coverage.

After completion of the first Inquiry, the inquired device changes its scan frequency and continues scanning for the next Inquiry and follows up with synchronization signaling.

8.2.7 Security

Bluetooth specifications provide usage protection and information confidentiality. Bluetooth has three modes of operation – non-secure, service-level, and link-level security. Devices also can be classified as trusted and distrusted. It makes use of two secret keys (128 bits for authentication and 8–128 bits long for encryption), a 128-bit-long random number, and the 48-bit MAC address of devices. Any pair of Bluetooth devices that wish to communicate will create a session key (called the link key) using an initialization key, the device MAC address, and a personal identity number. This protocol has been shown to have several vulnerabilities [WET01] by which a malicious entity could obtain the personal identity numbers and keys depending on how the session initialization of the communication protocol is performed.

8.3 IEEE 802.15.4 and ZigBee

ZigBee is a suite of protocols capable of mesh networking between nodes that are in range, some with multi-hop routing capability, defined for operation with the IEEE 802.15.4 standard that specifies the PHY and MAC layers for low data rate, low power, and low cost WPANs. The first ZigBee specification was ratified in 2004 and the latest version was released in 2007. The relation between IEEE 802.15.4 and ZigBee is also like the relation between IEEE 802.11 and Wi-Fi. We use the terms ZigBee and IEEE 802.15.4 interchangeably in many places.

In as much as the IEEE 802.15.1 Bluetooth standard was a low complexity, inexpensive, and low power *single-hop ad-hoc* network design influenced by the IEEE 802.11 FHSS standard, the design of the IEEE 802.15.4 standard has been influenced by the IEEE 802.11 DSSS standards. In comparison with the IEEE 802.11 Wi-Fi devices, ZigBee devices are designed for very low cost communications among scattered devices such as low-power sensors with minimal infrastructure. Many of the applications and the radio coverage of ZigBee devices are similar to those of Bluetooth. However, ZigBee intends to provide faster formation of a piconet, a larger number of active devices, a much longer battery life, but lower data rates of 20–250 Kbps. The flexible data rate and faster connection time make

ZigBee more desirable for connectionless data-oriented applications to enable communication between the sensors and the Internet.

To follow the same format of presentation that we have used to present the details of other wireless networking technologies, here, we start with the overall architecture and the general protocol stack followed by sections on the details of the physical layer, the medium access control layer, and how packets are formed at these layers.

8.3.1 Overall Architecture

One of the general architectural differences of the IEEE 802.15.4 standard is that it defines two types of nodes in the network. The two types of nodes are referred to as *full-function devices* and *reduced-function devices.* A full-function device is like Bluetooth devices with greater functionality in some respects and it can serve as a *coordinator* or master of a piconet. It can also serve as a common node or a full-function slave. A full-function device can communicate with any other device in the network (within its radio communication range) and it can further help in routing messages (packets or frames) throughout the ZigBee network. In contrast, reduced-function devices are defined to be extremely simple with very modest resources and sparse communication capabilities and are usually only used as slave nodes to communicate with a full-function device. Consequently, such devices are often in sleep modes most of the time and communicate only very infrequently (e.g., upon detecting an event or sensing some parameter), allowing their batteries to last for months at a time. This characteristic of reduced-function devices provides flexibility in the implementation of a variety of topologies that can address diversified applications. As an example, consider a typical example application of a wireless light lamp switch. The node at thc lamp can be a full function device since it is connected to the main power supply and does not have power constraints, while a battery-powered light switch would be a reduced-function device to conserve energy and increase the battery life. In fact, at the time of this writing, there is light emitting diode (LED) bulbs that have been introduced with ZigBee capability that communicates with a ZigBee bridge that can plug into a Wi-Fi router. Users can control the operation of the bulbs through the web or applications on their smartphones.

In a manner like Bluetooth and IEEE 802.11, IEEE 802.15.4 and ZigBee support both peer-to-peer and star network topologies. A new additional topology for the IEEE 802.4 is the cluster-tree topology. Figure 8.19 shows all

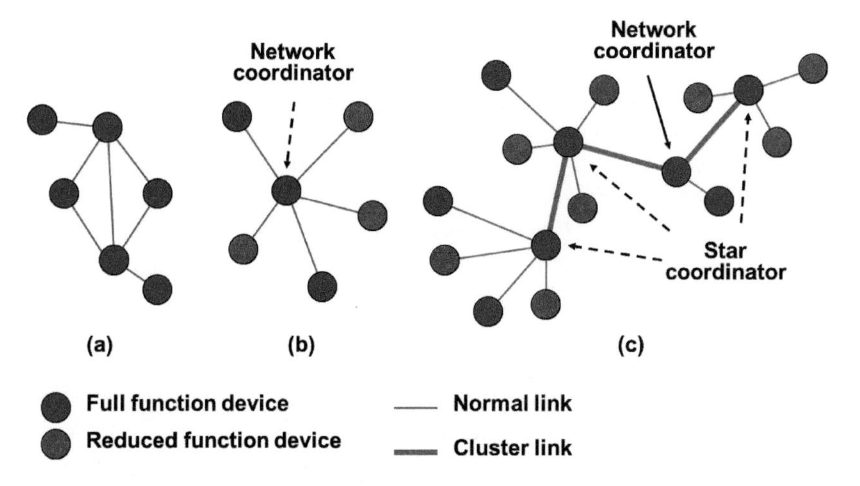

Figure 8.19 IEEE 802.4 ZigBee topologies. (a) Peer-to-peer. (b) Star for master–slave operation. (c) Clustered stars.

three topologies for IEEE 802.15.4 based ZigBee networks. The peer-to-peer networks, shown in Figure 8.19(a), form arbitrary patterns of connections. The extent to which the extension of these connections is possible depends on the distance between each pair of nodes. The nodes in Figure 8.19(a) are all full-function devices and are thus meant to serve as the basis for constructing on-the-fly networks that can perform self-management and organization. Figure 8.19(b) shows a simple star topology with a master–slave operation that is like that of Bluetooth. In the ZigBee network, however, as we mentioned earlier, we have two classes of nodes that provide more flexibility in forming a network to support an application. The network coordinator must be a full-function device and the other nodes in Figure 8.19(b) can be either full- or reduced-function devices depending on the application. Figure 8.19(c) shows an example of a more complex clustered tree topology formed by two different devices with three star coordinators and one network coordinator connecting stars together. The coordinators in this figure play the role of a router in the *ad-hoc* network, which means the network coordinator manages the entire network. This type of topology is unique to ZigBee and it allows the formation of a hierarchical *ad-hoc* network with simple end nodes which are in the sleeping mode most of the time, allowing extremely long battery lives.

In a cluster-tree topology, there are three distinct classes of node operation. The first class is the reduced-function or dumb end node devices, which are sleeping most of the time to extend the battery life. Each such end device allows up to 240 end points for separate applications sharing the same radio. For example, a three-gang light switch would have three distinct end points sharing the same radio electronics and battery. The second class of devices is the mid-level routers, which can stack up communication messages and respond to general enquiries about sleeping end devices in their vicinity. These routers are also responsible to find out the best way to pass on a message to a node that is not in the range. The third class of nodes at the top of the hierarchy is the network coordinator, which is always on and relies on connection to a good power source. In addition to being a router, the network coordinator sets the rules for basic network operation such as finding an appropriate frequency channel for the network.

In the case of the IEEE 802.15.4 standard (along with the ZigBee higher layer protocols), the setting up of the network is accomplished using WPAN coordinators. In the star topology, nodes communicate directly to a WPAN coordinator and all communications go through the coordinator. In many ways, this is like nodes communicating to an access point in wireless local area networks. In the peer-to-peer topology, devices can communicate directly with one another as long as they are in radio range.

In the star topology, a full-function device, when deployed, automatically becomes the WPAN coordinator. This is accomplished simply by the device, by transmitting a beacon and announcing itself as the coordinator, if it does not hear any other device when it powers up. In the peer-to-peer topology, a similar mechanism works and there is a WPAN coordinator (usually the first full function device to power up). However, devices may communicate directly with one another where allowed. If two or more full function devices attempt to become coordinators, some contention resolution beyond the scope of the base IEEE 802.15.4 standard becomes necessary.

A *cluster tree* is a generalized form of the peer-to-peer topology in IEEE 802.15.4 networks. Figure 8.19(c) shows an example of a cluster tree with three clusters. Here, the assumption is that most of the devices are full-function devices (although reduced-function devices can connect to clusters as leaf nodes). The first full-function device that announces itself (or a device with more power or capabilities) becomes the overall WPAN coordinator. It transmits beacon frames and provides other nodes (and other coordinators) synchronization. As the network grows, the overall WPAN coordinator instructs another device to become a WPAN coordinator of its own cluster. Clusters can develop this way into a large network. The standard specifies the resolution of conflicts between WPAN IDs, transmitting beacons, etc.

8.3.2 Protocol Stack and Operation

Though ZigBee WPANs are simple *ad-hoc* sensor networks designed to operate in short distances of up to 10 m, the network has several layers designed to enable communication within the network, connection to a network of higher level, and, ultimately, an uplink to the Internet. Like all other communications networks, IEEE 802.15.4 based ZigBee divides up the communications tasks into layers on a stack of protocols shown in Figure 8.20. The lower layers, i.e., the physical and medium access control layers, are defined by the IEEE 802.15.4 standard and the higher layers are specified by the ZigBee alliance. In addition, independent application developers can define their own application layers that can then communicate with the ZigBee defined layers and protocols. The top of the stack is the application layer and the bottom is the physical radio. The middle layers are used to glue the application to the actual transmission so that nodes can communicate reliably, efficiently, and securely and with interfaces that designers can employ to develop their applications in an easier supporting environment.

As shown in Figure 8.20, on top of the ZigBee defined protocols, we have a *ZigBee Device Object*, which is a special application object, available on all ZigBee nodes. The address of this application object is always "zero," and other application objects running on the ZigBee module are numbered 1–240. These form the various applications that an end device may be supporting. The Device Object is something that has its own profile, which other user application objects and other ZigBee nodes can access, and it is responsible for overall device management, security keys, and policies. Each profile in a ZigBee module includes a table of other ZigBee modules on the network and the services they offer. All other application end points use the available information in the table to discover other devices, manage to bind, and specify security and network setting. The *application support layer* protocol in Figure 8.20 routes messages on the network to different application end points running on a ZigBee module by maintaining a "binding table" and forwarding messages to the appropriate application.

The *network layer* provides the routing and multi-hop capability required to turn MAC-level communications into a full star, tree, or mesh network. ZigBee employs a distance-vector routing algorithm suitable for *ad-hoc*

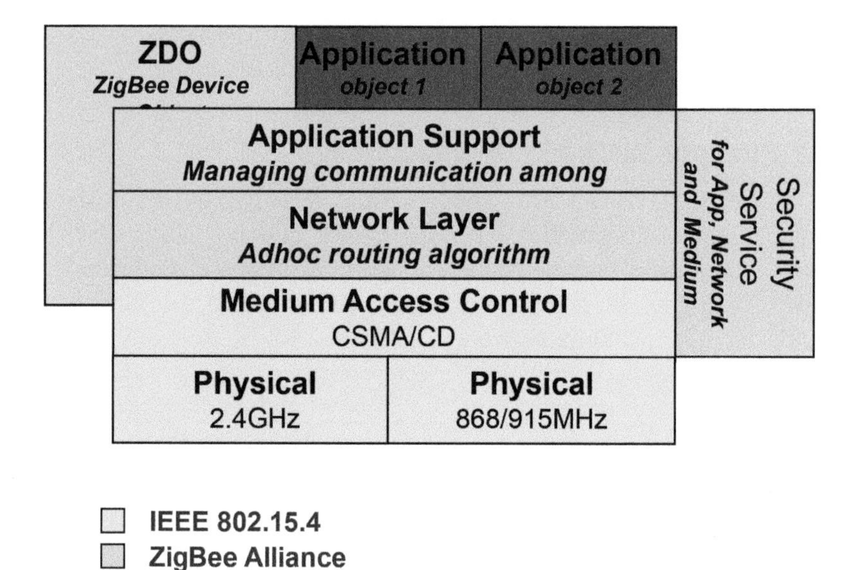

Figure 8.20 IEEE 802.15.4 ZigBee protocol stack.

networks called the *ad-hoc on-demand distance vector* (AODV) protocol. This algorithm automatically constructs a low-speed *ad-hoc* network of nodes by forwarding messages, discovering neighboring devices, and building up a map of the routes to other nodes. In the coordinator nodes, the network layer assigns network addresses to new devices when they join the network for the first time.

The *security service* protocol provides for establishing and exchanging security keys and using these keys to secure the communications link through encryption and message integrity checks. The security services work across three layers to provide security at each level. Like all other security services in wireless networks, this layer is responsible for encrypting and decrypting the data when it is generated or received and authenticating it (checking for integrity) when it is received. The ZigBee Device Objective layer dictates the security policies and configurations implemented by the security services.

8.3.3 Physical Layer

In a manner like IEEE 802.11, IEEE 802.15.4 operates in the unlicensed radio bands. In addition to the 2.4 GHz band used in IEEE 802.11b/g/n, which

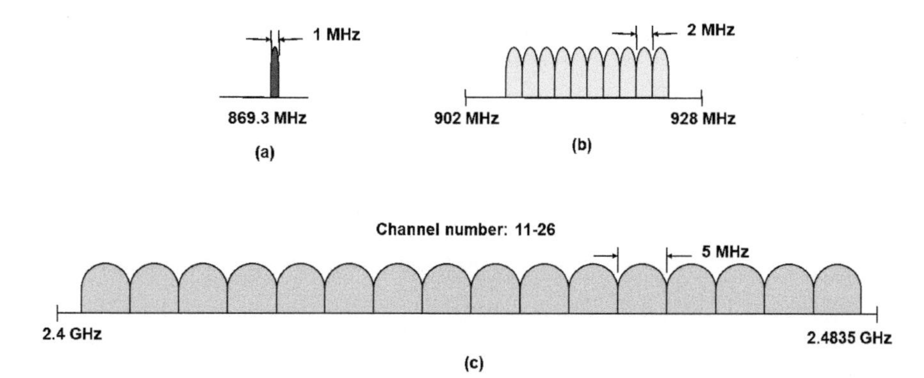

Figure 8.21 Frequency bands used by different PHY options in the IEEE 802.15.4 ZigBee. (a) One channel at 868 MHz. (b) Ten channels in 915 MHz. (c) Sixteen channels in 2.4 GHz.

is mostly available worldwide, the IEEE 802.15.4 standard also supports options for operation in the unlicensed 868 MHz bands in Europe and unlicensed 915 MHz bands in countries such as the USA and Australia. The physical layer of IEEE 802.15.4 defines 27 channels with three data rates of 20, 40, and 250 Kbps in these three different frequency spectrums. Figure 8.21 shows the 27 different channels and their associated bandwidth specified by the standard in the three different spectrums. The first channel with 0.6 MHz bandwidth, referred to as channel number "0," is at 868 MHz, 10 channels each with 2 MHz bandwidth are in the 915 MHz bands, and the remaining 16 channels each with 5 MHz bandwidth are in the 2.4 GHz bands. The IEEE 802.15 standards committee's task group 4c has been considering standards for the 779–787 MHz bands in China and task group 4d is looking at the 950–956 MHz bands in Japan. The standard defines *channel pages* in addition to the *channel numbers*. Channel pages are used to distinguish between the different possible modulation schemes used in the frequency channels.

The basic radios use DSSS, in a manner like legacy IEEE 802.11, but with different chip rates and modulation techniques. The binary phase shift keying (BPSK) modulation and a 15-chip linear feedback shift register (LFSR) pseudorandom sequence are used for the 868 and 915 MHz bands with chip rates of 0.3 and 0.6 Mcps, respectively. This results in a data rate of

$$\frac{0.3\ (\text{Mcps})}{15\ (\text{cpb})} = 20\ \text{Kbps}$$

for the single channel in 868 MHz and

$$\frac{0.6\ (\text{Mcps})}{15\ (\text{cpb})} = 40\ \text{Kbps}$$

for each of the 10 channels in the 8.15 MHz bands. An optional amplitude shift keying modulation scheme allows the data rates to be increased by using a form of code division multiplexing (where bits are sent in parallel by spreading them using almost orthogonal sequences).

The 2.4 GHz operation uses 16-ary orthogonal coding with code words of length 32 chips. The modulation technique used for the transmission of data is the offset quadrature phase shift keying (QPSK). This general structure is like the reverse channel of the IS-95 CDMA system, which also uses M-ary orthogonal coding and offset QPSK modulation. Offset QPSK provides a constant envelope, which reduces the power consumption of the last stage amplifiers in the radio. M-ary orthogonal coding reduces the error rate and adds to the integrity of the received bits. The receiver is implemented non-coherently, which does not need carrier synchronization between the transmitter and the receiver and, therefore, avoids excessive power consumption and complexity of the phase lock loops needed for coherent modulation. These measures are taken to satisfy the low cost and low power implementation of the radio. To form the transmitted symbols, the arriving raw data stream at a rate of 250 Kbps is used to create 4-bit blocks of data at a rate of

$$\frac{250\ (\text{Kbps})}{4\ (\text{bpS})} = 62.5\ (\text{Ksps})\,.$$

Each 4-bit block is then mapped to one of the 16-orthogonal symbols. The chips are then transmitted at the rate of 2 Mcps using offset QPSK modulation. Therefore, the net processing gain of this DSSS transmission scheme used in the IEEE 802.15.4 is

$$\frac{2\ (\text{Mcps})}{250\ (\text{Kbps})} = 16.$$

It is useful to remind the reader here that the processing gain in the basic IEEE 802.11 DSSS transmission scheme was $N = 11$, which is reasonably close to what is used in the IEEE 802.15.4 standard. It is also useful for the reader to remember that the complementary code keying (CCK) modulation scheme used in IEEE 802.11b is a different sort of M-ary orthogonal coding. These observations show the path of evolution of experimentally successful local radio networks.

The maximum transmission power of IEEE 802.15.4 radios is usually 0 dBm (1 mW) compared to a maximum transmit power of 20 dBm (100 mW) used in the IEEE 802.11 devices. This 20 dB difference, assuming free space propagation with a distance-power gradient of 2, accounts for one order of magnitude higher coverage for IEEE 802.11 devices. As we explained in Chapter 2, the coverage depends on the environment. But it is customary to assume that the coverage of WPANs is roughly 10 m and that of WLANs is approximately 100 m. The lower transmission power and power conscious design of the radio is one of the major differences between the design of IEEE 802.15.4 and IEEE 802.11 devices. The receiver sensitivity in the 2.4 GHz bands is specified to be at least -85 dBm. For BPSK modulation in the 868/916 MHz bands, the receiver sensitivity should be at least -92 or -85 dBm. This allows an extra 7 dB for operation in the 868/915MHz bands. In addition, since the path-loss in the first meter is calculated from $20\alpha\log(4\pi/\lambda)$, for the path loss in the first meter at 2.4 GHz and 868 MHz, we have another

$$20\log\left(\frac{2.4\ (\mathrm{GHz})}{868\ (\mathrm{MHz})}\right) = 8.8\ (\mathrm{dB})$$

difference between the two bands that adds up to a 15.8 dB edge for operation at lower frequencies. In free space, this accounts for approximately up to $10^{15.8/20} \approx 6$ times longer coverage for operation in lower frequency bands. In indoor environments, lower frequencies also penetrate the walls better, which increases the coverage at lower frequencies to even higher values.

8.3.4 MAC Layer

Medium access control for the IEEE 802.15.4 standard is based on CSMA/CA is also the main medium access control technique used in the IEEE 802.11 standard. Some details of CSMA/CA were provided in Chapter 4. In general, the CSMA/CA in IEEE 802.15.4 is a simpler version of the CSMA/CA used by the IEEE 802.11 standard which includes different options for conserving battery and reducing power consumption. The PCF option, a variety of inter-frame delays (PIFS, DIFS, and SIFS), and the RTS and CTS mechanisms which were included in the IEEE 802.11 MAC (see Section 7.4) are not included in 802.15.4. The functionality that the RTS/CTS and PCF mechanisms were providing in IEEE 802.11 is here provided by simpler and more practical guaranteed time slot transmission. The medium access control layer provides reliable communications between a node and

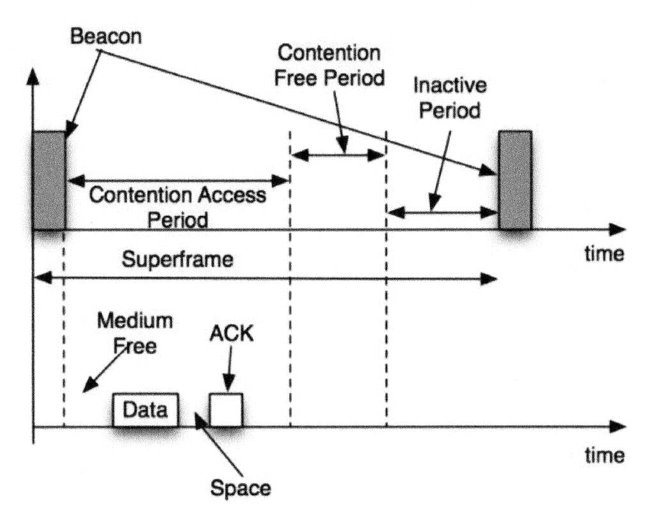

Figure 8.22 Illustration of medium access in IEEE 802.15.4.

its immediate neighbors (in radio range) and manages packing data into frames prior to transmission and then unpacks received packets to check them for errors. In addition, this layer provides beacons and synchronization to improve communications efficiency. Networks are formed either based on a beacon or without a beacon. Beacons do not follow carrier sensing and are sent on a fixed timing schedule. Acknowledgment packets are also sent without carrier sensing after the arrival of a packet or MAC frame. Two other important features of 802.15.4 MAC are (a) the allowance for guaranteed time slots and (b) integrated MAC support for secure communications. Devices that have low latency real-time requirements can use the so-called guaranteed time slots allocated to them by a coordinator without resorting to carrier sensing. The details are as follows.

The MAC protocol in 802.15.4 operates under a superframe structure (see Figure 8.22). A superframe is defined as the period between two *beacons*, which are special management packets transmitted by the coordinator. Beacons synchronize the WPANs and provide information about the network. Within the time between two beacons, i.e., the superframe, sensor nodes can have an active period and an inactive period. The active period can be divided into a contention period and a contention-free period. The contention period is slotted. In each slot, nodes use carrier-sensing multiple access with collision avoidance to access the channel. The process is quite simple. Each device waits for a random period to see if the channel is idle. If it is idle, it simply transmits. Otherwise, it backs off for another random period and tries again (see bottom of Figure 8.22). This access suffers from the disadvantages mentioned previously with carrier sense in Chapter 4 such as the hidden terminal problem. Scheduled access is possible through the use of the contention-free period where the WPAN coordinator creates guaranteed time slots that nodes can use without contention from other nodes. It is possible to have a WPAN without beacons in which case non-slotted CSMA/CA is adopted by all nodes.

The standard considers "transactions" that are initiated by the low power devices, which will otherwise have the choice to be in a low-power mode. In the star topology, nodes can send data to a coordinator or received data from a coordinator. In the former case, the node simply sends data to the coordinator using CSMA/CA and gets an acknowledgment if requested. In the latter case, the node should first request data from the coordinator and get an acknowledgment for its request followed by the data. Upon receipt of the data, it acknowledges it to the coordinator. Acknowledgments do not wait for the medium to be idle as in the case of data frames. Alternatively, the beacon can have information about whether or not there is pending data for a sensor node. Peer-to-peer transmissions occur similarly except that special steps may be necessary for synchronizing transmissions of two peer nodes. In order to allow the MAC layer to process frames, there must be some time that lapses between successive frame transmissions. The time between receipt of a frame and the transmission of an acknowledgment is the smallest while long and short inter-frame spaces are used to separate long or short frames.

In general, ZigBee protocols minimize the time the radio is "on" to reduce unnecessary power consumption. In non-beacon enabled networks, power consumption is decidedly mixed: some devices are always active, while others spend most of their time sleeping. In these networks, an unslotted CSMA/CA channel access mechanism is used and routers typically have their receivers continuously active, requiring a more robust power supply. These networks allow a bipolar consumption of energy. The router consumes substantially while end nodes only transmit when an external stimulus (e.g., an event such as a rise in temperature) is detected. The typical example of such an operation is the light lamp operation which we discussed earlier. In networks using a beacon, nodes are only activated after a beacon is transmitted. The special ZigBee router node transmits a beacon periodically to confirm its presence to other users. Other nodes may sleep between beacons to save energy consumption. For different options of the physical transmission, the beacon interval ranges from 15 to 24 ms. This option is more suited for higher traffic load and it is very similar to the operations in IEEE 802.11.

8.3.5 Frame Format

IEEE 802.15.4 uses four different types of frames for data, acknowledgment, beacon, and MAC commands. Figure 8.23 shows the frame format for all four types of packets. Every frame has a 4-byte preamble, a 1-byte start of the packet delimiter, and a 1-byte start of the frame which uses 7-bits to identify the length of the packet in bytes and 1-bit to identify the addressing mode because the device address is either 2-bytes (16-bits) or 8-bytes (64-bits). In addition to the payload, the physical service data unit for all packets has a 2-byte frame control to carry control messages, 1-byte sequence number to provide for tracking the sequence of the packets during fragmentation and reassembly process (for large application packets), up to 20-bytes of addressing, and 2-bytes of frame check code for error detection. The acknowledgment packet does not have any address field, but the other packets have source and destination addresses with 2-bytes for indicating the address of the coordinator node of the PAN plus 2- or 8-bytes for the device address. The payload for the data frame is the information to be delivered with a length that ensures an overall length of the physical service data unit to stay under 127-bytes. The payload for the beacon provides other devices the

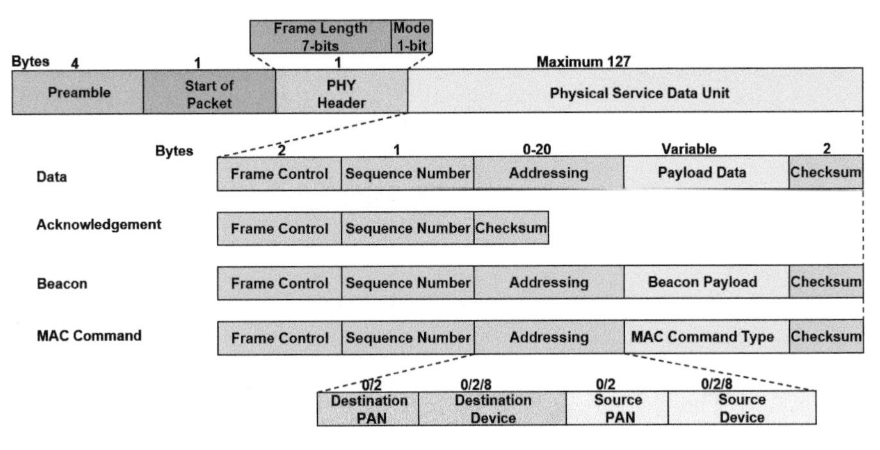

Figure 8.23 Frame format for IEEE 802.15.4 packets.

necessary information bits for synchronization and self-configuration. The MAC command control field carries different MAC control messages.

The 868/915 MHz bands which have better coverage for the same level of transmits power. The modulation techniques in the case of Bluetooth and ZigBee both use spread spectrum technology (but different types), which is a power efficient transmission technique. Both devices also implement non-coherent modulation to reduce the complexity and power consumption in the electronic design. The MAC of Wi-Fi and ZigBee are both based on CSMA/CA, which is a distributed MAC protocol suitable for data applications. The ZigBee implementation is a light version of the MAC protocol in Wi-Fi because it has fewer features. Each piconet in Bluetooth uses a centralized TDMA/TDD-like technique, based on polling, which is more appropriate for connection-based telephone applications. Different piconets are separated using FHSS/CDMA technique in the case of Bluetooth. The data rate of ZigBee networks is in the range of the data rates of Bluetooth which are suitable for *ad-hoc* low-rate sensor network applications. WLANs generally provide much higher data rates (54 Mbps for 802.11g) and are more suited to modern computer networking applications for home and small office networking and pervasive access in public buildings.

ZigBee allows two different types of terminals which allow the design of simple end nodes which may sleep most of the time to stretch the lifetime of the battery. Although the number of channels for Wi-Fi and ZigBee in 2.4 GHz looks similar, ZigBee channels are narrower (5 MHz) allowing the implementation of 16 non-overlapping channels. The IEEE 802.11g networks have only three non-overlapping channels. Bluetooth allows up to 79 piconets which are far beyond the other two, but the number of nodes per Master has a more practically important role because, usually, we have a few piconets but many nodes. In *ad-hoc* sensor networking, having only seven simultaneously connected nodes is rather limited and ZigBee's higher number of nodes (each with many applications) is very desirable. All the features of ZigBee are designed to minimize power consumption which results in a longer lifetime for the battery. The coverage of Wi-Fi is better (with a higher transmit power) and that is important for computer networking. Bluetooth and ZigBee are designed for low-rate *ad-hoc* sensor networking and the coverage can be extended by the selected topology. The ability of ZigBee to form cluster-tree networks increases the coverage of the *ad-hoc* network significantly making it a better choice for applications where we have a few sensor networks spread over a large geographical area.

8.4 IEEE 802.15.1 BLE and iBeacon

As a result of differences among Wi-Fi, Bluetooth, and ZigBee, summarized in Table 8.3, these technologies evolved for different market sectors. Wi-Fi dominated the WLAN market for a home, small office, and *ad-hoc* public building access. Bluetooth found several *ad-hoc* wire replacement applications for telephone connections inside the cars and audio devices. ZigBee originally emerged to address the market for WSN and IoT devices for a medical, power grid, home automation, and meter reading networking applications. As compared with Wi-Fi, ZigBee provides a long battery time, and, as compared with Bluetooth, it has a fast connection time and a better battery life. Bluetooth was designed for connection-based applications such a voice and credit card verification for which connection time is acceptable and it could accommodate only seven devices in a piconet. ZigBee enables the formation of large WSN, and it enables the implementation of IoT devices with battery life that can be stretched to a few months and more. Another extra feature of ZigBee is that it also operates at sub-GHz center frequencies allowing the wider area of coverage.

WSN and IoT devices also need wireless Internet access. Cellular and Wi-Fi have fixed infrastructure for wireless access for a device carrying cellular or Wi-Fi compatible chipsets and mobile devices can provide hot-spots for temporary access to the Internet to devices that are in their proximity and have Wi-Fi or Bluetooth compatible chipsets. When the first smartphone,

Table 8.3 Comparison of 802.11, 802.15.1, and 802.15.4.

Technology/feature	802.11g (Wi-Fi)	802.15.1 (Bluetooth)	802.15.4 (ZigBee)
Frequency	2.4 GHz	2.4 GHz	2.4 GHz–868/915 MHz
Modulation	OFDM	FHSS/BPSK	DSSS/QPSK
MAC	CSMA/CA	TDMA/TDD	CSMA/CA
Max. data rate	54 Mbps	1 Mbps	20/40/250 Kbps
Device types	One	One	Full/reduced function
No. of channels	11–14 (3 orthogonal)	79 (hopped)	26
Max. no. of devices	32	8	Up to 65,535
Battery life	Hours/days	Weeks	Months
Coverage	100 m	10 m	10/50 m
Topologies	Star, peer-peer	Star, peer-peer	Star and cluster-tree
Connection time	<10 s	3–5 s	30 ms

iPhone, was designed Wi-Fi and Bluetooth were integrated on top of the cell phone chipset, but ZigBee was not. Today, all smartphones have Wi-Fi and Bluetooth and billions of smartphones worldwide enable a hybrid network for access to the Internet. Since it is not practical to integrate ZigBee into smartphones on a large scale, but it is possible to design next generations of Bluetooth or Wi-Fi to imitate features of ZigBee, Wi-Fi, and Bluetooth community began to adopt low energy features. The IEEE 802.11ax, ah, aj, that we discussed in Section 7.5.6, addresses this issue for Wi-Fi. The Bluetooth V.4, known as smart or Bluetooth low energy (BLE) addressed this issue for IEEE 802.15.1 Bluetooth community.

8.4.1 Evolution of the Bluetooth PHY to BLE

The legacy Bluetooth v.1 was most successful for audio connection for headsets, speakers, and smartphones, although it was considered for the WSN and IoT devices. The early updates of this technology were to increase the quality of audio by increasing the transmission data rate of the original 1 Mbps. The legacy Bluetooth used the simple GFSK modem with two symbols separated with 1 MHz around two sides of the carrier frequency. FSK transmitter is easily implemented by applying the binary digital waveform from a device to an analog FM modulator. FM modulators map each voltage to a frequency; with two voltage levels from the data source, we will have two frequencies with their associated sidelobes. By passing the output of the modulator through a Gaussian filter, the sidelobes of the waveform are reduced before delivery to the antenna for transmission. A simple receiver can be designed by two filters at the two tones and a decision device to decide which level was transmitted. In digital communication literature, this simple transmission scheme is called GFSK. The receiver that we described does not need any synchronization like the one needed for QAM in the IEEE 802.11 OFDM signals. Synchronization circuitry will add to the complexity of the electronic and power consumption of the devices that relate to the life of the battery. GFSK is a binary communication system with 1-bit per symbol; to increase the data rate in a later version of classic Bluetooth, they adopted differential quadrature phase shift keying (DQPSK) with four points in the constellation and 2 bits per symbol to increase the data rate to 2 Mbps and eight phase differential phase shift keying (8-DPSK) with eight points in the constellation and 3 bits per symbol to go to 3 Mbps. Implementation of a higher data rate needs a compromise in battery life or in the coverage. The original 10 m coverage perceived for Bluetooth can be compromised in many short distance applications. The original Bluetooth was operating at 1 mW in the US; in the later versions, they left three transmission powers of (1, 25, and 100 mW) options. Availability of different options for data rate and coverage enables a wide variety of different scenarios for application development and the continual success of this technology in connection-based audio related applications.

The BLE resorts back to the GFSK binary modems with channel spacing of 2 MHz with two binary symbol transmission rates of 1 and 2 Mbps. In FSK, one can increase the symbol transmission rate by increasing the separation between the two tones that are kept the same as the incoming binary data rate. The transmission bandwidth of the 2 Mbps is double that of the 1 Mbps. Since channel spacing is 2 MHz, both symbol transmission rates are possible; but with a 2 Mbps transmission scheme, we will have more adjacent channel interference. For lower data rates, strong coding is added to the stream which allows flexibility to adjust with different application scenarios demanding a variety of power-coverage requirements. The difference in transmission between classical Bluetooth and BLE is that in classical Bluetooth, we resort to 2- and 3-bits per symbol transmission with DQPSK and 8-DPSK, while in BLE, we stay with binary transmission and increase the data rate to 2 Mbps by doubling the bandwidth. In classical Bluetooth, we have 79 channels with 1 MHz separation that cannot allow an increase in bandwidth and adjacent channel interference management.

8.4.2 BLE Technical Features

One of the unique features of the legacy Bluetooth was that it was introduced by a software CD carrying the protocol stack that facilitated application development. This was complemented gradually by integration in a variety of operating systems. The BLE is supported by all popular operating systems.[1] In 2013, Apple introduced the iBeacon protocol to create a class of BLE devices that broadcast their identifier to smartphones and tablets in Proximity of an iBeacon. The iBeacon uses BLE proximity sensing to transmit several bytes of the message to be picked up by a compatible app or operating system in its coverage area. This information includes RSS and other physical information. The broadcast channels host several innovative applications for proximity advertisements. Originally, these applications were like a bar that broadcasts its happy hours for pedestrians passing by.

[1] iOS5+, Android 4.3+, Apple OS X 10.6+Windows 8+, GNU/Linux Vanilla BlueZ 4.93+

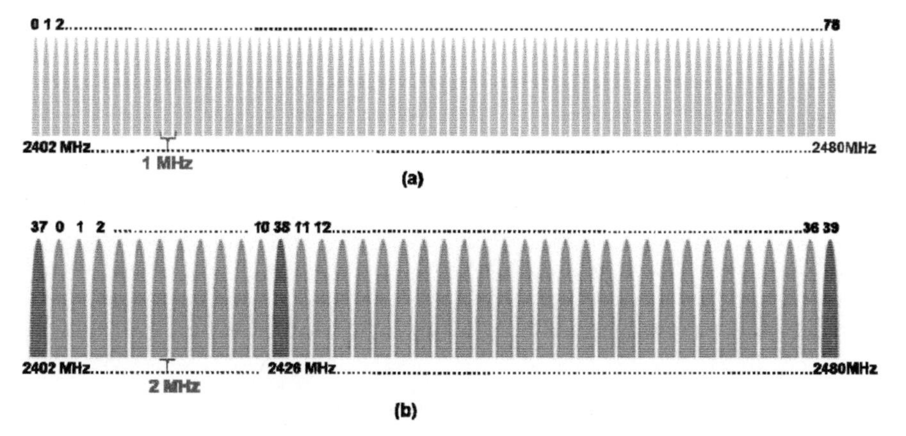

Figure 8.24 Channel number and frequency of operation for (a) 79 identical classical Bluetooth and (b) 3 advertising and 37 data channels in BLE.

Later, it became popular in other proximity check applications in security and social distancing measurements. As shown in Figure 8.24, unlike the classical Bluetooth with 79 identical channels (Figure 8.24(a)), BLE supports three broadcasting channels with 37 data channels for communications (Figure 8.24(b)). BLE also supports adjustable power levels from −20 dBm (0.01 mW) to 10 dBm (10 mW) allowing considerable flexibility for data rate and coverage compromises in different applications with a broadcasting ability that may extend to years.

Figure 8.25 shows the BLE star bus topology that is different from the scattered *ad-hoc* topology of classical Bluetooth is shown in Figure 8.7. In BLE, a device broadcast periodically and the piconet devices scan the broadcast channels periodically; when the two scans coincide, the connection is established, and data is exchanged if needed. The connection time for the BLE is less than 3 ms as compared with 100 ms connection time for classical Bluetooth. This change in connection time makes the data communications more efficient. Figure 8.26 shows the packet format of the classic Bluetooth, ZigBee, and the BLE. In legacy Bluetooth (Figure 8.26(a)), the packet format is divided into voice (Figure 8.14) and data packets (Figure 8.16) each with several options. This type of definition is like packets in cellular networks (Chapter 9) that are designed for a crowded domains with integrated voice and data users. ZigBee packets are shown in Figure 8.26(b) and their details are described in Figure 8.23. These packets are more like the Ethernet and Wi-Fi packets. BLE packets do not support voice and follow the format of

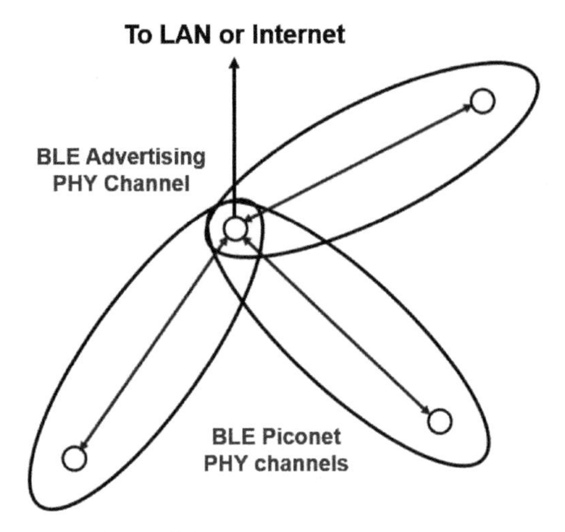

Figure 8.25 BLE star bus topology.

the ZigBee and Ethernet packets designed for access to the Internet. BLE and ZigBee provide similar services. BLE is better for smaller IoT home networks and ZigBee with wider coverage and ability to better connect many users that suit metropolitan area WSN. Table 8.4 summarizes the differences in features of the classic Bluetooth and BLE that we discussed in this section.

Preamble (0.5)	PN-Sequence (8)	Trailer (0.5)	Header (6.75)	Data (0-343)

(a)

Preamble (4)	Access Code (1)	Header (1)	Data (0-127)

(b)

Preamble (1)	Access Code (4)	Header (1)	Length (1)	Data (2-37/259)	CRC (3)

(c)

Figure 8.26 Comparison of packet formats in (a) classical Bluetooth, (b) ZigBee, and (c) BLE.

Table 8.4 Comparison of features of classic Bluetooth and BLE.

Technical specification	Classic Bluetooth	Bluetooth low energy
Distance/range	1 as reference	0.5
Over the air data rate	1–3 Mbps	125 Kbps–2 Mbps
Number of channels	79 with 1 MHz spacing	40 with 2 MHz separation
Active slaves	7	Not restricted to 7
Modulation	GFSK with 1–2 MHz spacing	GFSK, DQPSK, 8-DPSK
Connection time	100 ms	Less than 3 ms
Voice capable	Yes	No
Network topology	Scatter *ad-hoc* (Figure 8.5)	Star-bus (Figure 8.25)
Power consumption	1 as the reference	0.01 to 0.5
Service discovery	Yes	Yes
Tx power options	0–20 dBm	–20 to 10 dBm

8.5 Other IEEE 802.15 Standards for WSN and IoT

IEEE 802.15 has pioneering work in several other interesting areas related to the WSN and IoT that have not been as successful as Bluetooth and ZigBee. In this section, we introduce three of these standards, the IEEE 802.15L for wireless smart utility network (Wi-SUN), the IEEE 802.15.6 on wireless body area networking (WBAN), and the IEEE 802.15.7 on visible light communication networks are known as Li-Fi.

8.5.1 IEEE 802.15.4 for Low-Energy Wide Area Networking

To compliment low energy proximity wireless networking technologies such as ZigBee and BLE, there is a need for low energy wireless wide-area networking technologies for applications such as utility metering, disaster prevention and civil infrastructure monitoring, and agricultural information monitoring and management. In these applications, a very low speed low energy network is needed to connect the wireless sensors and IoT devices to the Internet. The wireless smart utility network (Wi-SUN), defining higher layers for the PHY and MAC of the IEEE 802.15.4e, is a good example of these technologies. The IEEE 802.15.4e PHY adopts the power efficient 2FSK and 4FSK and its MAC has coordinate sample listening (CSL) and receiver initiated transmission (RIT) features for saving in transmission power. This technology is designed for utilities, smart cities, and IoT devices to form an outdoor wireless wide area network. Figure 8.27 [Har14] illustrates an application scenario for this standard. In this scenario, utility

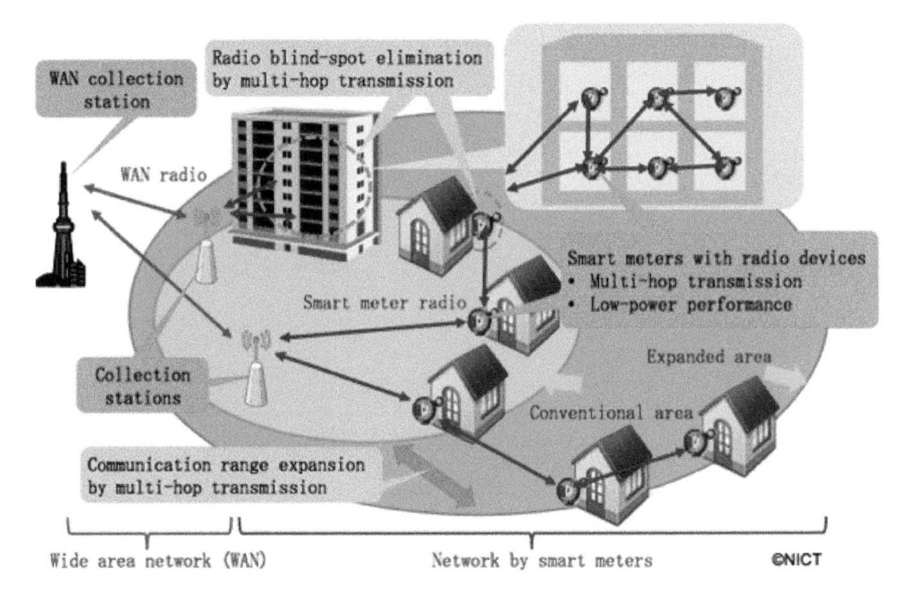

Figure 8.27 Wi-SUN and 802.15.4e application scenario for smart utility network [Har14].

meters with radio devices adopt multi-hop transmission for low power consumption to connect to a collection station with the capability to connect to the backbone wireless wide area network infrastructure. The multi-hop transmission capability allows the accommodation of utility devices in radio blind spots, where the coverage of the collection station is not available. This technology addresses a very important need and it has been adopted by the Tokyo electric company for energy management.

Another interesting technology for these types of applications is the IEEE 802.15.4k that covers up to 5 Km with 20 mW power. The PHY of this standard adopts low energy BPSK and offset or staggered QPSK modulation with DSSS. The chip rate or nominal bandwidth of this PHY is 800 Kcps and the processing gain of the DSSS has a wide range of 16–32,768 allowing huge flexibility for data rate and coverage compromises. This standard is defined at 902–928 MHz unlicensed ISM bands that help further expansion of the coverage. The MAC of this standard adopts a sleeping option to further extend the battery life. The choice of the band is like NB-Wi-Fi (IEEE802.11ah/ax), which covers up to 1 Km with bandwidths of 1–16 MHz [Wil17].

8.5.2 IEEE 802.15.3/4, UWB, and mmWave Technologies

The UWB was used in radars since WW-2; a new hype in this technology began around the turn of the 21st century. In this hype, UWB was considered for its low energy, capabilities for achieving Gbps wireless, and for precise indoor positioning. In UWB transmission, we can transmit very narrow pulses directly without modulation over a carrier signal by an extremely fast on–off switching of an antenna, and at the receiver, as soon as received power is passing a threshold, we can detect the time of arrival of the pulse. This way, electronic for implementation is very simple and power consumption is minimal and it is an excellent low energy transmission for short-range communications. The time of arrival is also used for precise ranging and positioning. If we switch at GHz, we can achieve Gbps, but we need UWB bands for this transmission. In early 2003, FCC released UWB unlicensed bands, and IEEE 802.15.3a began discovering technologies for short-range Gbps communication applications in these bands. The high light of communication application for this technology could be wireless USB or wireless HDMI cable replacements for storage to computer interfacing and multi-channel video transmission to connect TV distribution boxes to the large TV monitors. The IEEE 802.15.3 attracted considerable attention from consumer electronic giants such as SONY and Intel. The precise positioning feature was attractive for military applications as well as for applications with high precision in commercial stores for isle-by-isle positioning of inventory with a few cm accuracies.[2]

The IEEE 802.15.3a began its activities by focusing on 3.1–10.6 GHz portion of FCC's unlicensed UWB spectrum. Two leading 802.15 proposals brought forward after the 2003 FCC announcement are known as direct sequence UWB (DS-UWB) and multi-band OFDM (MB-OFDM). The DS-UWB technique is an RF impulse radio technique, coding RF pulses with direct sequence orthogonal codes. The transmission data rate is adjusted by changing the processing gain to provide flexibility in data-rate coverage choices for accommodating different applications. The DS-UWB solution divides the UWB spectrum into two channels, shown in Figure 8.28(a). The upper channel occupies approximately 2 GHz of the UWB spectrum and it is centered around 4.1 GHz. The lower band occupies approximately 4 GHz of the UWB spectrum centered at around 8.2 GHz. The region between the two bands is 5–6 GHz, and it is left empty to avoid interference with

[2] The traditional real-time location systems using Wi-Fi signals for inventory check have an accuracy of a few meters and they need fingerprinting of the area (see [Pah19]).

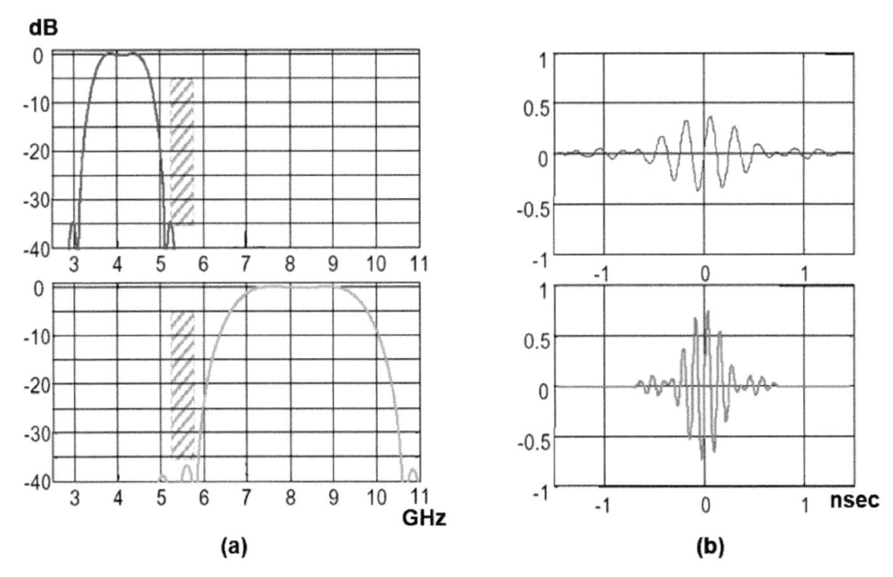

Figure 8.28 Two bands for DS-UWB IEEE 802.15.3a. (a) Frequency band. (b) The pulses as they are seen in time response [Koh04].

IEEE 802.11 devices operating in 5−6 GHz ISM bands. Figure 8.2(b) shows the RF pulses in the time domain (the inverse Fourier transform of the frequency domain). The narrower band on top generated RF pulses with a duration of approximately 2 ns with variations at the rate of 4.1 GHz. The wider bandwidth created narrower pulses with duration of approximately 1 ns modulated over 8.2 GHz. The chip rates in the low and high bands are 1.368 and 2.736 Gcps, respectively. Different data rates are derived from these basic rates and three different forward error control coding schemes. M-ary biorthogonal keying (MBOK) direct sequence spreading codes in this standard is implemented on a trinary (−1, 0, 1) voltage level. The standard also adds error correcting codes with three different coding rates. Different lengths of the code spreading and error correcting codes enable flexible implementation of a wide range of data rates from few tens of Kbps up to values close to a Gbps [Koh04]. The MB-OFDM approach divided the band into 15 non-overlapping 500 MHz channels providing an opportunity to the designer to implement a multi-band operation with data rate flexibility.

The IEEE802.153a activity lost its momentum as a wireless communication standard in 2009 when FCC released the unlicensed mmWave bands between 57 and 64 GHz. The mmWave bands provide the same UWB spectrum of approximately 7 GHz without overlay with the ISM bands. In addition, moving to higher center frequencies enabled the design of smaller antenna sizes opening a window of opportunity for massive MIMO implementation to support a higher number of streams to achieve higher bit rates. However, the legacy of precise positioning features of the UWB moved the DS-UWB physical layer specification to the IEEE 802.15.4a [Ye11]. At the time of this writing, DecaWave and a few other companies have introduced UWB precise positioning systems to the market. The mmWave technology was pursued by IEEE 802.15.3c [Zhu10], but it was overshadowed by IEEE 802.11ad and 5G activities in mmWave.

8.5.3 IEEE 802.15.6 WBAN

In the past decade, miniaturization and cost reduction of semiconductor devices have allowed the design of small low cost computing and wireless communication devices used as sensors in a variety of popular wireless networking applications, and this trend is expected to continue over the next two decades. It is expected that a myriad of new applications designed around sensor technologies will emerge to stimulate the world economy for another round of industrial growth. One of the most promising areas of economic growth associated with this industry is that of body sensor networks that are also referred to as BANs [Yan06]. These networks are expected to connect wearable and implantable sensory nodes together and with the Internet to support numerous applications ranging from traditional externally mounted temperature meters or implanted pacemakers to emerging blood pressure sensors, eye pressure sensors for glaucoma, and smart pills for health monitoring and precision drug delivery. Several technical challenges regarding size and cost, energy requirements, and wireless communication technology is under investigation, and at the heart of these investigations is the understanding of radio propagation in and around the human body.

To support the growth of this industry, the FCC has allocated specific bands for Medical RadioCommunication Services (MedRadio) bands [FCC09] and the IEEE 802.15.6 standard addresses standardization of these emerging technologies. The IEEE 802.15.6 standard intended to define the technologies and models for characteristics of the medium for wearable and implanted sensor networks [Aoy08][Kim08][Hag08a][Hag08b]. They classified the BAN application into scenarios and tried to define standard channel models, PHY specifications, and MAC layers for them. Figure 8.29 shows the overall architecture of an IEEE 802.15.6 BAN connecting implant and body mounted sensor devices to a body mounted base station and an external access point connecting to the Internet. The characteristics of the implant and body mounted devices are different because they may have different sizes, communications, localization, and power consumption requirements and capabilities. The body mounted base station is usually mounted on the belt above the hips of a person and carries larger batteries and has extended computational capabilities and onboard memory. The results of the sensing and other data can be delivered to the Internet either in real time or with opportunistic timing. As an example, in capsule endoscopy, the capsule inside the human digestive system sends the video to the body mounted base station that saves the information. Later, when the patient visits the doctor, the video is transferred to a computer for further processing and diagnosis. In a pacemaker application, a doctor may ask the patient to get close to a Wi-Fi access point at certain times of the day to send a sample of the heart signal for remote monitoring of a patient's heart. From the wireless networking point of view, the system should be designed to accommodate all these situations. The proper design of the network involves understanding and modeling the channel behavior and the design of the physical and MAC layer to support envisioned applications.

Radio propagation measurements and modeling for wireless networking applications begin with defining measurement scenarios and frequency bands. For each scenario, a measurement campaign is conducted, and the results of measurements are postprocessed and analyzed to develop statistical models for the behavior of the channel in each scenario. These models are then used for performance measurement and evaluation of alternative wireless networking solutions. Figure 8.29 illustrates a generic scenario involving implant and body-surface sensors and base stations and how they get connected to an external access point. IEEE 802.15.6 considers seven measurement scenarios for BAN channel modeling. We have four different classes of radio propagation models: implant to implant, implant to body surface and external access point, body surface to body surface, and body surface to the external network connection. These classes are further categorized into scenarios; we have one scenario for an implant to implant, two scenarios for each of the other channel model classes, which form a total of seven scenarios, summarized in the left top part of Figure 8.29. In Section 3.4.3, we presented the implant to body-surface models recommended by this standard.

The unlicensed frequency bands considered by IEEE 802.15.6 for BAN applications include the MedRadio bands [FCC09] from 401–406 MHz for measurement scenarios with implant sensors (the first two

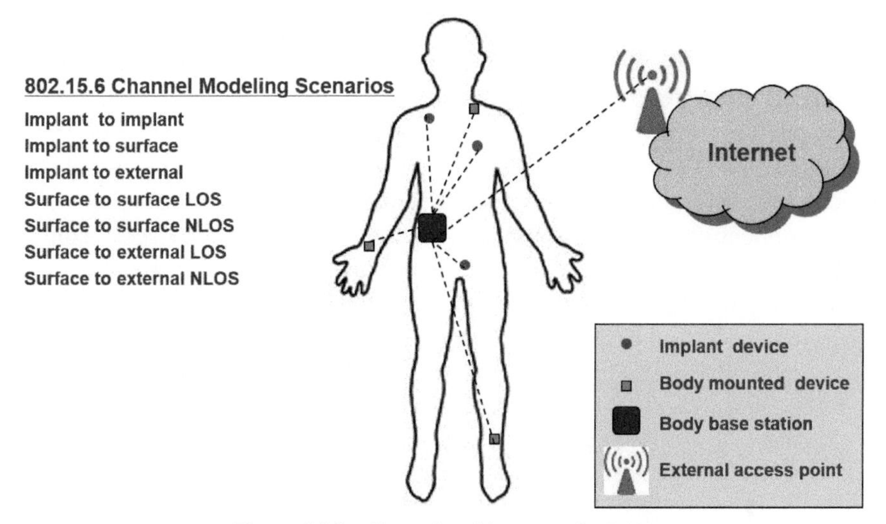

Figure 8.29 General architecture of a BAN.

802.15.6 scenarios) and industrial, scientific and medical (ISM) band and UWB frequency bands for non-implant sensors (last five scenarios). The MedRadio bands were released by the FCC to replace the medical implant communication system and the medical implant telemetry system bands from 402–405 MHz [FCC03]. All these bands are allocated by the FCC for low power and unlicensed BAN implant applications. Medical implant communication bands were originally used for data transmission to help diagnostic or therapeutic functions using external medical processing transceiver and implanted medical devices or between implanted medical devices. The medical implant telemetry system bands, which operate between 403.5 and 403.8 MHz, were used to provide transmission of one-way non-voice data periodically from an active implant to an external receiver. The MedRadio band is 2 MHz wider than medical implant and provides wider flexibility to accommodate a new types of devices. The ISM bands between 902–928, 2400–2483.5, and 5.15–5.35 MHz are the first unlicensed bands released for wireless local communication since 1985. The UWB 3.1–10.6 GHz band, released by FCC in 2003, is also an unlicensed band devoted to low power short distance communications. The IEEE 802.15.6 standard defines channel models for BAN applications at these frequencies and a few other frequencies as well [Yaz10].

The design of the physical and MAC layer for BANs is very much application dependent. For example, continual monitoring of the heart-beat may need a real-time low speed link to the Internet, while transmission of a few pictures every second for endoscopy capsule diagnostics may need a reliable low power transmission using strong error control coding, but processing of the data may be done after data collection. Certainly, for all BAN applications, power efficient modulation, and medium access control methods are needed in principle, and several researchers are working on these topics. The important fundamental issue is also localization of objects inside the human body and discovery of methods for localization that is suitable for body applications [Pah15].

8.5.4 IEEE 802.15.7 and Optical Wireless

The legacy IEEE 802.11 diffuse infrared (DFIR) wireless optical solution was a completed standard that did not meet a commercial success comparable to the RF spread spectrum options for the legacy 802.11. However, research in wireless optics was continual to take advantage of the unique features of wireless optical communications [Pah85]. Optical wireless signals for indoor applications are contained in the room that provides physical security against eavesdropping from outside of the room. Although, optical wireless signals are electromagnetic signals like F, but their frequency of operation is above THz that is well beyond 300 GHz regulated by FCC. As a result, the entire bandwidth is available for networking without any frequency regulation. The latest hype in the application of optical wireless is communications using visible light, like what we use in our rooms. The new marketing label for this movement is Li-Fi possibly because it rhymes with Wi-Fi and reminds success. Figure 8.30(a) shows the spectrum of visible and infrared lights. Figure 8.30(b) shows a typical light emitting diode (LED) that is a wireless optical transmitter and a typical photo sensitive diode (PSD) that is the receiver in wireless optics.

The DFIR option of the legacy 802.11, described in Section 7.5.1, used the very inexpensive LED devices and silicon detectors, which had emerged in the evolution of remote control technology for short-range wireless control of TV and other appliances. More expensive wireless LED and photo sensitive devices are used for focused beam laser communications which support much higher data rates. Laser beams do the same thing that massive MIMO does to RF; they focus the beams allowing multiple streaming to enable spatial diversity to increase the capacity of wireless to a theoretical

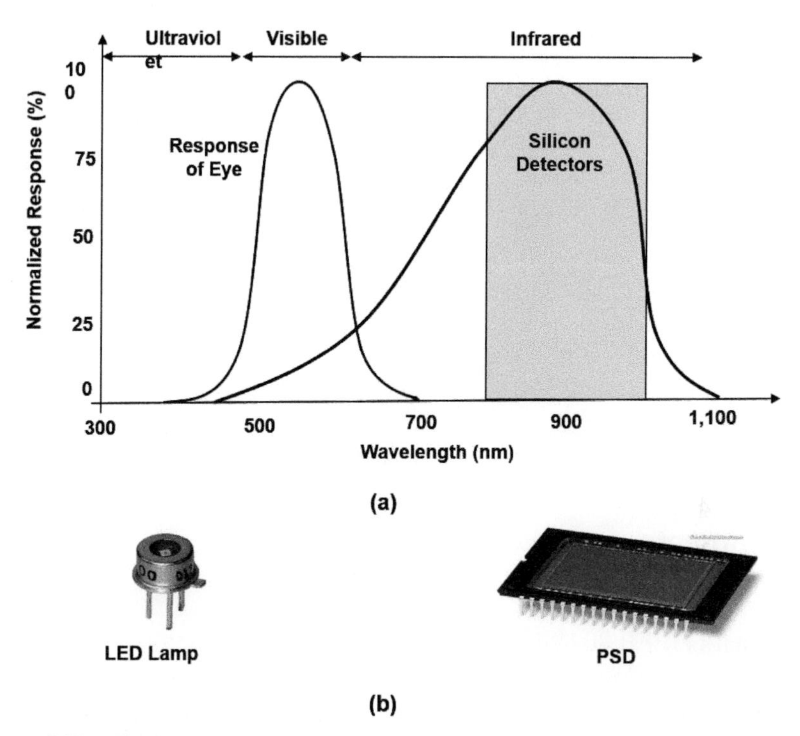

Figure 8.30 Elements of wireless optical communications. (a) Visible and infrared spectrums. (b) Sample light emitting diode (LED) lamp and photo sensitive diode (PSD).

infinite. A focused beam is indeed a "wireless" "wire"; it is a guided media, and guided media has infinite capacity. Implementation of the laser beam can be less expensive than massive MIMO and that opens the new marketing name of THz communications for 6G and beyond. The diffused infrared (DFIR) optical signals were first adopted for wireless local area network around 1980 [Pah85], but Li-Fi gained momentum in mid-2010's, after the invention of inexpensive LED/PSD technology for visible lights [Tso14]. The IEEE 802.15.7 began in January 2009 to develop a standard for Li-Fi. Figure 8.31 shows a basic indoor application scenario of Li-Fi for a PC, laptop, and a smartphone communicating with ceiling mounted boxes and light bulbs. The bulb changes its glowing intensity at the communication rate and that is transparent to the eye. This way, a high-speed communication link overplayed an appliance light.

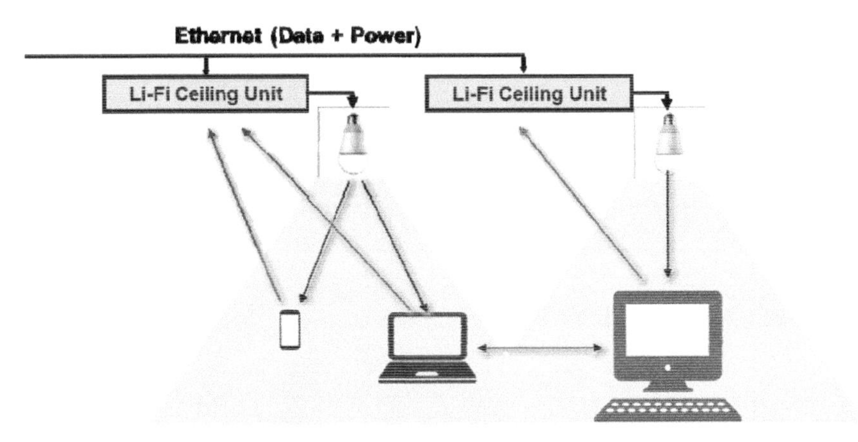

Figure 8.31 The basic application scenario for Li-Fi.

Assignments

Questions

1. What is IEEE 802.15 and what is its relation to Bluetooth, ZigBee, and BAN technologies?
2. What are the differences between IEEE 802.15 device specifications and the device specifications of IEEE 802.11 WLAN devices?
3. Name the four states that a Bluetooth terminal can take and explain the difference among these states.
4. Name the three classes of applications that were originally considered for Bluetooth technology and identify those which are similar to 802.11 WLAN technologies.
5. What are the similarities and differences between the FHSS transmission scheme used in IEEE 802.11 and that used in Bluetooth in terms of data rate, modulation technique, available frequencies for hopping, speed of the hop, and the number and pattern of the hops?
6. What are the differences between *ad-hoc* networking solutions offered by 802.11 and 802.15?
7. What is the difference between a full-function device and a reduced-function device in IEEE 802.15.4?
8. What is the difference between a channel page and a channel number in IEEE 802.15.4?
9. Differentiate between the star topology and the cluster-tree topology in IEEE 802.15.4 based sensor networks.

10. What are the differences between the frame format and MAC protocol of Bluetooth and IEEE 802.11 FHSS?
11. How many different voice services do Bluetooth support and how are they differentiated from one another?
12. How many different symmetric and asymmetric data services does Bluetooth support?
13. What is the maximum supported asymmetric packet data rate in Bluetooth? How many slots per hop does it use? What is its associated data rate in the reverse channel?
14. Compare the header and access code of Bluetooth with the physical layer convergence protocol (PLCP) header of FHSS-based IEEE 802.11.
15. What are the differences between the implementation of paging and inquiry algorithms in Bluetooth?
16. What are the differences between the frame format and MAC protocol of ZigBee and IEEE 802.11 DSSS?
17. What are the fundamental differences between supported data rates, packet format, and MAC layer of Bluetooth and ZigBee?
18. What are the differences between the protocol stack of IEEE 802.11 and that of ZigBee?
19. What is a BAN? Which standardization activity regulates it and what are the typical applications supported by this technology?
20. What are the computational methods used for radio propagation analysis inside the human body?

Instructor's solution available on River Publishers' website:
https://www.riverpublishers.com/book_details.php?book_id=919

Problem 1

Draw and explain the complete protocol stack for the implementation of e-mail application over Bluetooth.

Problem 2

Consider that encoded voice in Bluetooth is at 64 Kbps in each direction.

1. Using the packet format for the HV1 channels, show that these packets are sent every six slots.
2. Using the packet format for the HV2 channels, calculate how often these packets are sent.
3. Repeat (b) for HV3 packets.

Problem 3

1. What is the hopping rate of frequencies in Bluetooth? How many bits are transmitted in each one-slot packet transmission?
2. If each frame of an HV3 voice packet in Bluetooth carries 80 bits of the samples of speech, what is the efficiency of the packet transmission (ratio of the overhead to the overall packet length)?
3. Determine how often an HV3 packet has to be sent to support 64 Kbps voice in each direction.
4. The DH5 packets carry 2712 bits for every five-slot packet. Determine the effective data rate in each direction in this case.

Problem 4

Repeat Examples 9.7 and 9.8 for all other data rates supported by Bluetooth (see Table 9.1).

9

Cellular Wireless 2G–6G Technologies

9.1 Introduction

Cellular and Wi-Fi are the dominant wireless infrastructures for Internet connection worldwide. Wi-Fi supports a higher data rate per user and does not involve monthly fee payment. As a result, it is the first choice for devices and applications. Cellular networks provide comprehensive coverage and support mobile connectivity. Traditionally, cellular antennas are installed outdoors with a wider area of coverage connecting many users to a cell tower and they operate in licensed bands that are more expensive and scarce. In addition, they are designed to support both circuit-switched and packet-switched networks. Cellular networks evolved from telephone applications and they still support connection-based telephone with ISDN addressing. Since the analog plain old telephone service (POTS) are gradually disappearing from home and office markets, a cellular telephone is surviving as a technology supporting connection-based telephone services with ISDN addressing.

Wi-Fi was originally designed for data and then it began integrating voice through voice over IP (Vo-IP). Today, both Wi-Fi and cellular networking industry use orthogonal frequency division multiplexing (OFDM)/multiple-input multiple-output (MIMO) for transmission of data over the air interface. The difference between the two technologies is in the method they form and send the packets of information, signaling, and management at the medium access control (MAC) protocol. Traditionally, the connection-based cellular networks have user data plane, signaling plane, and management planes implemented on different types of packets and sent through different logical channels. The flow of data and control and management of packets in cellular networks are implemented on a centrally controlled MAC. With a large number of users per cell, expensive licensed bands, and higher mobility in the cars, cellular infrastructure is much more complex. This complexity lies in the need for a more comprehensive deployment strategy, frequency channel management, mobility management for handoff and roaming, security to avoid fraud, and an accounting system to charge the subscribers. The cellular network technologies have evolved to address these challenges.

To understand the differentiation of the cellular networking technology from Wi-Fi, it is helpful to review the evolution of cellular networks. The details of this evolution demonstrate the differences between circuit-switched and packet-switched network infrastructures as well as a need for wireless access to them.

9.1.1 Evolution of Cellular Networks

So far in this book, we have described the Internet, and we introduced the Ethernet, Wi-Fi, and sensor networks to connect to the Internet. The Internet is a connectionless network with an IP/MAC addressing method. We differentiated the core of the Internet with the adoption of routers as interconnecting elements and we discussed hubs, bridges, and LAN switches as interconnecting elements of the Ethernet. We also referred to the PSTN as a connection-based network using ISDN addressing and adopting long-haul switches or virtual switching as their interconnecting elements. The PSTN core was a digital network, but it was originally designed to connect analog POTS. The analog signal arriving at the end office was converted to a 64 Kbps digital signal to be carried over the core. The core PSTN indeed supported two planes for information transmission, one for voice and one for data control and connection management.

The first generation (1G) analog systems imitating a POTS with wireless access. Around 1990, 2G digital wireless cellular technology introduced the first end-to-end digital service to connect to the PSTN. GSM was the most successful 2G technology and it was called a second generation (2G) technology because it was replacing the 1G analog cellular telephone technology. In addition to digital telephone service, GSM was also supporting connection-based data of up to 9600 bps plus a short messaging system (SMS) as the first packet switching service offered in wireless networks. The packet-switched SMS service was running on the data control and management network and the voice was on the traffic plane. The unexpected success of the Internet and its penetration in homes increased the popularity of data applications and packet switching. In the mid-1990s, intermediate packet-switched data services, such as general packet radio service (GPRS), emerged to connect a cellphone to the Internet over the wireless digital cellular infrastructure. These services are sometimes referred to as 2.5G and they are connected to both PSTN and the Internet core.

At the turn of the 21st century, the 3G cellular network emerged with the 3rd Generation Partnership Project (3GPP) as the core of that wireless infrastructure. The 3GPP was an IP network carrying both circuit-switched and packet-switched user data. The 3G interface to 3GPP was the code division multiple access (CDMA) technology originally invented by QUALCOMM and later adopted in the Pan European 3G universal mobile telephone services (UMTS). The next step in the evolution of cellular was the emergence of 4G and 5G access to 3GPP core to support higher data rate using OFDM and MIMO technologies, like the Wi-Fi.

9.1.2 The Cellular Concept

Cellular topology is a special case of an infrastructure multi-base station (BS) network configuration that exploits the *frequency reuse* concept. Radio spectrum is one of the scarcest resources available and every effort must be made to find ways of utilizing the spectrum efficiently and to employ architectures that can support as many users as theoretically possible with the available spectrum. This is extremely important especially today considering the huge demand for capacity. Spatially reusing the available spectrum so that the same spectrum can support multiple users separated by a distance is the primary approach for efficiently using the spectrum. This is called *frequency reuse*. Employing frequency reuse is a technique that has its foundations in the attenuation of the signal strength of electromagnetic waves with distance. For instance, in a vacuum or free space, the signal strength falls as the square of the distance. This means that the same frequency spectrum may be employed without any interference for communications or other purposes, provided the distance separating the transmitters is sufficiently large, and their transmit powers are reasonably small (depending on the reuse distance). This technique has been used, for example, in commercial radio and television broadcasts where the transmitting stations have a constraint on the maximum power they can transmit so that the same frequencies can be used elsewhere. The cellular concept is an intelligent means of employing frequency reuse. Cellular topology is the dominant topology used in all large-scale terrestrial and satellite wireless networks. The concept of cellular communications was first developed at the Bell Laboratories in the 1970s to accommodate many users with limited bandwidth.

By cellular radio, we mean deploying much low-power BSs for transmission, each having a limited coverage area. In this fashion, the available capacity is multiplied each time a new BS or transmitter is set up since the same spectrum is being *reused* several times in a given area. The fundamental principle of the cellular concept is to divide the coverage area into a number of contiguous smaller areas each served by its own radio BS. Radio channels are allocated to these smaller areas in an intelligent way to minimize the interference, provide adequate performance, and cater to the traffic loads in these areas. Each of these smaller areas is called a *cell*. Cells are grouped into *clusters*. Each cluster utilizes the entire available radio spectrum. The reason for clustering is that adjacent cells cannot use the same frequency spectrum because of interference. So, the frequency bands must be split into chunks and distributed among the cells of a cluster. The spatial distribution of chunks of radio spectrum (which are called sub-bands) within a cluster has to be done in a manner such that the desired performance can be obtained. This forms an important part of network planning in cellular radio.

Two types of interference are important in such a cellular architecture. The interference due to using the same frequencies in cells of different clusters is referred to as *co-channel interference*. The cells that use the same set of frequencies or channels are called *co*-channel cells. The interference from frequency channels used within a cluster whose sidelobes overlap is called *adjacent channel interference*. The allocation of channels within the cluster and between clusters must be done to minimize both.

The cellular concept can increase the number of customers that can be supported in the available frequency spectrum as illustrated by the following examples by deploying several low-power radio transmitters (see Section 5.2).

9.1.3 Cellular Hierarchy

There are three reasons to use a hierarchical cellular infrastructure supporting cells of different sizes. One is to extend the coverage to the areas that are difficult to cover by a large cell. For example, cells designed to cover suburban areas have antennas with tall towers and cover a large area. Signals from these antennas, however, cannot propagate sufficiently into urban canyons or indoor environments. For urban canyons, we need to install antennas at lower heights, and in indoor areas, we may mount the antennas on walls to provide comprehensive coverage. Antennas mounted in these locations are of low power and cover a smaller area resulting in the creation of a smaller sized cell. The second reason to have a cellular hierarchy is to increase the capacity of the network for those areas that have a higher density of users. Imagine the number of cellular phone users in the world trade center and compare it with the number of mobile users on an interstate highway. To support the larger subscriber demand and higher traffic in smaller areas, we need to increase the number of cells by reducing their sizes. The third reason is that, sometimes, an application needs certain coverage. Consider the increasing number of wireless devices that we are carrying in our bags these days and the increasing need for communication between these devices. This necessitates extremely small sized cells that provide a wireless network for connecting laptops or notepads to cellular phones.

In a modern deployment of a cellular network, several cell sizes are used to provide a comprehensive coverage supporting traffic fluctuations in different geographical areas and supporting a variety of applications. One way of dividing the cells into a hierarchy is to define the following cell sizes:

Personal-cells: These are the smallest unit of the cellular hierarchy used for connection of personal equipment such as laptops, notepads, and cellular telephones. These cells need to cover only a few meters where all these devices are in the physical range of the user.

Pico-cells: These are small cells inside a building that support local indoor networks such as wireless LANs. The size of these networks is in the range of a few tens of meters.

Micro-cells: These cells cover the inside of streets with antennas mounted at heights lower than the rooftop of the buildings along the streets. They cover a range of hundreds of meters and are used in urban areas to support personal communication services (PCS).

Macro-cells: Macro-cells cover metropolitan areas and they are the traditional cells installed during the early phases of cellular telephony. These cells cover areas on the order of several kilometers and their antennas are mounted above the rooftop of typical buildings in the coverage area.

Mega-cells: Mega-cells cover nationwide areas with ranges of hundreds of kilometers and are mainly used with satellites.

Figure 9.1 illustrates the relationship between different cells with example applications. An ideal network has a hierarchy of these cells to cover airplane travelers with mega-cells, car drivers in sub-urban areas with macro-cells, pedestrians in the streets via micro-cells, indoor users with pico-cells, and connect personal equipment with femtocells. The focus of this chapter is on cellular telephone networks that typically deploy macro-cells to cover their service areas.

The cellular infrastructure with a good deployment ensures comprehensive coverage for the cellular networks on highways, city streets, and inside a commercial and residential buildings. To ensure mobility of a device with

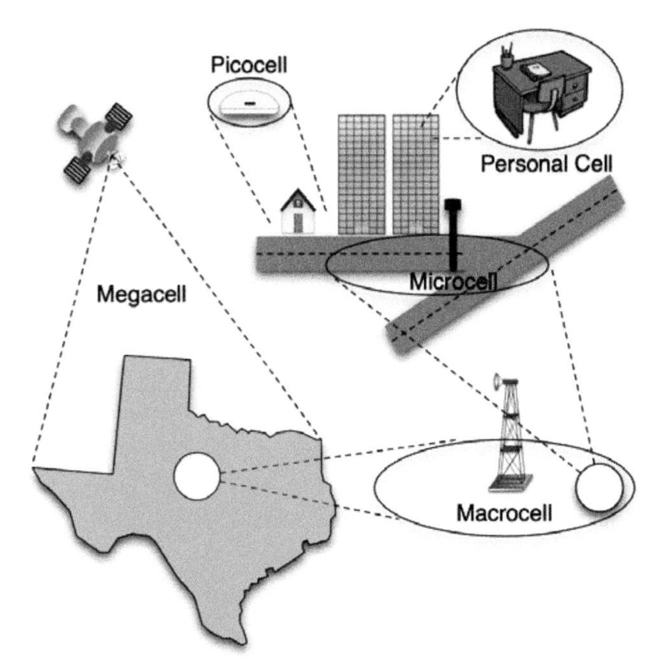

Figure 9.1 Cellular hierarchy.

a wireless connection to this network, we need serious attention to a few mechanisms to maintain the mobility. To explain these mechanisms, we need to address the infrastructure of cellular networks first, and, for that, we need an example. We give the 2G GSM cellular infrastructure as the first digital infrastructure for wireless connection to the PSTN.

9.2 Connection-Based 2G Cellular Network Infrastructure

The 2G digital cellular was the first end-to-end digital service over the PSTN. Specifications of these cellular wireless networks are complex with detailed descriptions of the terminal, fixed hardware backbone, and software databases that are needed to support the operation. To describe such a complex system, in the beginning, a reference model or overall architecture is needed to provide an understanding of the network elements and operation, and, later, it is possible to divide the system into subsystems. Figure 9.2 presents an overall view of the GSM as an example of a 2G reference model with typical hardware and software elements used in a cellular network

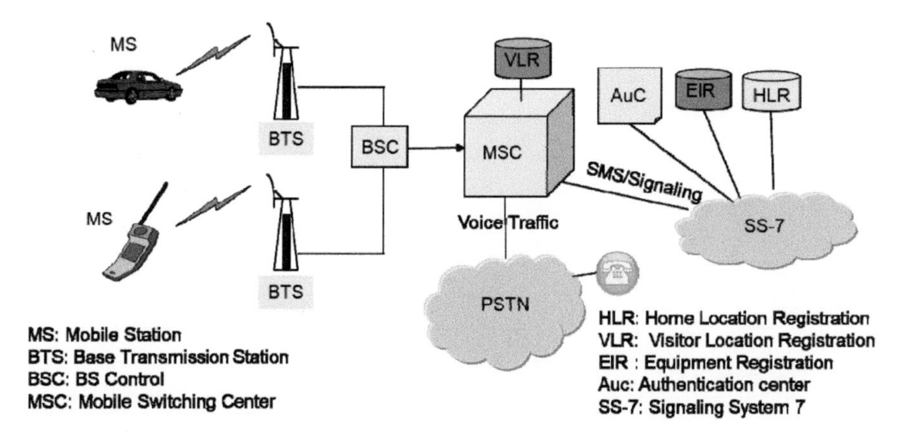

Figure 9.2 General 2G architecture for access to the PSTN.

divided into hardware and software elements. These elements can be further classified into three subsystems: mobile station (MS), base station subsystem (BSS), and network and switching subsystem (NSS). This division of the architectural elements was adopted from [Hau94] and we will follow that for the description of the system elements. The description of the elements of this reference model is as follows.

9.2.1 Mobile Station

MSs form the interface for exchanging information with the user and modify the information onto the transmission protocols of the air interface to communicate with the BSS. The user information is communicated with the MS through a microphone and speaker for speech, keypad and display for short messaging, and cable connections to data terminals for other applications. The MS has two elements. The first element is the ***mobile equipment (ME)*** which is a piece of hardware that the customer purchases from the equipment manufacturer or their dealers. This hardware piece contains all the components needed for the implementation of the protocols to interface with the user and the air interface to the BSS. The components include speakers, microphones, keypad, and the radio modem. Therefore, the ME is an expensive piece of hardware. To encourage more users to subscribe to the wireless services, several service providers in the early days of the cellular industry, and even today subsidize the price of the MEs.

The second element of the MS in the reference model of Figure 9.3 is the ***subscriber identity module (SIM)*** that is a smart card issued at

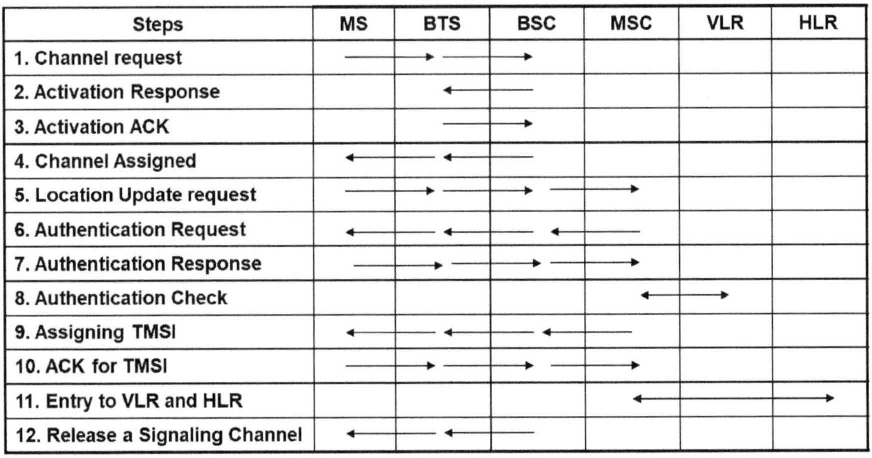

Steps	MS	BTS	BSC	MSC	VLR	HLR
1. Channel request	→	→				
2. Activation Response		←				
3. Activation ACK		→				
4. Channel Assigned	←	←				
5. Location Update request	→	→	→			
6. Authentication Request	←	←	←			
7. Authentication Response	→	→	→			
8. Authentication Check				↔		
9. Assigning TMSI	←	←	←			
10. ACK for TMSI	→	→	→			
11. Entry to VLR and HLR				↔	↔	
12. Release a Signaling Channel	←	←				

Figure 9.3 Registration procedure in a typical digital cellular network

subscription time identifying the specifications of a user such as address and type of service. The calls in the system are directed to the SIM rather than the terminal. Short messages are also stored in the SIM card. Although implementing a SIM is a simple concept, it has a significant impact on the way that a user transacts with the service provider. A SIM card carries every user's personal information that enables several useful applications. For example, people visiting different countries but not keen on making calls at their home number can always carry their own terminal and purchase a SIM card in every country that they visit. This way, they avoid roaming charges at the expense of having a different contact number. Since SIM cards carry the private information for a user, a security mechanism is implemented in the network that asks for a four-digit PIN number to make the information on the card available to the user. This can sometimes play a negative role because users are not keen on remembering too many passwords. Using SIM cards was not a possibility with the analog cellular systems, and some systems such as the North American digital cellular standards have not implemented this option.

9.2.2 Base Station Subsystem

The BSS communicates with the MS through the wireless air interface and with the wired infrastructure through wired protocols. In other words, it translates between the air interface and fixed wired infrastructure protocols.

The needs of the wireless and wired media are different since the wireless medium is unreliable, bandwidth limited, and needs to support mobility. As a result, protocols used in the wireless and wired mediums are different. The BSS provides for the translation of these protocols. For example, consider speech conversion. The user's speech signal at the MS is converted into a voice encoding scheme around 10 Kbps (for bandwidth efficiency) with a speech coder and is communicated over the air interface. The backbone wired network uses 64 Kbps PCM digitized voice in the PSTN hierarchy. Conversion from analog to 10 Kbps voice takes place at the MS and the change from 10 to 64 Kbps coding takes place at the BSS. As another example, the signaling format to establish a connection in wired networks is the multi-tone frequency scheme used in POTS. Digital cellular systems, on the other hand, establish the call through the exchange of several packets. The translation of this communication into a dialing signal is made in the BSS.

As with speech coding and dialing, explained in the above examples, data transmission protocols over the air interface are different from those in the wired infrastructure. All these translations are performed at the BSS. To implement packet data services on the same air interface, the BSS also separates packet switching data from the circuit-switched PSTN traffic to direct them toward the PSTN and the Internet separately.

There are two architectural elements in the BSS. The ***base transceiver system (BTS)*** is the counterpart of the MS for physical communication over the air interface. The BTS components include a transmitter, a receiver, and signaling equipment to operate over the air interface and it is physically located at the center of the cells where the BSS antenna is installed. The second architectural element of the BSS is the ***base station controller (BSC)*** that is a small switch inside the BSS in charge of frequency administration and handover among the BTSs inside a BSS. The hardware of the BSC in a single BTS site is located at the antenna and in multi-BTS systems in the switching center where other hardware elements of the NSS are located. One BSS may have anywhere from one up to several hundred BTSs under its control [RED95].

9.2.3 Network and Switching Subsystem

The NSS is responsible for network operation. It provides for communications with other wired and wireless networks as well as support for registration and maintenance of the connection with the MSs. The NSS could be interpreted as a wireless specific switch that communicates with other switches in the

PSTN and, at the same time, supports functionalities that are needed for a cellular mobile environment. The NSS is the most elaborate element of the cellular network and it has one hardware, MSC, and four software elements: visitor location registration (VLR), home location register (HLR), equipment identification register (EIR), and authentication center (AUC).

Mobile switching center (MSC) is the hardware part of the wireless switch that can communicate with PSTN switches using the signaling system-7 (SS-7) protocol, commonly used in the PSTN, as well as other MSCs in the coverage area of a service provider. The MSC also provides the network specific information on the status of the mobile terminals.

HLR is database software that handles the management of the mobile subscriber account. It stores the subscribers' address, service type, current location, forwarding address, authentication/ciphering keys, and billing information. In addition to the telephone number for the terminal, the SIM card is identified with an international mobile subscriber identity (IMSI) number that is totally different from the ISDN telephone number. The IMSI is used totally for internal cellular networking applications. For example, the telephone number of a subscriber in Finland could be 358-40-770-5246. The first three digits are the country code, the next two are the digits for the specific MSC, and the rest are the telephone number. The IMSI of the same user can be 244-91 followed by a 10-digit number that is totally different from the actual telephone number. The first three digits of the IMSI identify the country, viz. Finland and the next two digits, the billing company.

VLR is a temporary database software similar to the HLR identifying the subscribers that are visiting inside the coverage area of an MSC. The VLR assigns a temporary mobile subscriber identity (TMSI) that is used to avoid using IMSI on the air to protect the user billing number. Maintenance of two databases at home and at the visiting site allows a mechanism to support call routing and dialing in a roaming situation where the MS is visiting the coverage area of a different MSC. The mechanism of holding two databases (home and visiting) to support mobility is used almost in all mobile networks.

AUC holds different algorithms that are used for authentication and encryption of the subscribers. Different classes of SIM cards have their own algorithms and the AUC collects all these algorithms to allow the NSS to operate with different terminals from different geographical areas.

EIR is another database managing the identification of the ME against faults and theft. This database keeps the international ME identity that reveals the manufacturer, country of production, and terminal type. Such information can be used to report stolen phones or check if the phone is operating according to the specification of its type. The implementation of the EIR is left optional to the service provider.

9.3 Connection-Based 2G Cellular Network Operation

Now that we have described all the hardware and software elements of a typical circuit-switched 2G digital cellular network, we can describe how different functionalities of the network are implemented with these elements. Four mechanisms are embedded in all circuit-switched voice-oriented wireless networks that allow mobile to establish and maintain a connection with the network. These mechanisms are *registration*, *call establishment*, *handoff* (or handover), and *security*. Registration takes place as soon as one turns the mobile unit on in a new environment, call establishment occurs when the user initiates or receives a call, handover helps the MS to change its connection point to the network, and security protects the user from fraud and eavesdropping. In this section, we describe the details of their implementation with examples using the reference architecture that was described in the last section. To illustrate the complexity of cellular wireless networks, when we discuss registration and call establishment, we compare these mechanisms with their counterparts in POTS.

9.3.1 Registration

When we subscribe to a POTS service, the telephone company brings a pair of wires to our home that is connected to a port of a switch in a PSTN end office. Then our telephone number is registered in a database in the network and our registration is fixed. Therefore, the connection and registration process for wired access to the network is a one-shot operation, and, after that, the connection is active and registration is valid as long as the subscription to the service is valid. With wireless access to a cellular network, each time that we turn on the MS, we need to establish a new connection and possibly establish a new registration with the network. We may connect to the network at different locations through a BS that may not be owned by our service provider. Therefore, a wireless network needs a registration process that is far more complex than the registration in wired networks.

Technically speaking, as we turn on an MS, it passively synchronizes to the frequency; bit and frame timings of the closest BS to get ready for information exchange with the BS. After this preliminary setup, the MS reads the system and cell identity to determine its location in the network. If the current location is not the same as before, the MS initiates a *registration* procedure. During a registration procedure, the network provides the MS with a channel for preliminary signaling. The MS provides its identity in exchange for the identity of the network, and, finally, the network authenticates the MS. The simplest connection takes place if the MS is turned on in the previous area and the most complex registration process occurs when the mobile is turned on in a new MSC area which needs changes in the entries of the VLR and HLR. The following example illustrated the complexity of the registration process when mobile is turned on in a new MSC.

Figure 9.3 shows the 12-step registration process in a typical digital cellular network that takes place when an MS is turned on in a new MSC area. During each step, a message is carried through a certain elements of the overall infrastructure of the network. In the first four steps, a radio channel is established between the MS and BSS to process the registration. In the next four steps, the NSS authenticates the MS. In the next three steps, a TMSI is assigned and adjustments are made to the entries in the VLR and HLR. In the final step, the temporary radio channel for communication is released and transmission starts over a traffic channel.

9.3.2 Call Establishment

Call establishment in POTS starts with a dialing process that transfers the number to the nearest PSTN switch where a routing algorithm finds the best connection through intermediate switches to the destination. After the establishment of the link, the last switch (end office) at the destination sends a signal back to the source to announce whether the destination is available or busy that is signaled to the user at the source. When the destination POTS terminal is off-hook, another signal is sent to the source end office to stop the waiting tone and establish the traffic line. In the mobile environment, we have two separate call establishment procedures for mobile-to-fixed and fixed-to-mobile calls. Mobile-to-mobile calls are a combination of the two. The following two examples provide the detailed procedure in a typical network for both types of call establishment.

The five-step procedure in POTS for call setup changes to a 15-step mobile originated call establishment procedure in a typical digital cellular network. As shown in Figure 9.4, the first five steps are like the registration process, except that these are done to prepare for call establishment. The next two steps start ciphering (encryption) to provide protection against

Steps	MS	BTS	BSC	MSC
1. Channel request		→	→	
2. Channel Assigned	←	←		
3. Call Establishment Request		→	→	→
4. Authentication Request	←	←	←	
5. Authentication Response		→	→	→
6. Ciphering Command	←	←	←	
7. Ciphering Ready		→	→	→
8. Send Destination Address		→	→	→
9. Routing Response	←	←	←	
10. Assign Traffic Channel		→	→	
11. Traffic Channel Established	←	←		
12. Available/Busy Signal	←			
13. Call Accepted	←	←	←	
14. Connection Established		→	→	→
15. Information Exchange	←			→

Figure 9.4 Steps in establishing a mobile originated call.

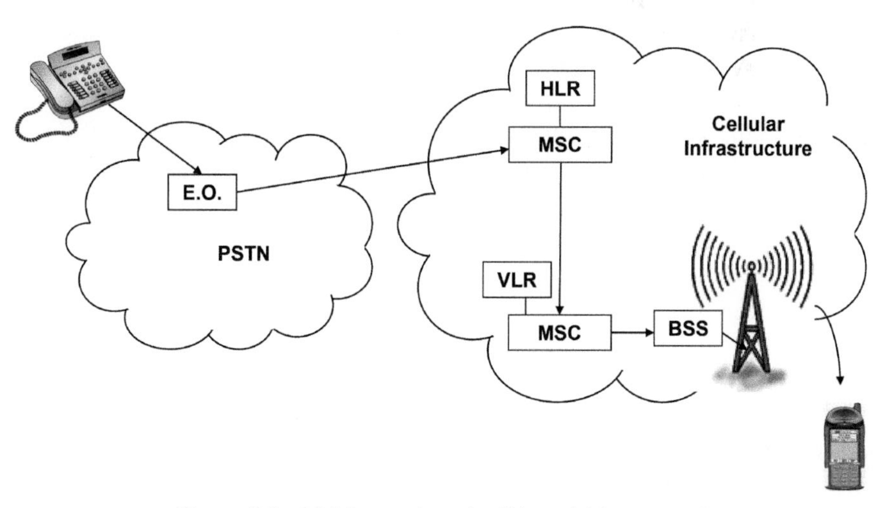

Figure 9.5 Mobile terminated call in a visiting network.

eavesdropping. The rest of the steps are like those in wired networks except that we have an additional traffic channel assignment procedure.

The most complicated call establishment is for the situation where a fixed telephone dials a mobile visiting another MSC. As shown in Figure 9.5, after dialing, the PSTN directs the call to the MSC identified by the destination address. The MSC requests routing information from the HLR. Since, in this case, the mobile is roaming around a different MSC, the address of the new MSC is given to MSC and it contacts the new MSC. At the destination MSC, the VLR initiates a paging procedure in all BSSs under the control of the MSC holding the registration. After a reply from the MS, the VLR sends the necessary parameters to the MSC to establish the link to the MS.

9.3.3 Handoff

Handoff refers to the switching of connections by an MS from one point of connection to the fixed network to another. The MS is initially connected to a BS and then it moves out of the service area of the BS to that of another BS. When this occurs, there must be protocols in place to *detect* that the MS is in a new service area, to *initiate* the handoff, and, finally, to *complete* the switch from the old BS to the new BS. There are two types of handoffs – internal and external. Internal handoff is between BTSs that belong to the same BSS and external handoffs are between two different BSSs belonging to the same MSC. Sometimes, there are handoffs between BSSs that are controlled by two different MSCs. In such a case, the old MSC continues to handle call management. Roaming between two MSCs in two different countries is prohibited and the call simply drops.

Handoff is initiated because of a variety of reasons. Signal strength deterioration is the most common cause for handoff at the edge of a cell. Other reasons include traffic balancing where the handoff is network oriented to ease traffic congestion by moving calls in a highly congested cell to a lightly loaded cell. The handoff could be synchronous where the two cells involved are synchronized or it may be asynchronous. Since the MS does not have to resynchronize itself in the former scenario, the handoff delay is much smaller (100 ms against 200 ms in the asynchronous case). The decision to make a handoff typically is the responsibility of the BSC or the MSC.

Figure 9.6 shows the details of messages for a handoff procedure between two BSSs that are controlled by one MSC in a typical cellular network. The details of handoff in different cellular networks vary substantially and depend on the technology. For example, GSM using time division multiple access

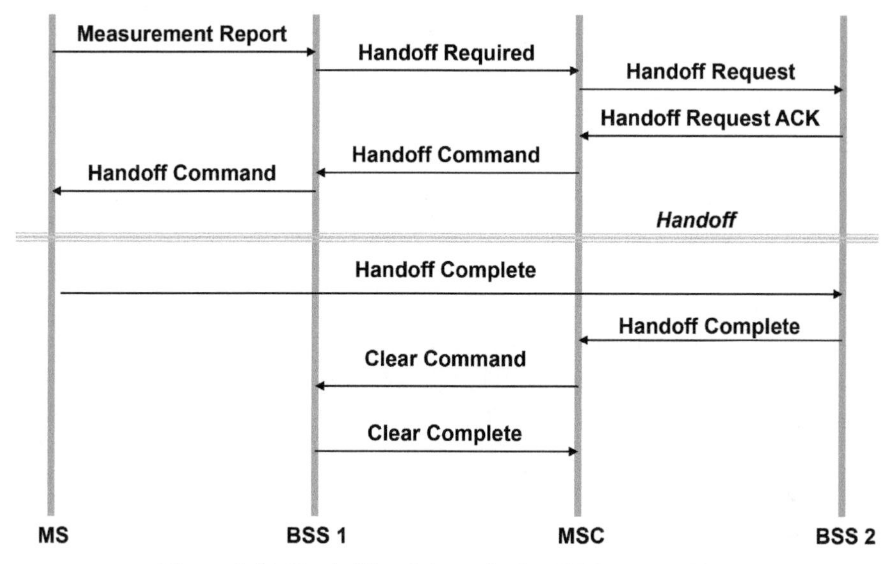

Figure 9.6 Handoff involving a single MSC but two BSSs.

(TDMA) technology adopts mobile assisted handoff. The BTS provides the MS a list of available channels in neighboring cells. The MS monitors the RSS from these neighboring cells and reports these values to the MSC. The BTS also monitors the RSS from the MS to make a handoff decision. Proprietary algorithms are used to decide when a handoff should be initiated. If a decision to make a handoff is made, the MSC negotiates a new channel with the new BSS and indicates to the MS that a handoff should be made using a handoff command. Upon completion of the handoff, the MS indicates this with a handoff complete message to the MSC.

The CDMA networks employ soft handoff, which refers to the process by which an MS is in communication with multiple candidate BSs before finally communicating its traffic through one of them. The reason for implementing soft handoff has its basis in the near–far problem and the associated power control mechanism. If an MS moves far away from a BS and continues to increase its transmit power to compensate for the near–far problem, it will very likely end up in an unstable situation. It will also cause a lot of interference to MSs in neighboring cells. To avoid this situation and ensure that an MS is connected to the BS with the largest received signal strength (RSS), a soft handoff strategy is implemented. An MS will continuously track all BSs nearby and communicate with multiple BSs for a short while, if necessary, before deciding which BS to select as its point of attachment.

The soft handoff procedure involves several BSs. A controlling primary BS coordinates the addition or deletion of other BSs to the call during soft handoff. The MS detects a pilot signal from a new BS and informs the primary BS. After a traffic channel is set up with the new BS, a frame selector join message is used to select signals from both BSs at the BSC/MSC. After a while, the pilot signal from the old BS starts falling and the MS will request its removal, which is achieved via a *frame selector remove* message. The pilot channels of each cell are involved in the handoff mechanism. The reason behind this is that the pilot channel provides MSs with a measure of the RSS. The MS maintains a list of pilot channels that it can hear and classifies them into different sets. Based on RSS values, thresholds, and timers, a handoff algorithm will decide which BSs or cell sectors to connect to at which time.

9.3.4 Security

Security in cellular systems is usually employed to prevent fraud via authentication, avoid revealing the subscriber number over the air, and encrypt conversations where possible. All of these are achieved using proprietary (secret) algorithms. Security requirements for wireless communications are very similar to the wired counterparts but are treated differently because of the applications involved and the potential for fraud. Different parts of the wireless network need security. Over the air, security is usually associated with the privacy of voice conversations in cellular systems. This is changing with the increasing use of wireless data services. Message authentication, identification, authorization, etc., also become issues with cellular networks and wireless local area networks (WLANs). Wireless networks are inherently insecure compared to their wired counterparts. The broadcast nature of the channel makes it easier to be tapped. Analog telephones are extremely easy to tap and conversations can be eavesdropped on using a radio frequency (RF) scanner. Digital cellular systems (2G and up) are much harder to tap and RF scanners cannot do the trick anymore, but since the circuitry and chips are freely available, it may not be hard for someone to break into a system that does not employ security.

Privacy Requirements of Cellular Networks:

A variety of control information is transmitted over the air in addition to the actual voice or data. These include call setup information, user location, user

ID (or telephone number) of both parties, etc. These should all be kept secure since there is a potential for misusing such information. Calling patterns (traffic analysis) can yield valuable information under certain circumstances. A flurry of calls between the CEOs of two major companies may indicate certain trends if it was discovered, even if the actual information in the calls was secure. Hiding such information is also important. In [WIL95], various levels of privacy are defined for voice communications.

We commonly assume that all telephone conversations are secure. While this is not true, it is possible to detect a tap on a wireline telephone. It is impossible to detect taps over a wireless link because of the broadcast nature of the medium. To provide the privacy that is equivalent to that of a wired telephone, for routine conversations, it may be sufficient to employ some sort of encryption that will take more than simply scanning and decoding to decrypt. Wilkes [WIL95] defines two levels of security as levels: zero and one. Level-0 privacy is when there is no encryption employed over the air so that anyone can tap into the signal. Level-1 privacy provides privacy equivalent to that of a wireline telephone call, one possibility being encrypting the over the air signal. To alert wireline callers about the insecure nature of a wireless call that is *not at all* encrypted, a "lack-of-privacy" indicator was suggested. For commercial applications, a much stronger encryption scheme would be required that would keep the information safe for more than several years. Secret key algorithms with key sizes larger than 80 bits are appropriate for this purpose. This is referred to as Level-2 privacy in [WIL95]. Encryption schemes that will keep the information secret for several hundreds of years are required for military communications and fall under Level-3 privacy. For wireless data networks, a bare minimum level would be to keep the information secure for several years. The primary reason for this is that wireless electronic transactions are becoming common. Credit card information, dates of birth, social security numbers, e-mail addresses, etc., can be misused (fraud) or abused (junk messages, for example). Consequently, such information should never be revealed easily. A Level-2 privacy will be essential for wireless data networks. In certain cases, Level-3 privacy is required. Examples are wireless banking, stock trading, mass purchasing, etc. Most cellular systems and WLANs employ strong encryption today and this is becoming a moot point.

Authentication and Integrity:

While privacy and confidentiality continue to be important issue in wireless networks, other security requirements are becoming significant in recent times. There has been widespread fraud and impersonation of analog cellular telephones in the past. Although this is more difficult with digital systems, it is not impossible. There is thus a need to correctly *identify* and *authenticate* a mobile terminal. Control messages need to be checked for integrity to ensure that spoofed messages do not cause the network to behave abnormally leading to widespread disruption of communications.

Implementation:

The SIM cards have a microprocessor chip that can perform the computations required for security purposes. Figures 9.7 and 9.8 show the principles of operation for authentication and ciphering in a typical cellular network. We have used GSM as an example, but similar procedures are adopted in other systems. A secret key K_i is stored on the SIM card and it is unique to the card. This key is used in two algorithms, A1 and A2, for authentication and confidentiality, respectively. For authentication purposes, shown in Figure 9.7, the secret key K_i is used in a challenge response protocol between the BSS and the MS. The secret key K_i is used to generate a privacy key K_c that is used to encrypt messages (voice or data) using the A2 algorithm. The control channel signals are encrypted using a third encryption algorithm. The size of a typical secret key K_i and the response to the challenge play an important role in the robustness of the security. Figure 9.8 illustrates the basic principles of ciphering in a typical cellular network. The challenge random string and the secret key K_i are used with Algorithm A2 to generate a new ciphering key K_c which is used with a third algorithm, A3, to cipher the data. Another aspect of security in cellular networks is that the secret key information is not shared between systems. Instead, a triple consisting of

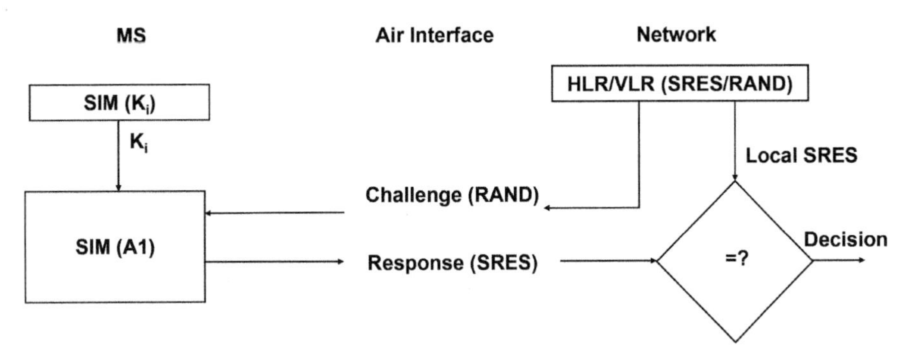

Figure 9.7 Basic principles of authentication in a cellular network.

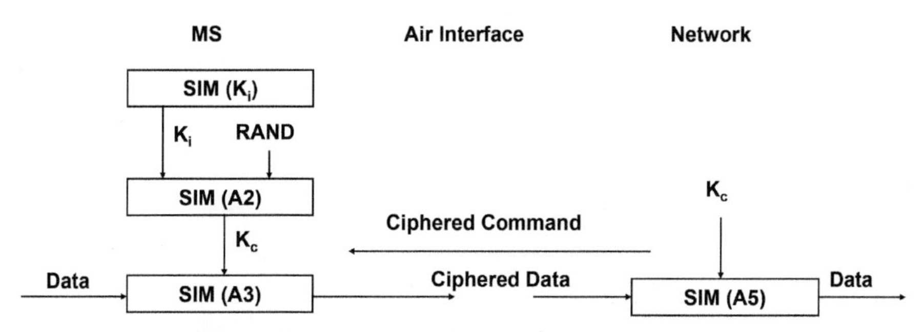

Figure 9.8 Basic principles of ciphering in a typical cellular network.

the random number used in the challenge, the response to the challenge, and the data encryption key K_c is exchanged between the VLR and the HLR. The VLR verifies if the response generated by the MS is the same. The algorithms A2 and A3 are secret and not shared between different systems. In 3G systems, message authentication is used to ensure the integrity and authenticity of control messages.

9.4 2G Connection-Based TDMA Cellular

In the previous sections of this chapter, we introduced the infrastructure elements as well as an overview of the mechanisms that allow this infrastructure to support mobile operation for a 2G cellular network. In this section, we will provide a description of how these elements and mechanisms communicate with each other using communications protocols. We begin with the description of protocol stacks and we continue radio transmission specifications. Our example for 2G remains as GSM, the most successful 2G TDMA standard.

9.4.1 Protocol Stack of a 2G Cellular Network

Elements of a network communicate with each other through a protocol stack that is specified by a standardization committee. The standard specifies the interfaces among all the elements of the architecture. As an example, Figure 9.9 shows the typical protocol architecture for communication between the main hardware elements and the associated interfaces in the GSM.

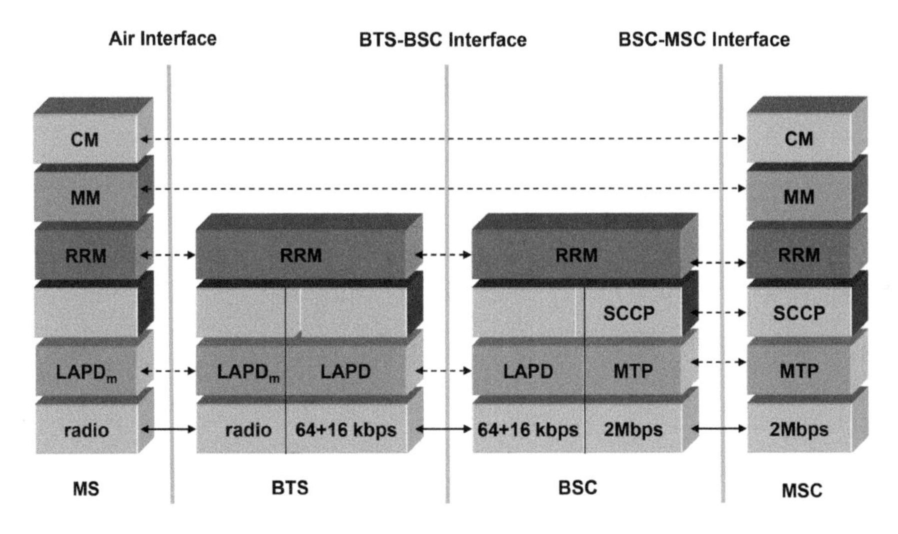

Figure 9.9 A typical protocol stack for a connection-based cellular network.

The air interface, which specifies the communication between the MS and BTS, is the most detailed and wireless-related interface. The interface between the BTS and BSC and the interface between BSC and MSC draw on existing wired protocols. In Figure 9.9, for example, ISDN protocols are used for physical wired connections.

The support for the interface between the BTS and BSC is for voice traffic at 64 Kbps and data/signaling traffic at 16 Kbps. Both types of traffic are carried over LAPD which is the data link protocol used in ISDN. The interface between different BSCs to the MSC takes place on a physical layer at 2 Mbps and employs PSTN's signaling system-7 (SS-7) protocols for communication. The message transport protocol (MTP) and the signaling connection control part (SCCP) of SS-7 are used for error-free transport and logical connections, respectively. The applications that employ the SS-7 protocols deal with direct transfer of data and management information for radio resource handling and operation and maintenance information for the operation and maintenance communication messages. To further simplify the description of the protocols, they are divided into three sub-categories or sub-layers: radio resource management (RR), MM, and CM messages or layers shown in the upper part of Figure 9.9.

9.4.2 Wireless Communications in 2G TDMA

Radio transmission in the air interface is the most complex part of a cellular network. This layer specifies how the information from different voice and data services are formatted into packets and sent through the radio channel. It specifies the radio modem details, structure of traffic and control packets in the air, and the packaging of a variety of services into the bits of a packet. This layer also specifies modulation and coding techniques, power control methodology, and time synchronization approaches that enable the establishment and maintenance of the channels. This section provides the details of the implementation of these functionalities in a cellular TDMA network using GSM as an example.

Since 2G cellular networks cover longer distances and were primarily designed for the telephone application with a fixed rate and quality of service (QoS), they use simpler modulation techniques than multi-rate wireless data networks in Wi-Fi. For example, GSM uses the Gaussian minimum shift keying (GMSK) modulation for implementation of the modem, like the Gaussian frequency shift keying (GFSK) used in Bluetooth. GMSK occupies half of the spectrum of GFSK, but it needs synchronization, which demands more complex electronic circuitry consuming more battery power. The bit transmission rate of this modem in GSM is 270.33 Kbps and it occupies a spectrum of 200 kHz per carrier. The MAC of the GSM is TDMA and it accommodates eight users per carrier. The standard organization recommends 124 carriers in 25 MHz of a band with two 100 KHz guard bands at each side of the spectrum (see Figure 9.10).

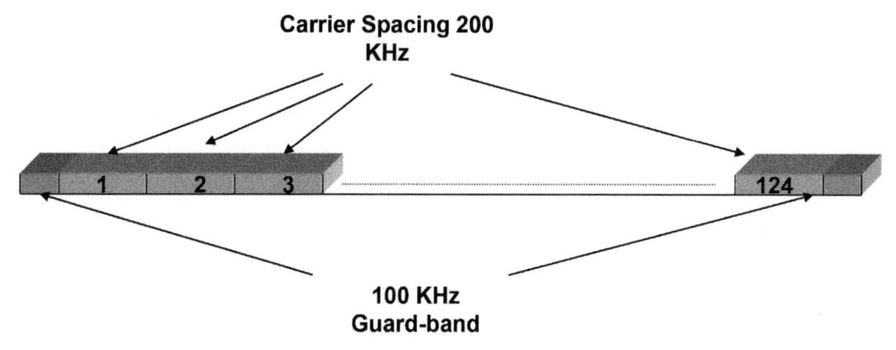

Figure 9.10 Division of spectrum into carriers for GSM cellular networks.

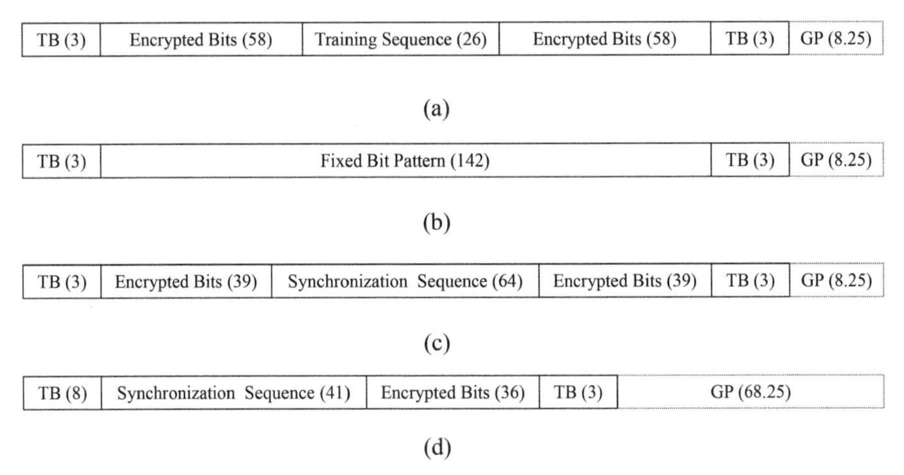

Figure 9.11 Four frame types used in GSM (a) Normal Burst, (b) Frequency Correction Burst, (c) Synchronization Burst, (d) Random Access Burst.

Connection-based networks have traffic planes and control and management planes. Control and management packets establish and maintain the link and traffic packets carry the voice or data traffic. The packet format in GSM includes four types of data bursts for traffic and control signaling. Figure 9.11 shows all four data packet types. The ***normal burst (NB)***, shown in Figure 9.11(a), consists of three tail bits (TB) at the beginning and at the end of the packet, equivalent to 8.25 bits of gap period (GP), two sets of 58 encrypted bits (a total of 116 bits), and a 26 bits training sequence. The TBs are three zero bits providing a gap time for the digital radio circuitry to cover the uncertainty period to ramp on and off for the radiated power and to initiate the convolutional decoding of the data. The 26 bits training sequence is used to train the adaptive equalizer at the receiver. Since the channel behavior is constantly changing during the transmission of the packet, the most effective place for the training of the equalizer is in the middle of the burst. The 116 encrypted data bits include 114 bits of data and two flag bits at the end of each part of the data that indicates whether data is user traffic or for signaling and control.

The other three types of bursts are designed for specific tasks. The simplest of all the remaining bursts is the ***frequency-correction burst***, shown in Figure 9.11(b). It has three TBs at the start and the end of the burst. The rest of the packet contains all "0"s that allow simple transmission of the carrier frequency without any modulated information. An equivalent of 8.25-bits duration is used as the GP between this burst and others. This burst is used to implement the ***frequency control channel*** used by the BTS to broadcast carrier synchronization signals. An MS in the coverage area of a BTS uses the broadcast frequency control channel to synchronize its carrier frequency and bit timing.

Synchronization burst, shown in Figure 9.11(c), is very similar to the NB except that the training sequence is longer, and the coded data are used for the specific task of identifying the network. The BTS broadcasts the frequency-correction and synchronization bursts and the MSs use it for initial frequency and time slot synchronization as well as training of the equalizer and initial learning of the network identity. This burst is used for the implementation of a ***synchronization logical channel*** used by the BTS to broadcast frame synchronization signals to all MSs. Using this channel, MSs will synchronize their counters to specify the location of arriving packets in the TDMA hierarchy.

The ***random access burst*** is used by the MS to access the BTS as it registers to the network. The overall structure is like NB except that a longer startup and synchronization sequence is used to initiate the equalizer. Another major difference is the length of the much longer GP that allows rough calculation of the distance of the MS from the BTS. This calculation is possible by determining the arrival time of the random access burst. A GP of 68.25 bits translates to 252 μs. The signal transmitted from an MS should travel more than 75.5 km (at the speed of 300,000 km/s) before arriving at the BTS to exceed this GP. After calculation of the distance of the mobile when it first sends a packet to the BTS, the BTS calculates a time advance for the user so that packets arriving from different MSs are better aligned. In another word, the time jitter of arriving packets from different stations caused by the difference among their distance reduces from 68.25 to 8.25 bits using the time advance calculated at the BTS.

Using this radio transmission infrastructure, GSM could support eight voice or data channels. The voice channels had a 13 Kbps digital voice and the data channels could support up to 9600 bps of connection-based data. On top of that, one of the innovative contributions of GSM was to offer short messaging systems (SMS) on traffic and management channels.

9.5 3G CDMA and Internet Access

The transition from 2G to 3G had several highlights in the evolution of wireless data applications and their relation to cellular infrastructure and the emergence of CDMA as a universal cellular radio networking technology. The trademark of 2G cellular can be labeled as integrated voice and data, a wireless ISDN, allowing end-to-end digital services on PSTN. The core of the PSTN was digital PCM for many years and ISDN intended unsuccessfully to extend that to home and office. But the 2G GSM implemented that on mobile platforms. However, in the early 1990s when GSM emerged, voice services were the dominant applications generating income for the cellular industry. The income of connection-based data services supported by GSM was negligible with the income from mobile telephone applications. The SMS, however, became a revolutionary technology engaging many users to a new application like the old telegraph for fast communications. The advantage of this new telegraph was that you did not need to contact an intermediary telegraph office and you could send your message from your personal pocket phone.

It is important to follow modern communications events in parallel with computer application technologies. Between the emergence of GSM in the early 1990s and the emergence of UMTS in 2000, the most important event in the evolution of computer communications applications was the Internet penetration of homes in the mid-1990s. America Online (AoL) started that revolution by attracting over forty million subscribers over two years. It took AT&T almost 50 years to attract that many subscribers. This rapid penetration of the internet in homes helped the growth of wireless access. Wi-Fi became popular for home networking and the cellular industry began to discover wireless data access for data rates well above 9600 bps and the goal of UMTS became the support of 2 Mbps data services on 3G cellular telephone networks and wireless access to the Internet. To make this a reality, we needed a new wireless infrastructure architecture and a new transmission technology with a wider bandwidth. As Shannon–Hartley equation directs us, higher data rates always need higher bandwidths. The CDMA technology uses direct sequence spread spectrum (DSSS) for voice and the same encoded voice has a much higher bandwidth than GSM TDMA with traditional GMSK modulation. Just as an example, the QUALCOMM's cdmaOne had a bandwidth of 1.25 MHz as compared with the 200 kHz bandwidth of the GSM. The 3G Pan-European cellular, UMTS, adopted 3 MHz and above. CDMA was to allow multiple voice users to overlay; for the data applications, we could give all the voice channels to a single user to enhance the data rate.

The 2G digital cellular was designed to connect to PSTN; to facilitate data applications, we need to change the infrastructure to connect to the Internet as well. Then, if we have access to the Internet, why should not we use voice-over IP rather than circuit-switched cellular networks? This need to an all IP networks initiated the 3GPP standardization that has stayed as the wireless backbone today. In the next two sections, we first address changes in the infrastructure, and then we discuss 3G CDMA radio transmission.

9.5.1 Evolutional of Infrastructure

The infrastructure of 2G digital cellular networks shown in Figure 9.2 supports only circuit-switched voice and data services at maximum data rates of approximately 10 Kbps and it does not have any direct access to the Internet for packet switching. The only packet data service is the SMS that is superimposed on the SS-7 data network for connection control and management. Indeed, the connection-based PSTN has a low-speed data network for the implementation of the SS-7 protocol for CM. SMS services in 2G were carried by that data network, and, naturally, it had serious restrictions for large data applications. With the sudden jump in popularity of the internet after its penetration in homes in the mid-1990s, the cellular telephone industry began to discover methods for high data rate wireless access to the Internet. The first step in this direction was to extend the existing infrastructure to accommodate Internet access. The first of these technologies building on the GSM TDMA network was the GPRS. Figure 9.12 shows the GPRS infrastructure built on Figure 9.2 to allow access to both PSTN and the Internet. In this architecture, we add a serving GPRS support node (SGSN) to communicate with the BSC and direct the packet data to the internet as well as a gateway GPRS support node (GGSN) that is a mobility aware router to connect to the Internet and manage changes in connection point of the MS.

The next step in the evolution of infrastructure was the emergence of an all-IP network. Figure 9.12 has two different data networks, the Internet using IP protocol and the SS-7 with its own proprietary protocols. By turning SS-7 protocols to the IP equivalent, we can integrate the two networks. All IP network protocols evolved under 3GPP, which was an umbrella term for several standards organizations, devoted to the Pan European 3G cellular networks. Figure 9.13 shows the general block diagram of the 3G cellular networks. Since the functionality of the devices and the protocols are modified, new names are adopted to the radio interface elements. The MS is referred to as user equipment (UE), the BS with new functionalities is called Node-B, and the BSC with new functionalities is called radio network controller (RNC). The remainder of the elements remains the same, except that the SS-7 data network for the control and management of connection-based services is integrated on the Internet. In this new formation

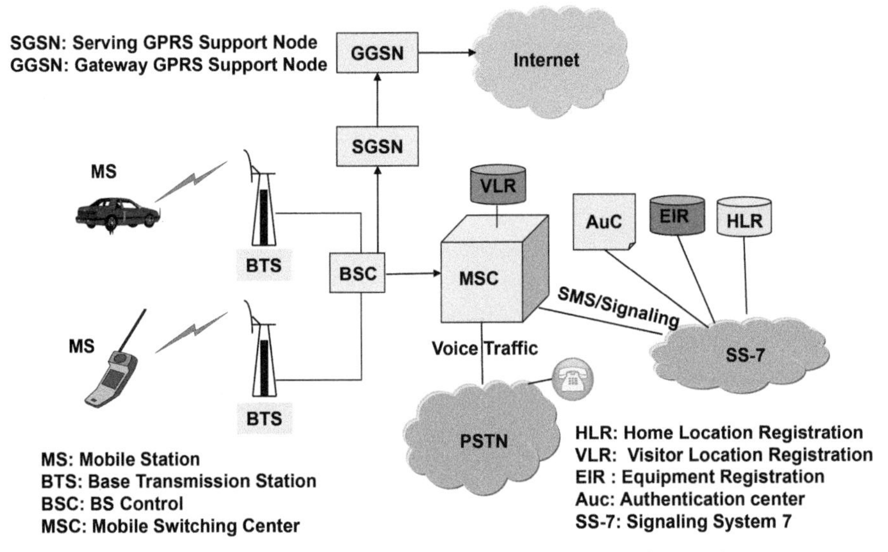

Figure 9.12 Extension of 2G Architecture to accommodate both PSTN and Internet access.

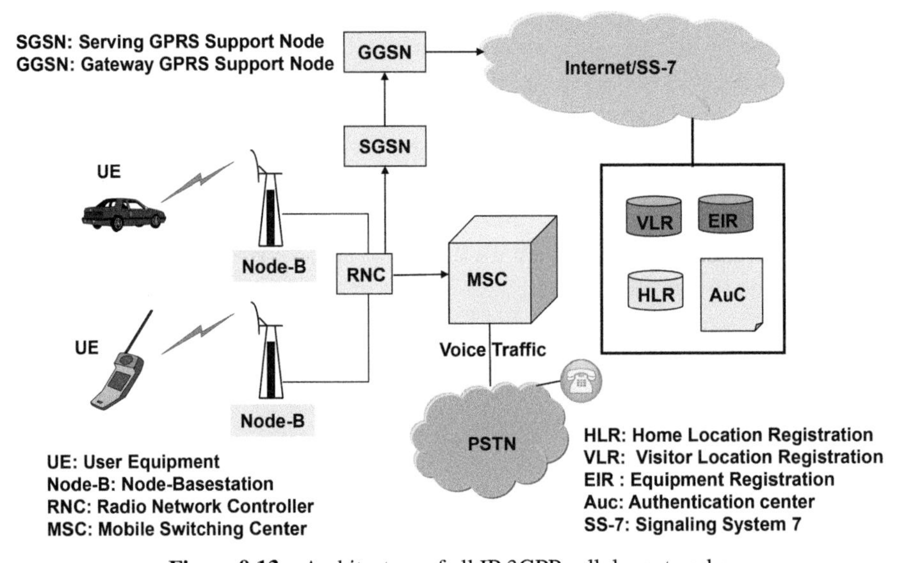

Figure 9.13 Architecture of all IP 3GPP cellular networks.

of the network, all databases communicate with each other through the Internet.

9.5.2 Wireless Communications in 3G CDMA

High data rate packet transmission needs a new radio access protocol for the air interface. The GSM radio interface supports a 270.833Kbps per carrier occupying 200 MHz of bandwidth using the GMSK. Each carrier supports eight parallel connection-based voice or data services. The user data for voice encoder is 13 Kbps, and for the data, it can go up to 9600 bps. For packet-switched transmission, the GPRS could increase the radio channel data rate up to 114 Kbps by allowing a single user occupying up to eight TDMA time slots; each channel carries 114/8 = 14.5 Kbps that was possible because it was in the rage of existing services. This is a simple solution without changing the modulation and remaining with GMSK modems. The next generation of packet data service over 2G TDMA GSM was the enhanced data rate for GSM evolution (EDGE) that adopted more efficient modems with three symbols per bit to achieve up to 236 Kbps. The objective of the ITU 3G cellular for International Mobile Telecommunications for the year 2000 (IMT-2000) was to achieve 2 Mbps packet switching rate for the cellular networks. GPRS and EDGE build on GSM with 200 MHz of bandwidth could not achieve this goal. Therefore, inspired by QUALCOMM's cdmaOne standard that occupies 1.25 MHz of bandwidth, the IMT-2000 adopted DSSS and CDMA technology to completely revolutionize the 2G GSM TDMA technology.

As we discussed in Sections 5.2.3–5.2.4, the CDMA air interface became the popular choice for 3G cellular networks because it provided inherent flexibility for multimedia traffic, provided a better quality voice, consumed less power (about 10% of analog or early TDMA phones), and does not require frequency planning since all cells employ the same frequency at the same time. In a manner like TDMA networks, CDMA carriers occupy a portion of the overall bandwidth available for a cellular operator, but the *carrier bandwidth* is much wider in the CDMA. For example, each carrier of the cdmaOne or IS-95 occupies 1.25 MHz of the total available band while carriers of GSM occupied 200 KHz. The carrier bandwidth of the 3G networks are even wider, for example, cdma2000 can potentially use multicarrier CDMA with at least three times the bandwidth of cdmaOne, and the pan European CDMA system, UMTS, occupies 5MHz of bandwidth. Within each carrier is the implementation of different logical channels.

Different channels in CDMA are separated using *orthogonal* codes. As a result, we can easily give all codes of a carrier to a single user to send its packet-switched data. As an example, QUALCOMM's cdmaOne had 64 channels each operating at 9600 bps with a 1-user bit per symbol transmission. If we assign all channels to one user without changing the modem, we have 64×9600 = 614.4 Kbps. By changing the model to a 3-bit per symbol modem and playing with the encoding technique, we can achieve 2 Mbps objective of the IMT-2000. The Pan European UMTS with 5 MHz of bandwidth can achieve the IMT-2000 goal simpler.

9.6 4G LTE and OFDM/MIMO for Asymmetric Data

Long-term evolution (LTE) was developed by the 3GPP standards organization to migrate from CDMA-based 3G UMTS systems to a cellular OFDM/MIMO system like Wi-Fi is suited for data traffic and voice-over IP. The entire communications in an LTE network employ IP packets and everything is packet switched. We discussed the PHY aspects of OFDM/MIMO in-depth in Sections 4.5.2–4.5.4 and we discussed their implementation in Wi-Fi devices in Sections 7.5.2–7.5.3. What is unique in LTE is the new asymmetric MAC with orthogonal frequency division multiple access (OFDMA) in downlink and single-carrier FDMA (SC-FDMA) in uplink. In this section, we first present architectural changes in the infrastructure of the cellular networks to transfer from 3G UMTS to 4G LTE. Then we review the protocol stack and the PHY and MAC specifics of the LTE.

9.6.1 Elements of LTE Architecture

One of the main objectives of LTE was to reduce the latency for user data (so-called user plane latency) i.e., the time it is transmitted by the MS to the time it leaves the radio access network, to something that is as small as 5 ms or less. For this reason, the network architecture in LTE was flattened as shown in Figure 9.14. As compared to the 3G or 2G cellular architectures, there is little hierarchy in the network architecture with most entities performing the needed functionalities. This infrastructure is becoming closer to that of Wi-Fi that has only an AP to connect to the Internet. However, in cellular networks, we have a large number of channels and scarce resources demanding seriously regulated frequency resource management as well as more advanced MM to accommodate mobile operation in high-speed vehicles. In addition, cellular networks are operated by service providers and they need to regulate the operation to enforce charging mechanisms to generate revenue to maintain the network.

As shown in Figure 9.14, the network is divided into an *evolved packet core* or EPC and an *evolved* UMTS terrestrial *radio access network (e-UTRAN)*. Notice that the radio access network consists of the UE or MS and an enhanced BS that is called *evolved Node-B* or *eNode-B*. If there is a potential for handoff between two e-NodeBs, there is a connection between them to enable them to communicate. The functions of the SGSN are incorporated into the mobility management entity (MME) shown in Figure 9.14. The serving gateway (S-GW) may be physically co-located with the MME or be separate. The S-GW lies between the radio access part and the core network. All the radio transmissions end in the S-GW. It forwards packets to the packet data network gateway (PDN-GW), which interfaces to the Internet. The PDN-GW could also be an IP multimedia subsystem (IMS) that supports voice over IP calls in an LTE network. The home subscriber server (HSS) handles the authentication of users and authorization of services provided by the IP network. In 2G and 3G cellular, authentication is performed at the MSC. Figure 9.15 summarizes how the elements of 3G UMTS architecture are mapped into the 4G LTE. In Wi-Fi networks, all three elements are combined in a corporate AP or a home router. These additional infrastructures in cellular networks are needed to

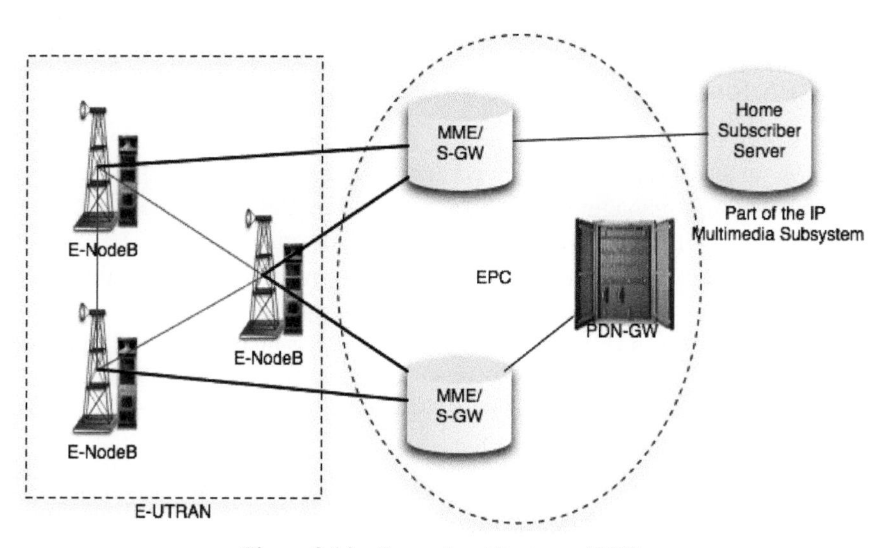

Figure 9.14 General architecture of LTE.

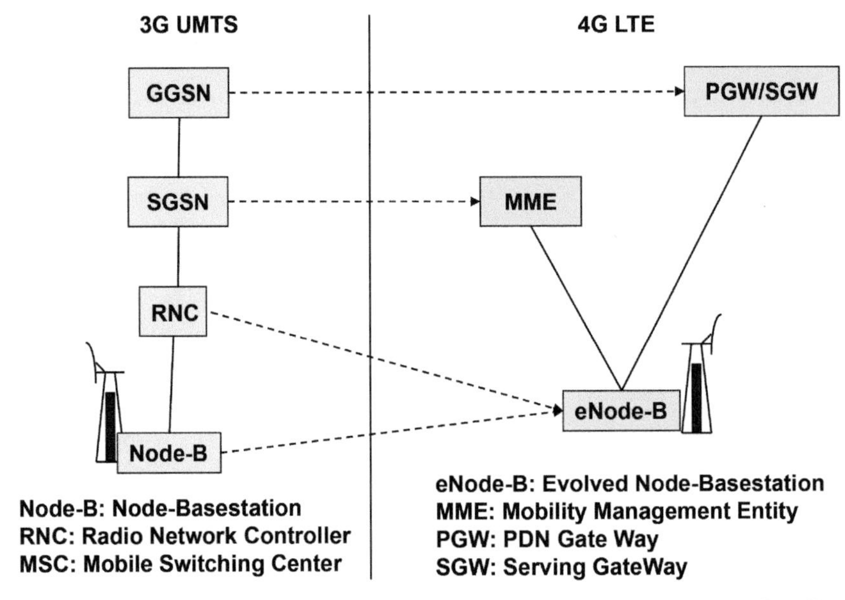

Figure 9.15 Mapping the architectural elements of the 3G UMTS to the 4G LTE.

maintain a practical distribution among many users operating in a fractured spectrum owned by a cellular company to maintain fair resource management and monitor distribution of resources for accounting purposes.

9.6.2 Protocol Stack for Data Flow in the 4G LTE

Figures 9.16 and 9.17 show simplified ways in which control data and user data flow in an LTE network. The control messages such as paging, security, mobility, session management, etc., are carried by the radio resource control (RRC) layer in Figure 9.16. The RRC layer exists only in the MS UE and the e-NodeB. Bearers are created between an MS and the PDN-GW that carry IP packets with specific QoS which may vary based on the application (e.g., the voice may have a different QoS compared to e-mail). Bearer IP packets and RRC packets are carried by a packet data convergence protocol (PDCP) layer that performs functions such as compression, encryption, and message integrity over the air. The PDCP layer also terminates in the e-NodeB. The radio link control (RLC) layer fragments the packets and works with the MAC layer to deliver packets in sequence to the PDCP layer. It performs automatic repeat requests for lost packets in an acknowledged mode, while an unacknowledged mode (where corrupted packets are not retransmitted) is

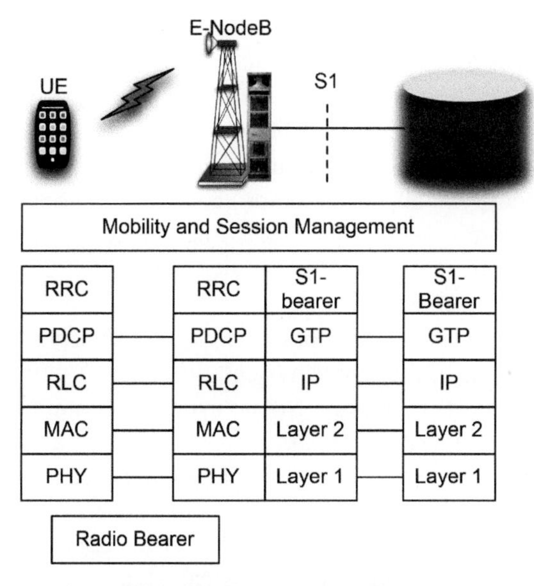

Figure 9.16 The flow of control data in LTE.

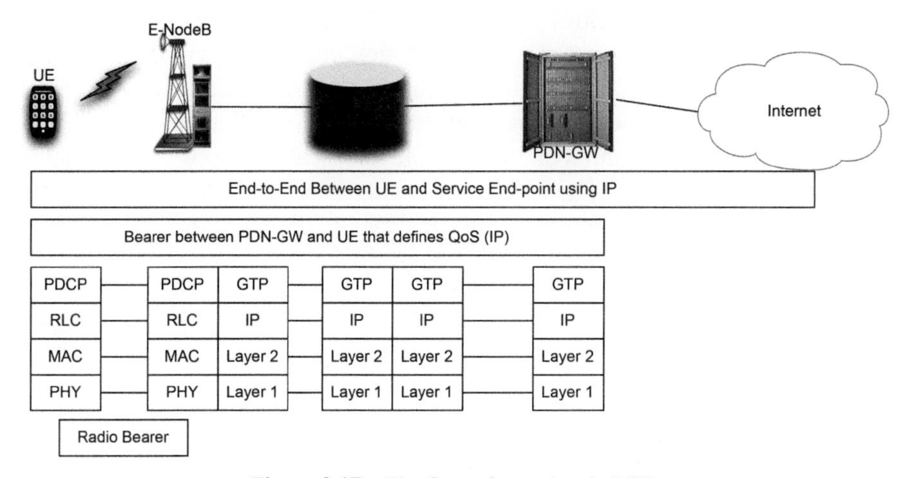

Figure 9.17 The flow of user data in LTE.

also possible. A transparent mode is used for random access where the RLC is not really used. The MAC layer handles the selection of modulation format, coding, MIMO scheme, transmit power levels, and error correction for a given packet. The physical layer transmits the packet using physical resource blocks, described later. The GPRS tunneling protocol (GTP) is an IP-based communication protocol carrying GPRS data within the GSM, UMTS, and LTE networks. The S1-bearer carries data along the S1-interface between an e-NodeB and an MME/S-GW. For the bearer IP packets that have the MS and the PDN-GW as the endpoints (see Figure 9.17), there may be a guaranteed bit rate with different packet error rates and latency specifications or a best effort service where the bit rates are not guaranteed. The guaranteed bit rate is used for applications such as real-time voice/video conversations, gaming, and streamed media. The best effort service is used for e-mail, file sharing, signaling for voice over IP, etc.

As in UMTS, LTE specifies logical channels, transport channels, MAC layer control information, and physical channels. Logical channels are created between the RLC and the MAC layer and contain the functionality needed for network operation. Transport channels that are created between the MAC and the physical layer include modulation, coding, and other details. MAC layer control information, also carried by specific physical channels, includes scheduling information, power control commands, etc., for physical layer procedures. The physical channels carry the transport channel data or MAC layer control information on the air. As an example, the logical broadcast control channel that contains system information is mapped into two transport channels – a downlink shared channel and a broadcast channel. The downlink shared channel is carried by a physical downlink control channel on the air, while the broadcast channel is carried by a physical broadcast channel on the air. A few logical, transport, and physical channels are defined in LTE, which we do not discuss in detail in this book. These logical channels enable communications in the network to implement radio registration, call-establishment, handoff, and security needed for the operation of the network.

9.6.3 Wireless Communications in 4G LTE

LTE is designed for access to Internet applications as well as traditional voice. Internet applications are very diversified and non-symmetric, and they have a wide spectrum from SMS short messaging demanding transmission of a burst of data up to video streaming on YouTube and other mediums which are extremely asymmetric and demand very high bandwidth for long sessions. Voice traffic is a two directional time division duplexed information stream at data rates of approximately 10 Kbps for long sessions that may take a few minutes. On the other hand, on the downlink of the cellular, the BSs can be made complex and can employ expensive techniques to handle a number of users with different bandwidth and connection time needs at the same time, but the reverse uplink from a mobile device can be very simple and power efficient to accommodate for the batter power restrictions of the mobile. As a result, LTE adopts quite different wireless communications techniques in the downlink. The downlink of the LTE employs a traditional simple and power efficient SC-FDMA and the downlink adopts a complex and innovative OFDMA with a complex MIMO antenna system. The SC-FDMA technique essentially retains the advantages of OFDM but, in effect, uses a *single carrier* instead of multiple carriers for transmitting data in a clever manner. Adoption of the OFDMA for wireless communications in the downlink allows an innovative view of RR management allowing a flexibility in accommodating multiple simultaneous mobile users with a wide variety of demand for their applications.

The 1G–3G cellular networks were fundamentally designed for mobile telephone circuit-switched applications with a growing ability to support packet-switched data. These networks had a fixed allocation of resources to the users throughout a session. For example, in the North American FDMA AMPS, the carrier bandwidth was fixed at 30 kHz, and in the TDMA GSM, the carrier bandwidth was fixed at 200 kHz to support eight users. In 2G QUALCOMM cdmaOne, the carrier bandwidth was 1.25 MHz accommodating 64 channels, while in 3G UMTS, the carrier bandwidth was 5 MHz to accommodate a higher number of users. In all these cases, once a carrier was assigned to an MS for any given duration, the bandwidth allocated for that duration was fixed, irrespective of whether it was necessary or not.

Downlink in LTE:

On the OFDMA downlink of the LTE networks, the medium access technique allows for extremely flexible allocation of resources to individual MSs at a granularity that is not possible in legacy 1G-3G systems. Figure 9.18 shows an example of how resources in OFDMA can be allocated on a per-slot basis. For simplicity, in this figure, we are assuming that *every* subcarrier can be flexibly allocated to any MS in each time slot. Usually, cellular systems such as LTE define what is called a *physical resource block* comprising a set of subcarriers and a transmission time interval, which can be flexibly allocated between different MSs.

The use of OFDMA in a downlink, which comes with the flexibility in allocating subcarriers on the downlink, allowed 4G LTE systems to exploit what is called multi-user diversity in both the frequency domain and time domain. The general idea behind multi-user diversity is that channel

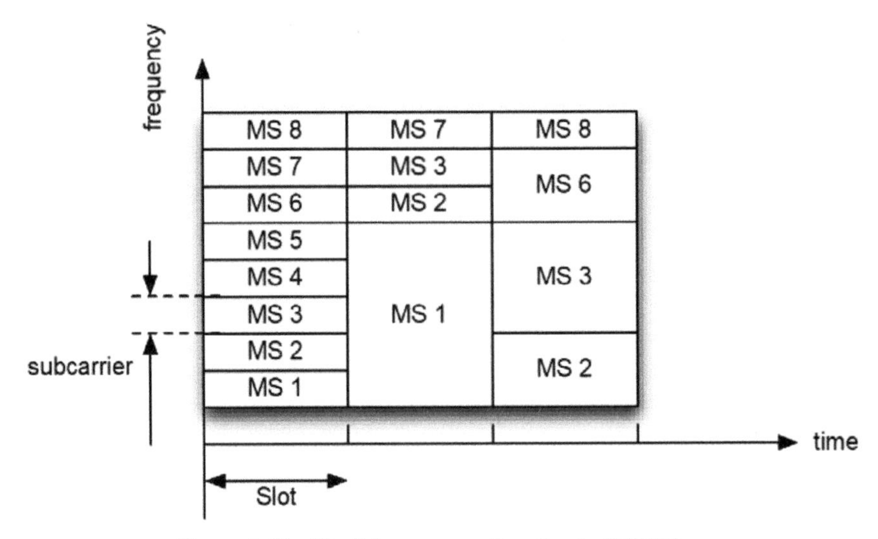

Figure 9.18 Flexible resource allocation in OFDMA.

conditions can be very different for different users or MSs due to their locations within the cell. By scheduling transmissions of MSs when they have good channel quality, the aggregate throughput of the network can be increased. In 3G systems, since the channel is constrained by the frequency carrier, the only variations observed in channel conditions are over time and an MS with "good channel conditions" can be scheduled, but only in time. In OFDMA systems, such scheduling can be done across frequency sub-carriers and in time.

Figure 9.19 shows an example of multi-user diversity as it applies to OFDMA systems. As the figure shows, different MSs at different locations may see different channel conditions as a function of frequency. Thus, by allocating the best sub-carriers to MSs in each time slot, the diversity in the channel can be exploited. The benefits of multi-user diversity diminish with the number of users, the amount of coding and interleaving, and other forms of reliability employed in the transmission scheme. For example, if transmit diversity with MIMO is employed, it provides resilience against narrowband small-scale fading making the quality of the channels of various MSs approximately the same. Thus, multi-user diversity, while useful, provides benefits only in certain scenarios.

As mentioned previously, LTE supports a variety of bandwidths and data rates. Obviously, smaller bandwidths can only support smaller data rates,

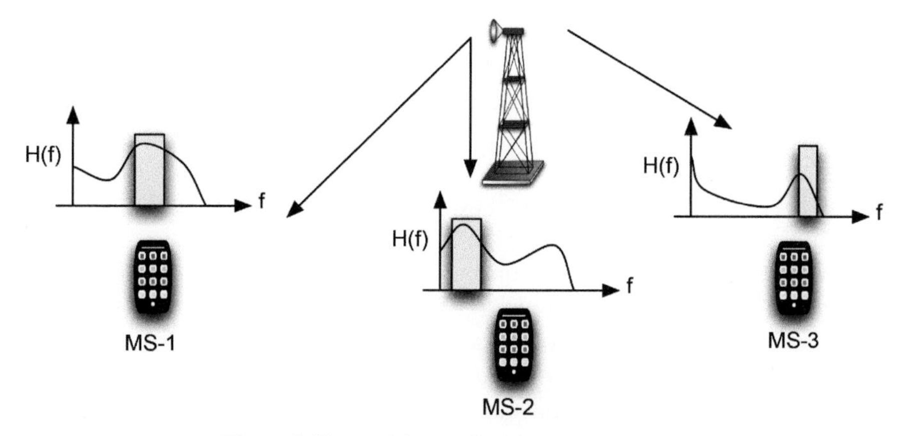

Figure 9.19 Multi-user diversity with OFDMA.

but the smallest bandwidth supported (1.4 MHz) also has a higher overhead in terms of a guard band, which is 22.8% of the bandwidth. Table 9.1 shows the bandwidths supported on the downlink and some of the associated parameters, the fast Fourier transform (FFT) size, number of sub-carriers, number of physical resources per block (PRB). Each sub-carrier in LTE is 15 kHz wide and, like the IEEE 802.11a,g, they are modulated BPSK, QPSK, 16-QAM, or 64-QAM.

The bandwidth shown in Table 9.1 includes the guard bands that are overhead. As an example, let us find the guard band in LTE for 1.4 MHz transmission bandwidths. For this bandwidth, the number of sub-carriers is 72, and the bandwidth occupied by the sub-carriers is 72 $\times$ 15 kHz = 1.08 MHz. The rest of the bandwidth of 0.32 MHz is used as a guard band for a total overhead of 0.32/1.4 = 22.8%. On the other hand, with a 20 MHz transmission bandwidth, there are 1200 sub-carriers that occupy 1200 $\times$ 15 kHz = 18 MHz, the guard band occupies 2 MHz for an overhead of 2/20 = 10%.

Table 9.1 Transmission bandwidths supported on the downlink in LTE.

Bandwidth	1.4 MHz	3 MHz	5 MHz	10 MHz	15 MHz	20 MHz
FFT size	128	256	512	1024	1536	2048
Number of sub-carriers	72	180	300	600	900	1200
Number of PRBs	6	15	25	50	75	100

Uplink in LTE:

SC-FDMA is used on the uplink in LTE. SC-FDMA still maps data to only a few of the sub-carriers out of the entire block of sub-carriers in the allocated bandwidth. But the way each of the sub-carriers carries the symbol is different. The use of SC-FDMA implies that it appears as if a wideband signal is being transmitted for a shorter duration in time although the signals are not generated this way.

We describe this in an informal way here (see Figure 9.20). Let us suppose that an OFDM symbol lasts for T_S seconds and comprises N sub-carriers. Each sub-carrier carries itself one data symbol (which depends on the modulation scheme used on that sub-carrier). In other words, the symbol and modulation scheme is constant for T_S seconds. This is implemented by taking N symbols in parallel, computing their Inverse FFT (IFFT), and then transmitting the IFFT samples in serial. The SC-FDMA "symbol" also lasts for T_S seconds. However, on any one sub-carrier, the data symbol lasts for roughly $1/N$ of the time. In other words, each sub-carrier has many data symbols throughout for the SC-FDMA symbol. However, the data symbol *across* a group of sub-carriers will be the same. Such sub-carriers may be contiguous or distributed in the allocated spectrum. If the sub-carriers are contiguous, the amount of frequency diversity seen by one MS may be limited, but multi-user frequency diversity is exploited. If the sub-carriers are distributed, each MS may benefit from frequency diversity. Note that the SC-FDMA symbol still has a concatenated cyclic prefix for protection against excess multipath delay.

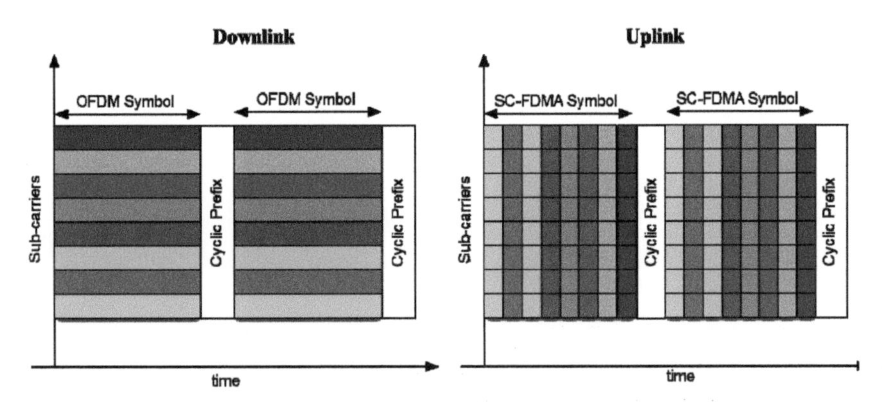

Figure 9.20 Illustration of OFDM and SC-FDMA symbols.

Achieving this requires the N data symbols to be pre-coded before the IFFT is taken. This pre-coding is performed using an FFT operation in SC-FDMA along with frequency shifting so that the signal occupies the right piece of the spectrum on the uplink. We recall here that multiple MSs will be transmitting at the same time on the uplink and they must be separated appropriately in frequency before transmission. At the BS receiver, separate IFFTs must be performed after the usual FFT operation to distinguish the signals that belong to different MSs. Although it appears from Figure 9.20 that the data symbols have been broken up into very small pieces, we emphasize here that the rectangular block of sub-carriers over time comprises the actual transmitted "symbol," whether it is OFDM or SC-FDMA. Thus, the block is a unit although it is easier to conceptualize the behavior of OFDM and SC-FDMA by referring to the sub-carriers and the data symbols that they carry over time.

On the uplink, LTE employs SC-FDMA rather than OFDMA as described previously. In the case of SC-FDMA, it may be possible to allocate sub-carriers in a contiguous manner (which is the option that is currently used) and to (uniformly) distribute the sub-carriers. In the case of contiguous sub-carriers, it is possible to view SC-FDMA as being like FDMA, in that separate chunks of spectrum are allocated to mobiles that are transmitting at the same time. Once again, the allocation is made in units of the physical resource block which is 180 kHz wide and comprises 7 OFDM symbols in the typical case. This allocation is made by the e-NodeB which ensures that no MS will interfere with another MS on the uplink within the cell. The physical downlink control channel carries the information about the allocated PRBs to the MSs. To exploit the benefits of frequency diversity, the channel quality may be a consideration in the allocation, as well as the use of frequency hopping (e.g., shifting the allocated PRBs in frequency every few time units or so).

Unlike the downlink, most MSs are expected to have only a single RF chain (although multiple antennas may be present on a device). Thus, transmission using multiple antennas is not an option, although it is possible for the e-NodeB to indicate which of the antennas provides the best quality of the signal. This is like selection diversity, and, in this situation, the MS uses the best antenna to transmit the signal to the e-NodeB. Sometimes, two MSs may be allocated the same PRBs at the same time. In this scenario, which is called *multi-user MIMO*, the e-NodeB uses its multiple antennas to separate the signals from different MSs. It is possible to view this as being like spatial multiplexing, although the transmissions are from different transmitters.

The next set of improvements on the LTE is sometimes referred to as LTE-advanced. Some of the primary features for this standard are:

1. To support data rates on the order of a Gbps, it becomes necessary to have more bandwidth. Toward this, LTE-advanced supports carrier aggregation, like IEEE 802.11n. Instead of stopping at 100 PRBs in 20 MHz, it can aggregate five carriers of 20 MHz for a total of 500 PRBs per link. This also increases the benefits of frequency and multi-user diversity.
2. Support for small cells such as femtocells to increase reliability and capacity is being actively considered. The coordination between small cells and macro-cells is considered as part of LTE-advanced.
3. LTE mostly supports two transmit and two receive antennas. LTE-advanced can support up to eight antennas on the downlink.
4. The architecture of the network includes relay nodes and heterogeneous entities with varying capabilities and cooperative transmissions to enhance reliability or capacity.

9.7 5G and mmWave Technology

5G is the natural evolution of 4G and an attempt to increase the capacity of cellular wireless all-IP networks. The fundamental innovative technologies considered for the 5G are [Nor17]: mmWave communications, small cell operation, massive MIMO, and full-duplex operation. The idea of a small cell with outdoor antennas has been examined for Wi-Fi in the past in a variety of different business models without any popular commercial success. Small cells for indoor operation directly compete with the Wi-Fi and its existing and very successful commercial success in that domain. A feasible new scenario for outdoor scenarios is deployment on utility posts or on the rooftops of a building using relays to cover up to 250 m without any wired connection to take advantage of solar energy to recharge the batteries. Full-duplex using time division duplex (TDD) with one carrier has been successful for Wi-Fi and was an option for the 3G; it may help the 5G as well. Beamforming with more precision enables the implementation of higher degrees of spatially separated beams for multiple streaming, but it needs an increase in the number of antenna elements above the 4G 4–8 antenna elements. In the 5G literature, antennas with a higher number of elements are called massive antenna arrays.

mmWave transmission at 30–300 GHz (1–10 mm wavelengths) facilitates smaller antenna arrays and well suited the implementation of massive antenna arrays. For these reasons, mmWave transmission became the corner stone and hype for the 5G technologies. mmWave were used for radars and satellite communication for many decades, but they were new for the cellular industry. The mmWave radio signals, like optical wireless signals, cannot penetrate through the walls and can be absorbed by the foliage. However, massive MIMO antennas can find other reflected routes to connect to the receiver. Laser beams are even more focused than mmWave beams, but they do not benefit from MIMO technology with adaptive beamforming capabilities for the time being.

In Chapters 6 and 7, we briefly discussed IEEE 802.11ad and IEEE 802.15.4c and their attempts for implementation of mmWave technology for indoor networking. The mmWave technology is a good solution for in-room applications such as multi-channel video transmission or wireless USB. But for the in-building applications involved in obstructed line-of-sight situations, walls block the signal and restrict its coverage making this technology look like optical communications with all its pros and cons. The situation outdoors is different, and a smart antenna using reflected paths can curb the blocked LOS scenarios. In addition, the small size of antennas enables the implementation of large antenna arrays for narrow beam antennas.

The main incentive to go to higher frequencies for data transmission is the availability of wider spectrums of bandwidth. The first challenge in operating at higher frequencies is the need for higher transmission power levels because the path loss in the first meter is inversely proportional to the wavelength (see Sections 3.3.1–3.3.2):

$$\begin{cases} L_p = L_0 + 10\alpha \log d \\ L_0 = 20 \log \frac{4\pi}{\lambda} = 20 \log \frac{4\pi f}{c} \end{cases} .$$

As an example, the path loss at a first meter for a popular 60 GHz mmWave is 68 dB, while for the 2.4 GHz popular ISM bands, it is approximately 40 dB. This means, in an environment with the same distance-power gradient, α, to maintain the same coverage, we need an additional 28 dB transmission power to compensate for the extra path loss at the first meter at 60 GHz. In practice, this difference is compensated by the antenna gain obtained by focusing the propagated energy. The transmitted power and antenna gain at mmWaves are regulated by government agencies such as FCC because extremely focused antennas are like lasers and they create too much heat on the surface of the body. The difference between the transmitted power in dBm and antenna gain in dBi is referred to as the equivalent isotropic radio propagation (EIRP) and it is specified with frequency administrators of government bodies in different

Table 9.2 Transmitted power, antenna gain, and EIRP for 60 GHz mmWaves in different regions.

Region of operation	Tx Power dBm	Antenna gain dBi	EIRP
US/Canada	27	15	43
Japan/Korea	10	48	58
EU	13	44	57

regions. Table 9.2 compares these regulations in the US/Canada, Japan/Korea, and in EU. It is interesting for the reader to notice that in all these regions, EIRP is much more than the 28 dB path loss from omni-directional antennas at 2.4 GHz, making mmWave transmission power reasonable for wireless communications.

Figure 9.21 shows an example design for an 8×8 antenna array. Figure 9.21(a) [Qua15] shows the pattern of radiation of different beams from a mmWave patch antenna array. Figure 9.21(b) shows the implementation of the antenna pattern and a physical picture of the antenna. Since the wavelength at 60 GHz is $\lambda = 5\,\mathrm{mm}$ and the size of the antenna elements is $\lambda/4 = 1.25\,\mathrm{mm}$, the size of the antenna with eight elements on each side is on the order of $1.25\,\mathrm{mm} \times 8 = 2\,\mathrm{cm}$ that is less than one inch (size of a quarter coin in the US). The picture in the middle of Figure 9.21 shows two different designs for the beamform and how a new design by FUJITSU reduces the sidelobes of the beam, interfering with adjacent beams. From this discussion, we can argue that design at 60 GHz mmWave provides higher bandwidths

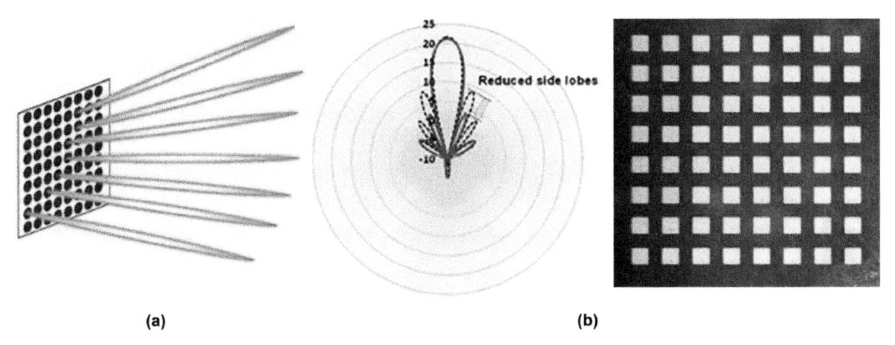

Figure 9.21 8×8 60 GHz patch antenna arrays for multiple streaming. (a) Multiple streaming concepts from QUALCOMM [QUA15]. (b) Two FUJITSU implementations of antenna patterns (black and red). (c) The designed array to achieve 12 Gbps.

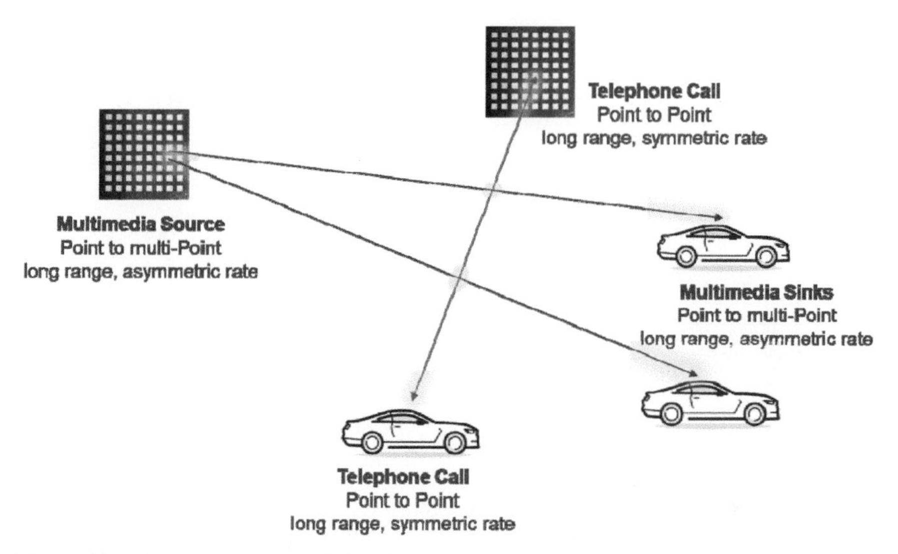

Figure 9.22 Directional antennas allow for spatial reuse using directional antennas with minimal interference.

and better coverage at the expense of the design of a more sophisticated smart antenna system.

Another advantage of the mmWave with focused antenna beams is its spatial reuse enabled by minimal interference among different beams that facilitate deployment planning and increases the capacity of the network close to infinity. Figure 9.22 shows a scenario for a video streaming broadcast to two users overplayed a telephone call application. With very narrow beams interference is minimal and multiple BSs can operate in an area. Focused beams operate close to wires. We can call them wireless wires; they create guided point-to-point connections with the flexibility of changing the distance and data rate. As shown in Figure 9.23, multiple beams can be assigned to a single link to increase its data rate. The key benefit of the smart antennas with focused beam forming in this application is to enable the support of data rates beyond that of a single path between the transmitter and the receiver. Examples in Figures 9.22 and 9.23 demonstrate a practical example of format flexibility of massive antenna arrays to accommodate different application scenarios.

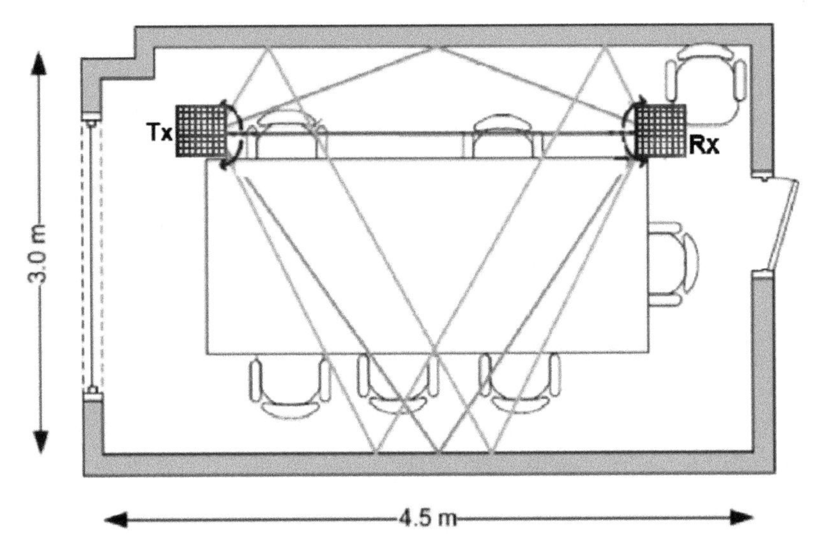

Figure 9.23 An office setting scenario with multiples stream from several paths increase the transmission rate between a transmitter (Tx) and a receiver (Rx).

9.8 6G and Future Directions

As the future is shaping with the penetration of artificial intelligence (AI) in our daily lives, we are witnessing growth for wireless networks to accommodate these needs. In Chapter 7, 8, and 9 we showed how IEEE 802.11, IEEE 802.15, and cellular networks evolved to support the evolution of computer, sensors, and vehicular industries. The IEEE 802.11 evolved into IEEE 802.11ax and the application of mmWaves and massive antenna arrays to accommodate smart word applications and the Internet of Things (IoT) and Bluetooth low energy (BLE) and ZigBee evolved to complement IEEE 802.11 technologies. The focus of the IEEE 802.11 and IEEE 802.15 was indoor networking where antennas are primarily deployed in indoor areas. The cellular industry evolved for primarily outdoor antenna deployments to cover wide areas. From a business model point of view, IEEE 802.11 and IEEE 802.15 networks were generating their revenue primarily from selling the equipment and networks were owned by individuals, corporates, and Internet service providers. The cellular industry involves in huge investment for deployment of the infrastructure and their revenue model is thought of as the subscriber monthly fees. From the technical point of view, a wide-area cellular and local and personal area IEEE 802.11 finally merged to use the same PHY but different MACs. The IEEE 802.15 completed these technologies. One distinct feature of the cellular and IEEE 802.11 industry is the ability of cells to provide wide area coverage for the vehicles.

In the 2020s, automated vehicles are becoming popular, and AI algorithms governing their operation need end-to-end precision latency (delay), which became a goal for 6G cellular networks. Researchers have already begun to create a wider image for these networks [Zha19]. In addition to the support of autonomous driving, these visions include integration of IoT, support of extremely higher data rates for interactive multimedia, and expansion of coverage to extreme environments such as in deep-sea and space. These will demand innovations in transmission, spectrum management, and medium access. Among technologies that have attracted attention for 6G networks are Terahertz (THz) communications, supermassive MIMO, large intelligent surfaces, holographic beamforming, orbital angular momentum multiplexing, laser and visible-light communications, blockchain-based spectrum sharing, and quantum communications and computing.

Assignments

Questions

1. What are the differences between a mobile digital telephone and POTS?
2. What are VLR and HLR, where are they physically located, and why do we need them?
3. What is the difference between registration and call establishment?
4. What are the reasons to perform handoff?
5. What is the difference between network decided and mobile assisted handovers?
6. What are the incentives for power control in a TDMA network?
7. Why are both architectural changes and changes in transmission schemes necessary for data transmission on cellular networks that were designed for voice?
8. What were the incentives to move from 2G to 3G cellular networks?
9. What were the incentives to move from 3G to 4G cellular networks?
10. What were the incentives to move from 4G to 5G cellular networks?
11. What are the incentives for 6G cellular networks?

10

Deployment of Communications Networks

10.1 Introduction

In this chapter, we discuss issues related to the deployment of wireless networks. We divide these networks into cellular and Wi-Fi because these are the two wireless networks with substantial infrastructure to cover metropolitan and indoor areas, respectively. Wireless service providers deploy a cellular infrastructure in an area to provide their services in licensed bands that they own in those areas. Deployment of cellular infrastructure needs maintenance and it changes as the wireless technology evolves in time. Wireless technologies supported by a service provider generate revenues based on the applications, which change with the communications habitual of people. The 1G–3G cellular services were focusing on the connection-based cellular telephone networks as their main source of revenue. The 4G technology and beyond generates their main revenue from smartphones which are small computing devices enabling millions of data applications as well as the telephone services. In terms of technologies, 1G, 2G, and 3G used frequency division multiple access (FDMA), Time division multiple access (TDMA), and code division multiple access CDMA technologies with their special needs for deployment. The wireless mobile data services beginning from 4G up to 6G that is under development at the time of this writing use orthogonal frequency division multiplexing (OFDM)/multiple-input multiple-output (MIMO) technologies affecting deployment strategies. Wi-Fi devices operate in unlicensed bands and they are most popular for indoor applications. Deployment of Wi-Fi infrastructure can be divided into a cellular and single antenna. Cellular deployment of the Wi-Fi devices is usually in university campuses, corporates, and other indoor areas where a small network operation group installs and maintains the network. In residential areas and in a small office and home office (SOHO) deployment, often, the owner or a service provider installs a single Wi-Fi router, and if the coverage is not adequate, they add a Wi-Fi relay or extender to expand the coverage. In this chapter, we provide an overview of these deployment technologies. To understand the deployment issues, we need to understand the interference in wireless networks.

10.2 Interference in Wireless Networks

When two wireless networks overlap in their coverage and operate at the same frequency at the same time without any access coordination, they will obviously interfere with one another. For example, several cordless phones, Bluetooth, ZigBee wireless personal area networks (WPANs), and Wi-Fi wireless local area networks (WLANs) operate at the 2.4 GHz unlicensed bands and they are subject to interference from each other. The literature on military communication systems offers many detailed analyses of the performance of communication systems in the presence of various intentional interferers or jammers [Sim85]. These jammers are designed to disrupt the operation of a system and they can employ relatively sophisticated techniques such as multi-tone jamming and pulsed jamming. In civilian applications, interference is neither intentional nor sophisticated. Most often, the interferer is simply another system designed to operate in an adjacent band or a portion of or the entire band of operation of a given system and the users are generally willing to cooperate so as to minimize the mutual interference. Cellular networks operating in licensed bands must maintain specific quality of service due to regulatory restrictions and satisfy paying subscribers. As we will see in Chapter 9, deployment of such networks is more complex and involves careful frequency reuse and interference management. WLANs and WPANs, due to the use of unlicensed bands and the relative simplicity of architecture, are usually deployed with minimal planning, unlike cellular networks. Depending on the level of coordination of the overlapping wireless network since the early days of the IEEE 802.11 [HAY91], the WLAN industry has specified three types of overlapping networks: interference, coexistence, and interoperation.

Multiple wireless networks are said to *interfere* with one another if collocation causes significant performance degradation of any of the devices. Multiple wireless networks are said to *coexist* if they can be collocated without significant impact on the performance of any of the devices. *Coexistence* provides for the ability of one system to perform a task in a shared frequency band with other systems that may or may not be using the same set of rules for operation. *Interoperability* provides for an environment for multiple overlapping wireless systems to perform a given task using a single set of rules. In an interoperable environment, multiple wireless networks exchange and use information among each other. Interoperability is an important issue for wired as well as wireless networks.

Coexistence and interference are issues that are significant primarily for wireless network designers and it becomes more important for the case of *ad hoc* networks. This terminology for unlicensed bands was first discussed in the IEEE 802.11 community [HAY91]. Later, when FCC released unlicensed PCS bands, the issue of *etiquettes* or rules of coexistence in unlicensed PCS bands attracted attention [PAH97] that ultimately lost its momentum, as the unlicensed bands did not gain significant popularity. Around the year 2000, the IEEE 802.15 WPAN group was engaged in interference analysis in one of its task groups. They performed preliminary interference analysis between Bluetooth and IEEE 802.11 devices operating in 2.4 GHz unlicensed bands working on practical coexistence and interoperability methods [IEE01][ENN98]. Bluetooth is a fast frequency hopping (1600 hops per second at 1 Mbps) wireless system operating in the 84 MHz of bandwidth that is available in the 2.4 GHz unlicensed bands that are also used for IEEE 802.11 systems mostly using a variety of technologies and IEEE 802.15.4 ZigBee using direct sequence spread spectrum (DSSS). Therefore, the interaction between a Bluetooth system and a co-located 802.11 WLAN or 802.11.4 ZigBee system needs an analysis of the interference between the different radio systems. In what follows, we use randomly deployed devices of different types in the 2.4 GHz to illustrate the general concept of interference in wireless networks.

10.2.1 Interference Range

The first issue in interference is the *interference range*, that is, the distance between two terminals that may cause them to interfere, in case they operate at the same frequency and at the same time. The range of interference is related to the radio propagation characteristics of the environment, the transmitted power from different devices, and the sensitivity level of the transmission technique to the interference. Figure 10.1 illustrates a general interference scenario between an interference source, the desired transmitter, and a target receiver co-located around coverage of the interference source. The interference takes place when the target receiver is receiving information from the desired source while the interference source is transmitting for its own objective of communicating with its paired receiver.

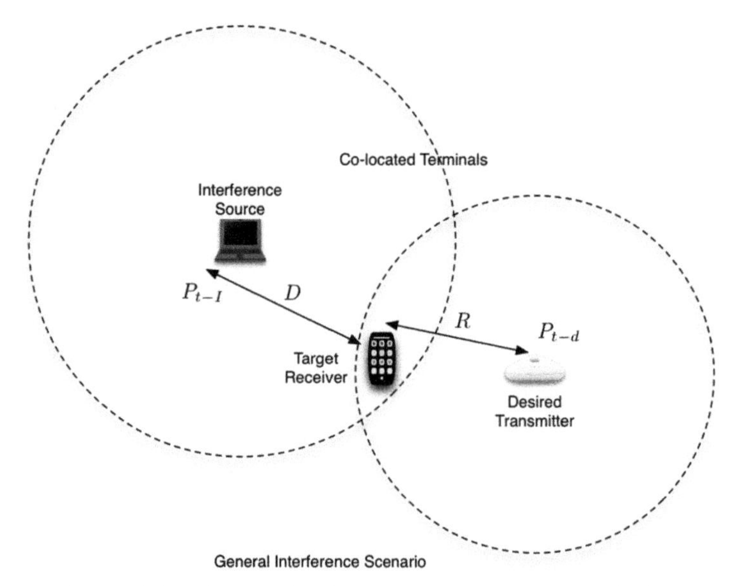

Figure 10.1 The basic interference scenario between two devices.

Considering the general Friis equation for open space propagation given by Equation (3.7), the received signal strength is

$$P_r = P_t \left(\frac{\lambda}{4\pi d}\right)^2 = \frac{KP_t}{d^2}; \qquad K = \left(\frac{\lambda}{4\pi}\right)^2, \tag{10.1a}$$

where *K* is a constant related to the wavelength of the signal and antenna gains. In all path loss models we presented in Section 3.3.2, we changed the exponent term in this equation from 2 to α and we named it the distance-power gradient; in this format, the generalized received signal strength will be:

$$P_r = \frac{KP_t}{d^\alpha} = KP_t d^{-\alpha}. \tag{10.1b}$$

Then, at the time that the target receiver is receiving a signal from the desired source and the interference source is also transmitting, *the signal-to-interference level* at the target receiver is

$$S_r = \frac{P_{r-d}}{P_{r-I}} = \frac{KP_{t-d}R^{-\alpha}}{KP_{t-I}D^{-\alpha}} = \frac{P_{t-d}}{P_{t-I}}\left(\frac{D}{R}\right)^\alpha, \tag{10.2}$$

where *R* and *D* are the distances between the desired receiver device and the target and interfering transmitters, respectively. Also, P_{t-d} and P_{t-I}

represent the transmitted power by the desired transmitter and the interference source, respectively. Therefore, the *range of interference* between the interference source and the target receiver is given by

$$D = R \times \left(\frac{P_{t-I}}{P_{t-d}} S_r \right)^{1/\alpha}.$$

If we define D_{int} as the maximum distance at which the two terminals interfere and $S_{\min}$ as the *minimum* acceptable *received signal-to-interference ratio* needed for proper operation of the target receiver, we have

$$D_{\text{int}} = R \times \left(\frac{P_{t-I}}{P_{t-d}} S_{\min} \right)^{1/\alpha}. \tag{10.3}$$

In other words, the range of interference of the interfering device to the target receiver is directly related to the distance to the desired transmitter, required signal-to-noise ratio (SNR) for proper operation of the target receiver, and transmit power of the interfering source, and it is inversely related to the transmit power of the desired transmitter. In general, as we discussed in Chapter 2, the value of α may change from less than 2 in hallways and open areas and up to around 6 in buildings with metal partitioning. Depending on the location of the devices, the path loss gradients between the interference source and target receiver may be different from α as well. In open areas with no walls, that include several scenarios involved with short-range devices, the environment is close to free space propagation and α is often close to 2.

Example 10.1. Range of a Bluetooth Interfering with a Wi-Fi: Consider a Wi-Fi access point (AP) with a transmitting power of 20 dBm and a minimum signal to noise requirement of 10 dB is located at 20 m of a Wi-Fi enabled device (Figure 10.2). If a Bluetooth device with transmitted power of 0 dBm is operating in that area, what is the minimum distance between the target Wi-Fi device and the interfering Bluetooth so that the signal from Bluetooth can intrude the Wi-Fi communication signal?

From Equations (10.2) and (10.3) for Figure 10.2, we have

$$\begin{cases} S_r = \frac{P_{t-d}}{P_{t-I}} \left(\frac{D}{R}\right)^{\alpha} = \frac{P_{\text{WL}}}{P_{\text{BT}}} \left(\frac{D}{R}\right)^{\alpha} \\ D_{\text{int}} = R \times \left(\frac{S_{\min} \times P_{t-I}}{P_{t-d}}\right)^{1/\alpha} = R \times \left(\frac{S_{\min} \times P_{\text{BT}}}{P_{\text{WL}}}\right)^{1/\alpha} \end{cases}.$$

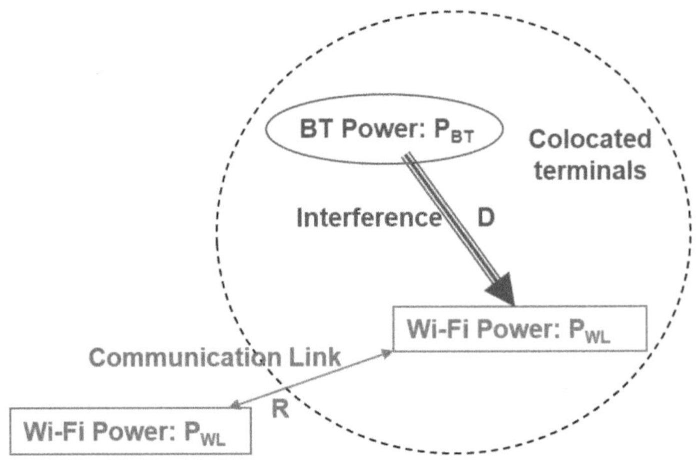

Figure 10.2 Example of Bluetooth interfering with Wi-Fi.

The parameters of the problem are

$$\begin{cases} P_{\mathrm{WL}} = 100\,\mathrm{mW}(20\,\mathrm{dBm}) \\ P_{\mathrm{BT}} = 1\,\mathrm{mW}(0\,\mathrm{dBm}) \\ S_{\mathrm{min}} = 10\,\mathrm{dB} \\ R = 20\,\mathrm{m} \end{cases}$$

The intrusion distance D_{int}depends on the distance-power gradient, α, of the environment. Following IEEE 802.11 model for path loss in line-of-sight (LOS) and obstructed line-of-sight (OLOS) environments, we have

$$\begin{aligned} &\mathrm{LOS}: \quad \alpha = 2 \Rightarrow D_{\mathrm{int}} = 20\ \times \left(\tfrac{10\times 1}{100}\right)^{1/2} = 6.3\,\mathrm{m} \\ &\mathrm{OLOS}: \quad \alpha = 3.5 \Rightarrow D_{\mathrm{int}} = 20\ \times \left(\tfrac{10\times 1}{100}\right)^{1/3.5} = 10.4\,\mathrm{m}. \end{aligned}$$

In LOS environments, interference is more effective than that in OLOS.

To cover large areas, wireless networks are deployed in a cellular manner, each cell covering a portion of the entire area of coverage. When the cells are far apart, we use the same frequency again because the interference is negligible. Considering that the government agencies assigned limited bands to service providers, understanding of the relationship between the number of users and available bandwidth in a cellular network first evolved for connection-based 1G cellular networks and then extended to other generation of connection based and packet switched cellular networks. We begin our study of deployment for wireless cellular networks with these classical deployments for connection-based services and then we discuss the deployment of packet switched wireless data networks.

10.2.2 Concept of Cellular Wireless

Cellular topology is a special case of an infrastructure multi-base station (BS) network configuration that exploits the *frequency reuse* concept. Radio spectrum is one of the scarcest resources available and every effort must be made to find ways of utilizing the spectrum efficiently and to employ architectures that can support as many users as theoretically possible with the available spectrum. This is extremely important especially today, considering the huge demand for capacity. Spatially reusing the available spectrum so that the same spectrum can support multiple users separated by a distance is the primary approach for efficiently using the spectrum. This is called *frequency reuse*. Employing frequency reuse is a technique that has its foundations in the attenuation of the signal strength of electromagnetic waves with distance. For instance, in a vacuum or free space, the signal strength falls as the square of the distance. This means that the same frequency spectrum may be employed without any interference for communications or other purposes, provided the distance separating the transmitters is sufficiently large, and their transmit powers are reasonably small (depending on the reuse distance). This technique has been used, for example, in commercial radio and television broadcast where the transmitting stations have a constraint on the maximum power; they can transmit so that the same frequencies can be used elsewhere. The cellular concept is an intelligent means of employing frequency reuse. Cellular topology is the dominant topology used in all large-scale terrestrial and satellite wireless networks. The concept of cellular communications was first developed at the Bell Laboratories in the 1970s to accommodate many users with a limited bandwidth [Mac79].

Example 10.1: Cellular Concept

Consider a single high power transmitter (see Figure 10.3) that can support 35 voice channels over an area of 100 km^2 with the available spectrum. If seven lower power transmitters are used so that they support 30% of the channels over an area of 14.3 km^2 each, a total of $\approx$80 voice channels are now available in this area instead of 35. Channels will have to be allocated to BSs in such a way as to prevent interference between one BS and another. In Figure 10.3, BSs 1 and 4 could use the same channels, as their coverage areas are sufficiently far apart and so also BSs 3 and 6. Suppose the cells labeled

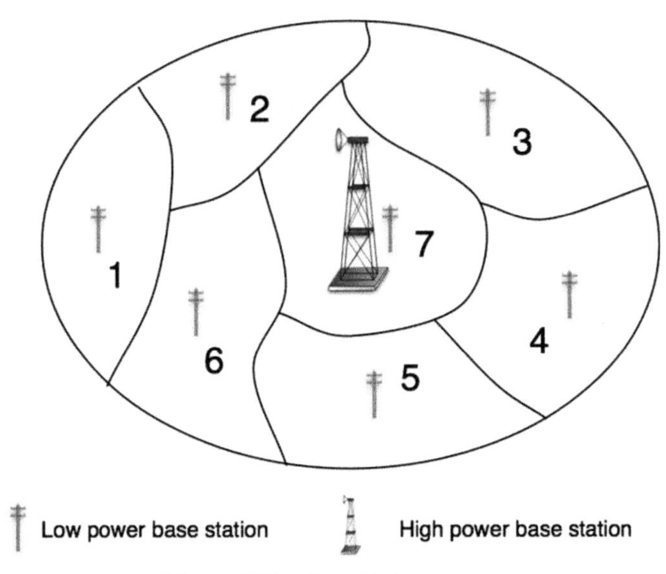

Figure 10.3 The cellular concept.

1, 2, 5, 6, and 7 use disjoint frequency bands, and the channels used in 1 and 6 are reused in 3 and 4. The set of cells {1, 2, 5, 6, 7} forms a cluster. Cells 3 and 4 form part of another cluster. In the limiting case, the density of BSs can be made so large that the capacity is infinite. However, in practice, this is impossible for several reasons that include drastic increases in the network and signaling load, the number and frequency of handoffs, and the cost of infrastructure and planning.

By cellular radio, we mean deploying many low-power BSs for transmission, each having a limited coverage area. In this fashion, the available capacity is multiplied each time a new BS or transmitter is set up since the same spectrum is being *reused* several times in each area. The fundamental principle of the cellular concept is to divide the coverage area into several contiguous smaller areas each served by its own radio BS. Radio channels are allocated to these smaller areas in an intelligent way to minimize the interference, provide adequate performance, and cater to the traffic loads in these areas. Each of these smaller areas is called a *cell*. Cells are grouped into *clusters*. Each cluster utilizes the entire available radio spectrum. The reason for clustering is that adjacent cells cannot use the same frequency spectrum because of interference. So, the frequency bands must be split into chunks and distributed among the cells of a cluster. The spatial distribution of chunks of radio spectrum (which are called sub-bands) within a cluster must be done in a manner such that the desired performance can be obtained. This forms an important part of network planning in cellular radio.

Two types of interference are important in such a cellular architecture. The interference due to using the same frequencies in cells of different clusters is referred to as *co-channel interference*. The cells that use the same set of frequencies or channels are called co-channel cells. The interference from frequency channels used within a cluster whose side-lobes overlap is called *adjacent channel interference*. The allocation of channels within the cluster and between clusters must be done to minimize both.

10.2.3 Cell Fundamentals and Frequency Reuse

Having looked at the cellular topology and the concept of employing a cellular architecture to increase the communications capacity and to cater to a large subscriber demand in hotspots, we now consider quantitative means to characterize the interference in a cellular topology. This, in turn, leads to quantitative means for determining the best cluster size and simple techniques for allocating the sub-bands of the spectrum and within a cluster.

Even though, in practice, cells are of arbitrary shape (close to a circle) because of the randomness inherent in radio propagation, it is easier to obtain insight and understanding for system design by visualizing all cells as having the same shape. Also, it is easier to mathematically analyze a cellular topology by assuming a uniform cell size for all cells. Once some insight is obtained as to what the effects of interference are, measurements, simulation, and a combination of these can be employed in determining the planning of a network.

For cells of the same shape to form a tessellation so that there are no ambiguous areas that belong to multiple cells or no cell, the cell shape can be of only three types of regular polygons: equilateral triangle, square, or regular hexagon as shown in Figures 10.4 and 10.5. In most of the literature and behind the envelope design, the hexagonal cell shape is chosen as the default cell shape. In cases that consider continuous distributions of traffic load and interference between different transmission schemes, a circular cell shape is employed for tractable calculation.

To investigate the effects of interference, which changes with distance, there is a need to come up with an elegant way of determining distances and identifying cells. Fortunately, it is possible to do this easily in the case of hexagonal cells [Mac79]. To maximize the capacity, co-channel cells must

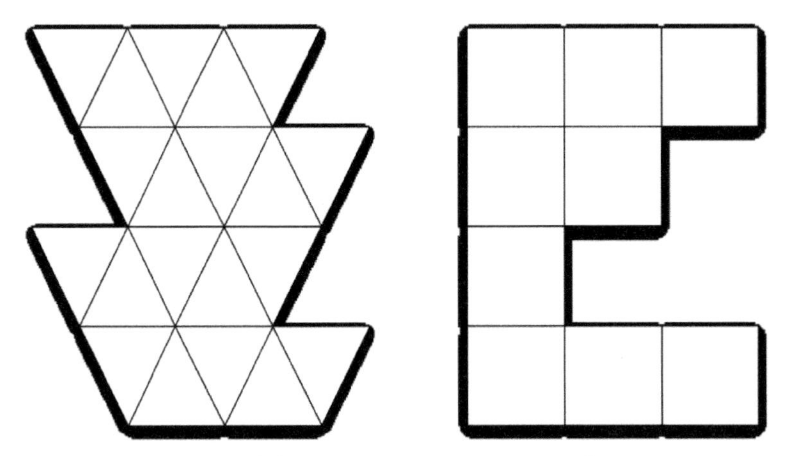

Figure 10.4 Triangular and rectangular cells.

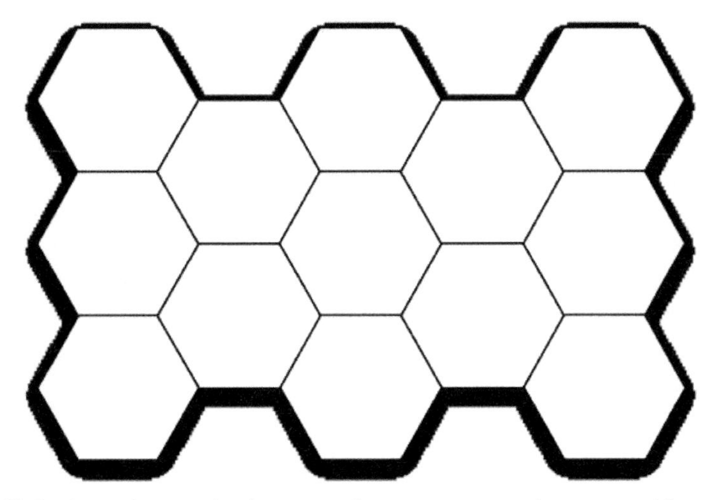

Figure 10.5 Arranging regular hexagons that can cover a given area without creating ambiguous regions.

be placed as far apart as possible for given cluster size. It can be shown that there are six co-channel cells for a given reference cell at this distance. The distance between the co-channel cells can be shown to be $D = \sqrt{3}R$. Here, R is the radius of a cell. The relationship between the distance between co-channel cells, the cluster size, and the cell radius is given by

$$\begin{cases} \frac{D}{R} = \sqrt{3N_f} \\ Nf = i^2 + j^2 + ij \end{cases}, \tag{10.4}$$

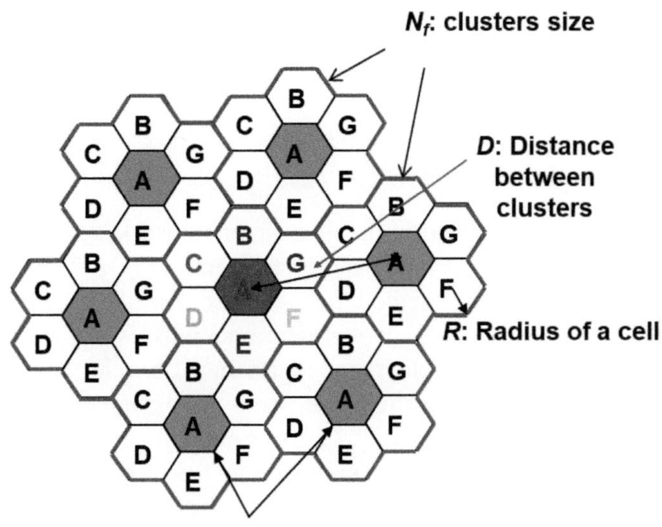

Figure 10.6 Hexagonal cellular architecture with a cluster size of $N_f = 7$.

where N_f is the frequency reuse factor or cluster size, and it can only take on values of the form $i^2 + ij + j^2$ where i and j are integers. In order words, cluster size can only take values $N_f = 1, 3, 4, 7, 9, 12, 13, 16, 19, 21, ...$ to guarantee that we can repeat the clusters with the same form to expand as much as we want.

Example 10.2: Cluster size of N_f = 7:

As described above, i and j can only take integer values. If we take $i = 2$ and $j = 1$, we see that $N_f = 4 + 2 + 1 = 7$. Selecting a cell, A, we can determine its co-channel cell by moving two units along one face of the hexagon and one unit in a direction 60° or 120° to this direction. Proceeding in this fashion, clusters of size $N_f = 7$ can be created as shown in Figure 10.6. A value of $N_f = 7$ is employed in the USA in the advanced mobile phone service (AMPS) 1G analog cellular networks with omni-directional antennas.

The number of cells in a cluster N_f determines the amount of co-channel interference and the number of frequency channels available per cell. Suppose there are N_c channels available for the entire system. Each cluster uses all the N_c channels. With fixed channel allocation, each cell is allocated N_c/N_f channels. It is desirable to maximize the number of channels allocated to a cell. This means that N_f should be made as small as possible. However, reducing N_f increases the signal-to-interference ratio (as discussed in the following section). There is, thus, a tradeoff between the system capacity and performance.

10.3 Cellular Deployment for Assigned Access

Classical 1G/2G/3G connection-based cellular network deployments were based on frequency reuse and interference control in licensed bands. A systematic deployment method has evolved for these networks, and understanding these classical deployment methods is beneficial for the understanding systematic deployment issues in all wireless networks. The 1G FDMA and 2G FDMA cellular networks user channels are separated in frequency and time, respectively. Cellular deployment of these networks requires a calculation of cluster size or the frequency reuse factor for the transmission technology adopted by these systems. In the 3G CDMA networks, the frequency reuse factor is one, and user channels are separated by the code they use to communicate with the network. In this section, we begin by describing how we can find the frequency reuse factor of the 1G/2G cellular networks followed by the description of architectural methods evolved to expand the capacity of these networks. Then, we address issues related to the deployment of CDMA networks.

10.3.1 Frequency Reuse Calculation for 1G/2G Cellular

In Section 10.1, we mentioned that a cellular architecture was essential to reuse the available spectrum while reducing interference caused by reusing the frequency spectrum. In this section, we will look in detail at the performance measures that are useful in system design, the signal-to-interference ratio and its relationship with the path loss, and the grade of service. Recall the signal-to-interference ratio calculations in the unlicensed frequency bands in Section 10.1 here.

In general, the signal-to-interference ratio is calculated in a manner similar to that in Section 5.3 and can be written as follows:

$$S_r = \frac{P_{\text{desired}}}{\sum\limits_i P_{\text{intereferer}-i}}. \tag{10.5}$$

The signal strength falls as some power of the distance α called the power-distance gradient or path loss gradient. That is if the transmitted power is

P_t, after a distance d in meters, the signal strength of a radio signal will be proportional to $P_t d^{-\alpha}$. In its most simple case, the signal strength falls as the square of the distance in free space (α = 2). Suppose there are two BS transmitters BS_1 and BS_2 located in an area with the same transmit power P_t and a mobile terminal is at *R* from the first and *D* from the second. If the mobile terminal is trying to communicate with the first BS, the signal from the second BS is interference. The *signal-to-interference* ratio for this mobile terminal will be

$$S_r = \frac{KP_t R^{-\alpha}}{KP_t D^{-\alpha}} = \left(\frac{D}{R}\right)^{\alpha}. \tag{10.6}$$

The larger the ratio *D/R* is, the greater is S_r (the smaller is the interference) and the better the performance. The objective in a cellular radio system is to allocate frequencies or channels to cells within a cluster so that the distance between interfering cells (co-channel or adjacent channel) is as large as possible. For urban land mobile radio, the distance power gradient increases from 2 (in the case of free space) to roughly 4 so that the received signal strength falls as the fourth power of the distance. This further improves the signal-to-interference ratio. If there are J_s interfering BSs surrounding a given BS, the general form of the SNR will be

$$S_r = \frac{P_{\text{desired}}}{\sum_i P_{\text{intereferer}-i}} = \frac{d^{-\alpha}}{\sum_{i=1}^{J_s} d_i^{-\alpha}}, \tag{10.7}$$

where the distance of the mobile from the given BS is *d* and its distance from the *n*th BS is d_n.

Recalling that there are exactly six co-channel cells with a hexagonal cellular structure, they will all cause similar levels of interference to a mobile terminal in the given cell, so J_s = 6 here. Also, the distance at which the co-channel cells are located depends on the size of the cluster from Equation (10.4). The farthest distance a mobile terminal can be from the BS of a given cell is the cell radius *R*. The approximate distance of the mobile terminal from the BSs of each of the co-channel cells is *D*.

$$S_r = \frac{KP_t R^{-\alpha}}{J_s KP_t D^{-\alpha}} = \frac{1}{J_s}\left(\frac{D}{R}\right)^{\alpha} = \frac{1}{J_s}\left(\sqrt{3N_f}\right)^{\alpha} \tag{10.8}$$

In cellular networks operating in urban areas, the distance-power gradient is assumed to be $\alpha = 4$, with six co-channel cells that make up the first tier

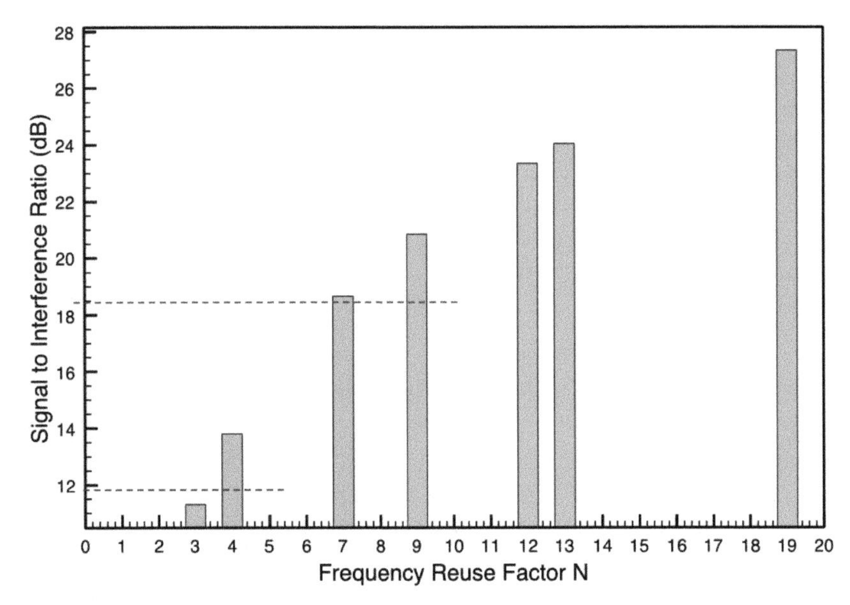

Figure 10.7 $S_{r-\mathrm{dB}}$ Sr as a function of N_f.

of interferers are considered, $J_s = 6$, the signal-to-interference ratio can be approximated as

$$\begin{cases} S_r = \frac{1}{J_s}\left(\sqrt{3N_f}\right)^{\alpha} = \frac{1}{6}\left(\sqrt{3N_f}\right)^{4} = \frac{3}{2}\left(N_f\right)^2 \\ S_{r-\mathrm{dB}} = 10\log\frac{3}{2} + 20\log N_f = 1.76 + 20\log N_f \end{cases}, \tag{10.9}$$

where $S_{r-\mathrm{dB}}$ is the signal-to-interference requirement in dB as a function of cluster size.

Figure 10.7 shows how the signal-to-interference ratio given by Equation (10.9) varies with the cluster size *N*. Equation (10.9) is commonly used to determine the cluster size for adequate performance. Note that the signal-to-interference ratio is influenced by the co-channel reuse ratio *D*/*R*, in that a given *D*/*R* has to be maintained for a particular S_r. However, it is an approximation since different BSs may employ different transmit powers and the path loss model may not be as simple as the d^{-4} model used here. The S_{r-dB} calculation will be different for the uplink (mobile terminal to BS communication) compared to the downlink (BS to mobile terminal communication).

Example 10.3: Cellular Architecture of AMPS:
As an example of cellular architecture, we consider the 1G US analog cellular AMPS that is based on an analog frequency modulation (FM) scheme. Each voice channel in AMPS occupies 30 kHz of bandwidth and it needs a minimum SNR of 18.6 dB. Figure 10.7 shows that to maintain this level of signal-to-interference, we need a minimum cluster size of 7.

A bandwidth of 25 MHz is allocated for both the uplink and downlink of the AMPS so that transmission is full duplex. The 25 MHz of spectrum is divided into two blocks of 12.5 MHz each. Block A was allocated to carriers who are not traditional telephone service providers. Block B was allocated to traditional telephone service providers. Every 12.5 MHz of the spectrum can support 416 channels each of which is 30 kHz wide. Of these, 395 are dedicated channels for voice and 21 are dedicated to call control.

Based on subjective voice quality tests, it was determined that a signal-to-interference ratio of 18.6 dB can be tolerated while providing a good voice quality to the user. From Equation (10.7), this means that the cluster size must be $N = 7$. Figure 10.6 shows the cellular architecture with this cluster size. Cells with the same label use the same frequency spectrum. They are separated by a distance $D = 4.58\ R$ in this case which ensures that the signal-to-interference ratio is better than18.6 dB.

Let the 395 voice channels available for a service provider be numbered from 1 to 395. For example, on the downlink, 869-869.030 corresponds to channel 1, 869.030-869.060 to channel 2, and so on. Channels 1, 8, 15, etc., are allocated to cells labeled A. Channels labeled 2, 9, 16, etc., are allocated to cells labeled B and so on. This ensures that there is sufficient separation between channels used within a cell so that adjacent channel interference is minimized. In practice, the numbering scheme is different since the entire 25 MHz of bandwidth was not available for AMPS initially. However, a separation of seven adjacent channels is maintained between channels used within a cell.

Example 10.4: Cluster Size for 2G GSM Digital Cellular:
The 2G digital cellular GSM uses GMSK modulation that needs 12 dB SNR to operate properly. As shown in Figure 10.7, a frequency reuse factor of $N_f = 4$ is adequate to maintain this level of signal-to-interference ratio.

10.3.2 Expansion Methods for 1G and 2G

In the past decade, the dominant source of income for the wireless telecommunication industry has been the cellular telephone service. This industry has grown exponentially during the last decade of the past millennium. Numerous companies are in fierce competition to gain a portion of the income of this profitable and prosperous industry. The main investment in deploying a cellular network is the cost of the infrastructure that includes the cost of BS and switching equipment, property (land for setting up the cell sites), installation, and links connecting the BSs. This cost is proportional to the number of BS sites. The income of the service is directly proportional to the number of subscribers. The number of subscribers should grow with time and a cellular service provider must develop a reasonable deployment plan that has a sound financial structure to account for many of these aspects. All service providers start their operation with the minimum number of cell sites to cover a service area that requires the least initial investment. As the number of subscribers increases, it generates a source of income for the service provider. At such a point in time, they can increase the investment in the infrastructure to improve service and increase the capacity of the network to support additional subscribers. Therefore, several methodologies have evolved to facilitate the expansion of cellular telephone networks.

There are basically four methods to expand the capacity of a cellular network. The simplest method is to obtain an additional spectrum for new subscribers. This is a very simple but expensive approach. The so-called PCS bands were sold in the USA for around $20 billion. If we assume that each new subscriber generates a profit of approximately $1000 per year, we will still need 20 million additional subscribers to recover this amount in a year. With the fierce competition to provide the lowest cost to the customer, this has proved to be suicidal. A case in example is that the top three companies that purchased the PCS bands have already filed for bankruptcy. The reader should not, however, conclude that this is not an acceptable method. With our pessimistic scenario, we are accentuating the vital importance of the need for other alternatives to expand capacity in addition to this simple approach of getting an additional spectrum.

The second method to expand the capacity of a cellular network is to change the cellular architecture. Architectural approaches include cell splitting, cell sectoring using directional antennas, Lee's micro-cell zone technique [Lee91], and using multiple reuse factors (called reuse partitioning [Hal83]). These techniques, described in detail in the rest of this section, change the size and shape of the coverage of the cells by adding cell sites or modifying the nature of antennas to increase the capacity. These techniques do not need additional spectrum or any major changes in the wireless modem or access technique of the system that will require the user to purchase a new terminal. These features of architectural approaches distinguish them as one of the more practical and less expensive solutions to expand the network capacity.

The third method for capacity expansion is to change the frequency allocation methodology. Rather than distributing existing channels equally among all cells, it is possible to use a non-uniform distribution of the frequency bands among different cells according to their traffic need. The traffic load of each cell is dynamically changed by the geography of the service area and with time depending on the traffic load. In most downtown areas, we have the largest traffic loads during rush hours and a relatively light traffic load in the evening hours and weekends. This situation is reversed in residential areas. If channels are allocated dynamically to different cells, we can increase the overall capacity of the network. These techniques do not need any change in the terminal or physical architecture of the system, and they are implemented somewhere inside the computational devices used for network control and management.

The fourth and the most effective method to expand the network capacity is to change the modem and access technology. The cellular industry started with analog technology using FM modulation and has now evolved toward TDMA and then a CDMA air interface using digital modems. Digital technology increases the network capacity and provides a fertile environment for the integration of voice and data services. However, this migration requires the user to purchase new terminals and the service provider to install new components in the infrastructure.

10.3.3 Architectural Methods for Capacity Expansion

Architectural methods interfere with the cellular physical deployment and modify it to increase the capacity of the network in supporting more subscribers. Here, we introduce a few of these techniques evolved for classical connection-based cellular networks. As we will see in Section 10.8, many of these methods are also applicable to the cellular packet radio data networks.

Cell Splitting:

As the number of subscribers increases within a given area, the number of channels allocated to a cell is no longer sufficient for supporting the subscriber demand. It then becomes necessary to allocate more channels to

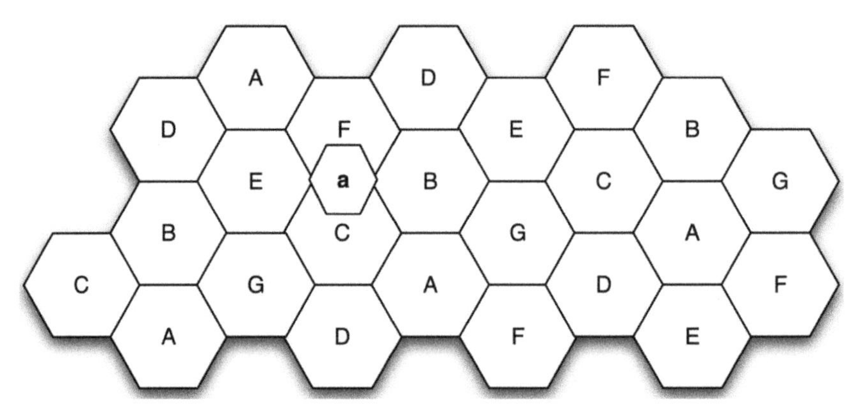

Figure 10.8 The idea of cell splitting.

the area that is being covered by this cell. This can be done by *splitting* cells into smaller cells and allowing additional channels in the smaller cells.

Consider Figure 10.8. In this figure, we have a cellular architecture where a cluster size of seven is employed. When the traffic load increases, a smaller cell is introduced such that it has half the area of the larger cells. This will ultimately increase the capacity fourfold (since the area is proportional to the square of the radius). However, in practice, only a single small cell will be introduced such that it is midway between two co-channel cells. In this case, these are the larger cells labeled **A**. It is logical too, thus, reuse the channels allocated to these cells in the smaller cell to minimize the interference.

This approach gives rise to some problems. Let us suppose that the radius of the smaller split cell (labeled **a**) is $R/2$. Let the transmit power of the BS of the small cell be the same as the transmit power of the larger cells. As far as the smaller cell is concerned, the signal-to-interference ratio is maintained because the maximum distance the mobile can be from the BS in this cell is $R/2$. So, though the distance between this cell and the co-channel cells **A** is reduced by half, the value of S_r remains the same. On the other hand, this is not the case for the cells labeled **A** since the co-channel reuse ratio for these cells is now $D_L/2R$ with respect to the smaller split cell. To maintain the same level of interference, the transmit power of the BS in the smaller cell should be reduced. But this will increase the interference observed by the mobiles in the smaller cell. The other alternative is to divide the channels allocated to cells labeled **A** into two parts: those used by **a** and those not used by **a**. The channels used by **a** will be used in the larger cells only within a radius of $R/2$ from the center of the cell so that the co-channel reuse ratio will be maintained as far as these channels are concerned. This is called the *overlaid cell concept* where a larger *macro-cell* co-exists with a smaller *micro-cell*.

The downside of this approach is that the capacity of the larger cells is reduced which will ultimately lead to introducing split cells in their area, till such time a chain reaction will result in the entire area being served by cells of a smaller radius. Also, the BSs in cells labeled **A** will become more complex and there will be a need for handoffs between the overlays.

Using Directional Antennas for Cell Sectoring:

The simplest and the most popular scheme for expanding the capacity of cellular systems is cell sectoring using directional antennas. This technique attempts to reduce the signal-to-interference ratio and thus reduce the cluster size, thereby increasing the capacity. The idea behind using directional antennas is the reduction in co-channel interference that results by focusing the radio propagation on only the direction where it is required. To achieve this, the coverage of a BS antenna is restricted to part of a cell called a *sector* by making the antenna directional. In implementing this technique, cell site locations remain unchanged and only the antennas used in the site will be changed. The main objective here is to increase the signal-to-interference ratio to a level that enables us to use a lower frequency reuse factor. A lower frequency reuse factor allows a larger number of channels per cell increasing the overall capacity of the cellular network.

As we discussed earlier (see Equation (10.9)), the signal-to-interference ratio for cellular networks is

$$S_r = \frac{1}{Js}\left(\frac{D_L}{R}\right)^4 = \frac{9}{J_s}N^2, \tag{10.10}$$

where J_s is the number of interfering cell sites. Using a sector antenna reduces the factor J_s resulting in the interference and an increase in S_r. The most popular directional antennas employed in cellular systems are 120° directional antennas. In some cases, 60° directional antennas are also employed. In the following two examples, we evaluate the impact of these antennas that enables the reuse factor to be reduced from $N = 7$ to $N = 4$ and $N = 3$, respectively.

Example 10.5: Three-Sector Cells and a Reuse Factor of $N = 7$:

Consider a seven-cell cluster scheme with 120° directional antennas shown in Figure 10.9. Channels allocated to a cell are further divided into three parts, each used in one sector of a cell. As shown in the figure, the

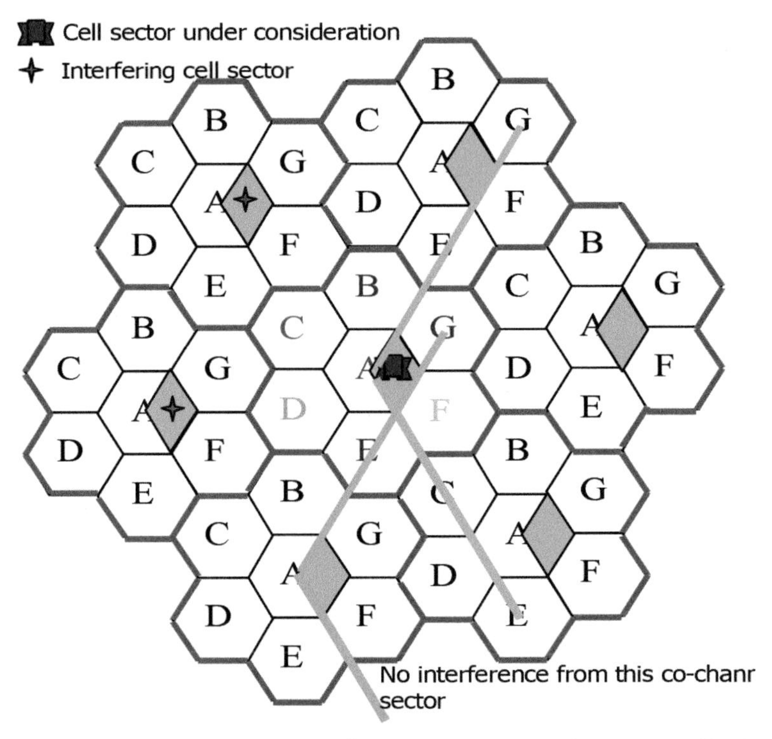

Figure 10.9 Seven-cell reuse with 120° directional antennas (three-sector cells, J_s = 2).

number of co-channel interfering cells is reduced from 6 to 2. Thus, there is an improvement in the signal-to-interference ratio. For omni-directional antennas (see Examples 10.3), the value of S_r for a cluster size of N_f = 7 is 18.66 dB. In this case, in a manner like (7), the signal-to-interference ratio is given by

$$S_r \approx \frac{R^{-4}}{J_s D^{-4}} = \frac{R^{-4}}{6D^{-4}} = \frac{1}{2}\left(\frac{D}{R}\right)^4 = \frac{9}{2}N^2. \tag{10.11}$$

For N = 7, this will give us S_r = 23.43 dB. To see the importance of this gain, note that the required SNR for AMPS systems is 18 dB which suggests N = 7. However, a larger S_r is required because of non-ideal situations.

In practice, we cannot ideally sector a cell because ideal antenna patterns cannot be implemented. Therefore, the numbers obtained in the above examples for ideal cell sectors are optimistic. However, our conclusion from the above example is that the use of sectoring increases the signal-to-interference ratio at the terminal. We should emphasize that, in the example, we could reduce the frequency reuse factor from $N = 7$ to $N = 4$. This reduction in frequency reuse from 7 to 4 would result in a capacity increase of 1.67, allowing an equal increase in the number of subscribers and consequently income of the service provider. The service provider needs to add these antennas to the BSs in the desired area. Compared to the cell-splitting method, using directional antennas is less effective in increasing capacity, but it can be significantly less expensive. The cost of additional cell sites, needed in cell splitting, includes costs of the property and installing the antenna mounting tower that is usually far expensive compared to deploying directional antennas. Cell splitting also requires additional planning efforts to maintain interference levels in the smaller cells. If directional antennas are used without reduction in the frequency reuse factor, the average required transmitted signal power from the mobile stations (MSs) will be reduced that can potentially result in longer battery life for the user.

Lee's Micro-Cell Method:

The disadvantage of using sectors is that each sector is nothing but a new cell with a different shape since channels must be partitioned between the different sectors of a cell. The network load is substantially increased since a handoff must be made each time a mobile terminal moves from one sector of a cell to another. Also, in all the discussions in the previous sections, it has been assumed that the BS antenna is located at the center of a cell, whether directional antennas are employed. In practice, employing directional BS antennas at the corners of cells can reduce the number of BSs [MAC79]. Lee's micro-cell zone technique [LEE91] exploits corner excited BSs to reduce the number of handoffs and eliminate the partitioning of channels between sectors of a cell.

Figure 10.10(a) shows Lee's micro-cell zones concept. In this case, there is one BS per cell, but there are three "zone-sites" located at the corners of a cell. Directional antennas that span 135° are employed at these zone sites. All three zone sites act as receivers for signals transmitted by a mobile terminal. The BS determines which of the zone sites has the best reception from the mobile and uses that zone site to transmit the signal on the downlink. The zone sites are connected to the BS by high-speed fiber links to avoid congestion and delay. Since only a single zone-site is active at a time, the interference faced by a mobile terminal from a co-channel zone site is smaller compared with what would be the interference with an omni-directional antenna. Consequently, the cluster size can be reduced to 3 and a capacity gain

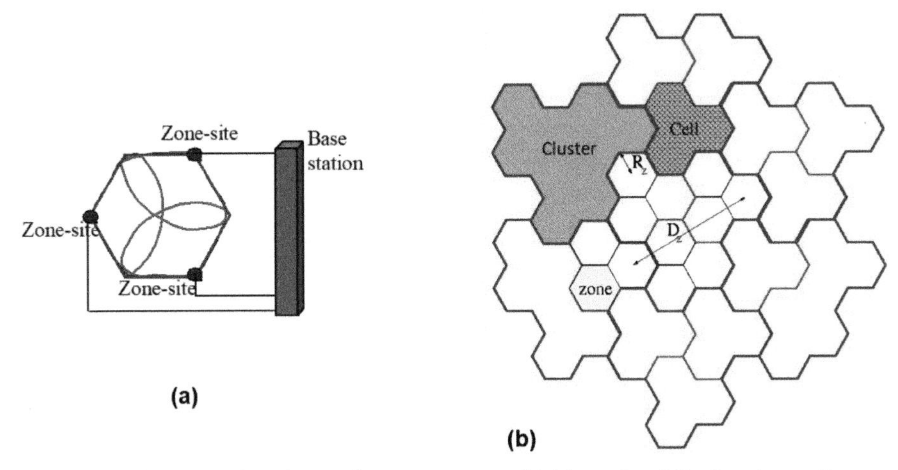

Figure 10.10 (a) Lee's micro-cell zone concept. (b) Example of Lee's micro-cell zone concept.

of 2.33 is obtained over a seven-cell cluster scheme. Consider the following example.

Figure 10.10(b) shows a cluster size of $N = 3$ for Lee's method. Each "cell" is divided into three "zones." On the downlink, only one of the zones is active. Since the zone sites are directional, they cause interference only in corresponding zone sites in another cluster. The co-channel reuse ratio D_z/R_z in Figure 10.10(b) is clearly $6 \times \sqrt{3}/2 = 5.196$ which is larger than $D/R = 4.6$ for traditional AMPS with omni-directional antennas. Even if all six co-channel zones cause interference, the capacity is still larger by a factor of 2.33 since the cluster size is now $N = 3$ as compared to the usual value of $N = 7$.

Using Overlaid Cells:

The *overlaid cell concept* introduced in the section on cell splitting can be used to increase the capacity of a cellular network. Here, channels are divided among a larger macro-cell that co-exists with a smaller micro-cell contained entirely within the macro-cell. The same BS serves both the macro- and micro-cells. Figure 10.11 illustrates the basic concept for overlaid cell concept. There are four parameters R_1 and D_1 representing the radius of coverage and distance among co-channel cells for the macro-cells and R_2 and D_2 denoting the radius of coverage and the distance among co-channel cells for the micro-cells. The design is made such that D_2/R_2 is larger than D_1/R_1, and from Equation (10.8), the signal-to-interference ratio for the micro-cells

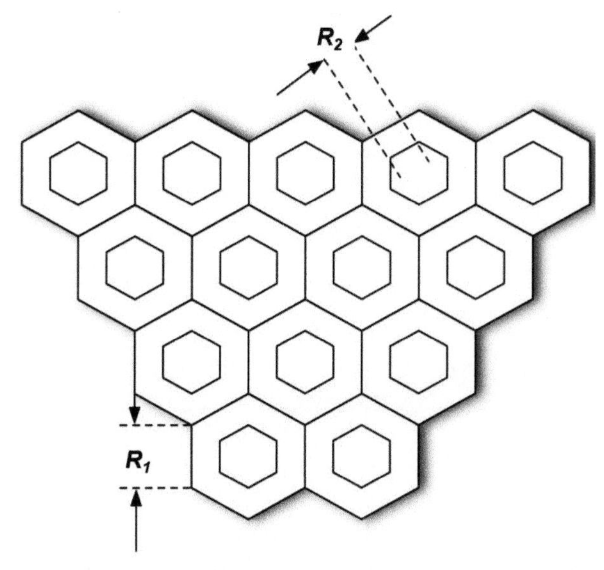

Figure 10.11 Reuse partitioning in a split band 1G cellular network.

will be substantially greater than that of the macro-cells. There are two methods to exploit this situation to increase the capacity of the network: using split-band analog systems and reuse partitioning. Often the micro-cells are said to belong to an *overlay* network that is overlaid on top of an underlying macro-cellular network referred to as the *underlay* network.

The split-band analog systems use a more bandwidth efficient modulation within the overlay cells. This technique is applied in analog cellular systems using FM. We have considered several examples of analog cellular systems in Chapter 1. In FM, the signal-to-noise requirement is inversely proportional to the square of the bandwidth. If we reduce the bandwidth to half the original value, the SNR requirement will be increased four times (by 6 dB). If we arrange R_2 and D_2 to have a co-channel reuse ratio that is four times larger than usual, we end up with a signal-to-interference ratio (from Equation (10.6)) that remains unchanged. The overlay system can then use FM with half the bandwidth of the underlaying system, doubling the capacity within the overlay part of the network. An example will further clarify this situation.

Example 10.6: Band-Splitting in 1G Analog AMPS:

The AMPS system uses a 30 kHz band for FM signals used for communication between the MS and the BS. As discussed earlier, the minimum required signal-to-interference ratio for this system is 18 dB. If we develop an overlay system with a 15 kHz bandwidth, the required S_r is 24 dB, that is, 6 dB more than the system employing a bandwidth of 30 kHz. From Equation (10.6), we have

$$10 \log \frac{\left(\frac{D_2}{R_2}\right)^4}{\left(\frac{D_1}{R_1}\right)^4} = 6\_\mathrm{dB}.$$

If we employ the same frequency reuse factor of $N = 7$ for the overlay and underlay networks, $D_1 = D_2$ and solving for the above equation, we have $R_2 = 0.7079\ R_1$. Since the area covered by each cell is proportional to the square of the cell radius, the area of the overlay cell, A_2, will be half of the area of the underlay cell, A_1. The overlay is responsible for terminals within the smaller hexagon, while the underlaying system supports users in the layer between the boundary of the overlay cell and the boundary of the underlay cell. These two areas are the same in our example. Therefore, the number of channels available to the overlay and underlay cells remains the same. If we represent this number by *M*, then the total bandwidth used by the system is $M(15 + 30)$ kHz.

In the original AMPS network, each service provider has 12.5 MHz of bandwidth that is divided into 416 channels from which 395 channels are used for voice and 21 channels for control signaling. Therefore, 395 × 30 kHz of bandwidth was used for actual traffic. If we replace that system with a split-band underlay–overlay network, we have

$$M(15 + 30) = 395 \times 30 => M = 263.$$

The total number of channels $M = 263$ for each of the overlay and underlay cells and $263 \times 2 = 526$ will be the total number of channels available. This number is 1.34 times larger than the original system, improving the capacity of the system by 34%.

Compared to cell splitting or using sectored cells, this technique provides a smaller improvement in capacity. However, it does not need any change in the hardware infrastructure. However, the MS and BS need minor changes to cope up with multiple bandwidths. To further improve the capacity of this technique, it is possible to use another layer of overlay system using even smaller cells. As we saw in Chapter 1, the Japanese analog systems use 25 kHz per user for underlay networks and 12.5 kHz (and even 6.25 kHz) for the overlay networks. The downside of underlay–overlay networks is the increased complexity at the BS for keeping track of which channel belongs to which overlay and the increased number of handoffs when a mobile move from one overlay to the next (or from a micro-cell to a macro-cell). This requires additional complexity at the BS and handoffs when a mobile terminal moves from a micro-cell to a macro-cell.

The overlaid cell concept described above can be used to increase the capacity of a cellular network through what is called the *reuse partitioning* concept [HAL83]. Here, channels are divided among a larger macro-cell and a smaller micro-cell contained entirely within the macro-cell. The bandwidth in both cells remains the same. Since the radius of the micro-cell is smaller, the signal-to-interference ratio (S_r) for the overlay is larger and it can employ a smaller co-channel reuse distance compared to the underlay or macro-cell. The channels allocated to the micro-cell, for instance, maybe reused in every third or fourth micro-cell, whereas the channels allocated to the macro-cells can be reused in only every 7th or 12th cell. This requires additional complexity at the BS and handoffs when a mobile terminal moves from a micro-cell to a macro-cell. To explain this situation, consider the following example.

In overlay–underlay with a split band, the bandwidth of the carrier is reduced to half to increase the minimum signal-to-interference ratio. The same goal can be achieved in any digital system with a fixed carrier bandwidth by change using two different frequency reuse factors in the overlay and underlay cells. Then the concept of overlay–underlay can be extended to any system with fixed bandwidth. Assume that in Figure 10.12, the radius of the underlay macro-cells is R_1 and the radius of the micro-cells of the overlay is R_2. If we have an AMPS network operating on this infrastructure, the required S_r for both networks is 18 dB. From Equation (10.6), both underlay and overlay networks should have $D_1/R_1 = D_2/R_2 = 4.6$. Since R_2 is smaller than R_1, D_2 can be made smaller than D_1 by a factor equal to the ratio of $R_1 - R_2$. The improvement in co-channel reuse ratio comes from the fact that the micro-cells in the overlay are not contiguous to one another.

Suppose the co-channel reuse ratio without reuse partitioning was $D_L/R = Q$. The cluster size N, in the case, is $Q^2/3$ (from Equation (10.6)) and the number of channels available per cell is $N_c/N = 3N_c/Q^2$. With reuse partitioning, let the ratio of the macro-cell radius to the micro-cell radius be $\kappa = R_1/R_2$. From Example 10.6, the cluster size for the micro-cells can be reduced by a factor of κ^2 since the micro-cells are non-contiguous.

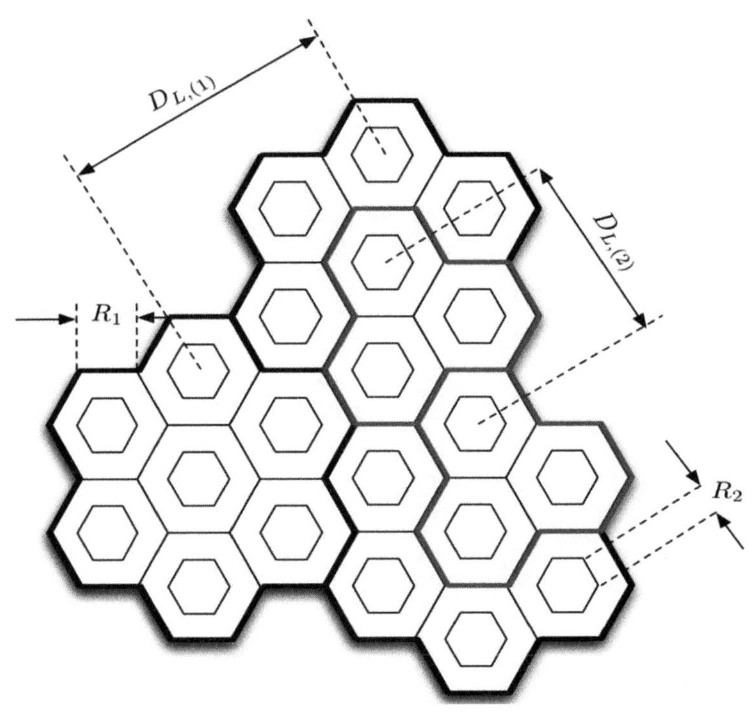

Figure 10.12 Reuse partitioning with a cluster size of 7 for the macro-cells and 3 for the micro-cells.

Example 10.7: Channel Allocation to Underlay and Overlay Cells: Consider Figure 10.12. Here we are using a cluster size of $N_1 = 7$ for the underlay macro-cells to ensure that $D_1/R_1 = 4.6$ to provide a suitable S_r for AMPS. We now overlay micro-cells with a radius R_2 such that the cluster size of the micro-cells is $N_2 = 3$. If $N_2 = 3$, we can see from Figure 10.12 that $D_2 = 3R_1$. Clearly, $3R_1/R_2 = 4.6$ or $R_2 = 0.652R_1$. One way of allocating channels to the micro-cells and macro-cells is to distribute them by the area occupied. This may not be the best case. However, for this example, we employ this technique. The area of a cell is proportional to the square of its radius. We see that the area of a micro-cell is 0.652^2 times the area of a macro-cell or $0.425 \times$ area of macro-cell. Let the total number of channels available be N_c. If channels are distributed according to the area and there are L channels available per cell, let us assume that $0.425L$ channels are allocated to the micro-cell and $0.575L$ channels are allocated to the macro-cell.

Since the cluster sizes are 7 and 3, respectively, we have

$$N_c = 7 \times 0.575L + 3 \times 0.425L => L = N_c/5.3.$$

The total number of available channels for an AMPS operator is 395. Therefore, L = 75. The inner overlay uses approximately 32 channels and the underlay uses 43 channels. Originally, we had 395 channels with N = 7, providing approximately 56 channels per cell. The increase in the capacity is 75/56 = 1.34, a 34% increase in capacity. A larger capacity can be expected since the channels allocated to the macro-cells may also be used within the micro-cells.

10.3.4 Network Planning for 3G CDMA

CDMA presents some unique features that are not present in traditional TDMA and FDMA systems. In TDMA and FDMA systems, the users operating in one channel are completely isolated from the users operating in other channels. The only interference comes from the fact that the same frequency bands are employed in spatially separated cells and this interference is the co-channel interference. Of course, leakage of signal from adjacent bands causing adjacent channel interference is also a factor but intelligent design can reduce this effect greatly. However, in the case of CDMA, all users are operating on the same frequency channel at the same time resulting in everyone causing co-channel interference. This problem is reduced on the downlink by employing time synchronized orthogonal codes. On the uplink, a combination of convolutional coding, spreading, and orthogonal modulation is employed to combat the effects of this interference. Network planning in the case of CDMA is far more complicated than in the case of TDMA/FDMA in that sense, but, at the same time, using CDMA completely eliminates the concept of conventional frequency reuse since the same frequencies can be deployed in all cells.

Instead of defining an acceptable signal-to-interference ratio, in CDMA, it is necessary to define the *quality of the signal* [HAL96]. Usually, this is expressed in terms of the acceptable energy per bit to total interference ratio E_b/I_t that results in roughly a 1% data frame error rate. The E_b/I_t is used in CDMA instead of the signal-to-interference ratio S_r. In this section only, we use E_b/I_t and S_r interchangeably. The reason for selecting this as a measure is that this frame error rate results in acceptable speech quality at the vocoder output. The value of E_b/I_t is usually between 6 and 11 dB depending on the speed of the mobile terminal, propagation conditions, the number of multipath signals that can be used for diversity, etc. The value of I_t depends on the number of interfering signals and the transmit powers of the interfering users. Consequently, power control and thresholds play a very important role in the coverage of a CDMA cell and the soft handoff process associated with it.

Many of the principles that apply to TDMA/FDMA systems also apply to CDMA systems, but there are important differences. For example, the path loss is very similar to TDMA systems in that the signal strength drops roughly as the fourth power of the distance in macro-cells and is quite a site specific and terrain dependent. However, the design issues that differ are described below.

In CDMA, managing the noise floor is very important. If the number of users in a particular area increases beyond a certain level, interference from other users dominates the background noise, the system is interference limited, and increasing the transmit power will not benefit any user or set of users as the total interference also increases. It is quite possible that interference from many cells can raise the noise floor to such a level that holes may be created in the region where the coding/spreading gain is not sufficient to overcome the interference levels. This is illustrated in Figure 10.13 [HAL96]. If there is an isolated three-sector cell, most of the cells have a signal-to-interference ratio of larger than 7 dB, and in the regions where there is a soft handoff (where the mobile terminal can connect to more than one BS), the signal-to-interference ratio value from each BS is around 3 dB providing sufficient diversity gain to allow communication. If too many cells are deployed as shown in the figure, there may be some regions where the noise level is so high that it is impossible to communicate. It is

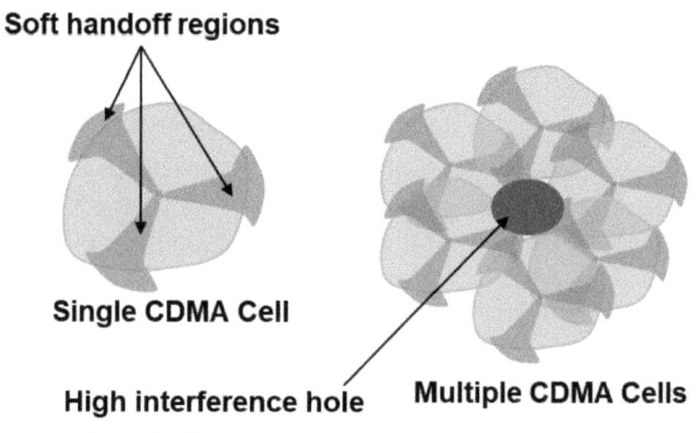

Figure 10.13 Noise floor management in CDMA.

often possible to cover the same area with fewer cells to reduce the total interference levels and it is usually not a very good idea to cover an area with more than three cells or cell sectors. The problem becomes more severe when terrain plays a role, and, in addition to site selection, it will be important to use the down tilt of antennas and the use of minimum radiated power levels to manage the noise floor.

In CDMA, the boundary of a cell is not fixed and depends on where the E_b/I_t value is reached. For example, consider the uplink signal-to-interference level value that is observed at a BS. As the number of traffic channels on the uplink is increased, this value also increases. This effect is called cell breathing. To ensure that a correct handoff is performed, the transmit power of the pilot channel of the BS must also be reduced so that the forward link handoff boundary is also maintained at the same level as the reverse link boundary. In some cases, cell breathing can have a deleterious impact on the system performance, and this should be considered while planning the system, either by deploying more cells or offloading capacity to other carriers.

10.4 Cellular Deployment for Random Access

Deployment of the 1G/2G/3G cellular networks was focused on connection base services for circuit switched mobile telephone applications. The 4G cellular networks were deployed as wireless mobile data services for packet switching like Wi-Fi. Mobile telephone application is a connection-based service with long statistically symmetric sessions demanding a low data rate (approximately 10 Kbps). The information packets in these connections cannot tolerate delays of more than 100ms, and they should support operation in vehicles with high-speed motion. Packet switched data networks support connectionless media access control (MAC) with bursts of often unbalanced data demanding as high a data rate as possible. Most applications are for semi-stationary scenarios and many of these applications are more tolerant to delays. Quality of service in connection-based networks is very sensitive to handoff delays, while data services are more tolerant to handoff delays. Circuit switched cellular networking methods evolved around the MAC options to support a higher number of telephone users. Deployment of packet switched networks evolved around physical (PHY) to maximize the data rate for user experiments. The PHY of wireless data networks began with omni-directional antennas and spread spectrum technology to take advantage of time diversity to combat with the server multipath. These transmission techniques shifted to OFDM to benefit from frequency diversity. The introduction of MIMO techniques opened a new horizon for beamforming and multiple spatial streaming to benefit from space-diversity. Deployment of cellular networks for these systems evolved over the same concept of overlay–underlay and is still evolving. We begin by describing the fundamentals of cellular mobile packet switched data networks, 4G, 5G, and beyond to build on our existing discussions. Then, we address the deployment of the local Wi-Fi data networks.

10.4.1 Deployment of 4G LTE Networks

Multiple overlay methods became popular in the deployment of long-term evolution (LTE) because they could provide an additional increase in capacity. As compared to the other expansion techniques, the advantages and disadvantages of frequency reuse partitioning are very similar to frequency splitting. However, reuse partitioning does not need modification in the BS or MS radio equipment, and it can be easily applied to other modulation techniques. The derivation of S_r for frequency splitting was highly correlated with how FM works, and it cannot be extended to digital systems in a straightforward way. We refer to a couple of these reuse partitioning methods for the 4G LTE networks.

Figure 10.14 [Yiw10] shows a deployment method called *soft frequency reuse* for the LTE. In this approach, the inner cells use all frequencies with low transmission power when they are close to the BS, while outer cells operate at higher power levels in non-overlapping bands. If an outer cell is not using a spectrum, the inner cell can use it.

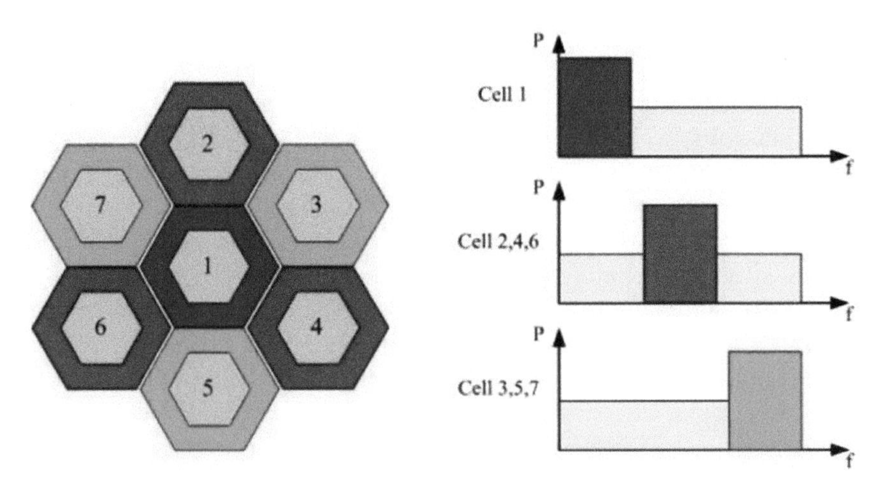

Figure 10.14 Soft frequency reuse partitioning for deployment of the LTE [Yiw10].

Reuse partitioning can be employed differently in 4G LTE systems where orthogonal frequency division multiple access (OFDMA) is the access method. In OFDMA, frequency sub-carriers and time slots are allocated to individual users for transmissions. Through what is called *fractional frequency reuse* or FFR [Saq13], certain frequency sub-carriers are used at lower transmission power so that they only service mobile devices that are at a closer distance to a BS, while other frequency sub-carriers are employed throughout a cell. This way, those low-power sub-carriers that are used in the center of a cell do not interfere with transmissions in neighboring cells at the same frequencies. The way these sub-carriers are selected depends on the cell sizes, loads in cells, tolerable interference, and other such factors. The primary difference between the way this is deployed in 4G LTE systems and in the past is that the power assignment and reuse can be done on a fine-grained time and frequency scale in 4G systems, especially because there are more powerful ways of predicting and coordinating interference in these systems. This is often called *inter-cell interference coordination* or ICIC in LTE systems [Ger10]. In contrast, the reuse partitioning approaches in 2G/1G cellular systems were mostly done on much longer time-scales and the frequency allocations were static.

Figure 10.15 shows two approaches for FFR in LTE systems [Gho11]. In the case of strict FFR, the interior of a cell uses its own frequencies which are not reused in the cell exterior. As shown in Figure 10.15(a), the interior of every cell uses frequencies f_1. The cell exteriors use frequencies f_2, f_3, f_4, with a reuse factor of 3 for a total of four different frequencies. Thus, the transmit power with frequency f_1 is smaller than the transmit power associated with the other frequencies. In the case of soft FFR, the interior of a cell can use the frequencies that are not used in its own cell's exterior but are used in the exterior of neighboring cells. For example, in the cell that uses f_4 in its exterior with a higher power, the frequencies f_2 and f_3 are used with a lower power so that they can be used only in the cell interior and not cause interference to the users at the edges of neighboring cells. We note here that we have referred to frequencies, but, as mentioned previously, it is actually the so-called physical resource blocks (that consist of a set of sub-carriers) that use different transmit powers in the case of FFR.

10.4.2 5G/6G Deployment, beamforming, and SDMA

At the turn of the 21st century, using smart antennas for capacity expansion for wireless networks attracted attention [LEH99]. Traditionally, FDMA,

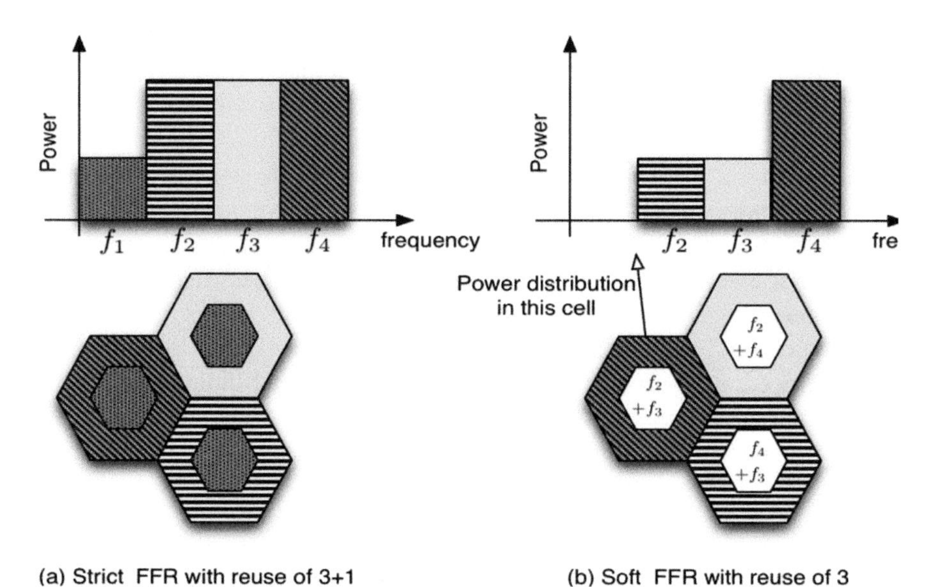

Figure 10.15 Fractional frequency reuse deployment in LTE systems [Saq13].

TDMA, and CDMA have been employed for cellular communications. Using smart antennas, users in the same cells can use the *same* physical communication channel if they are not located in the same angular region with respect to a BS. Such a multiple access scheme referred to as space division multiple access (SDMA), can be achieved by the BS *directing* a narrow antenna beam toward a mobile communicating with it. In addition to SDMA, interference between co-channel cells is greatly reduced since the antenna patterns are extremely narrow. Previously, we saw that using sectored cells, the reuse factor can be vastly reduced. Even larger advantages can be obtained with smart antennas. Simulations on a frequency-hopped GSM system reported a capacity increase of 300%. A fivefold (500%) increase in capacity was reported for CDMA [LEH99].

The breakthrough in wireless communications in the turn of the 21st century was the discovery of multiple antenna streaming using MIMO technology. The foundation of MIMO is based on two technologies: adaptive antenna arrays to focus the beam pattern of antennas and space time coding (STC) which enable the creation of multiple streams of data between two devices. The benefits of multiple transmitting and receiving have antennas existed in the antenna and propagation society since the 1930s [Jen16]. Seminal work on STC [Ala89][Tar89][Ala01] enabled multiple streams of data, and that is why it is considered as one of the most important worldwide innovations around the turn of the 21st century. Multiple streams of data using MIMO technology and benefitting from STC opened a new horizon in scaling the physical layer transmission rates in multipath fading channels [Fos89]. Multiple streaming enables SDMA and it began to emerge in 4G wireless cellular data networks. This trend continued into 5G and 6G cellular networks by resorting to massive- and super-massive MIMO antenna arrays, respectively, and emergence of orbital angular momentum (OAM) multiplexing for 6G to benefit from special diversity.

Very focused antenna beams controlled with artificial intelligence algorithms to adapt to the mobility of the environment open a new horizon for the deployment of 5G cellular networks and beyond. Figure 10.16 [Tao15] shows a 5G cellular deployment scenario benefitting from spatial diversity and focused beam to connect cellular macro-cells to small cells or relays deployed on top of utility posts with a wireless backhaul link. The small cell covers its proximity with broadband access. Figure 10.17 [Sha17] shows another scenario for a 5G user equipment (UE) to communicate at a high data rate with a small cell while it is receiving control data at a low speed channel from the macro-cell. In this scenario, the UE is benefitting from an unbalanced connection to establish a high speed wireless communication

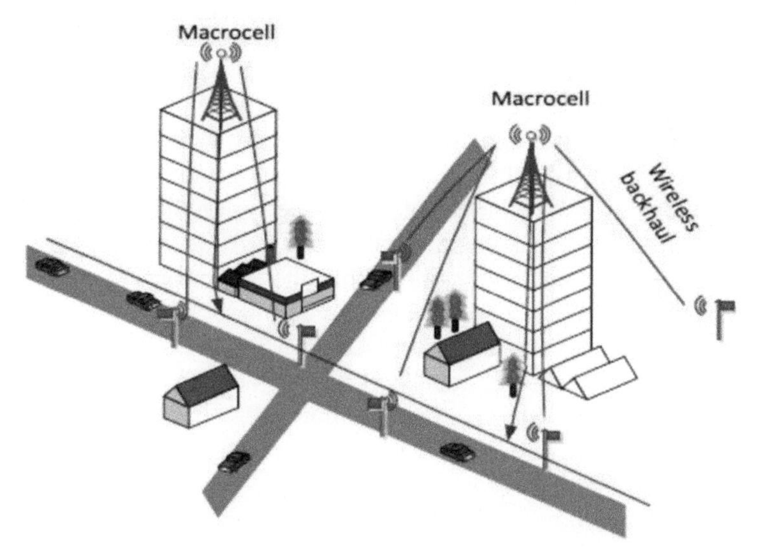

Figure 10.16 5G cellular deployment with wireless backhaul benefitting from massive MIMO for implementation of focused beams [Tao15].

link for data applications. The macro-cell link could be operating in lower frequencies and omni-directional antennas, while the small cell can operate at mmWaves with massive antenna arrays with focused beams. Similar arrangements are implemented in IEEE 802.11ad in which the 2.4/5.2GHz bands are used for control signaling while the actual data transfer at 60 GHz mmWave takes place.

10.4.3 Femtocells

Over the last decade or so, cellular network service providers have engaged in allowing the deployment of very small BS-like devices in the residences of subscribers or in businesses. These devices are generally called *femtocells* although different service providers refer to their devices by different names such as micro-cells, miniature towers, or wireless network extenders. These devices are being marketed as solutions to either improve the quality of reception or provide coverage in otherwise dead spots inside and around buildings. Femtocells are connected to the Internet through wired services that may be available in a residence or a business through coaxial cable, digital subscriber lines, or fiber. Femtocells can communicate with a service provider's network through the Internet. Femtocells typically can provide coverage over an area of 5000 square feet. The reader is referred to [CLAU08] for an overview of the femtocell concept.

Femtocells are installed by subscribers without the careful network planning aspects that accompany the deployment of BSs by service providers. Consequently, the deployment of femtocells requires that such devices be capable of self-configuration. The process usually works as follows. Upon connecting to the Internet through an existing wired connection such as cable, a femtocell device will communicate with a server on the service provider's network. This server can authenticate the femtocell and upgrade its firmware and capabilities without intervention from the subscriber. Most femtocells are equipped with GPS to obtain an estimate of their location (to ensure that it is allowed to be used there), also based on which the service provider can assign some parameters to the femtocell, such as the frequencies to use and in the case of CDMA networks, the pseudo noise (PN) codes to be used by the femtocell. The femtocell monitors the environment to detect what network deployed BS transmissions can be heard by it. Based on such measurements, the femtocell can create a neighbor list for handoffs and adjust its transmit power levels.

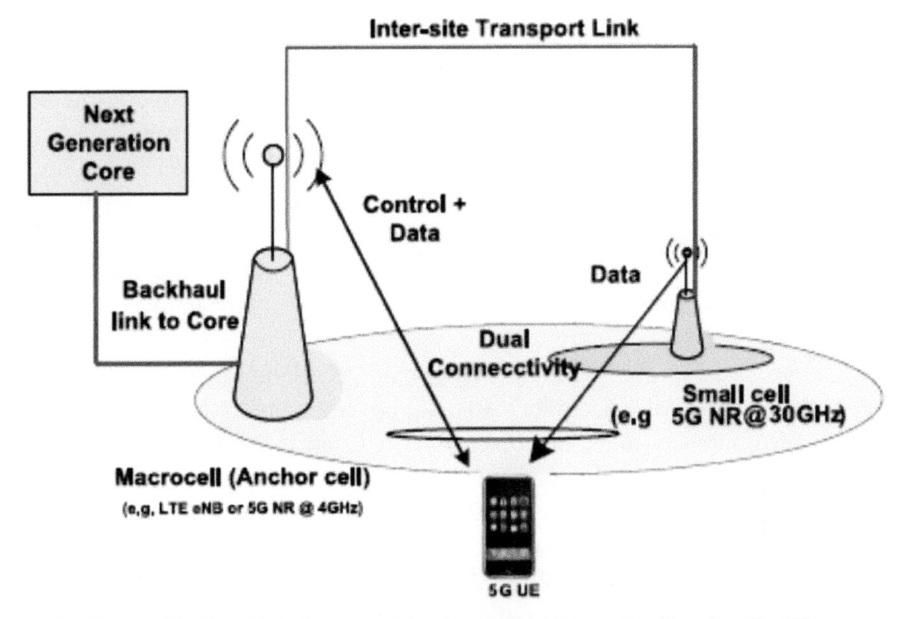

Figure 10.17 5G dual connectivity deployment for spatial diversity [Sha17].

There are a few challenges in the deployment of femtocells. There are many types of interference factors – femtocell downlink interfering with a regular BS's downlink and vice versa, a femtocell's downlink interfering with another femtocell's downlink, a mobile device connecting to a femtocell interfering on the uplink with transmission to a regular BS and vice versa, and a mobile device connecting to a femtocell interfering with another mobile's transmission on the uplink to another femtocell. Further, the dynamic range of femtocell devices must be large to account for the minimum transmit power limits on mobile devices and the short distances between mobile devices and femtocells.

10.5 Deployment of Wi-Fi Infrastructure

We begin deployment of Wi-Fi by comparing deployment of local and metropolitan area wireless cellular networks because this difference provides insight to understanding the economic and regulatory aspects of deployment of wireless networks. Wi-Fi APs operate in unlicensed bands while cellular wide-area networks use licensed bands. As we discussed in Chapters 5, the network capacity of cellular networks depends on the frequency reuse factor during deployment, which is determined by calculations of the interference from the neighboring cells. All these calculations assume that the band is licensed to one operator, which technically means that the network planner has control of the interference. Wi-Fi APs operate in unlicensed bands in which a network planner does not have control of the interference. Network managers in a university campus or a corporation may restrict students or employees in the deployment of Wi-Fi other than those owned by the university or the corporation to control interference, which does not comply with the government regulations on using these bands and can be considered illegal.

Unlike Wi-Fi, for optimal deployment, cellular metropolitan area networks use relatively accurate statistical coverage prediction models, such as the Okumura–Hata model discussed in Chapter 3. Statistical channel models for coverage in indoor areas are much less accurate and this poses a challenge in the analysis of coverage of Wi-Fi unless we resort to labor intensive empirical measurements or use computationally intensive ray tracing algorithms. However, Wi-Fi APs are very inexpensive, and they do not need expensive antenna towers and site landscape for installation. Cellular BSs are orders of magnitude more expensive than APs, and for large area coverage (macro-cells), they need expensive towers and appropriate land for cell sites. Radio resource management in Wi-Fi networks is much simpler because Wi-Fi devices have limited numbers of non-overlapping frequency bands (for example, a maximum of three non-overlapping channels in 2.4 GHz and 12 channels in 5.2 GHz ISM bands), and cellular networks have an order of magnitude more channels and users/cell/carrier to handle. Mobility management for cellular networks is much more complex than Wi-Fi because most popular Wi-Fi applications are quasi-stationary while cellular networks were designed to operate inside a vehicle. In addition, traditional Wi-Fi data traffic is in bursts; it is non-symmetric and location- and time-selective. Therefore, the medium access control is not performed centrally through radio resource management techniques. As a result of all of these differences, a Wi-Fi AP is simple and connects to the Internet backbone directly through a wired LAN while a cellular BS is connected to the public switched telephone network (PSTN) using a hierarchy that includes a radio network controller (RNC) and a public data network interface. These additional infrastructures are needed because in cellular networks, we have more complex radio resource management, mobility management, and connection management techniques. Cells are larger, user connections need more quality control, and cellular telephones have higher mobility. Cellular networks operate under the supervision of a large service provider company such as Verizon, while Wi-Fi devices in residential areas are deployed by users and in corporation with small groups in the IT service management division of the corporates. As a result, cellular network planning is much more complex and service providers spend considerable efforts to deploy and maintain them, while Wi-Fi devices are deployed with less care.

Wi-Fi deployments have been very successful and essential in residential and corporate indoor environments with local coverage. Over many years, they have been deployed in a variety of outdoor experiments as a complement to cellular networks for metropolitan area coverage. We can categorize these installations into two categories, random deployments, and quasi-grid deployments. Figure 10.18 provides examples of these deployment strategies, quasi-grid in a corporate, a shopping mall, and in an outdoor area as well as a random deployment of hot-spot in stores of a shopping mall. In random deployment for residential buildings and hotspots, people usually locate the

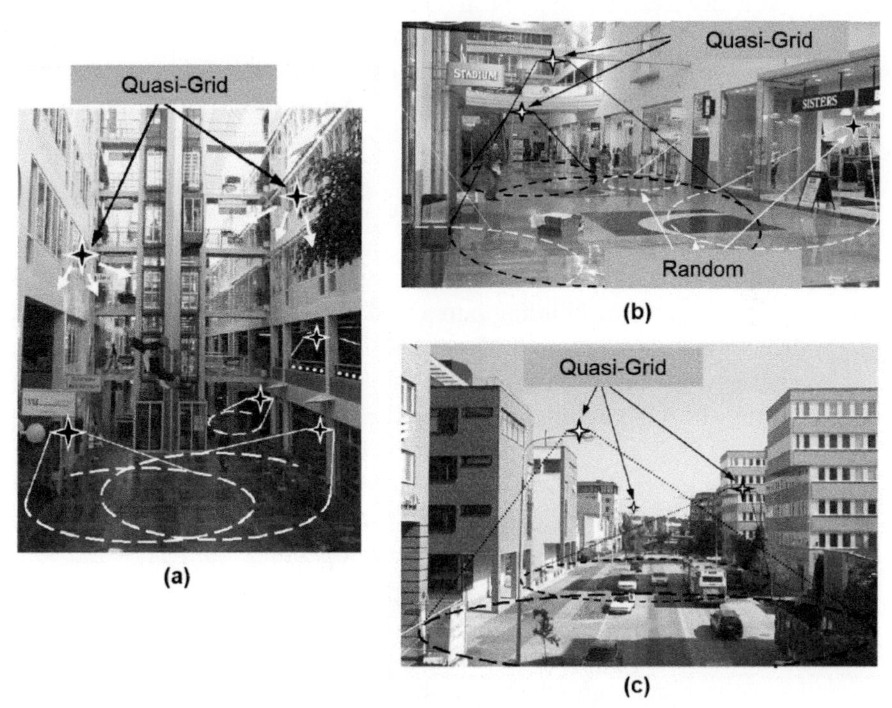

Figure 10.18 Wi-Fi deployment strategies in, (a) a corporate, (b) a shopping mall with hot-spots, and (c) an outdoor setting (adopted from [Unbo3].

AP close to the arriving Ethernet cable from the service providers; to optimize the location more, technical people try to install the AP at the center of the residence to provide a better coverage in the entire area or in the working office of the house where residents normally perform their technical work with their personal computing. For the areas that are not well covered, people use relays or extenders, which we will discuss in the next section. Service providers often sell multiband APs, which use low frequency standard such as 802.11ah at frequencies below 1 GHz, for wider coverage, and 802.11 ac at 2.4 GHz for medium distances and 5.2 GHz or even 802.11ad at 60 GHz for short distances with wideband needs such as connecting to large TV screens.

10.5.1 Cellular Deployment of Corporate Wi-Fi

Wi-Fi AP manufacturers such as Cisco usually provide guidelines for the deployment of corporate Wi-Fi networks. These corporate networks have proprietary features for mesh networking and optimization of the performance to manage the handoffs and maintenance of connectivity as a user connects from different locations to the network. The manufacturer guidelines for optimal deployment are very simple and are illustrated in Figure 10.19. In the first step, they recommend the user to calculate the coverage R for the highest data rate with a high percentage of certainty (for example, 95%). Details of these calculations are provided in Example 3.5 of Section 3.3.2 on the calculation of fade margin and coverage. After calculation of desired coverage for a given certainty, they recommend dividing the layout of the building into a grid and install APs at the center of the grids, as shown in Figure 10.19. In practice, corporate network planners deploy the network based on this guideline and the availability of Ethernet access wirings of the building. In crowded areas such as lobbies, conference rooms, and class rooms, network managers add additional APs as needed. Like residential areas, corporate APs are also multiband to cover better and support higher data rates for access. The difference between the residential and corporate Wi-Fi APs is their ability in operating as a network with multiple APs. In residential areas, we often have one AP that we commonly refer to as a home router because it connects directly to the Internet through a cable and has multiple Ethernet ports to connect desktops or other devices, as well as connecting with smartphones, tablets, security systems, TVs, and other Wi-Fi devices using a wireless connection.

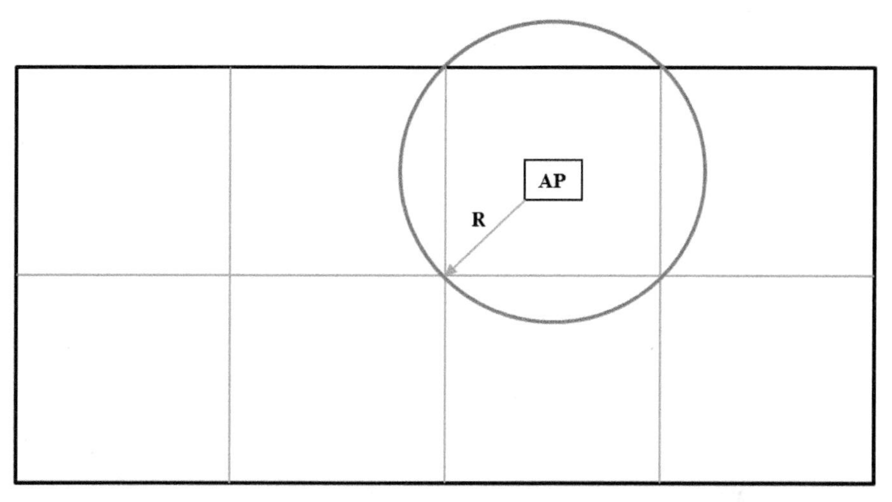

Figure 10.19 Ideal grid installation in a corporate and its relation to AP coverage.

Example 10.8: Cisco Recommended Deployment at 2.4 GHz: Figure 10.20[1] provides a summary of Cisco's recommendation for deployment for IEEE 802.11g/n, which can be extended to 802.11ac if the bandwidth is kept at 20 MHz like n/g. Figure 10.20(a) shows the channel spacing in 2.4 GHz unlicensed ISM bands recommended by the IEEE 802.11g/n standard with 22 MHz separation between the three non-overlapping channels (1, 6, and 11). Only non-overlapping channels are used for deployment to minimize the interference and these channels are assigned in sequence in neighboring cells (Figure 10.20(b)). The ideal scenario for this grid deployment based on the desired coverage calculation is illustrated in Figure 10.20(c). In practice, buildings are not rectangular; Figure 10.20(d) shows a practical quasi-grid deployment in an example layout of a building. In quasi-rectangular grid, distances are kept close to the desired value and the location is picked based on the availability of space and wiring to connect to the Ethernet.

Example 10.9: Cisco Recommended Deployment at 5 GHz: Figure 10.21 shows a summary of Cisco recommended deployment at 5.2 GHz unlicensed national information infrastructure (UNII) bands. Figure 10.21(a) shows the IEEE 802.11a/n/ac channel numbering.

[1]Source: Cisco: Channel Planning Best Practices: https://documentation.meraki.com/MR/WiFi_Basics_and_Best_Practices/Channel_Planning_Best_Practices

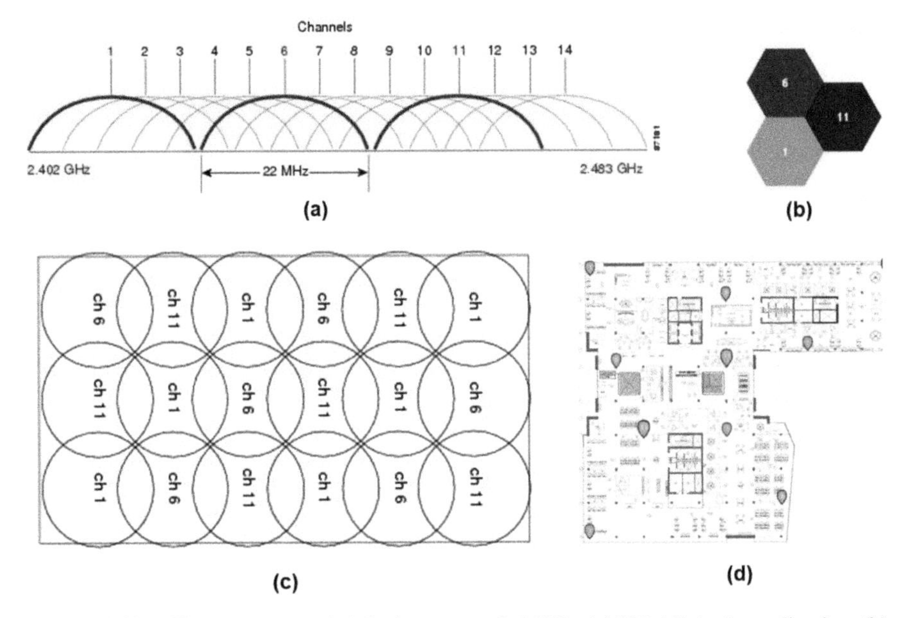

Figure 10.20 Cisco recommended deployment at 2.4 GHz. (a) 802.11b/g channelization. (b) Non-overlapping channels in neighboring cells. (c) Ideal setting. (d) Deployment in practice.

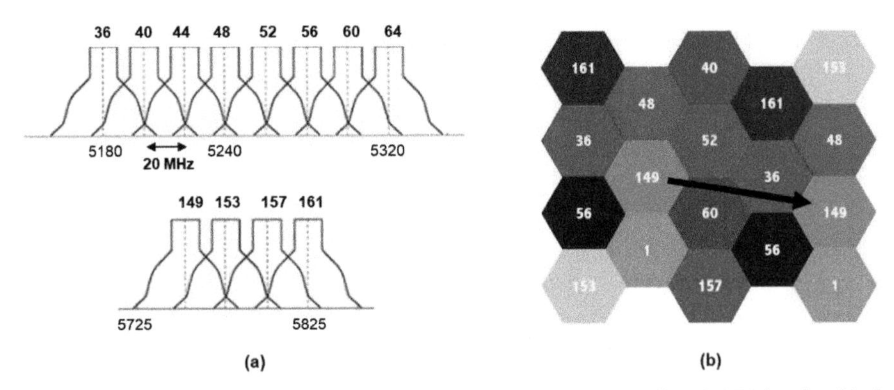

Figure 10.21 (a) IEEE 802.11 non-overlapping channels for 5 GHz UNII bands. (b) A sample neighboring cell assignment with reduced interference.

Figure 10.21(b) shows a sample that recommended channel numbers that include channel 1 from 2.4 GHz and ten of these channels.

In Chapter 9 on a cellular network, we discuss systematic neighboring channel assignment and relate that to the interference controlled environment for licensed band operations. For indoor deployments, Examples 7.6 and 7.7 are adequate for understanding the issues related to Wi-Fi deployment in practice. At this point, we begin addressing the spatial throughput of a wireless data device.

10.5.2 Throughput Analysis for Wi-Fi Deployment in SOHO

In SOHO deployment of the Wi-Fi routers in residential homes, stores, or small offices, the service providers or the property owner deploy a single Wi-Fi router to cover the property. Deployment planning for these applications is simple and often makes a compromise on location by the ease of connection to the backbone wired network provided by the service provider. If the coverage of routers is not satisfactory, we use a Wi-Fi relay or extender to cover all devices on the property. In these networks, it is valuable to analyze the spatial throughput of the network to understand the relation between the data rate and distance from the router and how a relay impacts that relationship. These are the topics that we discuss in this section.

All the Wi-Fi APs are multi-rate systems for which the data rate is adjusted depending on the received SNR and packet losses. In multi-rate systems, a Wi-Fi device can operate at one of the multiple choices of the data rates according to the value of its received SNR. In a single user environment, the data rate of an MS is the average of all data rates it operates at while moving in the area. Different Wi-Fi technologies have different coverage–data-rate relationships. For example, IEEE 802.11b has data rates of 1, 2, 5.5, and 11 Mbps, while IEEE 802.11g has data rates of 6, 9, 12, 18, 24, 36, 48, and 54 Mbps. The coverage of an AP using any of these data rates is different, and as a user moves in coverage of these APs, it observes different data rates. In this section, we introduce a method to calculate the average data rate of a mobile device moving in a uniformly distributed manner in the coverage area of an AP and we refer to that as the spatial throughput or average observed data rate in the coverage area.

In communication with Wi-Fi APs, at short distances from the AP, a multi-rate Wi-Fi device has its highest data rate. As the distance from the AP increases, the received signal strength and, consequently, SNR reduces until a point where the necessary SNR for the highest data rate is not available and the Wi-Fi device has to be switched to the next lower data rate. As the distance continues to increase, the data rate continues to fall to lower rates until the signal strength falls below the coverage of the AP at the lowest allowed rate. In an infrastructure Wi-Fi network with multiple APs providing comprehensive coverage, we expect that when the signal falls below one of the lower thresholds, there is another AP to connect to. Therefore, if we consider an area that is covered by a Wi-Fi AP, the data rate available to the user has a spatial distribution in which associated with any location is one of the available multiple rates of the system. In other words, the data rate in a random location around the coverage forms a *discrete random variable*. One way to define a capacity for this multi-rate system with a statistical data rate is to define the *spatial capacity* as the *average* of the data rates that a user randomly located around the coverage of the AP observes. With this definition, the spatial capacity will be given by

$$R_{\mathrm{av}} = \sum_{n=1}^{N} p_n R_n, \tag{10.12a}$$

where R_{av} is the average spatial data rate, R_n is one of the available multi-rates, and p_n is the probability of occurrence of that data rate, which is the ratio of the areas in which we have that specific data rate to an overall area of the coverage of the AP or the BS.

Example 10.10: Spatial Throughput of IEEE 802.11b

IEEE 802.11b supports four data rates, namely 11, 5.5, 2, and 1 Mbps. According to one of the manufacturers, in a semi-open indoor area, these data rates can be used up to distances of 50, 70, 90, and 115 m, respectively. Figure 10.22 shows the details of the calculation of the average spatial throughput of this technology. Figure 10.22(a) plots the relation between data rate and range of coverage of each data rate. Figure 10.22(b) shows the coverage area for different data rates, for example, $A_1 = \pi {D_1}^2 = 7850\,(\mathrm{m}^2)$ or $A_2 = \pi {D_2}^2 - A_1 = 7536$, etc. Knowing the areas, we can calculate the probability of observation of each data rates from

$$p_n = \frac{A_i}{\pi {D_4}^2}. \tag{10.12b}$$

Figure 10.22(c) is a table summarizing all these calculations, and Figure 10.22(d) shows a plot of the distribution function of data rates. If we substitute the data rates and their probabilities from the density function in Figure 10.22 into Equation (10.12), the average data rate or the spatial throughput of the AP is 2.584 Mbps, which is well below the expected 11 Mbps.

This type of calculation provides a good insight for single AP installations, for example, for most home routers, to compare different

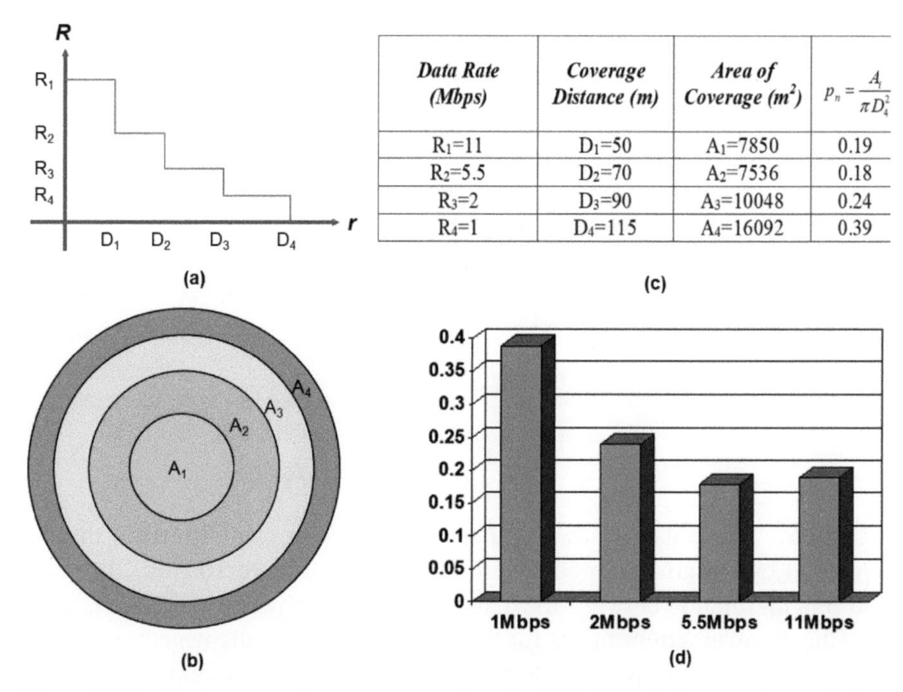

Data Rate (Mbps)	*Coverage Distance (m)*	*Area of Coverage (m²)*	$p_n = \frac{A_i}{\pi D_4^2}$
R_1=11	D_1=50	A_1=7850	0.19
R_2=5.5	D_2=70	A_2=7536	0.18
R_3=2	D_3=90	A_3=10048	0.24
R_4=1	D_4=115	A_4=16092	0.39

Figure 10.22 Data rates and coverage areas for the IEEE 802.11b in a semi-open indoor area. (a) Relation between data rate and distance. (b) Coverage area for different data rates. (c) Calculation of probabilities for data rates. (d) Probability density function of the data rates.

technologies. In corporate-network deployment, as we discussed in Examples 10.8 and 10.9, we calculate the coverage to achieve the highest data rates in the entire area.

Continuous Relation Between Data Rate and Distance:

The relation between the data rate and the distance shown in Figure 10.22(a) is a step function assumed to be common in industrial deployment guidelines. This model assumes that up to a certain distance, we have a data rate, and, after that, we switch to a different data rate. This assumption is not correct, and, at any distance, we may observe any data rate but the probability of observing higher data rates decreases continuously as we go away from an AP. We can explain this phenomenon under the light of effects of shadow fading on a received signal strength of a device located at a certain distance from an AP.

If we represent the received signal strength by p, the received signal strength in a distance, x, is

$$p(x) = f(x) + \eta, \tag{10.13a}$$

where η is the zero mean Gaussian shadow fading with variance of σ, and using IEEE 802.11 path loss model, we have

$$f(x) = P_t - L_0 - \begin{cases} 20\log(x), & x < x_{\text{bp}} \\ 20\log(x_{\text{bp}}) + 35\log(x/x_{\text{bp}}), & x \geq x_{\text{bp}} \end{cases}. \tag{10.13b}$$

Then, the data rate as a function of p is

$$R_p = \begin{cases} R_1; & p < P_1 \\ R_i; & P_{(i-1)} \leq p < P_i \end{cases}, \tag{10.13c}$$

where R_i, P_i are data rates and their power requirements in dBm that are provided by manufacturers of the Wi-Fi devices. Figure 10.23(a) illustrates the general relation between the data rate and their power requirements. Note that these power requirements for Wi-Fi devices are usually negative numbers in the ranges of -70 to -95 dBm; therefore, the horizontal axis in the figure only has a left side. Equation (7.2a) shows that in a given distance x, the received signal strength $p(x)$ is a Gaussian random variable with the mean of $f(x)$ and variance of σ. Therefore, the data rate can be any of the available data rates with different probabilities. Figure 10.23(b) shows the method for the calculation of the probability of having a specific data rate. We proceed with the details of this calculation by explaining the analytical mathematics

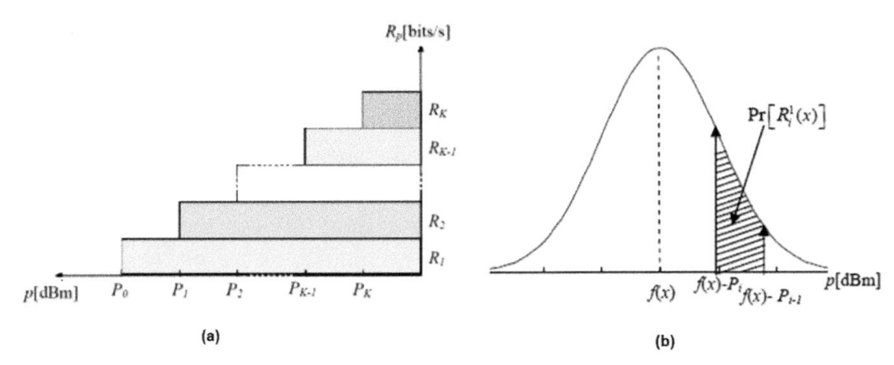

Figure 10.23 (a) Data rates and power requirement in dBm. (b) Calculation of probability of a data rate.

of this derivation for different data rates. As shown in Figure 10.23(b), the probability of observing the data rate of R_1at the distance *x*, as shown by Equation (10.13c), is the same as the probability of received power is less than P_1

$$\begin{aligned}\Pr\left[R_1(x)\right] &= \Pr\left[p(x) - f(x) < P_1\right] \\ &= \int_{-\infty}^{P_1} \frac{1}{\sqrt{2\pi}\sigma} e^{\frac{-[p(x)-f(x)]^2}{\sigma^2}} dp = 1 - \tfrac{1}{2}\mathrm{erfc}\left(\tfrac{f(x)-P_1}{\sqrt{2}\sigma}\right).\end{aligned} \tag{10.14a}$$

The probability that the data rate is any other data rate, R_i, as shown by Equation (10.13c), is that the received power is between $f(x) - P_i$ and $f(x) - P_{i-1}$

$$\begin{aligned}\Pr\left[R_i(x)\right] &= \int_{P_{(i-1)}}^{P_i} \frac{1}{\sqrt{2\pi}\sigma} e^{\frac{-[p(x)-f(x)]^2}{\sigma^2}} dp \\ &= \tfrac{1}{2}erfc\left(\tfrac{f(x)-P_{(i-1)}}{\sqrt{2}\sigma}\right) - \tfrac{1}{2}\mathrm{erfc}\left(\tfrac{f(x)-P_i}{\sqrt{2}\sigma}\right).\end{aligned} \tag{10.14b}$$

Also, always, there is change that power is not enough to connect to the AP and that probability is

$$\Pr\left[R_{\mathrm{NC}}(x)\right] = 1 - \sum_i \Pr\left[R_i(x)\right]. \tag{10.14c}$$

Figure 10.24 shows the probability of observing any of the four data rates supported by the IEEE802.11b and the probability of having no connection at all for different distances. On the right-down corner is a typical manufacturer table relating required power in dBm to the four possible data rates. These values are inserted in Equations (10.14a), (10.14b), and (10.14c) and (10.15b) to generate the plots. Up to 40 m distance from the AP, the Wi-Fi device is more likely to observe 11 Mbps. After 40 m, the probability of having no connection passes the probability of any of the other data rates availability. At this distance the probability that the terminal is connected (1-no-connection) is approximately 65%. For a distance of approximately 110 m, the probability that we can get a connection is approximately 10%. When distances increase beyond that, the probability of cover decreases exponentially to get closer.

Equations (10.14a) and (10.14b) calculate the probability of having different data rates among all possible rates of a Wi-Fi device at a distance *x* from an AP. If we substitute these probabilities in Equation (10.12), the

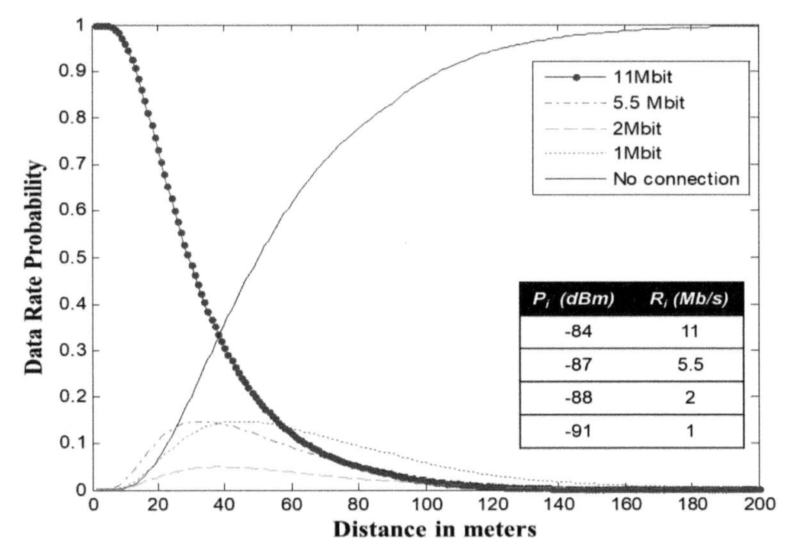

Figure 10.24 Probability of observing a data rate at different distances for the IEEE 802.11b.

average expected data rate at distance x is

$$\bar{R}(x) = \sum_i R_i \Pr\left[R_i(x)\right]. \tag{10.15}$$

Example 10.11: Average Data Rate Versus Distance for IEEE 802.11b:

Figure 10.25(a) provides the MATLAB code for calculation of Equation (10.15) for average expected data rate from an IEEE 802.11b device at different distances when a device is connected to the AP. Figure 10.25(b) provides the resulting plot from the MATLAB code. Near the AP, the data rate is close to 11 Mbps, which is consistent with observations from Figure 10.24. As the distance increases beyond 40 m, the data begin to decrease, around 110 m, it reduces to half, at around 180 m, it goes below 2 Mbps, and around 220 m, it gets close to the minimum supportable data rate of 1 Mbps.

Equation (10.15) averages the data rates including the data rate of zero when there is no connection. If we eliminate that probability, we need to scale Equation (10.14) and calculate conditional probability, given that the device is connected to the AP. The result is not much different, and we leave that as an exercise for the readers.

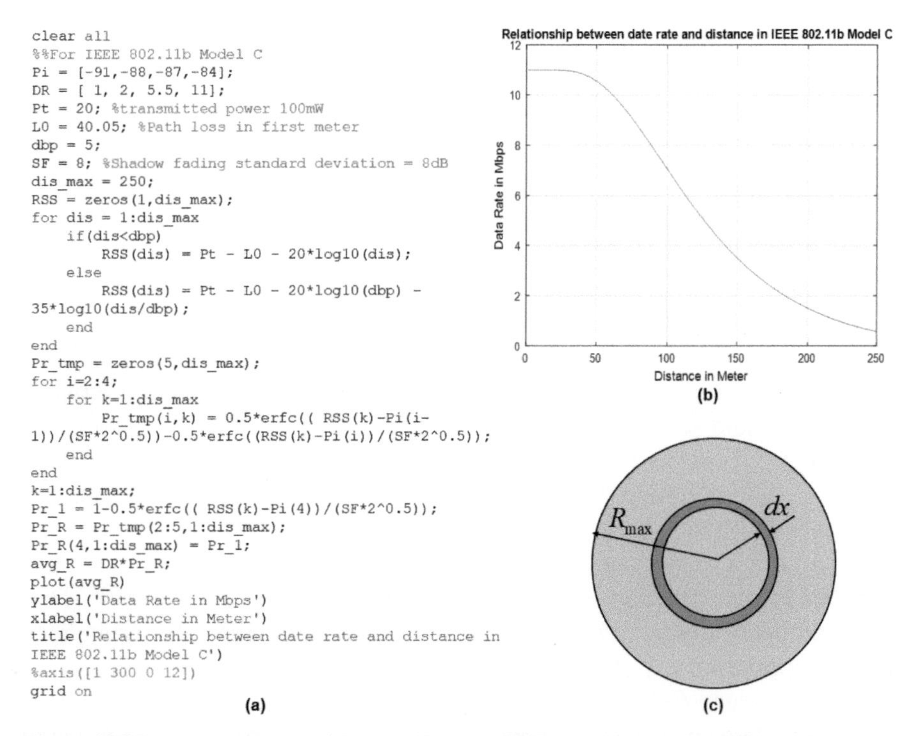

Figure 10.25 Average expected data rate from an IEEE 802.11b device at different distances. (a) The MATLAB code. (b) The performance. (c) Spatial distribution of data rate.

Figure 10.25(b) demonstrates that the relation between average data rate observed by a device and distance is not a step function like Figure 10.22(a), and, indeed, it is a monotonically decreasing continuous function of the distance. This will change the calculation of spatial throughput from discrete mathematics to continuous. We use Figure 10.25(c) to explain this situation and calculate the spatial throughput for a continuous relation between the average user data rate and the distance. When the average data rate, $\bar{R}(x)$, is a function of the distance, x, we can assume that in a rim with a thickness of dx, it holds its value. Then the probability of holding that value is the ratio of the area in the rim, $2\pi\, x\, dx$, to the area of the coverage of the device, $\pi R_{\max}^2$

$$p(x) = \frac{2\pi x dx}{\pi R_{\max}^2}, \tag{10.16a}$$

where $R_{\max}$ is the maximum coverage of the AP that we accept.

Then the average spatial throughput is

$$R_{\text{av}} = \int_0^{R_{\max}} p(x)\,\bar{R}(x)\,dx = \frac{2\int_0^{R_{\max}} x\,\bar{R}(x)\,dx}{{R_{\max}}^2}\,. \qquad (10.16b)$$

Equations (10.16a) and (10.16b) for a continuous distribution of the average data rate as a function of distance are like Equations (10.12a) and (10.12b) when the relation between data rate and distance is approximated by a step function. Both methods provide the average expected data rate for users in the coverage area of an AP. All examples for calculation of the average data rate that we provided are based on the PHY transmission data rates. The user observed data rates include overheads from logical link control (LLC), medium access control (MAC), transmission control protocol (TCP), and Internet protocol (IP) headers as well as details of the implementation of MAC. We discussed these issues separately in our performance analysis section of Chapter 4 on MAC. In general, packet overheads reduce the average user experiment with a fixed scale factor of around 50% in Wi-Fi devices. For example, the maximum user data rate experiment with 802.11b is approximately around 5.5 Mbps and 802.11g users observe a maximum data rate of approximately 27 Mbps. As an alternative to analytical approaches, we can measure $\bar{R}(x)$ empirically at different distances and fit the results with a function and use that function in Equation (10.26b) to calculate spatial throughput.

Relays Deployment to Extend the Coverage:

In the previous section, we used IEEE 802.11b as an example because it has only four data rates and it is easier for explaining the concepts. Analysis of coverage and data-rate–distance relationship, and spatial throughput follows the same pattern with more detailed tables. For example, Table 10.1 shows the IEEE 802.11g manufacturer supplied tables to relate discrete step relation

Table 10.1 Relation between data rate distance and power for IEEE 802.11g (Cisco).

Data Rate (Mbps)	Indoor distance (m)	Outdoor Distance (m)	P_i (dBm)
54	27	76	–72
48	29	76	–72
36	30	84	–73
24	42	124	–77
18	54	183	–80
12	64	203	–82
9	76	247	–84
6	91	396	–90

between the data rate, coverage, and minimum RSS requirement to achieve a data rate, and it has eight data rates. Like the table shown in Figure 10.22(c), we can use Table 10.1 to plot the discrete relation between the data rate and distance, like Figure 10.22(a), for calculation of the discrete spatial throughput of 802.11g. We can also use the relation between data rate and power requirements P_i, similar to Figure 10.23(a) to plot the continuous relation between the average data rate and distance by a simple modification on the code provided in Figure 10.25(a). The result would be a monotonically decreasing curve like Figure 10.25(b). This can be repeated for any other Wi-Fi technology and the results are similar; first, there is a relatively flat area close to the AP, followed by a sharp decrease up to around reasonable coverage area and decaying slowly after that (see Figure 10.25(b)). This analysis is insightful for residential area deployments in which we often have one AP or Wi-Fi router. In residential areas at the edge of the cell coverage when the probability of coverage falls short and/or the data rate falls significantly, we often use Wi-Fi relays, which are sometimes referred to as Wi-Fi extenders.

Figure 10.26 provides the basic concepts behind the operation with and without relays. Figure 10.26(a) shows a basic communication between a Wi-Fi AP and a device (DV). Packets sent for the AP propagate through the air one after another in sequence to arrive at the destination DV. The maximum distance between the AP and the DV is the coverage of the AP, $R_{\max}$, and the average data rate that application on DV experiences at a distance x is $\bar{R}(x)$ calculated from Equation (10.15) for different Wi-Fi technologies. Figure 10.26(c) illustrates this average data rate for IEEE 802.11g as a function of distance in blue color. When the distance between AP and the DV exceeds $R_{\max}$, the probability of connecting to the AP becomes very small, and even when we have the connection, the data rate is very low. At that point, we consider adding a relay (RL) or repeater to extend the coverage, shown in Figure 10.26(b). The RL receives the packets designated by "1" and propagates them in the air as a new packet, designated by "2." The physical data rate of the packets arriving at the RL is $\bar{R}(r_1)$, and the physical data rate of the connection between RL and DV is $\overline{R}(r_2)$, where the distance between the AP and the DV is $x = r_1 + r_2$. Since packets with relay occupy the space two times, the user data rate is reduced to half, and the DV net application data rate with an RL in between is $\bar{R}_1(x) = \bar{R}(r_2)/2 = \bar{R}(x - r_1)/2$, shown by the green line in Figure 10.26(c). The maximum DV application data rate is achieved when the RL is in the middle of AP and the DV, where $r_1 = r_2 = x/2$. As a general guideline, RL should be installed in the middle

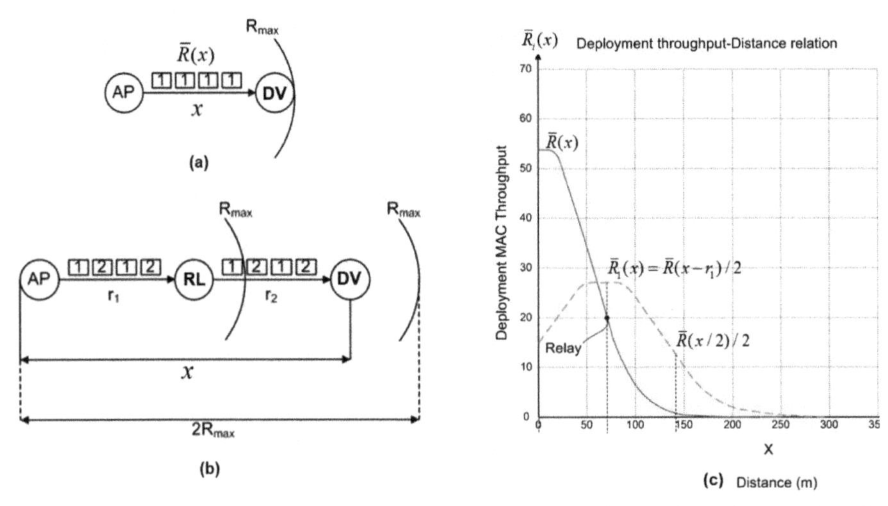

Figure 10.26 Relay and PHY/MAC throughput. (a) Normal operation without relay. (b) Operation with relay. (c) Throughput data rate and distance relation with and without relay.

or closer to the device, and if it is not in the middle, the user throughput at the DV is calculated from

$$\bar{R}_1(x) = \frac{1}{\dfrac{1}{\bar{R}(r_1)} + \dfrac{1}{\bar{R}(x - r_1)}}.$$

The detailed practical application and simulation for relays in cellular networks and analytical solutions to this problem are available in [Pao04] and [Alk15], respectively.

10.6 Assignments

Questions

1. Name the five different cell types in the cellular hierarchy and compare them in terms of coverage area and antenna site.
2. Why is hexagonal cell shape preferred, over square or triangular cell shapes, to represent the cellular architecture?
3. What are the most popular frequency reuse factors for 1G FDMA, 2G TDMA, and 3G CDMA?

4. Of the following, what values are possible for a cluster size in a cellular topology? Why? Assume a hexagonal geometry: 8, 21, 23, 30, 47, 61, 75.
5. Name five architectural methods that are used to increase the capacity of an analog cellular system without increasing the number of antenna sites.
6. Explain why band splitting is not used in the second-generation cellular networks.
7. Explain why reuse-partitioning can be used for both 1G and 2G cellular networks.
8. What is the difference between band-splitting and underlay–overlay techniques for increasing the capacity of cellular networks? What is the effectiveness of each in improving the capacity and how do they differ from one another?
9. Explain how smart antennas can improve the capacity of a cellular network.
10. How are the high interference holes in CDMA deployment created?
11. What is the difference between the deployment of 4G LTE and earlier generation of cellular networks?
12. How do massive and super-massive MIMO antenna systems help the deployment and capacity of cellular networks?

Instructor's solution available on River Publishers' website:
https://www.riverpublishers.com/book_details.php?book_id=919

Problem 1

Consider the general interference scenario of Figure 10.2 in which a Bluetooth device is collocated with an IEEE 802.11 device and interferes with its operation.

a) Shannon–Hartley bound gives the relation between minimum signal to noise, $S_{\min}$, and the number of bits per transmitted symbol, m. Using that $S_{\min}$, calculate the minimum distance for interference, D_{int}, as a function of distance-power gradient, α, power of the Bluetooth device, P_{BT}, the power of the desired AP, P_{AP}, and the distance between the AP and the 802.11 device, R.
b) Plot the minimum distance for interference, D_{int}, as a function of the transmission rate, $R_b = m\,R_s$, $\alpha = 2$, $P_{\text{BT}} = 0$ dBm, and $P_{\text{AP}} = 20$ dBm.
c) Repeat for $\alpha = 3$ and $\alpha = 4$.

Problem 2

Assume that you have six-sector cells in a hexagonal geometry. Draw the hexagonal grid corresponding to this case. Compute S_r for reuse factors of 7, 4, and 3. Comment on your results.

Problem 3

Assume that we wanted to deploy an analog FM AMPS system with a half band of 15 kHz rather than the existing 30 kHz. Also assume that in analog FM carrier-to-interference ratio (C/I), requirement is inversely proportional to the square of the bandwidth (4 times increase in C/I for dividing the band into two).

a) What is the required C/I in dB for the 15 MHz per channel if the required C/I for the 30 kHz systems was 18 dB?
b) Determine the frequency reuse factor K needed for the implementation of this 15 kHz per user analog cellular system.
c) If a service provider had a 12.5 MHz band in each direction (uplink and downlink) and it would install 30 antenna sites to provide its service, what would be the maximum number of simultaneous users (capacity) that the system could support in all cells? Neglect the channels that are used for control signaling.
d) If we use the same antenna sites but a 30 kHz per channel system with $K = 7$ (instead of the 15 kHz system), what would be the capacity of the new system?

Problem 4

We have an installed cellular system with 100 sites, frequency reuse factor of $K = 7$, and 500 overall two-way channels.

a) Give the number of channels per cell, total number of channels available to the service provider, and the minimum carrier-to-interference ratio (C/I) of the system in dB.
b) To expand the network, we decide to create an underlay–overlay system where the new system uses a frequency reuse factor of $K = 3$. Give the number of cells assigned to inner and outer cells to keep a uniform traffic density over the entire coverage area.

Problem 5

a) What is the number of radio frequency (RF) channels per cell in the GSM network described in Chapter 7? The frequency reuse factor of the GSM is $K = 4$.

b) What is the maximum number of simultaneous users per cell in this system?

c) Assume that we want to replace this GSM system with an IS-95 spread spectrum system in the same frequency bands. What is the maximum number of users per cell? Assume an ideal power control and use the practical considerations for the IS-95 system.

Problem 6

a) Determine the carrier-to-interference ratio, in dB, of a cellular system with a frequency reuse factor of $K = 7$.

b) Repeat (a) for $K = 4$.

c) If we consider multi-symbol quadrature amplitude modulation (QAM) for the digital transmission of the information, how many more bits per symbol can be transmitted with $K = 4$ as it is compared with $K = 7$ architecture?

11

RF Cloud and Cyberspace Applications

11.1 Introduction

Wireless information networks have become a necessity of our day-to-day life. Over a billion Wi-Fi access points (APs), hundreds of thousands of cell towers, and billions of Internet of Things (IoT) devices, using a variety of wireless technologies creates the infrastructure that enables this technology. The radio signal carrying the wireless information propagates from antennas through the air and creates a radio frequency (RF) cloud carrying a collection of data that is commonly accessible by anyone. The Big Data of the RF cloud contains information about the transmitter type and address, as well as signal features such as received signal strength (RSS), time of arrival (TOA), direction of arrival (DOA), channel impulse response (CIR), and channel state information (CSI). We can benefit from the contents of the RF cloud and their temporal and spatial variations to engineer intelligent cyberspace applications. Opportunistic cyberspace Wi-Fi positioning is the most popular application emerging in this domain, and, today, each of the three major Wi-Fi positioning systems receives over a billion hits per day and their databases enable location intelligence with a variety of applications. Other opportunistic RF positioning systems use cell tower, Bluetooth, ultra-wideband (UWB), and ZigBee signals. More recently, new opportunistic applications of the RF cloud, such as motion, activity, and gesture detection, as well as physical layer authentication and security, have emerged to host numerous innovative applications. In this chapter, we provide a holistic view of emerging cyberspace intelligence applications benefitting from the Big Data embedded in the RF cloud of wireless devices [Pah20].

The holistic view of wireless data communications for office information networking emerged in the mid-1980s [Pah85, Pah88] and the IEEE 802.11 standardization activity for wireless local area networking, commercially known as Wi-Fi, began in the late 1980s to address this industry. Today, when we arrive at a hotel registration desk, the first fundamental questions we ask related to our basic needs are: Where is my room? Where is the restaurant? And how can I connect to the Wi-Fi? Over a billion Wi-Fi APs deployed worldwide connect our mobile, personal, and fixed devices to the Internet and cyberspace. They have become an essential part of our lives to the extent that some people take Wi-Fi as the foundation of human needs, where Maslow's hierarchy of human needs lands on (Figure 7.1).

In the late 1990s, the IEEE 802.15 standardization activities began and introduced Bluetooth, ZigBee, and UWB technologies for personal area networking [Sei00]. Radio frequency identification (RFID) technologies have emerged as the icon of supply chain management, inventory control, and many other applications [Wei05]. More recently, with the emergence of millimeter wave (mmWave) technology for Wi-Fi and cellular networks, leading manufacturers such as Texas Instruments have introduced short-range radar sensor devices employing this technology [Iov17]. At the time of this writing, the RF signal radiating from over a billion Wi-Fi APs, several hundred thousands of cell towers, and trillions of IoT devices using Bluetooth, ZigBee, UWB, mmWave, and RFID technologies invites innovative opportunistic Big Data application developments for cyberspace [Ban20]. The RF signals radiating from these devices create an RF cloud reachable to any device with an RF front end to sense their signals. The features of these RF signals, such as RSS, TOA, DOA, CIR, and CSI, provide a fertile ground for numerous innovative opportunistic cyberspace applications.

This chapter provides a visionary overview of these emerging cyberspace applications and explains how they benefit from RF cloud to operate. We first discuss the Big Data contents of features of the RF cloud. Then, we explain how innovative cyberspace applications are emerging to benefit from the Big Data in these features. We begin with explaining opportunistic wireless positioning benefitting from Big Data from the RF cloud. Then, we explain how researchers are studying applications of these features for motion, activity, and gesture detection as well as authentication and security to open a new horizon for human–computer interaction (HCI).

11.2 Big Data in the RF Cloud

Figure 11.1 explains the concept of RF cloud for Wi-Fi APs in a database of a Wi-Fi positioning system in the Bay Area, Manhattan, and Seattle [Pah10]. The Big Data embedded inside the RF cloud is divided into two

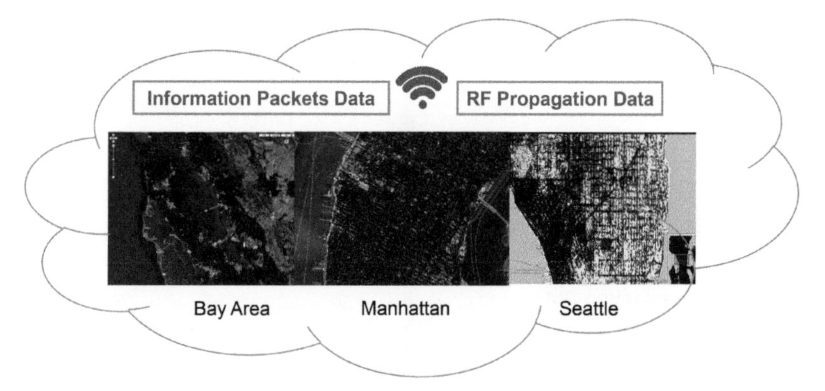

Figure 11.1 Access points in Manhattan creating the Wi-Fi RF cloud utilized for Wi-Fi positioning.

types: 1) the data in the information packets to exchange information among wireless devices, and 2) the data related to the multipath characteristics of RF signals carrying this information. The data embedded in RF propagation features reflect the structure of the environment surrounding the source and destination antennas of the RF devices.

We can also divide wireless devices into two general classes, wireless communication devices, and radars (Figure 11.2). Wireless communication devices (Figure 11.2(a)) transmit symbols, each carrying a limited number of bits of information in binary format. The transmitted packet of information consists of a bundle of these symbols carrying an information packet destined to a receiver with information about the system and the devices, which are beneficial for any receiver to gain cyber intelligence. These packets are broadcast, and they are accessible to all other devices in the coverage area of the transmitter. In indoor and urban areas where wireless communication devices operate, the received signal arrives through different paths, bouncing off objects between the transmitter and the receiver. As such, the signal contains information related to the objects in the environment, embedded in the characteristics of the RF propagation channel between the transmitter and the receiver. Modern wireless devices measure these characteristics to enhance the quality of the wireless communication link. That way, characteristics of the RF propagation channel are available to end users. Radars (Figure 11.2(b)), like communication devices, also transmit electronic waveforms. However, the transmitter and receiver are in the same location and the received waveforms are compared with the transmitted symbols to measure the characteristics of the paths reflected from surrounding objects in the environment.

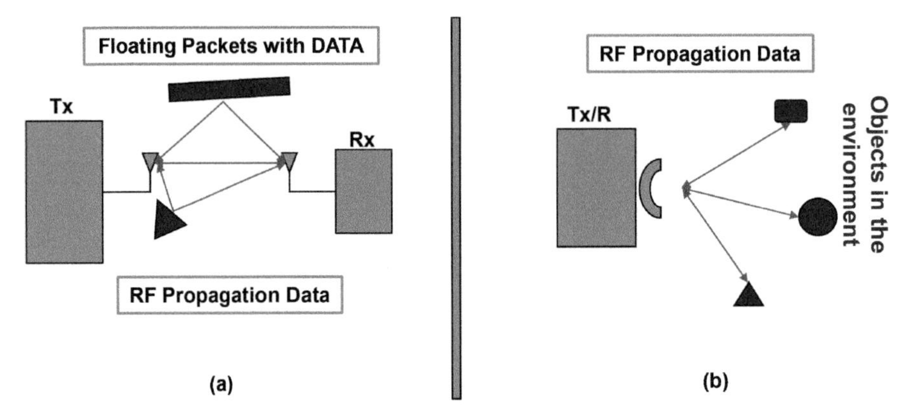

Figure 11.2 Two classes of wireless devices. (a) Wireless communication devices, with transmitter (Tx) and receiver (Rx) in different locations, and (b) radars with integrated Tx and Rx.

Receivers in both radars and wireless communication devices can measure the magnitude, phase, and time of flight of multiple paths reflected from surrounding objects in the environment. As objects move in the environment, the data associated with paths fluctuate and an intelligent receiver can use this to design motion-related cyberspace applications for positioning, tracking, motion and gesture detection, authentication, and security. In recent years, many cyber intelligent applications have evolved benefitting from the contents of data broadcast from wireless devices and the data associated with RF channel characteristics measured by RF receivers.

11.2.1 Data Contents of Floating Packets

Figure 11.3(a) shows typical fields in a packet used for wireless communications. It consists of a preamble, starting delimiter (SD), destination/source addresses (DA/SA), control bits, information data, and a checksum code. The length of the packet depends on the information length and the rest of the data is considered as the overhead of the packet. Figure 11.3(b) shows the type of data contents in each field of a packet. The header is different in different technologies and it contains data on the type of technology used for the packet communication. Addresses contain data about the source and destination and can associate the packet to the physical location of the source. In wireless communications, coverage of the devices is limited. As a result, when we read a packet from a transmitter, we know we

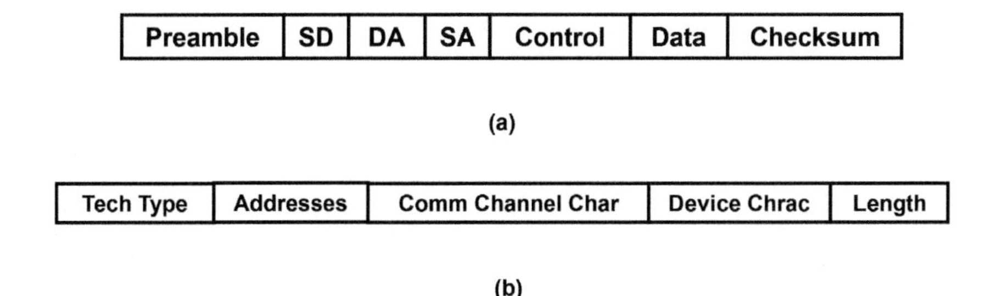

Figure 11.3 (a) Typical fields in a wireless communication packet. (b) Typical common data in floating packets.

are at a certain distance from its location. Control data contains information on communication links and sometimes channel information that can be used for environmental monitoring. The data itself and the checksum code are aimed at communication applications. This data does not contain any special information for intelligence; however, they affect the length of the packet, and variations of the length contain information. For example, variation of the length of data arriving from a specific device can reflect the unique behavior of the source as a measure for authenticity.

11.2.2 Data Contents in Features of RF Propagation

Motion in the environment affects RF propagation features including the RSS, embedded in the amplitude of the carrier of the received signal, and time of flight or TOA, which is embedded in the phase of the carrier of the received signal. The TOA can also be measured using the envelope of the carrier signal, but it is much less reliable than that obtained from the measurement of the phase of the signal. Using multiple antennas, we can also extract DOA by utilizing the differences among the TOAs in antenna arrays. The quality of TOA ranging for measuring the distance between a transmitter and a receiver is superior to RSS-based ranging. However, TOA-based ranging is extremely sensitive to excessive multipath propagation conditions, and if it is not controlled, it may perform worse than RSS-based ranging. Multipath conditions increase as we go into partitioned spaces: in open space areas, there is no multipath; in suburban areas, we have some multipath; in dense urban areas, multipath increases significantly; and in indoor areas, it is extensive. If the receiver can measure the characteristics of the individual multipath components, there is an opportunity to take care of multipath effects using signal processing algorithms [Pah19].

The CIR for wireless devices operating in multipath indoor and urban areas are commonly represented by

$$h(\alpha_i; \tau_i; \theta_i, \psi_i) = \sum_{i=1}^{N} \alpha_i e^{j\theta_i} \delta(t - \tau_i)\delta(\psi - \psi_i), \tag{11.1a}$$

where $(\alpha_i; \tau_i; \theta_i; \psi_i)$ are the magnitude, TOA, phase, and DOA of the *i*th path. In this equation, the TOA is related to the phase of the arriving path by

$$\tau_i = \frac{\theta_i}{2\pi f_c} = \frac{d}{c}, \tag{11.1b}$$

where f_c is the carrier frequency of the signal, *d* is the length of the path, and *c* is the speed of light.

We can calculate the RSS of the received signal from

$$\mathrm{RSS} = P_r = |r(t)|^2 = \left| \sum_{i=1}^{N} \alpha_i e^{j\theta_i} \delta(t - \tau_i)\delta(\psi - \psi_i) \right|^2 = \left| \sum_{i=1}^{N} \alpha_i \right|^2 . \tag{11.2}$$

We can easily measure the RSS from a transmitting wireless device without any synchronization with the source, while measurement of TOA needs tight synchronization between the devices as well as some additional signal processing.

As the objects or the wireless devices move in the environment or we change the frequency of operation, characteristics of the multipath features fluctuate drastically and cause fading in the received signal. In the wireless communication literature, this phenomenon is discussed under temporal, frequency-selective, and spatial fading (see Chapter 3). By taking the Fourier transform of these fluctuations, we can measure the speed of movement of the objects. Different wireless devices measure some of these parameters for enhancing their communication quality and those measurements are available for the development of other cyberspace applications, which we present in this chapter.

RF Data Content of Radars:

The popularity of mmWave technology operating at around 60 GHz for the 5G and 6G cellular networks has enabled the implementation of low-cost short-range radars at these frequencies. The Texas Instruments mmWave sensor radar device is a popular example of such devices operating at

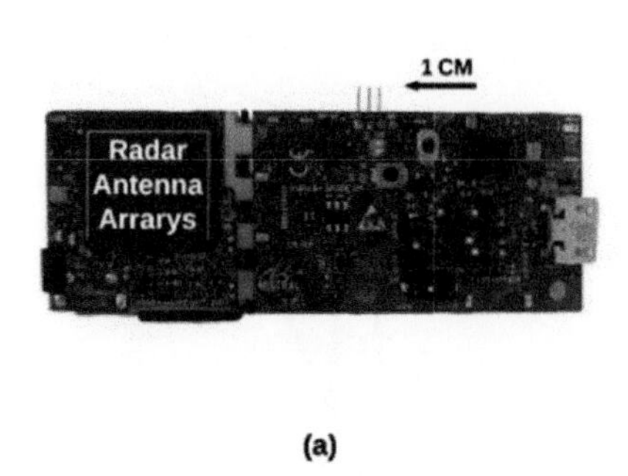

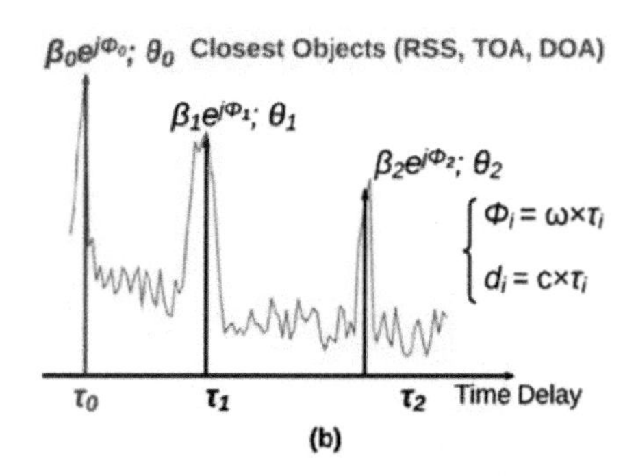

Figure 11.4 Overview of the TI's mmWave radar. (a) The physical appearance. (b) Abstraction of CIR. (c) A typical measure of range-amplitude profile.

76–81 GHz [Iov17]. This compact and low-cost radar, shown in Figure 11.4, emits chirp signals to capture the distance, velocity, and angle of objects surrounding the device. This information includes the RSS, TOA, DOA, and velocity of motion of these objects. This mmWave radar features a flat 8 × 8 multiple-input-multiple-output (MIMO) array antenna enabling the device to capture refined spatial information from detected objects. Operation at high GHz has enabled the device to have a small array and advancements in microelectronics have integrated this device in a finger-sized package. The availability of this device in the market initiated several interesting research projects in micro-gesture detection. We will discuss more details on the research on these topics in Section 11.3.2.

Figure 11.4 shows the basics of TI's radar characteristics. Figure 11.5(a) shows the physical appearance of the device with size metrics. Figure 11.5(b) illustrates a sample range-amplitude profile captured by the radar receiver from different surrounding objects, representing the CIR. In this measurement, the first peak associates with the gesture of a hand kept close to the device and other major peaks are a reflection from the environment located at longer distances.

RF Data Content of Wireless Communications:

The enormous success of the wireless communication industry has nurtured several successful technologies that include, Wi-Fi, cellular, Bluetooth, ZigBee, and UWB. In addition to the common data available in the floating packets, devices using these technologies also have access to data from features of RF propagation reflecting motions in the environment. All these devices support the measurement of the RSS. As a result, RSS of Wi-Fi,

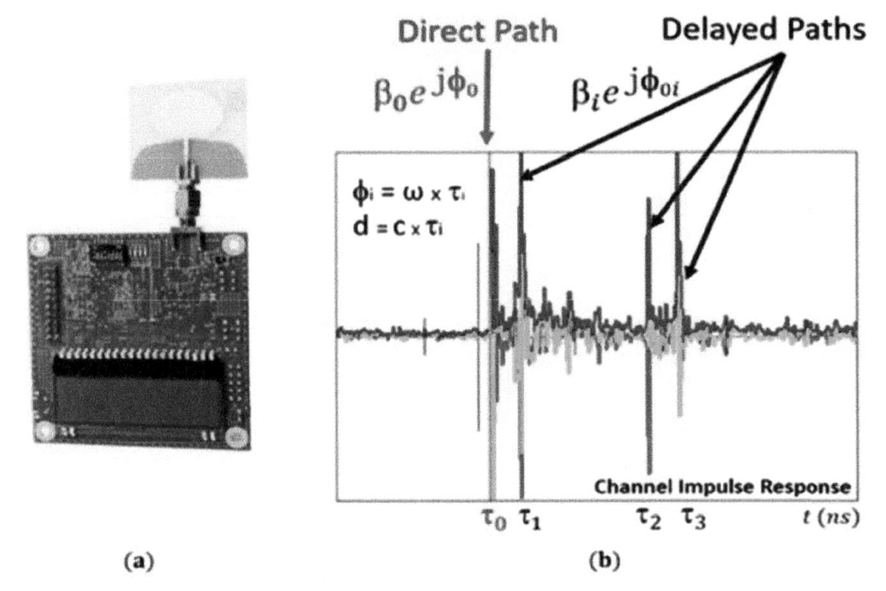

Figure 11.5 Overview of DecaWave EVK100 UWB wireless communication and ranging system. (a) Physical appearance. (b) A typical channel impulse response measurement, with abstraction of CIR.

Bluetooth, and ZigBee have found their ways in a variety of cyberspace applications.

Other devices measure the CIR with different levels of precision. UWB devices provide an accurate estimate of the CIR suitable for opportunistic applications in human–computer interfaces. The popularity of UWB technology operating at around 3–10 GHz for positioning and communication applications has enabled the implementation of low-cost UWB devices. The DecaWave's EVK1000 UWB positioning system is a good example of these devices [Rui17]. This small size, a low-cost accurate indoor positioning system (Figure 11.5) uses UWB signals to measure the CIR between a transmitter and a receiver and position a device in an unknown location using the known location of several reference devices. Figure 11.5(a) shows the physical appearance of the device, and Figure 11.5(b) illustrates a typical measurement of the CIR and detected direct and reflected paths. In addition to accurate positioning applications for a system consisting of several reference points and a tag, the CIR between any two transmitters and receivers provide a multi-channel data stream that is useful for HCI and other cyberspace applications.

At the time of this writing, orthogonal frequency division multiplexing (OFDM) is the most popular wireless communication technology for Wi-Fi and cellular networks. An OFDM signal consists of a large group of narrowband transmission systems modulated over neighboring carrier frequencies. In theory, if we have N-carriers, we have N-streams of magnitudes and phases. However, unlike the CIR multiple data streams, the multiple streams of OFDM data are highly correlated. It is possible to obtain CIR from OFDM signals and most OFDM receivers estimate the CIR to enhance the quality of transmission. However, users should notice that the quality of CIR estimates is proportional to the bandwidth and UWB systems provide a much better estimate of CIR.

Wireless communication systems with MIMO antennas, shown in Figure 11.6, is commonly used in Wi-Fi and cellular networking technologies. These systems can provide for multiple streams of CIR and DOAs. MIMO antenna systems transmit multiple streams through different paths at different arrival angles, each carrying the magnitude and phase of the signal. In the MIMO literature, these streams of information are referred to

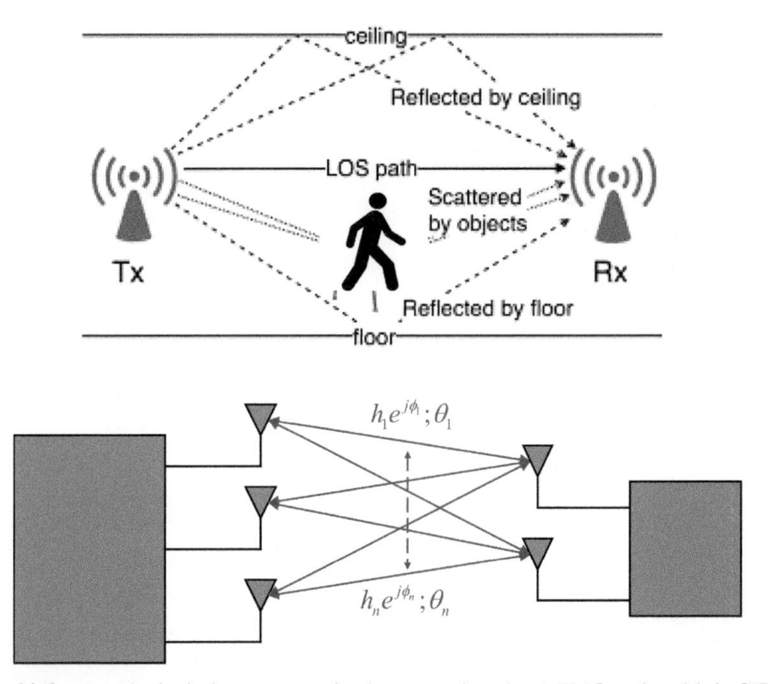

Figure 11.6 A typical wireless communication scenario using MIMO and multiple CIR and DOA information.

Table 11.1 Summary of signals and features in RF cloud.

	Time-Domain Features	Frequency-Domain Features
RSS	Mean, Standard Deviation (STD), Peak-to-Peak	Spectrum (mean, STD, entropy)
CSI (MIMO, OFDM)	Same as RSS with multiple streams and sub-carriers	Doppler Spectrum (spread, decay, entropy, n-th order moment)
CIR (UWB)	Mean, STD, Power of direct/**multi-path**, Time Delay, Root-Mean-Square Time Delay	Spectrum (centroid, n-th order moment, entropy)
CIR (Radar)	Mean, STD, Power of direct/**reflected path**, Time Delay, Root-Mean-Square Time Delay	Same as CIR (UWB)

as CSI [Hal11]. The CSI is another rich signal space with multiple streams, which has been popular in recent literature for motion-related cyberspace application development. Table 11.1 summarizes the features of signals embedded in the RF cloud of wireless devices.

RF Data Content of Communication Devices:

Digital wireless communications take place through symbol transmissions, each symbol carrying a group of information bits. As shown in Figure 11.7(a),

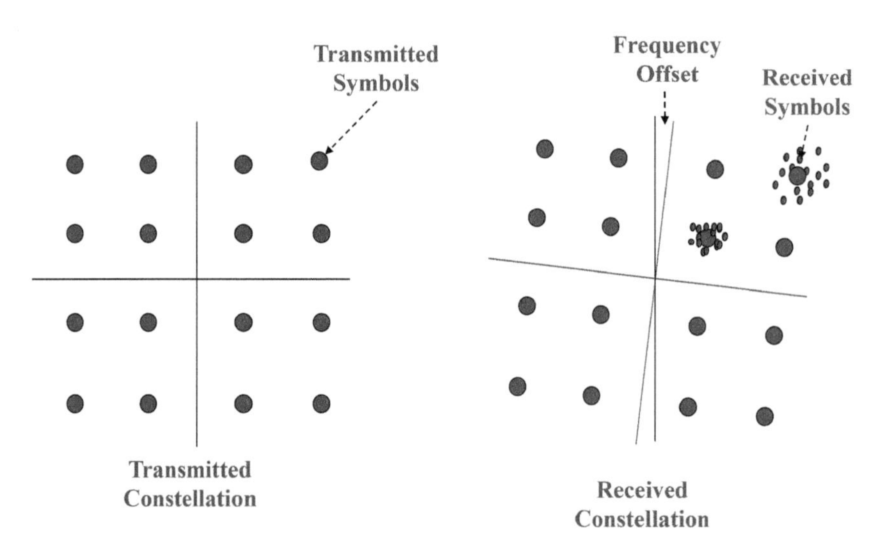

Figure 11.7 (a) Transmitted and (b) received signal constellation reflecting frequency offset and nonlinearities of the device.

transmitted symbols are represented by a signal constellation. Due to the thermal noise, carrier synchronization error, and nonlinearities of the receiver amplifiers, the received symbols arrive around the targeted transmitted symbol and the signal constellation has a frequency offset (Figure 11.7(b)). The statistical pattern of the noise around the transmitted symbol changes with non-linearity of the specific receiver electronic circuits that has unique features for any device. We can benefit from these unique electronic features of the communication devices obtained from statistical behavior of the received symbols in the signal constellation to identify a device type.

11.3 Opportunistic Cyberspace Applications of RF Cloud

Section 11.2 described the RF cloud and its Big Data contents. We showed that the RF cloud radiating from wireless devices surrounding is a valuable source of information. Each wireless device has a unique address and, if fixed, a unique location, and it radiates an RF signal with different coverage, which changes its features with motions. One can create a database of these addresses and the available signal features (RSS, TOA, DOA, CIR, and CSI) associated with the addresses to develop opportunistic cyberspace applications.

This section describes examples of these cyberspace applications, which have evolved around the RF cloud from wireless devices. The most widespread cyberspace applications of the RF cloud are related to indoor positioning using wireless signals of opportunity [Pah19]. Other applications using RF cloud include gesture and motion detection and using signals of opportunity for authentication and security. We provide an overview of these three categories of RF cloud applications in the next three subsections.

11.3.1 Wireless Positioning with RF Cloud Data

In the late 1990s, indoor geolocation science and technology began to evolve to extend the coverage of global positioning system (GPS) to indoor areas [Pah98, Pah02]. The high cost of dense infrastructure, needed for proper operation of these systems, moved this industry toward opportunistic positioning using RF cloud data from the existing Wi-Fi AP infrastructure [Li00, Bah00]. A real-time localization system (RTLS) industry, with a limited vertical market, evolved around this idea for applications in specific areas, such as museums, warehouses, and hospitals. Fingerprinting of the RF cloud for RTLS systems is done manually by surveying inside the building for the site of application. Manual sight survey is expensive and restricts scaling to large areas of coverage. In the mid-2000s, the wireless positioning system (WPS) industry evolved around the same idea with a new method for fingerprinting. In WPS, the RF cloud fingerprinting takes place by driving in the streets and tagging the collected data using a GPS receiver. This automated process enabled WPS systems to scale to metropolitan areas. For that reason, WPS was adopted for the original iPhone and it became integrated into with all smartphones and smart devices since [Pah10]. In the remainder of this section, we explain how WPS works and how it is evolving to enhance the opportunistic wireless positioning industry.

Wi-Fi RSS Positioning and WPS:
Today, the most popular positioning system is WPS, which is the main positioning engine for hundreds of thousands of applications on smart devices. Skyhook, Google, and Apple own the three major Wi-Fi location databases of APs for these systems. The database of Skyhook, the pioneer of the technology, receives over a billion hits per day and includes close to a billion Wi-Fi AP addresses with their estimated locations. In the original WPS systems, cars driving in the streets of a city collected the RSS fingerprint of Wi-Fi devices identified by their MAC addresses provided in the floating beacon packets and tag them with the GPS readings of the locations. Intelligent algorithms process the big database of these readings to estimate the location of any device from its Wi-Fi readings in an unknown location. Therefore, WPS relies on GPS because it is a database associating Wi-Fi addresses with GPS readings in the streets. The advantage of WPS is that it works indoors, where GPS does not work.

Initially, cars driving in the streets of different cities collected the database. Then, organic RSS reading data from devices searching for their unknown location augmented the database of AP addresses and locations. The accuracy of WPS systems are typically around 10–15 m [Pah10], which is on the order of the average coverage of Wi-Fi. This accuracy is adequate for turn-by-turn navigation of cars in streets to differentiate building addresses from each other in urban areas. To increase the precision of WPS for indoor positioning applications, demanding a few meters of accuracy to differentiate different rooms from each other, we need indoor manual fingerprinting, similar to RTLS, and that is expensive.

Location Intelligence: An Outcome of WPS:
GPS is a physical real-time system providing position information based on current readings of TOA from satellites. WPS is a cyberspace information system built on a big database and an intelligent search engine with intelligent algorithms. Each time we agree that an application on our smart device can use our location address, we send a packet to the WPS database and WPS knows our device location. With around one billion hits per day, WPS service providers can extract cyberspace intelligence about our location. We can use this new outcome of WPS technology to implement location-time traffic analysis, geo-fencing (for supporting elderly people, animals, prisoners, and suspicious people), real-world consumer behavior analysis, location certification for security and privacy, positioning IP addresses, and customizing content and experiences [10]. These are secondary outcomes of WPS technology, enabling other cyberspace applications for location intelligence.

Future Directions of WPS:
As we mentioned in Section 11.3.1, the current state-of-the-art WPS technology without indoor fingerprinting has 10–15 m accuracy. For accuracy in the range of meters, we need expensive indoor site surveys and fingerprinting. Typical smart devices carry several other sensors such as accelerometer, gyroscope, magnetometer, barometer, step counter, and compass. These devices provide information on the speed and direction of movements of the device. Using hybrid AI algorithms, we can integrate this motions-related information with the absolute position estimate from the WPS to enhance the positioning and to refine the tracking in indoor areas [Ye11, Bar16, Bar16].

Wi-Fi APs are installed in office buildings approximately 30 m apart. In a typical office building such as Atwater Kent Laboratory at the Worcester Polytechnic Institute (WPI) (approximately 50 m $\times$ 100 m), each floor is covered only with 3–7 Wi-Fi APs. That is why we need fingerprinting to increase the precision to a few meters to differentiate rooms from each other. With the increase in the "smartness" of office buildings, every room of this building has at least two IoT devices controlling the light and the temperature. IoT devices use Bluetooth low energy (BLE), ZigBee, or other active RFID technologies, which have smaller coverage than Wi-Fi. Smaller coverage indeed helps the precision. Imagine we have an RFID with coverage of 1 m; if we read its signal, we know our location with 1-m accuracy. With such density of deployment of small coverage IoT devices, we may not need indoor fingerprinting anymore. It can be shown that the precision of Wi-Fi positioning in a typical building (e.g., WPI's Atwater Kent Laboratory), with three Wi-Fi APs in 90% of locations is better than 15 m, while with only eight randomly distributed IoT devices in that floor, this precision comes close to 2 m [Yin17, Yin19]. In practice, the design of such systems is practical because all devices measure their RSS and they are connected to the Internet; therefore, they can pass that information to a positioning database to enhance the precision of positioning.

Cell Tower RSS Positioning:
RSS-based Wi-Fi positioning is a device-based positioning system. The metric data used for positioning is collected by the device independent from the communication network provider. We can apply this technology to cell tower positioning using fingerprinting of cell towers [Akg09]. The advantage of this approach for cell tower positioning is that the positioning system takes advantage of cell towers from all cellular providers without any specific coordination. The positioning service provider drives in the streets to identify cell towers and develop a database of their fingerprints tagged with the GPS location. Then using the RSS readings of the cell towers around a device, the service provider can come up with a position estimate for the device. The device needs to have a cellular chipset to read the RSS values of the cell towers.

As compared with Wi-Fi positioning, the density of cellular networks is far less: we have billions of Wi-Fi APs as compared with hundreds of thousands of cell towers worldwide. Therefore, the accuracy of these RSS based cell-tower positioning systems (CPS) is around 100−250 m, which is significantly lower than WPS [Akg09]. However, CPS has more comprehensive coverage, which includes highways as well as urban areas. The original iPhone did not include GPS and it used CPS as a backup for WPS for these areas. With the increase in density of deployment in 5G and 6G cellular networks, the gap between precision of WPS and CPS should reduce significantly. This intuitive observation needs to be justified by empirical research data.

Cell Tower TOA Positioning:
WPS, CPS, and GPS are device-based positioning systems, in which the device measures the features of the RF cloud for positioning. Another approach to positioning is network-based positioning, where cell towers or APs measure the features of RF signals from the device and send that

to a central computational server to locate the device. The first popular application of this approach was the uplink-time difference of arrival (U-TDOA) positioning systems, designed in 2G cellular networks to comply with FCC regulations for E911 services for cell phones [Pah19]. These TOA-based systems utilize the difference between arriving signals from a cell phone to locate the device. One of the advantages of this approach is that we can locate a device without its active participation in the positioning process.

The U-TDOA provides for approximately 100 m precision for E-911 service using existing cell tower signals [Mia12]. This level of precision is not adequate for many popular indoor and urban area positioning and navigation applications, but it has comprehensive coverage, which makes it appealing for emergency response.

The U-TDOA was a patch solution to the position because 2G standard organizations had not included positioning in their agenda. If we consider positioning as a part of the standardization of communication protocols, we should be able to achieve higher precisions using TOA and DOA technologies. The fundamental challenge for TOA-based systems is sensitivity to multipath effects and the need for atomic clock synchronization to achieve sub-meter precision. By integrating the GPS clock with the cellular system standards, we can have a practical solution for synchronization, but multipath effects are serious, with indoor areas [Rui17].

UWB transmission controls the effects of multipath arrivals by isolating them from one another; antenna beamforming focuses the transmission to a single path, and we can design algorithms for positioning in the absence of a direct path [Pah06]. The emerging 5G and 6G cellular systems with massive MIMO and mmWave technologies benefit from UWB transmission as well. In theory, these characteristics of 5G/6G technologies can enable high precision TOA-based positioning. However, implementation of these systems to make it available for precision sensitive positioning applications needs algorithm and system design with a focus on performance evaluation in realistic positioning application scenarios. In general, standards organizations are focused on the increase in capacity, which directly affects the user experience. They need to increase their attention to positioning and navigation as a fundamental enabling technology for millions of applications. More details on the design and performance evaluation of positioning systems are available in the lead author's recent book in this area [Pah19].

11.3.2 Motion, Activity, and Gesture Detection with RF Cloud

Motions of the wireless device or objects close to the antennas of the wireless devices cause temporal fluctuations of the characteristic of RF cloud features measured at the receiver antennas. Recently, several researchers have studied these characteristics of RF cloud from wireless devices for activity, motion, and gesture detection. This area of research expects to revolutionize HCI and introduce a variety of other cyberspace applications by taking advantage of the variations in RF cloud features due to motions in the environment.

Detection of RF Features Due to Motion:

Wireless communication receivers measure features of the RF cloud reflecting motions in the environment. Signal processing techniques help detect these motions and prepare them for cyberspace application development. Figure 11.8 illustrates the temporal variations of RSS of a receiver antenna in the proximity of a transmitting antenna. The figure also shows the Fourier transform of the signal representing the Doppler spectrum and the short-term Fourier transform representing its spectrogram. Figure 11.8(a) shows a situation with no-motion, Figure 11.8(b) shows a situation with a handheld between the two antennas, and Figure 11.9(c) shows

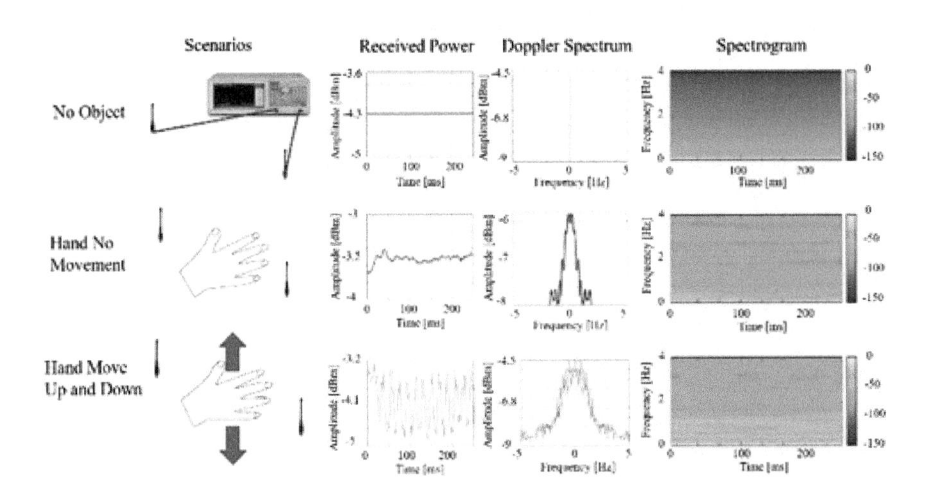

Figure 11.8 The temporal variations of RSS for a receiver antenna in proximity of a transmitting antenna and its Doppler spectrum and spectrogram (a) with no-motion, (b) with a hand with natural motions, and (c) with a moving hand between the antennas.

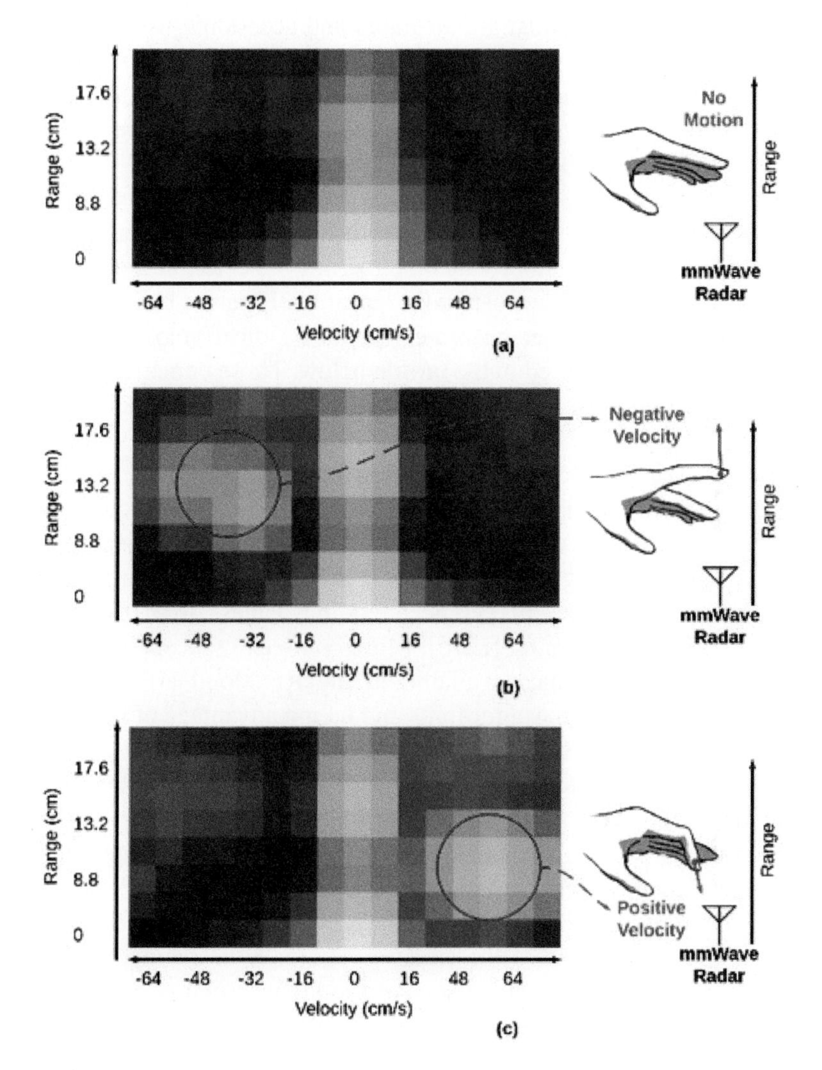

Figure 11.9 Range-velocity profile of TI's mmWave radar with (a) the hand staying still in front of the radar device, (b) a finger tilting backward, and (c) a finger tilting forward.

the results when the hand moves between the antennas. As the speed of motions increases, the bandwidth of the Doppler spectrum and the contrast of colors in the spectrogram increase. We can benefit from this change in the depiction of the RSS characteristics, to develop hand motion related applications. All modern wireless devices measure RSS and many other features of the RF cloud that are available and accessible with the software, opening an interesting area for motion-related cyberspace applications.

The mmWave radar development environment (Figure 11.4) also supports other aspects helpful in the classification of motions. Figure 11.9 shows the range-velocity profile of the device illustrating motions of the finger in different directions. The mmWave sensor extracts velocity information and consolidates it with the range data to form the range-velocity profile. Figure 11.9(a) shows a hand, which is a strong reflector, at a close distance from the radar and its corresponding profile. Figures 11.9(b) and 10(c) demonstrate that the finger movement creates radical velocities relative to the radar, and thus mirrored in the profile below. These depictions of motions open an opportunity for micro-gesture detection from finger motions.

Motion-Related Cyberspace Applications:
In recent years, several researchers have benefited from RF cloud features to introduce innovative cyberspace applications. As a simple example, using an algorithm measuring variations of the RSS above its average value, one could detect the number of people attending a class [Yan16] or monitor newborn babies in a hospital [Li16]. More complex cyberspace applications using opportunistic signals available in the RF cloud are achievable by using artificial intelligence algorithms and taking advantage of more complex features of the signal, such as CIR, CSI, TOA, and DOA. In recent years, several research laboratories have pursued this idea.

At the WPI, variations of the RSS of body-mounted sensors are used for activity monitoring of first responders to find out if a firefighter carrying a device is standing, walking, laying down, crawling, or running [Fu11, Yan16, Don18]. These states of motion reflect the temporal behavior of the firefighter, revealing the seriousness of the situation she or he is facing. The work in [Fu11] uses traditional characteristics of the fading, such as coherence time, rms Doppler spread, and threshold crossing rate of the RSS of simple devices such as Bluetooth, to differentiate different motions, and the work presented in [Yan16] integrates AI algorithms into the motion detection process. The work presented in [Don18] benefits from more complex CSI signals of Wi-Fi devices along with more complex AI algorithms, such as long short-term memory regressive neural network (LSTM-RNN), to increase the capacity of the system in differentiating different motions on a flat floor or when climbing the stairs. As we explained in Section 11.2, CSI provides multiple streams of RSS and more diversified variations of the signal. In [Li20], the research group demonstrates the use of mmWave radar in tracking the motion of a finger, opening further study in the gesture-based application controls in the HCI research area.

Researchers at the University of Washington [Pu13] have used Wi-Fi signals for hand gesture recognition to differentiate nine different hand motions. Multiple RSS streams from different channels of the OFDM signal of Wi-Fi are depicted by a spectrogram to generate frequency-time characteristics color images. The AI algorithm classifies the image to detect the nine gestures of the hand motion. At Michigan State University [Ali15], the CSI of a Wi-Fi signal is used for keystroke detection. When typing a certain key, the hands and fingers move in a unique formation and direction, and there is a unique pattern of CSI RF fingerprint. By training an AI algorithm, they have detected the keystrokes of the keyboard user. At the Massachusetts Institute of Technology [Zha18], researchers have used radar signals like the Wi-Fi signals with multiple antennas, for human pose estimation through walls and occlusions. They demonstrated the detection of multiple human postures through the walls using the RF signal and a neural network algorithm. They used visual data captured by a camera during the training period for the AI algorithm. At Stanford University [Kot17], commodity Wi-Fi signals are used for tracking hand motion for virtual reality applications to replace existing infrared devices.

In parallel with academic studies, practical applications of RF signals for motion and gesture detection and tracking are emerging in the industry. As an example, Google [Lie19] uses RF radar signals at mmWave frequencies obtained from antenna arrays, for micro-motion tracking of hand and finger gestures for applications such as winding a wristwatch without touching the surface. RF signal variations can replace any application using mechanical sensors. For example, interactive electronic games commonly use mechanical sensors such as an accelerometer, and an accelerometer mounted on the gait of a patient has been used to measure the extent of progress in Parkinson disease [Moo07, Abu17]. The RF cloud of UWB devices, measuring the CIR, can replace many of these mechanical sensors and be used in interactive electronic gaming [Zhe16], to help the visually impaired [Zan16] and to provide gait motion detection.

Building on the advances in motion, activity, and gesture detection using RF cloud, researchers have begun to explore the possibilities for future HCI applications. Early work explored using unmodified GSM signals to enable recognition of eight tapping gestures, four hover gestures, and two sliding gestures around a mobile device and to enable incoming call management as well as phone navigation from a distance [Zha14]. More recent work has demonstrated a mmWave gesture recognition pipeline [Lie19] as well as the recognition of 11 gestures with short-mmWave radar with a goal of them being used in HCI [Lie19a]. Other work explored mmWave gesture recognition for in-car infotainment control [Smi18]. Radar signals have also been explored for automatically classifying everyday objects to support various applications including a physical object dictionary that looks up objects that are recognized, context-aware interaction, as well as future applications such as automatic sorting of different types of waste, assisting the visually impaired and smart medical uses [Yeo16]. Using radio signals and one external sensor hanging on the wall, researchers have demonstrated that gait velocity and stride length, which are important health indicators, can be monitored, enabling health-aware smart homes [Hsu17]. Taking advantage of indoor Wi-Fi signals to identify motion direction, researchers have created a contactless dance "exergame" [Qia17] as well as sign language gesture recognition [Ma18]. Other work demonstrated that 5 GHz Wi-Fi can be used to achieve decimeter localization accuracy of up to four users as well as activity recognition of up to three users doing six different activities [Tan19].

Security and Authentication with RF Cloud:

In recent years, several researchers have shown interest in developing authentication and security applications benefitting from Big Data embedded in the RF cloud. These researchers investigate various kinds of devices, including Wi-Fi, Bluetooth, ZigBee, and RFID, to evaluate the threat, to assess the vulnerability of the systems, and propose frameworks for specific authentication and security schemes.

To analyze the security of the networks, it is customary to refer to a layered architecture [Zha13]. Figure 11.10 shows a general layered architecture and the relations among different layers. The architecture of the security system in this figure consists of three layers: perception layer, network layer, and application layer. The functionality of the perception layer is data collection, preprocessing of data, and secure transition of this data to the network layer. The network layer checks the security of data and transmits it to the application layer. The application layer analyzes and processes the data to support the application.

Since most of the RF data collection sensors are deployed in environments with no human supervision, and the data is collected through a wireless medium, this data can be easily monitored, intercepted, and modified. In these environments, an attacker can access the sensor and take control of the device or damage these sensors or physically remove them from their

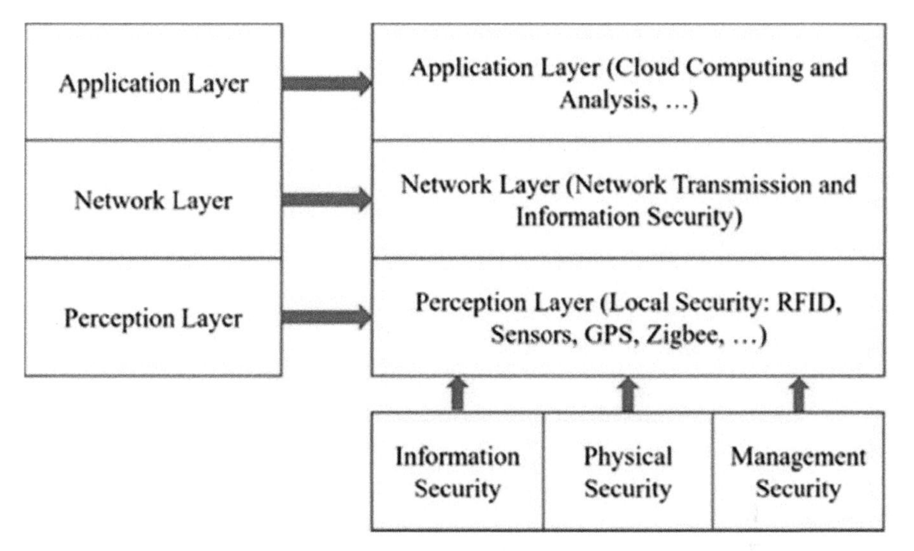

Figure 11.10 Security architecture for applications involved in the RF clouds.

assigned location. As a result, most of the security designers for RF cloud applications implement their measures at the perception layer.

The application of machine learning methods for the classification of devices for authentication and security has been very popular in the recent literature [Chat18]. The time-domain features of the RF cloud from Wi-Fi have been used to train a classifier to differentiate between trusted and untrusted devices operating in close vicinity of each other [Di16]. Researchers have also examined physical authentication using a unique coding technique to generate location-related public keys based on RF cloud signature in each location [Zha18].

In Section 11.2.2, we introduced the main features of the RF cloud, which includes RSS, TOA, CIR, and CSI, and how we can process them for extraction of traditional statistical features such as mean and standard deviation, as well as Doppler spectrum related features. At the perception layer of security systems, we can use the fingerprint of these features for RF authentication. Fingerprinting is the process of identifying radio transmitters by examining their unique transient characteristics at the beginning of transmission. A complete identification system has been presented, which includes data acquisition, transmission detection, RF fingerprint extraction, and a variety of classification subsystems [Ure07]. Following this pioneering work, several researchers have examined different machine learning methods for RF cloud related research in authentication and security.

Using non-parametric and multi-class ensemble classifiers for RF fingerprinting, researchers demonstrated improved ZigBee device authentication over the traditional algorithms [Pat14]. Other works extracted novel RF fingerprint features to design a hybrid and adaptive classification scheme adjusting to the environmental conditions and carry out extensive experiments to evaluate the performance of these systems [Pen18]. A low-cost system has been introduced for bit-level network security, benefitting from physical unclonable functions, which is challenging to replicate [Wil10]. A device recognition algorithm based on RF fingerprint has also been proposed [Wan16]. In this work, a Hilbert transform and principal component analysis are used to generate the RF data fingerprint of the device and traditional machine learning algorithms are used to classify the devices. The accuracy of RF fingerprinting employing low-end receivers has been evaluated showing that receiver impairment effectively decreases the success rate of impersonation attack on RF fingerprinting [Reh14]. Another area of emerging security and authentication research related to RF cloud applications are the design of testbeds for risk analysis for IoT-based physically secure systems. To assess security risks, researchers have proposed testbeds and methodologies for risk analysis and evaluation of vulnerability [Zha14][Mie17]. There are other works proposing a testbed for authentication of IoT objects benefiting from RF fingerprinting, along with a machine learning technique [Ali18, Dab19].

In this chapter, we explained how the success of wireless networks has resulted in the deployment of a huge infrastructure as well as the development of inexpensive wireless devices. Big Data from the RF cloud of the infrastructure and devices has enabled several intelligent cyberspace applications in positioning and tracking, motion and gesture detection, and security and authentication. These innovative cyberspace applications have the potential for creating a major paradigm shift of untethered human–computer interfacing and the development of popular applications in the health and gaming industries.

Research challenges facing this industry include learning how to integrate multiple sensors to enhance positioning and tracking for universal operation in all environments. Another challenge is in finding methods for systematic performance evaluation of alpha–beta classification capability of micro-gestures and performance evaluation of motion and micro-motion tracking techniques. Designing universal data acquisition interfaces for multiple RF sources is another technical challenge facing the existing devices for practical applications in health, interactive gaming, and HCI.

11.4 Assignments

a) What is RF cloud and which technologies are contributing to that?
b) What are the data contents of the RF cloud?
c) What are the differences between RF cloud of communication devices and those of radars?
d) How can a communication device create a unique ID?
e) How does Wi-Fi positioning work?
f) What are the differences between indoor and urban area RF signature data collection methods?
g) What are the differences among RSS and TOA cell tower positioning?
h) How do gesture and motion detection techniques using communication devices differ from those using radars?
i) Explain how the proximity check using RSS of communications devices can help in the authentication of a device.
j) What is the CIR and how can we benefit from that for security and authentication?

References

[Abb18] Abbas, N., Zhang, Y., Taherkordi, A. and Skeie, T., 2018. Mobile edge computing: A survey. *IEEE Internet of Things Journal*, *5*(1), pp.450-465.

[Ala98] S. Alamouti, "A Simple Transmit Diversity Technique for Wireless Communications," IEEE J. Sel. Areas. Comm., pp. 1451-1458, Oct. 1998.

[Aya96] E. Ayanoglu *et al.*, "Mobile information infrastructure," *Bell Labs Tech. J.*, pp. 143–63, Autumn 1996.

[Ber87] D. Bertsekas and R. Gallagher, Data Networks, Prentice Hall, 1987.

[Bla92] K. L. Blackard et al, "Path Loss And Delay Spread Models As Functions Of Antenna Height For Microcellular System Design," Proc. 42nd IEEE Vehicular Technology Conference, Denver, CO, 1992.

[Blu00] Bluetooth Special Interest Group, "Specifications of the Bluetooth System, vol. 1 v. 1.1, 'Core' and vol. 2 v. 1.0 B 'Profiles'," 2000.

[Cha99] Chaudhury, P., Mohr, W. and Onoe, S., 1999. The 3GPP proposal for IMT-2000. *IEEE Communications magazine*, *37*(12), pp.72-81.

[Che97] G. Cherubini, et.al., "100BASE-T2: A New Standard for 100 Mb/s Ethernet Transmission over Voice-Grade Cables", *IEEE Comm Mag.* Nov. 1997.

[Cro97] B. P. Crow, I. Widjaja, L.G. Kim and P.T. Sakai, "IEEE 802.11 Wireless Local Area Networks", *IEEE Communications Magazine*, Vol. 35, No. 9, pp. 116-126, Sept. 1997.

[Dem06] I. Demirkol et al, "MAC Protocols for Wireless Sensor Networks: A Survey," IEEE Communications Magazine, April 2006.

[Den96] L. R. Dennison, "BodyLAN: A wearable personal network", *Second IEEE Workshop on WLANs*, Worcester, MA, 1996.

[Dha02] Al-Dhahir, N.; Fragouli, C.; Stamoulis, A.; Younis, W.; Calderbank, R.; Space-time processing for broadband wireless access , *IEEE Communications Magazine*, Vol. 40 No. 9, pp. 136-142, Sep 2002

[Edn04] J. Edney and W.A. Arbaugh, Real 802.11 Security: Wi-Fi Protected Access and 802.11i, Pearson Education, 2004.

[Enn98] Greg Ennis, Doc: IEEE P802.11-98/319, *Impact of Bluetooth on 802.11 Direct Sequence*, September 15, 1998.

[Ert98] Richard B. Ertel, Paulo Cardieri, Kevin W. Sowerby, Theodore S. Rappaport and Jeffrey H. Reed, "Overview of Spatial Channel Models for Antenna Array Communication Systems", IEEE Personal Communications, Feb. 1998.

[Fos98b] G. J. Foschini and . M. Gans, "On Limits of Wireless Communications in a Fading Environment Using Multiple Antennas" *IEEE Wireless Communications*, 6: 311-315, 1998.

[Frii46] Harald T. Friis, A note on a Simple Transmission Formula, *Proceedings of the IRE and Wave and Electrons,* May 1946, pp 254-256.

[Gan91] R. Ganesh and K. Pahlavan, "Modeling of the Indoor Radio Channel," *IEE Proc. I: Commun. Speech Vision*, 138, 153-161, 1991.

[Gas02] M. S. Gast, *802.11 Wireless Networks: The definitive guide*, O'Reilly & Associates, 2002.

[Haa00] J.C. Haartsen and S. Mattisson, "Bluetooth-a new low-power radio interface providing short-range connectivity", *Proceedings of the IEEE*, Vol. 88, No. 10, pp. 1651-1661, Oct. 2000.

[Hal83] S.W. Halpern, "Reuse partitioning in cellular systems", *Proc. Of the IEEE Vehicular Technology Conference*, pp. 322-327, 1983.

[Hal96] C.J. Hall, and W.A. Foose, "Practical Planning for CDMA Networks: A Design Process Overview", *Proc. Southcon'96*, pp.66-71, 1996.

[Hal99] K. Halford, S. Halford, M. Webster, C. Ander, "Complementary code keying for RAKE-based indoor wireless communication", IEEE International Symposium on Circuits and Systems, Vol. 4, pp. 427-430, Orlando, FL, 1999.

[Ham86] J.L. Hammond and P.J.P. O'Reilly, *Performance analysis of local computer networks*, Addison-Wesley, Reading, MA, 1986.

[Hau94] T. Haug, "Overview of GSM: Philosophy and Results", *International Journal of Wireless Information Networks*, Jan 1994.

[Hay91] V. Hayes, "Standardization efforts for wireless LANS", *IEEE Network*, Vol. 5, No. 6, pp. 19-20, 1991.

[How90] S. J. Howard and K. Pahlavan, "Measurement and Analysis of the Indoor Radio Channel in the Frequency Domain," IEEE Trans. Instr. Meas., No. 39, pp. 751–55, 1990.

[HOW92] S.J. Howard and K. Pahlavan, "Autoregressive Modeling of Wideband Indoor Radio Propagation", *IEEE Trans. Communication*, Vol.40, pp. 1540-1552, 1992.

[IEE01] *Proc. IEEE Workshop on Wireless LANs*, Newton, MA (Sep. 2001).

[Kap02] S. Kapp, "802.11a. More bandwidth without the wires", IEEE Internet Computing, Volume:6, Issue:4, July-Aug. 2002.

[Kei89] G.E. Keiser, *Local Area Networks*, McGraw-Hill, New York, 1989.

[Ker00] J.P. Kermoal, L. Schumacher, P.E. Mogensen and K.I. Pedersen, "Experimental investigation of correlation properties of MIMO radio channels for indoor picocell scenarios", *52nd IEEE Vehicular Technology Conference*, 2000, pp. 14-21, 2000.

[Kle75] L. Kleinrock, Queing Systems: Volume 1: Theory, John Wiley, New York, 1975.

[Kni05] Knightson, K., Morita, N. and Towle, T., 2005. NGN architecture: generic principles, functional architecture, and implementation. *IEEE Communications Magazine*, *43*(10), pp.49-56.

[Koh04] R. Kohno, M. Welborn, and M. McLaughlin, "DS-UWB Proposal," *IEEE* 802.15-04/140r2, March 2004.

[Kot04] S. Kota., K. Pahlavan, and P. Leppanen, *Broadband Satellite Internet*, Kluwer Publishing Company, 2004.

[Lee06] M. J. Lee and J. Zheng, "Emerging Standards for Wireless Mesh Technology," *IEEE Wireless Communications*, pp. 56–63, April 2006.

[Lee91] W.C.Y. Lee, "Smaller Cells for Greater Performance", *IEEE Communications Magazine*, pp. 19-23, November 1991.

[Lee98] William C.Y. Lee, Mobile Communications Engineering : Theory and Applications (McGraw-Hill Series on Telecommunications), McGrawHill Professional Publishing, 1999.

[Leh99] P.H. Lehne and M. Pettersen, "An Overview of Smart Antenna Technology for Mobile Communications Systems", *IEEE Communications Surveys*, pp. 2-13, Vol.2, Fourth Quarter, 1999.

[Mac79] V. H. MacDonald, "The Cellular Concept", *The Bell System Technical Journal*, Vol.58, No.1, pp. 15-41, January 1979.

[Mar85] M.J. Marcus, "Recent US regulatory decisions on civil use of spread spectrum", *Proc. IEEE Globecom*, 16.6.1-16.6.3, New Orleans, December 1985.

[Mcd98] J.T. Edward McDonnell, "5 GHz indoor channel characterization: measurements and models", *IEE Coll. on Ant. and Prop. for future mobile communications*, 1998.

[Mcd98] McDysan, D.E. and Spohn, D.L., 1998. *ATM theory and applications*. McGraw-Hill Professional.

[Min89] Minzer, S.E., 1989. Broadband ISDN and asynchronous transfer mode (ATM). *IEEE Communications Magazine*, *27*(9), pp.17-24.

[Nag98] A. Naguib et al, "A Space Time Coding Modem for High Data Rate Wireless Communications," *IEEE J. Sel. Areas. Comm.*, pp. 1459-1477, October 1998.

[Pah00a] K. Pahlavan, P. Krishnamurthy, et. al., "Handoff in hybrid mobile data networks", *IEEE Personal Communications Magazine*, April 2000.

[Pah00B] K. Pahlavan, X. Li, M. Ylianttila, R. Chana, and M. Latva-aho, "An Overview of Wireless Indoor Geolocation Techniques and Systems", MWCN'2000, Paris, May 2000.

[Pah02] K. Pahlavan and P. Krishnamurthy, *Principles of Wireless Networks: A Unified Approach*, Pearson Ed., 2002.

[Pah05] K. Pahlavan and A. Levesque, Wireless Information Networks, 2nd Ed., John Wiley and Sons, 2005.

[Pah20] Pahlavan, K. and Krishnamurthy, P., 2020. Evolution and Impact of Wi-Fi Technology and Applications: A Historical Perspective. International Journal of Wireless Information Networks, pp.1-17.

[Pah85] K. Pahlavan, "Wireless Communications for Office Information Networks," *IEEE Communications Magazine*, September 1985.

[Pah88] K. Pahlavan and J. L. Holsinger, "Voice-band Data Communication Modems – a historical review: 1919-1988," *IEEE Communications Magazine*, Vol. 26, pp. 16-27, January 1988.

[Pah94] K. Pahlavan and A. H. Levesque, "Wireless data communications", *Proceedings of the IEEE*, Vol. 82, No. 9, pp. 1398-1430, Sept. 1994.

[Pah95] Kaveh Pahlavan and Allen Levesque, *Wireless Information Networks*, John Wiley and Sons, New York, 1995.

[Pah97] K. Pahlavan, A. Zahedi, and P. Krishnamurthy, "Wideband local access: wireless LAN and wireless ATM", *IEEE Communications Magazine*, Vol. 35, No. 11, pp. 34-40, Nov. 1997.

[Pah98] K. Pahlavan, P. Krishnamurthy and J. Beneat, "Wideband radio propagation modeling for indoor geolocation applications", *IEEE Comm. Magazine*, pp. 60-65, April 1998.

[Pat03] W. Pattara-atikom, P. Krishnamurthy and S. Banerjee, "Distributed Mechanisms for Quality of Service in Wireless LANs", *IEEE Wireless Communications: Special issue on "QoS in Next-generation Wireless Multimedia Communications Systems*, Vol. 10, No. 3, pp.26-34, June 2003.

[Pat96] Sandeep Patel, Howard W. Johnson, and J. R. Rivers, "Method and apparatus for implementing a type 8B6T encoder and decoder", *United States Patent 5525983*, Issued on June 11, 1996

[Ped00] K.I. Pedersen, J.B. Andersen, J.P. Kermoal, and P. Morgensen, "A stochastic multiple-input-multiple-output radio channel model for evaluation of space-time coding algorithms", *52nd IEEE Vehicular Technology Conference*, pp. 893 -897, 2000.

[Pir17] Pirandola, S., Laurenza, R., Ottaviani, C. and Banchi, L., 2017. Fundamental limits of repeaterless quantum communications. *Nature communications*, *8*(1), pp.1-15.

[Pro00] John G. Proakis, *Digital Communications*, 4th Ed., McGraw Hill, 2001.

[Rap96] S.Y. Seidel and T.S. Rappaport, "914 MHz path loss prediction models for indoor wireless communications in multifloored buildings", *IEEE Transactions on Antennas and Propagation*, Vol. 40, No. 2, pp. 207-217, Feb. 1992.

[Red95] Siegmund Redl, Matthias K. Weber, Malcolm Oliphant, Werner Mohr, *An Introduction to GSM,* The Artech House Mobile Communications Series, Artech House, 1995.

[Sex89] T. Sexton and K. Pahlavan, "Channel modeling and adaptive equalization of indoor radio channels", *IEEE JSAC*, Vol. 7, pp. 114-121, 1989.

[Sie00] T. Siep, I. Gifford, R. Braley, and R. Heile, "Paving the Way for Personal Area Network Standards: An Over View of the IEEE P802.15 Working Group for Wireless Personal Area Networks", *IEEE Personal Communications*, Feb. 2000.

[Sim85] M.K. Simon et al., *Spread Spectrum Communication*, Computer Science Press, 1985.

[Sko08] D. Skordoulis et al, "IEEE 802.11n MAC Frame Aggregation Mechanisms for Next Generation High Throughput WLANs," IEEE Wireless Communications, pp. 40-47, Fenruary 2008.

[Sta00] W. Stallings, Local and Metropolitan Area Networks, 6th Ed., Pearson, 2000.

[Sti02] D. Stinson, Cryptography: Theory and Practice, CRC Press, 2002.

[Tak85] H. Takagi and L. Kleinrock, "Throughput analysis for persistent CSMA systems", *IEEE Trans. Comm.*, Vol. 33, pp. 627-638, 1985.

[Tan03] A.S. Tanenbaum, Computer Networks, Fourth Edition, Prentice Hall, 2003.

[Tar98] V. Tarokh, N.Seshadri, and A. R. Calderbank, "Space-time codes for high data rate wireless communications," IEEE Trans. Inf. Theory, pp. 744-765, March 1998.

[Tel95] E. Telatar, "Capacity of Multiantenna Gaussian Channels," AT&T Bell Labs Tech. Memo, 1995.

[Tho06] R. Thompson et al, The Physical Layer of Communication Systems, Artech House, 2006.

[Tob75] F.A. Tobagi and L. Kleinrock, "Packet switching in radio channels Part II: The hidden terminal problem in CSMA and the busy tone solution", *IEEE Trans. Comm.*, Vol. 33, pp. 1417-1433, 1975.

[Tuc91] B. Tuch, "An ISM band spread spectrum local area network: WaveLAN", Proc. 1st IEEE Workshop on WLANs, pp. 103-111, Worcester, MA 1991.

[Val98] R.T. Valadas, A.R. Tavares, A.M.deO. Duarte, A.C. Moreira, and C.T. Lomba, "The infrared physical layer of the IEEE 802.11 standard for wireless local area networks", *IEEE Communications Magazine*, Vol. 36, No. 12, pp. 107-112, Dec. 1998.

[Vas05] D. Vassis, G. Kormentzas, A. Rouskas and I. Maglogiannis, "The IEEE 802.11g standard for high data rate WLANs", IEEE Network, pp. 21-26, May/June 2005.

[vNee99] R. van Nee et al., "New high-rate wireless LAN standards", IEEE Communications Magazine, Vol. 37, No. 12, pp. 82-88, Dec. 1999.

[Wet01] M. Jakobsson and S. Wetzel, "Security Weaknesses in Bluetooth", *RSA Conference'01*, April 8-12, 2001.

[Wil95] T. A. Wilkinson, T. Phipps, S. K. Barton, "A Report on HIPERLAN Standardization," International Journal on Wireless Information Networks, Vol. 2, pp. 99-120, March 1995.

[Xia05] Y. Xiao, "IEEE 802.11n: Enhancements for higher throughput in wireless LANs", IEEE Wireless Communications, pp.82-91, December 2005.

[Zha19] Zhang, Z., Xiao, Y., Ma, Z., Xiao, M., Ding, Z., Lei, X., Karagiannidis, G.K. and Fan, P., 2019. 6G wireless networks: Vision, requirements, architecture, and key technologies. *IEEE Vehicular Technology Magazine*, *14*(3), pp.28-41.

[Zha90] K. Zhang and K. Pahlavan, "An integrated voice/data system for mobile indoor radio networks", *IEEE Trans. Vehicular Technology*, Vol. 39, pp. 75-82, 1990.

Index

D

Q

R

S

T

About the Author

Kaveh Pahlavan is a Professor of ECE, a Professor of CS, and Director of the Center for Wireless Information Network Studies, Worcester Polytechnic Institute (WPI), Worcester, MA, USA. He is also a recipient of "overseas famous scholar award" from R.I. China and serves as a Visiting Professor with University of Science and Technology Beijing (2019–2021). He has also been a Visiting Professor of Telecommunication Laboratory and CWC, University of Oulu, Finland (1997–2007). He has spent sabbaticals at University of Oulu, Finland (1997), Olin College, Needham, MA, USA (2004) and Harvard University, Cambridge, MA, USA (2011). His areas of research include multipath radio frequency (RF) indoor propagation measurement and modeling for cyberspace applications for communications, positioning, motion and gesture detection, and physical layer security and authentication. He is the principal author of the seven textbooks including *Wireless Information Networks* (with Allen Levesque), John Wiley and Sons, 1995 and *Principles of Wireless Networks – A Unified Approach* (with P. Krishnamurthy), Prentice Hall, 2002. He has been a consultant to a number of companies worldwide including Verizon Laboratories, Jet Proportion Laboratories, Pasadena, CA, USA, United Technology Research Center in Connecticut, Honeywell in Arizona, Nokia, LK-Products, Elektrobit, TEKES, and Finnish Academy in Finland, and NTT in Japan. He is the Editor-in-Chief of the *International Journal on Wireless Information Networks*. He was the founder, the program chairman, and organizer of the *IEEE Wireless LAN Workshop*, Worcester, in 1991 and 1996, and the organizer and technical program chairman of the *IEEE International Symposium on Personal, Indoor, and Mobile Radio Communications*, Boston, MA, USA in 1992, 1998, 2011, 2014, and 2017. He has also been selected as a member of the Committee on Evolution of Untethered Communication, US National Research Council, 1997 and has led the US review team for the Finnish R&D Programs in 1999 and 2003. For his contributions to the wireless networks, he was the Westin Hadden Professor of Electrical and Computer Engineering at WPI during 1993–1996, was elected as a fellow of the IEEE in 1996, and become a fellow of Nokia in 1999. From May to December 2000, he was the first Fulbright-Nokia scholar at the University of Oulu, Finland. Because of his inspiring visionary publications and his international conference activities for the growth of the wireless LAN industry, he is referred to as one of the founding fathers of the wireless local area network (WLAN) industry, commercially known as Wi-Fi. His research work was the core for more than 25 patents by Skyhook Wireless where he acted as the chief technical advisor of the company (2004–2014). Details of his contributions to this field are available at www.cwins.wpi.edu.